Zoonoses

Bacterial Diseases

The Editor

Dr. Sudhi Ranjan Garg is a senior Professor of Veterinary Public Health and Epidemiology in the College of Veterinary Sciences, Lala Lajpat Rai University of Veterinary and Animal Sciences at Hisar (India). He has also worked as Head of the Department. Dr. Garg obtained the degree of Bachelor of Veterinary Sciences and Animal Husbandry in 1979. The Indian Council of Agricultural Research awarded Junior Research Fellowship to Dr. Garg for postgraduate studies. Dr. Garg completed his Master of Veterinary Sciences in Veterinary Public Health and Epidemiology in 1982. He was later awarded Russian Government Scholarship for doctoral studies and obtained Ph.D. from the prestigious All-Russian Institute of Experimental Veterinary Medicine in Moscow in 1991. Dr. Garg has over 30 years of experience of extensive graduate and postgraduate teaching and research on zoonotic diseases. He has guided many students for their Master's and Doctoral degrees. Dr. Garg works as a subject matter specialist for many organizations. Recognizing his contributions, the Indian Association of Veterinary Public Health Specialists has honoured Dr. Garg as Fellow of the Association.

Dr. Garg has been actively associated with the Alliance for Rabies Control (UK), Global Alliance for Rabies Control (USA), and the Commonwealth Veterinary Association in their campaigns for rabies prevention, control and elimination. Dr. Garg has been regularly coordinating mass education programmes for the schoolchildren, college students, pet owners, farmers and urban people. The Global Alliance for Rabies Control has commended his efforts and has cited him as an example of "Local Heroes" taking action towards rabies prevention and control efforts in the country. Dr. Garg also works as resource faculty in the Continuing Veterinary Education Programme run by the Veterinary Council of India and other training programmes. Apart from coordinating training programmes on Diagnosis of Rabies, Food Safety Standards, and Capacity Building of Decentralized Institutions for SPS Compliance in Raw Milk Sector, Dr. Garg has delivered lectures to the trainees in a number of such programmes. Dr. Garg is a Life-Member of the Indian Association of Veterinary Public Health Specialists, Association for Prevention and Control of Rabies in India, Indian Association of Veterinary Microbiologists, Immunologists and Specialists in Infectious Diseases, and the Association of Food Scientists and Technologists of India. He is currently working as a Member in the Editorial Board of the journal Microbes and Health published by the Bangladesh Society of Veterinary Microbiology and Public Health and in the Journal of Foodborne and Zoonotic Diseases which has been launched recently.

Dr. Garg is currently engaged in propagating the 'One Health' concept in the region. He has successfully organized a multidisciplinary symposium on Interdisciplinary Approach for Tuberculosis Control and a Veterinary-Medical Workshop on Rabies and its Prevention. Dr. Garg has attended a large number of scientific conferences and meetings to present invited and lead papers. He has over 100 scientific papers and several books to his credit. His earlier books include Understanding Rabies and its Prevention; Interdisciplinary Approach for Tuberculosis Control; Veterinary and Livestock Sector: A Blueprint for Capacity Building; Environmental Security: Human and Animal Health; Handbook of Quality Control of Dairy and Meat Products; Veterinary Diagnostics: Current Trends; and Human and Animal Health: Environmental Perspectives.

Zoonoses

Bacterial Diseases

Sudhi Ranjan Garg

College of Veterinary Sciences,
Lala Lajpat Rai University of Veterinary and Animal Sciences,
Hisar – 125 004, India

2014

Daya Publishing House®

A Division of

Astral International Pvt. Ltd.

New Delhi – 110 002

Published by : **Daya Publishing House®**
A Division of
Astral International Pvt. Ltd.
– ISO 9001:2008 Certified Company –
4760-61/23, Ansari Road, Darya Ganj
New Delhi-110 002
Ph. 011-43549197, 23278134
E-mail: info@astralint.com
Website: www.astralint.com

Laser Typesetting : **Classic Computer Services**, Delhi - 110 035

Printed at : **Replika Press Pvt. Ltd.**

PRINTED IN INDIA

This book is dedicated to my loving father,
Shri Narender Pal Garg
who has always been a source of inspiration.

Preface

Pathogens perpetuate by adapting themselves to a variety of environmental conditions and by moving from one infected host to another susceptible host. Many of these organisms affecting a large variety of animal species are also transmissible from these to man and vice versa. Such zoonotic pathogens circulating in animal population thus threaten not only animal health but also jeopardize public health. Zoonoses can be endemic, re-emerging or emerging diseases with epidemic potential. These affect the well-being of the societies worldwide but assume particular significance in the developing countries because of the close everyday interactions of people with animals, people's dependence on livestock for livelihood, and high relative density of human and animal populations. Urbanization of populations combined with the changes in people's food habits and lifestyles, has caused intensification of the livestock production systems resulting in greater human-animal linkages and threat of zoonoses.

Rapid growth of international trade in animals and animal products, combined with the unprecedented international travel, has substantially increased the possibility, territorial limits and speed of transmission of infections. The emergence and re-emergence of zoonoses is, therefore, a growing concern everywhere. The potential of zoonotic pathogens as biological weapons is another important facet of the problem. However, despite their potentially disastrous impact on public health and economy, these are too often forgotten or neglected, particularly in the developing world because of the attribution of low priority and general constraints in human and financial resources. During the recent years, public health has faced several major threats of zoonotic infections causing global scare among the populations in both developed and developing countries. While the industrialized countries have generally been able to respond rapidly to such outbreaks and contain them, many developing countries have not been able to do so adequately because of several reasons. Surveillance, early diagnosis and people's perception are crucial components of disease prevention, control and eradication strategies. The majority of developed countries have efficient surveillance systems in place to detect and control major zoonotic diseases but the developing countries are most likely to be deficient in this which may result in most of zoonoses going unrecorded at such places. Moreover, preparedness for outbreaks of zoonotic diseases is also suboptimal at many places due to which the diseases may often catch the animal health as well as public health systems by surprise.

Predicting and controlling diseases at the human-animal interface is a huge challenge, however, appropriate strategies can control and eliminate zoonotic diseases and improve the health status and quality of human life. The strategic action plans should aim at risk reduction, early disease diagnosis and containment, continuing disease monitoring and surveillance, and strengthening the preparedness and response mechanisms. Capacity building of individuals, groups, institutions, organizations and societies is essential to enhance their abilities in the area of surveillance, prevention and control of zoonotic diseases, agricultural development, food safety and food security. The infrastructure and veterinary public health services generally require rebuilding and strengthening in the developing nations and those in transition.

With One Health approach gaining momentum worldwide, integrated multidisciplinary coordinated approach assumes paramount importance in alleviating the sufferings and losses caused by zoonoses. The zoonoses control activities need to cover people, animals and their environments taking into consideration all epidemiological features of a disease in a holistic manner. It is important that veterinary and public health sectors work collectively to ensure successful control of zoonotic diseases. Close collaboration of wildlife management, environmental health and other key sectors is also important in controlling the animal reservoirs and niche of zoonotic diseases.

In the current scenario, the epidemiology, prevention and control of zoonotic diseases constitute an essential component of the curriculum of veterinary sciences, but the coverage of these issues is likely to be grossly overlooked in medical education, particularly in the developing countries. The present book is a humble attempt to fill up the gap and provide a rich resource for all those having stake in the zoonoses control activities. Particular emphasis has been given to the risk analysis and strategies for zoonoses management in the developing nations. I am confident that the book will be quite useful to the students, teachers, researchers, academicians, policymakers and other professionals in the fields of animal health, public health, wildlife management and environmental health. The book is also expected to be a handy tool with the professionals for taking up advocacy, public awareness and health education programmes concerning zoonoses.

I am extremely grateful to the contributors of different chapters for their painstaking efforts. I am also thankful to my colleague, Dr. Vijay J. Jadhav, for helping in this project. Finally, I express my deep sense of gratitude to my wife, Dr. Meenakshi Garg, for her unfailing support throughout the years I worked on this book.

Sudhi Ranjan Garg

Contents

List of Contributors

Abdul Rehman
Department of Epidemiology and Public Health, Faculty of Veterinary Science, University of Veterinary and Animal Sciences, Lahore, Pakistan

Abraham Ali
Ethiopian Health and Nutrition Research Institute, Zoonoses Research Team, Addis Ababa, Ethiopia

Asefa Deressa
Ethiopian Health and Nutrition Research Institute, Zoonoses Research Team, Addis Ababa, Ethiopia

B.C. Bera
Veterinary Type Culture Collection, National Research Centre on Equines, Hisar – 125001

B. Sunil
Department of Veterinary Public Health, College of Veterinary and Animal Sciences, Mannuthy, Thrissur – 680651

Bhoj Raj Singh
Centre for Animal Disease Research and Diagnosis, Indian Veterinary Research Institute, Izatnagar – 243122

Chandan Prakash
Centre for Animal Disease Research and Diagnosis, Indian Veterinary Research Institute, Izatnagar – 243122

D.K. Sinha
Centre for Animal Disease Research and Diagnosis, Indian Veterinary Research Institute, Izatnagar – 243122

D.N. Garg
Department of Veterinary Public Health and Epidemiology, Lala Lajpat Rai University of Veterinary and Animal Sciences, Hisar – 125004

Debabrata Mahapatra
Texas A&M Veterinary Medical Diagnostic Laboratory, Amarillo, TX 79106, USA

Fasil Mengistu
Ethiopian Health and Nutrition Research Institute, Zoonoses Research Team, Addis Ababa, Ethiopia

G.B. Manjunatha Reddy
Project Directorate on Animal Disease Monitoring and Surveillance, Bengaluru – 560024

Geetanjali Singh
Dr. G.C. Negi College of Veterinary and Animal Sciences, C.S.K. Himachal Pradesh Agricultural University, Palampur – 176062

H. Rahman
Project Directorate on Animal Disease Monitoring and Surveillance, Bengaluru – 560024

K.N. Viswas
Indian Veterinary Research Institute, Mukteswar – 263138

K. Vrinda Menon
Department of Veterinary Public Health, College of Veterinary and Animal Sciences, Mannuthy, Thrissur – 680651

Kuldeep Dhama
Division of Pathology, Indian Veterinary Research Institute, Izatnagar – 243122

M. Nagalingam
Project Directorate on Animal Disease Monitoring and Surveillance, Hebbal, Bengaluru – 560024

Mandeep Sharma
Department of Veterinary Microbiology, Dr. G.C. Negi College of Veterinary and Animal Sciences, C.S.K. Himachal Pradesh Agricultural University, Palampur – 176062

Mohd. Yaqoob Wani
Division of Veterinary Biotechnology, Indian Veterinary Research Institute, Izatnagar – 243122

Muhammad Athar Khan
Department of Epidemiology and Public Health, Faculty of Veterinary Science, University of Veterinary and Animal Sciences, Lahore, Pakistan

P.K. Kapoor
Department of Veterinary Public Health and Epidemiology, Lala Lajpat Rai University of Veterinary and Animal Sciences, Hisar – 125004

Pradeep Mahadev Sawant
Division of Veterinary Biotechnology, Indian Veterinary Research Institute, Izatnagar – 243122

R.K. Vaid
Veterinary Type Culture Collection, National Research Centre on Equines, Hisar – 125001

Rajesh Chahota
Department of Veterinary Microbiology, Dr. G.C. Negi College of Veterinary and Animal Sciences, C.S.K. Himachal Pradesh Agricultural University, Palampur – 176062

Rajeswari Shome
Project Directorate on Animal Disease Monitoring and Surveillance, Bengaluru – 560024

Ruchi Tiwari
Department of Microbiology and Immunology, College of Veterinary Sciences, Pandit Deen Dayal Upadhyaya Veterinary University, Mathura – 281001

S.K. Khurana
National Research Centre on Equines, Hisar – 125001

S.R. Garg
Department of Veterinary Public Health and Epidemiology, College of Veterinary Sciences, Lala Lajpat Rai University of Veterinary and Animal Sciences, Hisar – 125004

S. Rajagunalan
Division of Veterinary Public Health, Indian Veterinary Research Institute, Izatnagar – 243122

Sandeep Kumar
Centre for Animal Disease Research and Diagnosis, Indian Veterinary Research Institute, Izatnagar – 243122

Sanjay Kapoor
Department of Veterinary Microbiology, College of Veterinary Sciences, Lala Lajpat Rai University of Veterinary and Animal Sciences, Hisar – 125004

Sathish B. Shivachandra
Clinical Bacteriology Laboratory, Indian Veterinary Research Institute, Mukteswar – 263138

Senthilkumar Natesan
Institute of Human Virology and Department of Medicine, University of Maryland School of Medicine, Baltimore, MD 21201, USA

Subhash Verma
Dr. G.C. Negi College of Veterinary and Animal Sciences, C.S.K. Himachal Pradesh Agricultural University, Palampur – 176062

V. Balamurugan
Project Directorate on Animal Disease Monitoring and Surveillance, Hebbal, Bengaluru – 560024

V.J. Jadhav
Department of Veterinary Public Health and Epidemiology, Lala Lajpat Rai University of Veterinary and Animal Sciences, Hisar – 125004

V.M. Vaidya
Department of Veterinary Public Health and Epidemiology, Bombay Veterinary College, Mumbai – 400012

Younis Farooq
Centre for Animal Disease Research and Diagnosis, Indian Veterinary Research Institute, Izatnagar – 243122

2014, Zoonoses: Bacterial Diseases
Editor: **Sudhi Ranjan Garg**
Published by: **DAYA PUBLISHING HOUSE, NEW DELHI**

Pages **1–12**

1

Zoonoses: An Overview

Sudhi Ranjan Garg
Department of Veterinary Public Health and Epidemiology,
College of Veterinary Sciences, Lala Lajpat Rai University of
Veterinary and Animal Sciences, Hisar – 125 004

The relationship between man and animals has been in existence since antiquity due to the man's dependence on animals for food, fibre, clothing, draught power, fuel, medicines, etc. In the modern day context, the link between the human and animal populations may look closer in the developing countries, but in the industrially advanced countries too animals are widely used for intensive food and fibre production, sports and recreational activities, and scientific experimentations. While the people such as farmers, livestock handlers, veterinarians, slaughterhouse workers, zoo personnel, hunters, etc. routinely remain in direct contact with animals due to their occupation, many others are closely linked to their companion animals. Apart from these, human beings share the environment with different categories of animals that include pet animals, domesticated animals, domiciliated animals, wild animals and birds, which establishes an inextricable link between the human and animal populations even when there is no direct contact or closeness between the two. Because a number of communicable diseases (known as zoonotic diseases, or zoonoses) are transmitted from animals to humans, human-animal interactions can lead to serious public health problems. Likewise, humans can transmit infections to animals. Infectious agents that are transmissible under natural circumstances from vertebrate animals to humans are responsible for causing zoonoses.

Zoonoses: Definition and Historical Perspective

The WHO Expert Committee on Zoonoses at its third meeting in 1966 defined zoonoses as those infections which are naturally transmitted between vertebrate animals and humans. Zoonoses may arise from wild or domestic animals or from products of animal origin. It is not a new phenomenon; zoonotic diseases have affected human health all along. For example, bubonic plague has caused substantial illness and death around the world since ancient times (Wheelis 2002). There are biblical references to plague. A possible epidemic of bubonic plague was described in the *Old Testament*, in the First Book of Samuel. Similarly, rabies was described in Mesopotamia in hunting dogs, as early as 2,300 BC. Early Chinese, Egyptian, Greek, and Roman records also have descriptions of rabies (Blancou 2003). The studies suggest that Alexander the Great died of encephalitis caused by West Nile virus in Babylon in 323 BC (Marr and Calisher 2003).

Zoonoses Burden

In recent years, zoonotic diseases have acquired growing national and international significance with regard to health, food safety, trade, security and economics due to the increasing awareness about the direct relevance of animal diseases to human health. Further, the increasing population density, intensive livestock production systems and rapid movement of people, animals and animal products have increased the vulnerability to cross-species illnesses. About 75 per cent of the new diseases that have affected humans over the past 10 years have been caused by pathogens originating from an animal or from products of animal origin. Many of these diseases have the potential to spread through various means over long distances and to become global problems (WHO 2012). Grace *et al.* (2011) estimated that in least developed countries, 20 per cent of human sickness and death was due to zoonoses or diseases that recently jumped from animal species to people. Zoonotic diseases affect the production and safety of foods of animal origin too and have adverse effect on the international trade in animals and animal products, which negatively influences the overall socio-economic development.

According to a recent study by Grace *et al.* (2012), 56 zoonotic diseases selected in the study together are responsible for an estimated 2.7 human million deaths and around 2.5 billion cases of human illness a year. Thirteen most important zoonoses in terms of human health impact, livestock impact, amenability to agricultural interventions, severity of diseases and emergence include in descending order zoonotic gastrointestinal disease, leptospirosis, cysticercosis, zoonotic tuberculosis, rabies, leishmaniasis, brucellosis, echinococcosis, toxoplasmosis, Q fever, zoonotic trypanosomosis, hepatitis E and anthrax. These 13 prioritized zoonoses have been assessed to be responsible for causing 2.2 million human deaths and 2.4 billion cases of illness. Nine of these top-ranked zoonoses are known to have high impact on livestock, all have a wildlife interface, and all are amenable to agriculture-based interventions. The systematic literature review of these 13 zoonoses identified as important by Grace *et al.* (2012) revealed that eight of these are 'endemic classical zoonoses', that is, zoonoses that are typically present across a wide range of communities at most times (although perhaps showing annual and inter-annual variability).

Zoonoses Hotspots

Zoonotic diseases threaten the public health, livestock industry, and the food safety and security throughout the globe, especially in the developing world (WHO 2012). Though most common in rural areas, zoonotic diseases are encountered in urban areas as well. The impact of these diseases is, however, greatest on the poorest householders. There is a strong association between poverty, hunger, livestock keeping, and zoonoses (Grace *et al.* 2012). It is estimated that around 1 billion poor people (< $2 a day) depend on livestock, and around two-thirds of the rural poor and one-third of the urban poor depend on livestock. Livestock provide one-fifth to one-half of household income for the poor and in poor countries, livestock provide from 6 to 36 per cent of protein intake (Grace *et al.* 2012).

Four countries including India, Nigeria, Ethiopia and Bangladesh have 44 per cent of the world's poor livestock keepers. Based on various parameters such as zoonoses burden, poverty burden, and reliance on livestock, the top hotspots for poverty, emerging livestock systems and zoonoses are in South Asia which in descending order of importance include India, Bangladesh and Pakistan. These are followed by Ethiopia, Nigeria, Congo DR, Tanzania and Sudan in East and Central Africa; then China, Indonesia, Myanmar and Vietnam in South East Asia; and finally Burkina Faso, Mali and Ghana in West Africa (Grace *et al.* 2012). However, the existing disease reporting systems may not allow true assessment of the impact of zoonoses; the developing countries may have lot of unpublished information. According to the studies, the top 20 countries that bear the greatest burden of zoonoses are India, Nigeria, Congo DR, China, Ethiopia, Bangladesh, Pakistan, Afghanistan, Angola, Brazil, Indonesia, Niger, Tanzania, Kenya, Cote d'Ivoire, Uganda, Sudan, Burkina, Mali and Iraq in descending order of importance. The countries which top in the events of emerging zoonoses include USA, UK, Australia, France, Brazil, Canada, Germany, Japan, China, Sweden, Italy, Malaysia, Switzerland, Congo DR, Sudan, Argentina, India, Israel, Peru, Trinidad and Tobago, Uganda and Vietnam (Grace *et al.* 2012).

Zoonoses Risk Categories

The list of zoonotic diseases is quite long. The frequency and severity of different diseases may have great variations at different places. While attempting to measure the global burden of zoonoses and prioritizing different diseases, Grace *et al.* (2012) categorized the zoonoses in the following categories:

1. *Endemic zoonoses*: These include those diseases which are continually present to a greater or lesser degree in certain populations. Generally, these occur regularly at many places affecting many people and animals. Some of such diseases are brucellosis, bovine tuberculosis, leptospirosis, foodborne zoonoses, cysticercosis, etc. According to Grace *et al.* (2012), endemic zoonoses are responsible for the great majority of human cases of illness (estimated 99.9 per cent) and deaths (estimated 96 per cent) as well

as the greatest reduction in livestock production. They are common in poor populations and are responsible for around a billion illnesses and millions of deaths every year.

2. *Outbreak or epidemic zoonoses*: Cases of such diseases may not occur regularly; the occurrence is sporadic, but sudden in an unpredictable epidemic or outbreak form. In terms of spatial and temporal distribution, their appearance may vary to a great extent. Under certain circumstances and predisposing conditions, the endemic zoonoses may occur as severe outbreaks.

3. *Emerging zoonoses*: These include those diseases that are newly recognized or newly evolved in a population or in a geographical area. A disease which is endemic at one place can be an emerging disease at another place. A previously existing disease in an area that shows an increase in incidence or expansion in its geographical, host or vector range may also fall under this category.

4. *Old zoonoses*: Those diseases which were originally zoonotic but now spread mainly or entirely by human-to-human transmission may be termed as old zoonoses (Grace and McDermott 2012, Grace *et al.* 2012). HIV-AIDS, influenza, malaria, measles and dengue are some such diseases which have jumped species in many places.

Zoonoses Transmission

Zoonoses are a complex group of disease conditions involving a variety of pathogens, host species, reservoirs and vectors, their characteristics and the interplay of a number of environmental factors. Zoonotic pathogens can infect human beings in different ways. The major routes are:

1. By touch or close contact with animals or animal discharges, excreta or waste, for example, anthrax, brucellosis, leptospirosis, Ebola haemorrhgic fever, ringworm, etc.

2. Through bite of an animal, for example, rabies, cat scratch disease, etc.

3. Through food and water, for example, salmonellosis, *E. coli* infection, campylobacteriosis, listeriosis, staphylococcosis, taeniasis, echinococcosis, etc.

4. Through inhalation of contaminated air, dust, droplets of infected saliva or nasal discharge during sneezing, coughing, etc., for example, influenza, tuberculosis, Q fever, hantaviral disease, etc.

5. Through arthropod vectors like mosquitoes, fleas, ticks, etc., for example, Japanese encephalitis, Lyme disease, babesiosis, leishmaniasis, etc.

Classification of Zoonoses

Zoonoses are a biologically heterogeneous group of infections and infestations which differ greatly from each other in terms of their causative agents, hosts, reservoirs, vectors, etc. The epidemiological patterns of these diseases are also different. Hence, it is quite useful to group them according to their reservoir hosts, modes of transmission, causative agents, etc. to facilitate some kinds of collective preventive and control plans. To understand the epidemiology of zoonotic diseases in its right perspective, it is essential to consider human beings as one among the other animal species. It needs to be noted that human has evolved and attained the highest position in the evolutionary ladder but he belongs to the animal kingdom and is considered a social animal. Biologically, humans share many qualities and characteristics with animals.

Classification According to Disease Agent

This is probably the simplest mode of classification of zoonotic diseases. As different kinds of pathogens are involved in the disease causation, zoonoses may be classified in the following categories:

☆ **Bacterial zoonoses:** The zoonotic diseases caused by bacteria fall in this group. Major diseases include anthrax, brucellosis, tuberculosis, leptospirosis, salmonellosis, listeriosis, plague, etc.

☆ **Rickettsial zoonoses:** This category includes the zoonotic diseases caused by rickettsial organisms. Some examples are Q fever, scrub typhus, etc.

☆ **Viral zoonoses:** Zoonotic diseases caused by viruses belong to this category. Some major viral zoonoses are rabies, Japanese encephalitis, avian influenza, yellow fever, etc.

☆ **Fungal or mycotic zoonoses:** Fungal diseases of zoonotic nature such as dermatophytosis, histoplasmosis, candidiasis etc. fall in this group.

☆ **Parasitic zoonoses:** The zoonotic diseases caused by different types of parasites, namely, protozoa, round worms, tape worms, flat worms, ectoparasites, etc. are known as parasitic zoonoses. Major examples include amoebiasis, giardiasis, trichinosis, taeniasis, hydatidosis, schistosomiasis, scabies, etc.

Classification According to Reservoir Host

A reservoir host is a naturally infected animal, bird or human in which the causative agent of the disease usually lives, multiplies and depends primarily for survival and to maintain the infection in nature for long periods. Different diseases may have different reservoir hosts including domestic animals, wild animals, birds or human beings. The reservoir hosts of most of the important zoonoses are various species of domestic and wild vertebrate animals, but in some cases, humans are the major reservoirs or they can be as suitable as animals. Depending upon the involvement of animals, humans, or both as reservoir hosts for a particular disease, a zoonotic diseases can be classified as anthropozoonosis, zooanthroponosis, or amphixenosis (Schwabe 1969).

Anthropozoonoses

This group includes most of the important zoonotic diseases. Various species of domestic and wild vertebrate animals act as reservoir hosts in these diseases. The infections thus can exist in nature independently of man. Human involvement often occurs only through occupational or accidental exposures. Rabies, anthrax, brucellosis, glanders, leptospirosis, echinococcosis are some examples of anthropozoonoses. In this way, the direction of transmission of such diseases is from animals to humans (Figure 1.1).

In many zoononses of this type, infection in man is considered a dead end or cul-de-sac host as further transmission of the infection from an infected person may not occur due to man's certain social and cultural practices and behaviour. For example, hydatid disease (echinococcosis) and some other zoonoses end blindly in mai only because of the usual practice o disposal of dead bodies of humans b' deep burial or burning. However, it ma' not be so in certain tribal and othe communities who traditionally dispos of the dead in open, exposing them t wild animals and birds.

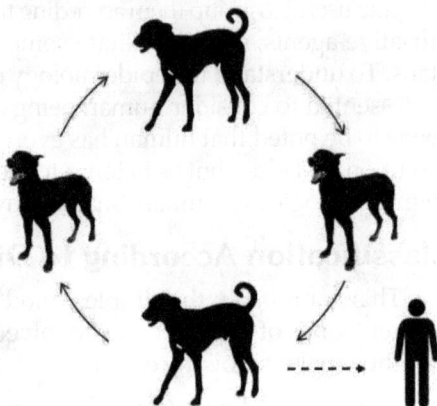

Figure 1.1: Anthropozoonosis (Example: Rabies)

Zooanthroponoses

This group has a small number o zoonoses which normally pass fron man to man but may occasionally infec other vertebrates. Thus, the direction o transmission of these infections is just reverse of that of anthropozoonoses (Figure 1.2). Tuberculosis of the human

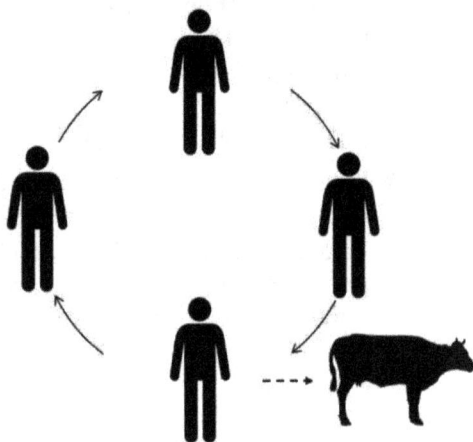

Figure 1.2: Zooanthroponosis (Example: Diphtheria)

type, amoebiasis, diphtheria, infectious hepatitis and poliomyelitis are some major zooanthroponoses.

Amphixenoses

This group includes several more ubiquitous zoonoses in which man and lower vertebrates are equally suitable reservoir hosts. Such diseases are transmitted in both the directions, *i.e.* animals to man as well as from man to animals (Figure 1.3). Such infections can persist in nature in the absence of any of these two categories of the reservoir hosts. Streptococcal, staphylococcal and salmonella infections constitute important amphixenoses.

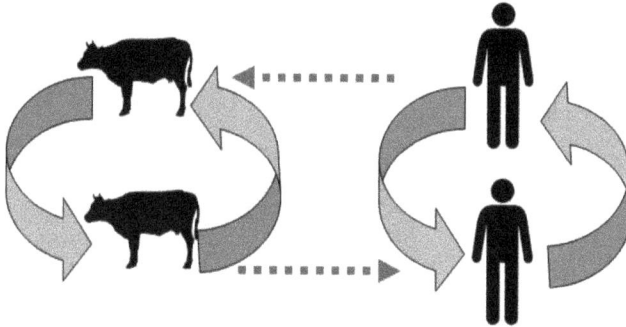

Figure 1.3: Amphixenosis (Example: Streptococcosis)

Classification According to the Cycle of Transmission

Due to the involvement of a vast variety of pathogens, host species, reservoirs and other factors, the epidemiological patterns of each disease differ to great extent, however, some features in the disease transmission process are common in many diseases. While many diseases can be perpetuated in nature by a single vertebrate species, others require more than one vertebrate species. Yet some other diseases essentially require the participation of invertebrate animals or inanimate reservoirs in their transmission cycles. The zoonotic diseases can be classified in the following categories according to their modes of transmissions.

Direct Zoonoses

The diseases in this category require only a single vertebrate species for their maintenance and perpetuation in nature (Figure 1.4) and are transmitted from an infected vertebrate host to a susceptible vertebrate host by contact, vehicle, or mechanical vector. The transmission of infection is thus not only through direct contact with the infected host but can also occur through infected food, air or vectors. Further, the disease causing agent itself undergoes little or no propagative change and no essential developmental change during transmission. A large number of

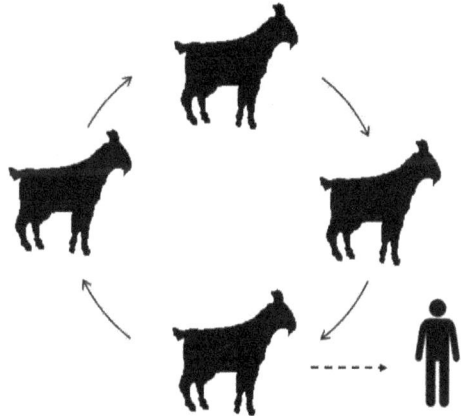

Figure 1.4: Direct zoonosis (Example: Brucellosis)

diseases such as anthrax, brucellosis, leptospirosis, staphylococcal infection, rabies, ringworm, scabies etc. fall in this group.

Cyclozoonoses

In these types of zoonoses, the infectious agent essentially requires more than one vertebrate host but no invertebrate host to complete its developmental cycle. Most

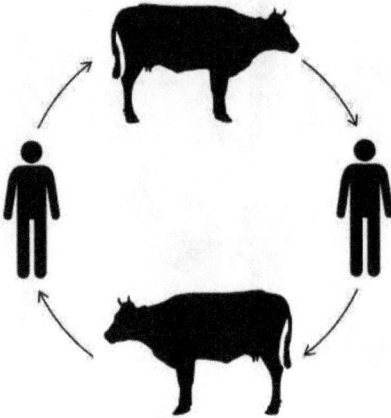

Figure 1.5: Obligatory cyclozoonosis (Example: Taeniasis)

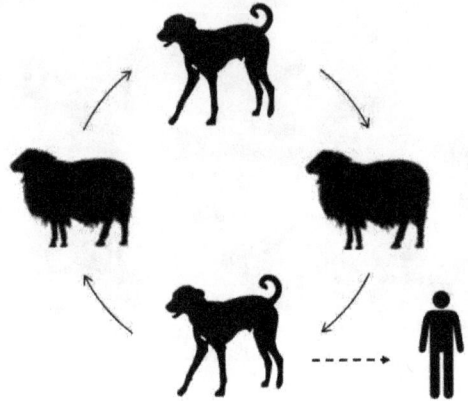

Figure 1.6: Nonobligatory cyclozoonosis (Example: Hydatid disease)

of these zoonoses are due to cestodes. In some diseases, like *Taenia saginata* and *T. solium* infections, man must be one of the vertebrate hosts to complete the life-cycle of the causative agent. These are called obligatory cyclozoonoses. Cattle and man are required in the life-cycle of *T. saginata* (Figure 1.5) and pig and man for *T. solium*. In non-obligatory cyclozoonoses, the life-cycle may be completed in animal hosts alone but man may acquire infection by getting involved accidentally in the transmission cycle. For example, in hydatid disease, the life-cycle of *Echinococcus granulosus* can be completed in sheep and dog but man may acquire infection accidentally (Figure 1.6).

Metazoonoses

This group includes those zoonoses that are transmitted biologically by invertebrate vectors. In metazoonoses, both vertebrate as well as invertebrate hosts are involved in the transmission of the causative organism (Figure 1.7). The role of the invertebrate may be different in different diseases. In some diseases, the causative agent multiplies in the invertebrate vectors, while the invertebrate also serves as a reservoir of infection (Schwabe 1969). This is known as propagative or cyclo-propagative transmission. In others, the infectious

Figure 1.7: Metazoonosis (Example: Yellow fever)

agent merely develops but the invertebrate is not a reservoir here. This is known as developmental transmission. In metazoonoses, there is always an extrinsic incubation period in the invertebrate host before the transmission to another vertebrate host is possible. Metazoonoses include important vector-borne zoonotic diseases like plague,

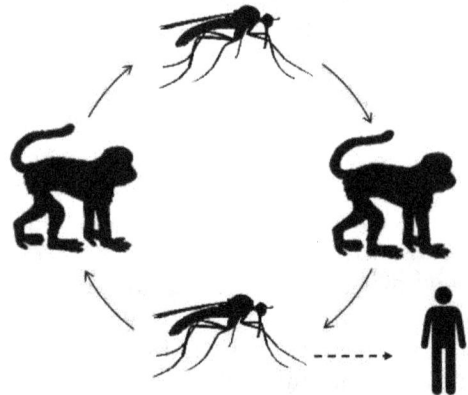

tick-borne encephalitis, yellow fever, etc. Depending upon the hosts required, at least four subtypes of meta-zoonoses may be distinguished.

⭐ Subtype I: requiring one vertebrate and one invertebrate host, for example, sylvatic yellow fever

⭐ Subtype II: requiring one vertebrate and two invertebrate hosts, for example, paragonimiasis

⭐ Subtype III: requiring two vertebrate and one invertebrate hosts, for example, clonorchiasis

⭐ Subtype IV: representing transovarian transmission, for example, tick-borne encephalitis

Saprozoonoses

Saprozoonoses are those zoonoses which require a non-animal site to serve either as a true reservoir of infection (not merely an optional medium) or as a site for an essential phase of development. Thus, the transmission cycle of the diseases in this group depends upon inanimate reservoirs or development sites as well as upon vertebrate hosts (Figure 1.8). Non-animal sites include organic matter, food, soil and plants. Fungal diseases like cryptococcosis and histoplasmosis are major examples of saprozoonoses.

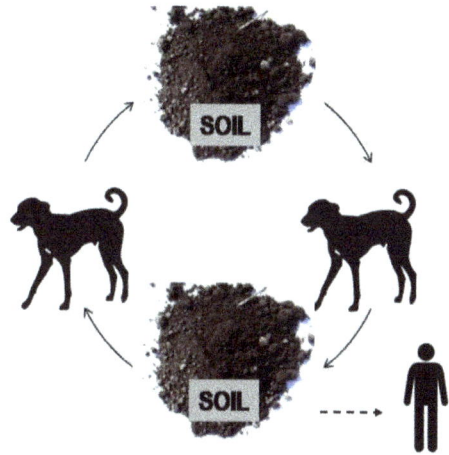

**Figure 1.8: Saprozoonosis
(Example: Cutaneous larva migrans)**

In different diseases, the agent may propagate or it may undergo essential development without propagating in the non-animal site. For example, in the case of histoplasmosis, soil enriched by bird or bat manure serves as a true reservoir but in *Ancylostoma brasiliense* infection, the agent undergoes essential development without propagating. The transmission is accordingly referred as propagative, cyclopropagative, or developmental as in the case of vector-borne infections in the metazoonoses category (Schwabe 1969).

The epidemiological patterns in the above-mentioned infections are quite difference. The mycotic saprozoonoses organisms are apparently not dependent for survival upon either man or another vertebrate, while the helminthic saprozoonotic organisms are dependent on some vertebrate host.

Other Zoonoses

Besides the above four categories, few diseases combine features of more than one of the above types in their transmission cycles. All of these are either flukes which utilize plants for metacercarial encystment or ticks with more than one hosts which spend portions of their life cycles of the ground (Schwabe 1969).

Classification Based on the Type of Animals

Different categories of animals such as wild animals, domiciliated animals and domestic animals may be involved in transmission of different diseases. The zoonoses can thus be categorized as: (1) wild animal-man shared infections; (2) domiciliated animal-man shared infections; and (3) domestic animal-man shared infections, although there may be some overlapping in these cases (Schwabe 1969). For prevention and control of zoonoses, different measures may be required depending on the role of different groups of animals.

Classification Based on Frequency and Severity of Disease

Zoonoses differ in the frequency of human involvement and the severity of human infection. Some zoonoses may be of great agriculture economic importance but may be of only minor public health significance due to very infrequent occurrence or very mild manifestations in man. On the other hand, some infections may be of great public health concern but may have little or no agricultural economic significance. Grouping of zoonoses according to frequency and severity of disease occurrence in human beings is of considerable practical value in allocating different priorities, medical importance and resources to different diseases (Schwabe 1969).

Emerging and Re-emerging Zoonoses

Many of the zoonotic diseases are well-known since the olden days, but a vast number of new or previously less familiar zoonoses have surfaced to the fore during the recent past. A new infection due to the evolution or change of an existing pathogen or parasite resulting in a change of host range, vector, pathogenicity or strain; or the occurrence of a previously unrecognised infection or disease is considered as an emerging disease. Most of the recently emerging diseases affecting public health, sometimes even at global level, have an animal origin. Similarly, there have been instances of the re-emergence of several previously known diseases due to the shifts in their geographical settings or expansion in host range, or significant increase in their prevalence.

The appearance of emerging and re-emerging diseases has substantially increased the impact of zoonoses worldwide, particularly due to the unprecedented levels of global travel, international trade, urbanization, intensive agricultural and livestock farming systems, and consumerism. Further, there might be many lesser known or obscure infectious agents in nature having the potential to cause zoonotic transmission and serious public health consequences in the future. This is supported by the examples of acquired immunodeficiency syndrome (AIDS) and influenza. These diseases have their origins as zoonoses, but they subsequently adapted to human-to-human transmission. AIDS originated from non-human primates, but has now developed into one of the most significant infectious diseases of man. Similarly, certain influenza virus strains have been found to cross the species barrier leading to large human pandemics.

Zoonoses Control

Despite a strong association between poverty, hunger, livestock keeping and zoonoses, and great variations in the incidence and distribution of zoonoses in different regions, zoonotic diseases attract worldwide concern. Because a relatively small number of countries including India bear the greatest burden of zoonoses, it would be a great idea to target such zoonoses hotspots to substantially alleviate the disease burden in the regions riddled with poverty and resource constraints. Endemic zoonoses, including the foodborne zoonoses, require particular attention in the poor countries to get maximum results in terms of lower human disease burden and improved livestock productivity. Further, to attain the cost benefits, it would be worthwhile to apply the scarce resources primarily to those particular diseases that are responsible for the majority of human and animal zoonoses burden (Grace *et al.* 2012).

The outbreaks of diseases such as avian influenza, severe acute respiratory syndrome (SARS), the Ebola virus, and bovine spongiform encephalopathy have pointed to the ease with which zoonotic diseases cross the species barriers between animals and human beings. As the causative agents of such diseases do not depend on human hosts for survival, these go on perpetuating in the nature far beyond the reach of medical intervention. Undoubtedly, a zoonotic disease cannot be controlled without tackling its causative agent at its animal source. The traditional approach of pathogen-specific measures that often operate within the public health department or animal health department in isolation does not help much. The fight against zoonoses requires coordinated actions between animal and public health authorities based on the principles of One Health. Prevention and control of a zoonotic disease requires a multi-dimensional, integrated approach keeping in mind the holistic picture because the wildlife, environment, socio-economic conditions and the interplay of several other factors also influence the disease occurrence.

References

1. Blancou J. 2003. *History of the surveillance and control of transmissible animal diseases*. Office International des Epizooties, Paris.

2. Grace D and McDermott J. 2012. Livestock epidemics. In: Ben Wisner B, Gaillard JC and Kelman I (eds.) *Handbook of hazards and disaster risk reduction*. Routledge, Abingdon Oxfordshire, UK.

3. Grace D, Jones B, McKeever D, Pfeiffer D *et al.* 2011. *Zoonoses: Wildlife/livestock interactions*. A report to the Department for International Development, UK submitted by the International Livestock Research Institute, Nairobi and Royal Veterinary College, London.

4. Grace D, Mutua F, Ochungo P, Kruska R, Jones K, Brierley L, Lapar L, Said M, Herrero M, Phuc PD, Thao NB, Akuku I and Ogutu F. 2012. *Mapping of poverty and likely zoonoses hotspots*. A Report to the Department for International Development, UK submitted by the International Livestock Research Institute, Vietnam and Hanoi School of Public Health, Vietnam.

5. Marr JS and Calisher CH. 2003. Alexander the Great and West Nile virus encephalitis. *Emerg Infect Dis* 9: 1599-1603.

6. Schwabe CW. 1969. *Veterinary medicine and human health*. 2nd ed. William and Wilkins Co., Baltimore.

7. Wheelis M. 2002. Biological warfare at the 1346 Siege of Caffa. *Emerg Infect Dis* 8: 971-975.

8. WHO. 2012. Veterinary public health. World Health Organization. http://www.who.int/zoonoses/vph/en/. Accessed 9 January 2013.

2014, Zoonoses: Bacterial Diseases
Editor: **Sudhi Ranjan Garg**
Published by: **DAYA PUBLISHING HOUSE, NEW DELHI**

Pages **13–37**

2

Anthrax

Sathish B. Shivachandra

Clinical Bacteriology Laboratory,
Indian Veterinary Research Institute, Mukteswar – 263 138

Anthrax, an acute and fatal bacterial disease, derives its name from the Greek word for coal – *anthrakis*, as it causes black coal like skin lesions. Caused by *Bacillus anthracis* organisms, the disease is known to affect herbivores predominantly, which is transmissible to humans. The knowledge of anthrax as a disease and its importance as a zoonosis in the Greco-Roman world had been revealed through several classical texts and mythological sources and until the nineteenth century, numerous names with different linguistic origins were given to the disease throughout history. Among them, most common synonyms around the world are charbon, wool sorters disease, rag pickers disease, malignant carbuncle, and malignant pustule. The descriptions of anthrax begin in antiquity, with the best ancient account by the Roman poet Virgil. During the 19th century, anthrax organisms were involved in several important veterinary medical developments (Sternbach 2003). In 1870s, Robert Koch cultured *B. anthracis* and first established the microbial origin of an infectious disease which led to famous 'Germ Theory'. In 1881, Louis Pasteur and Greenfield independently attenuated the organism and successfully developed vaccines for anthrax in animals (Turnbull 1991). Since the earliest historical records until the development of an effective veterinary vaccine (Sterne 1937), together with the subsequent development of effective antibiotics, the disease has been one of the foremost causes of uncontrolled mortality in herbivores (cattle, sheep, goats, horses and pigs) worldwide. Humans are almost invariably known to contract anthrax directly or indirectly from domestic animals.

In recent times, the disease has received an exceptional global attention mainly due to the possible use of bacillus spores in biological warfare and bioterrorism (Friedlander 2000). The use of an anthrax-based weapon as part of a military campaign was a principal perceived threat during the period of Cold War. In the post-Cold War world too, this remains a concern, but following the attacks conducted through the U.S. postal system in 2001 (Jernigan *et al.* 2001), this scenario has been eclipsed by the worry that anthrax spores might be used against the general population.

Anthrax is an archetype anthropozoonosis, which affects a wide range of animal species, especially herbivorous mammals, and is transmissible to humans through touching either animals killed by anthrax or infected products. Increased international trade in animals and animal products, as well as intense travel of people around the world, represent greater risks of dissemination of anthrax with a higher devastating global impact on public as well as livestock health scenario.

Geographical Distribution

Anthrax has worldwide distribution and has been reported from every country having livestock population. The disease is most prevalent in tropical and subtropical countries which practise largely pastoral agriculture. From the global disease prevalence data reported to the World Organization for Animal Health (OIE), it is evident that anthrax is enzootic in most countries of Africa and Asia, a number of European countries, countries/areas in the American continent and certain areas of Australia. In many other countries, the disease occurs sporadically.

Despite the global prevalence of anthrax, in many cases of incidences of disease outbreaks go unreported. Thus, the reported outbreaks only provide an index of the magnitude of the disease in particular region and could be an underestimate of the extent of the problem. Unfortunately, the mere absence of reported outbreaks is no proof of absence of the disease. Insufficient examination of unexpected livestock deaths and reporting deficiencies are worldwide surveillance defects. However, the understanding of the epidemiology of anthrax is growing better in recent times with the advent of new molecular systems for distinguishing different genotypes/phenotypes of the microorganism (Beyer and Turnbull 2009).

Etiological Agent

The causative agent of anthrax, *Bacillus anthracis*, is a Gram-positive, rod shaped, 2.5 x 10 µm in size, spore forming bacterium. Within the genus *Bacillus*, this one is the only obligate pathogen. Most other species of the genus are common ubiquitous environmental saprophytes such as *B. cereus*, *B. licheniformis* and *B. subtilis*, which are occasionally associated with food poisoning in humans and with other clinical manifestations in both humans and animals (Quinn and Turnbull 1998).

B. anthracis grows readily on nutrient agar, however, 5-7 per cent horse or sheep blood agar is the medium of choice. Following overnight incubation at 37°C, *B. anthracis* colonies are grey-white to grey, 0.3-0.5 mm in diameter, non-haemolytic, with a ground glass moist surface, and very tacky when teased with an inoculating loop. Tailing and prominent wisps of growth trailing back towards the parent colony, all in the same direction, are seen, which is characteristically described as a 'medusa head' appearance. Methionine and thiamine are required for growth, in addition to the growth medium supplemented with multiple amino acids, which most likely serve as sources of nitrogen (Thorne 1993). The organism hydrolyzes casein medium containing glucose, adenine, uracil, thiamine, tryptophan, cysteine and glycine. Confirmation of *B. anthracis* cells is generally accomplished by the demonstration of a capsulated, non-motile, spore forming, Gram-positive rod in blood culture (OIE 2008).

B. anthracis is one of the most monomorphic and homogeneous bacterial species. The chromosome structure of *B. anthracis* is relatively conserved with a few large rearrangements. The *B. anthracis* genome is tripartite and comprised of a single circular chromosome and two circular virulence plasmids; pXO1(182 kb) and pXO2 (96 kb). The genome is highly A+T rich, with only ~35 per cent of the bases from G+C. The prevalence of A+T means that this DNA has a higher buoyant density and lower melting temperatures than many others. The plasmids are relatively large and code for many different genes, but importantly they carry the toxin and capsule determinants. The pXO1 plasmid (Okinaka *et al.* 1999) genes mainly encode for exotoxins which include *pag*A (protective antigen and intra membrane toxin transporter), *lef* [lethal factor (LF), has Zn2+-dependent endoprotease activity] and *cyaA* [Oedema Factor (EF), has calmodulin-sensitive adenylate cyclase activity) (Leppla 1999). The pXO2 plasmid carries the capsule biosynthesis genes (*capBCADE*) which are found in a cluster and contribute to formation of capsule (Okinaka *et al.* 1999).

Virulence Factors of *B. anthracis*

The capsule and the toxin complex are considered as major virulence factors of *B. anthracis*. The poly-D-glutamic acid capsule is presumed to act by protecting the bacterium from phagocytosis. The exotoxin complex, which consists of three synergistically acting proteins, Protective Antigen (PA, 83kDa), Lethal Factor (LF, 87 kDa) and Oedema Factor (EF, 89 kDa), is produced during the log phase of growth of *B. anthracis*. LF in combination with PA (lethal toxin) and EF in combination with PA (oedema toxin) are now regarded as responsible for the characteristic signs and symptoms of anthrax (Leppla 1999). Currently, crystalline structures of all three exotoxins are available for better understanding of their role in pathogenesis (Petosa *et al.* 1997). According to the well-established PA model, PA binds to receptors (CMG2) on the host's cells and is activated by a host protease which cleaves off a 20 kDa fragment, thereby exposing a secondary receptor site for which LF and EF compete to bind. The PA+LF or PA+EF are then internalized and the LF and EF are released into the host cell cytosol. EF is an adenylate cyclase which, by catalysing the abnormal production of cyclic–AMP (cAMP), produces the altered water and ion movements that lead to the characteristic oedema of anthrax (Friedlander 2000). High intracellular cAMP concentrations are cytostatic but not lethal to host cells. EF is known to impair neutrophil function and its role in anthrax infection may be to prevent activation of the inflammatory process. LF appears to be calcium and zinc dependent metalloenzyme endopeptidase (Hammond and Hanna 1998). It has recently been shown (Duesbery *et al.* 1998) that it cleaves the amino terminus of two mitogen-activated protein kinases and thereby disrupts a pathway in the eukaryotic cell concerned with regulating the activity of other molecules by attaching phosphate groups to them. This signalling pathway is known to be involved in cell growth and maturation; the manner in which its disruption leads to the known effects of LF has yet to be elucidated. On the basis of mouse and tissue culture models, macrophages are a major target of lethal toxin which is cytolytic in them. The initial response of sensitive macrophages to lethal toxin is the synthesis of high levels of tumour necrosis factor and interleukin-1 cytokines and it seems probable that death in anthrax results from a septic shock type mechanism resulting from the release of these cytokines. The endothelial cell linings of the capillary network may also be susceptible to lethal toxin and the resulting histologically visible necrosis of lymphatic elements and blood vessel walls is presumably responsible for systemic release of the bacilli and for the characteristic terminal haemorrhage from the nose, mouth and anus of the infected animals (Friedlander 2000).

Host Range

Anthrax is primarily a disease of herbivorous animals, although all mammals, including humans, and some avian species can contract it. Mortality can be very high, especially in herbivores. According to Sterne (1959), amphibians and reptiles are naturally resistant but warming of cold-blooded animals allows them to be experimentally infected. The susceptibility to anthrax varies considerably among different animal species. Naturally, herbivores are particularly susceptible, and omnivores and carnivores are moderately resistant but still succumb. Generally, higher incidences have been reported in large ruminants (cattle and buffaloes), followed by small ruminants (sheep and goat) and a fewer outbreaks reported in pigs and elephants. Although reports of anthrax occurrence in dogs/carnivorous animals, wild animals scavenging anthrax carcasses have been reported from zoological gardens and wildlife sanctuaries or national parks, outbreaks affecting large number of carnivorous animals are very rare.

Previously published LD_{50} doses for anthrax by the parenteral route range from <10 spores for a guinea pig through 3×10^3 for Rhesus monkey, 10^6 for rat, 10^9 for pig, and 5×10^{10} for dog (Watson and Keir 1994). Minimum infectious dose (MID) estimates are only rarely available, but an aerosol MID of 35,000 spores for sheep has been recorded (Fildes 1943). Since it is thought that animals generally acquire anthrax by ingestion of spores, coupled with a lesion for establishment of infection, LD_{50} values, particularly parenterally determined, only provide a rudimentary guide to natural susceptibility. Hence, extrapolation of experimental findings to the natural situation need to be done cautiously, taking into consideration the factors influencing infectivity, such as the strain of *B. anthracis*, route of infection, and the species, breed, strain and state of health of the experimental animal as well as the times and sites at which such tests are done.

Disease Transmission

Although, the disease has been recognized for centuries, it has not yet been established scientifically on how grazing healthy animals contract it. The possible modes of disease transmission are depicted in Figure 2.1. The disease is primarily

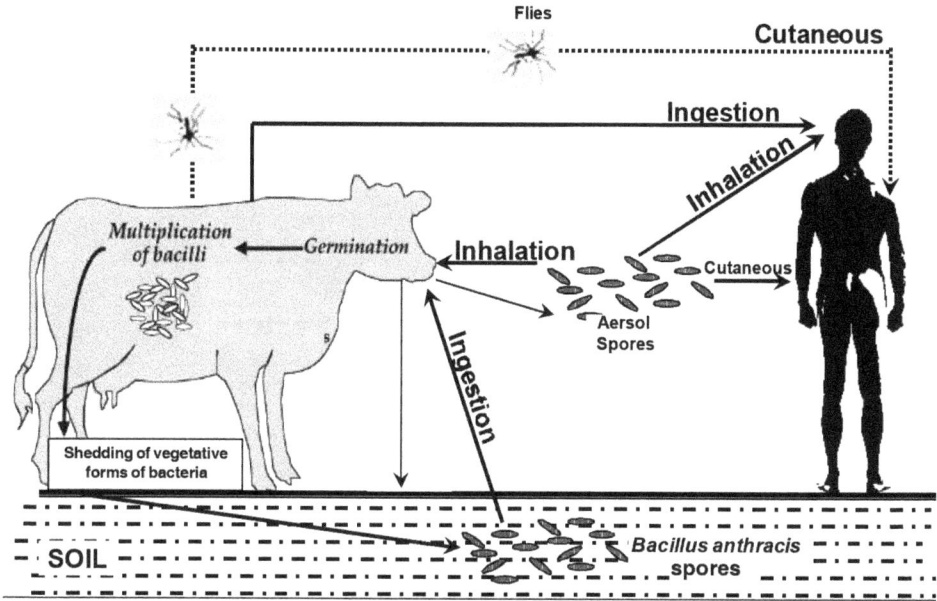

Figure 2.1: Transmission of anthrax between human and animal

noticed in warmer regions around the world, however, it has also been carried to cooler regions through the trade of infected animals or contaminated animal products. The importance of disease transmission lies in its ability to affect large number of livestock at one time. The sporulated forms of the organism shed by an animal dying or dead from anthrax generally provide the source of infection. It is generally assumed that ingestion of spores while close grazing to the soil is a frequent mode of uptake and a lesion is necessary for the initiation of infection. Generally, infective materials or carcasses of infected animals pose a greater hazard to humans as well as other animals in the vicinity and at a distance through their meat, hides, hair, wool or bones, which may be transported to large distances for use in industries, feedstuffs or handicrafts. Livestock may acquire infection through contaminated feedstuffs, or from spores that have reached fields (Turnbull 1998). The source can either be improperly treated locally produced meat and bone meals salvaged from moribund stock or imported infected bones/contaminated meat and bone meals. Cross-contamination can also occur during shipping if the containers are not cleaned out between shipments or if infected dry hides are placed in contact with feed. Similarly, cross-contamination following the reuse of contaminated feed bags has also resulted in outbreaks of anthrax. Moreover, contaminated pastures pose the risk of infection to the grazing stock in several countries. The seasonal occurrence of peak anthrax outbreaks indicate the association of the disease with climatic conditions in varied geographical regions, supporting the hypothesis that an animal can harbour the spores for long periods and manifests the disease only during seasonal stress or immuno-compromised conditions. Sometimes acquisition of the disease is by inhalation of spore-laden dust; however, it would vary with season. It is also widely accepted that in some regions certain types of flies transmit anthrax which is also attributed to particular season.

In recent times, intentional release of anthrax spores into the environment is of greater concern. The fear of terrorist attacks has increased since the events of 11 September 2001 in New York. In the weeks following the terrorist attack, letters containing anthrax spores were received at various locations in the USA (Jernigan *et al.* 2001). The virulent anthrax spores could be distributed via the air and drinking water and many more food sources.

Environmental and Other Predisposing Factors

B. anthracis is an obligate bacterial microbe. Its vegetative forms are more 'fragile' and die more spontaneously in simple environments such as water or milk (Turnbull *et al.* 1991, Bowen *et al.* 1992). The organism, which is in the vegetative form inside the infected host, when released into the outside environment, ceases its multiplication under adverse conditions and undergoes sporulation in the presence of free oxygen to form spores. The spores are markedly resistant to environmental extremes such as heat, cold, pH desiccation, etc. as well as to chemicals including disinfectants, irradiation, and other extreme conditions. Hence, spores constitute the predominant phase of the organism in the environment and can remain there for a longer duration. The spores germinate within the infected host to produce vegetative forms which multiply, eventually killing the host.

Based on some incidences, anthrax is also considered as a seasonal disease; its incidence in any one place is related to temperature, rains or drought; however, the literature shows that the conditions which predispose to outbreaks differ widely from location to location. Climate probably acts directly or indirectly by influencing the way in which an animal comes into contact with the spores during close grazing or water scarcity or varied levels of host resistance to infection.

Although several hypotheses are considered such as the effects of season, rainfall, temperature, soil, vegetation, host condition and population density on the epidemiology of anthrax, little agreement exists on the roles played by these factors in the incidence of anthrax. Some reports indicate that anthrax outbreaks are likely to develop mainly during the dry months that follow a prolonged period of rain when the spores are exposed and the ruminants have greater access to them. Hot and humid season facilitates germination of spores in the environment. However, there is a need to understand the multi-factorial role in persistence and precipitation of anthrax outbreaks.

Prevalence of Anthrax in Animals

It has been clearly noticed that inadequate reporting and insufficient examination of unexpected livestock deaths are common throughout the world. In warmer countries, especially in Africa and Asia, anthrax epidemics are more common in livestock. It has been hypothesised that as countries become free of anthrax or the annual incidence of outbreaks approaches near zero, the number of animals affected in an outbreak increases. This seems to be due to the decreasing veterinary experience in recognizing cases and in dealing appropriately with the outbreaks. The mere absence of reported livestock anthrax does not mean that a country is free from the disease.

Prevalence in India and Other Developing Countries

Anthrax continues to be endemic in most of the underdeveloped and developing countries, especially in the tropical regions, where animal husbandry practices are poor. The disease is enzootic in southern India, but it is less frequent to absent in the northern states where soil is more acidic. Many factors such as poor disease surveillance coverage, inadequate diagnostic capacity, and lack of clear strategies to address the plight of zoonotic diseases have been contributing to underdiagnosis and underreporting of zoonotic diseases.

In the data of the National Animal Diseases Referral Expert System (NADRES) of the Project Directorate on Animal Disease Monitoring and Surveillance (PDADMAS) for the last two decades (1991-2010), anthrax features as one of the top ten diseases reported in India and also a major cause of deaths in livestock (Gajendragad and Uma 2012). Anthrax has been reported in eighteen states, namely, Andhra Pradesh, Assam, Bihar, Chhattisgarh, Gujarat, Himachal Pradesh, Jammu and Kashmir, Jharkhand, Karnataka, Kerala, Madhya Pradesh, Maharashtra, Manipur, Meghalaya, Odisha, Rajasthan, Tamil Nadu and West Bengal during 1991 to 2010. Despite annual fluctuations in the frequencies in the states, there has been gradual increase in the number of outbreaks since 1992 with a peak between 2000 and 2002, however, subsequently, gradual decline has been observed. This has been attributed to the mass awareness programmes among the farming community as well as simultaneous implementation of control measures by the central and respective state governments.

In India, the incidence of anthrax is known to vary greatly in various agro-climatic zones of the country, with widely reported incidences in the South zone, followed by East, West and Northeast zones. Most of the districts of Andhra Pradesh, few districts of Karnataka, Tamil Nadu, Kerala, Gujarat, West Bengal and Assam have been identified as very high anthrax pathozones. The decadal and quinquenial trend of anthrax occurrence showed that there were periodic fluctuations in its occurrence in different states as well as considerable annual variations. The reports in the last two decades indicate an overall decrease in the occurrence at anthrax in 2001-2010, except in few states like Karnataka, Tamil Nadu, Odisha and Assam. The epidemics of anthrax are known to increase in some seasons, especially between July to September and in November and January, coinciding with the post-monsoon months across the country. Seasonal variations in anthrax occurrence have also been reported between the different zones of India (Gajendragad and Uma 2012).

Anthrax is also endemic in India's neighbouring country, Nepal. Some Southeast Asian countries such as Myanmar, Vietnam, Cambodia, Western China are severely affected; while Thailand used to be free, now it is endemic and continually threatened. Malaysia and Taiwan are free, and the disease is limited to single regions of the Philippines and Indonesia. All these countries differ from the rest of the world in that pigs and water buffalo, as opposed to ruminants, are the animals commonly affected.

Anthrax is enzootic in Peru, Bolivia and Venezuela. In South Africa, the annual number of outbreaks is less than five and occasionally zero, despite the continued occurrence of the disease in the wildlife in various parks. The disease remains and, reported in Botswana, Zimbabwe and Zambia despite the implementation of good control programme. In equatorial African countries, anthrax is enzootic or hyperenzootic.

Overall, the exact prevalence of anthrax in developing countries needs to be estimated systematically.

Prevalence in Developed Countries

With the advent of national programmes directed at progressive reduction in livestock population over the past three decades, anthrax in animals is now truly absent or only sporadic in the middle and higher latitudes of Europe. However, it is still common in some European countries such as Greece, Italy, Spain, Turkey and Yugoslavia, especially adjoining the Mediterranean. In Canada, apart from its continued incidence in bison in the MacKenzie Bison Range, North West Territory, and in the Wood Buffalo National Park in Northern Alberta, anthrax is sporadic in southern Alberta and Saskachewan, with a recent singular outbreak in Ontario. In the USA, the disease is confined to a few persistent pockets with sporadic cases in South Dakota, Nebraska, and Oklahoma, but probably a hyperendemic situation persists throughout the South-Western quadrant of Texas. In Latin America, the true situation is unclear due to underreporting and failure to diagnose unexpected livestock deaths that occur, especially in small ruminants. Anthrax is enzootic in Central America, Mexico and Guatemala, with decreasing occurrence in Belize, the Caribbean with the exception of Haiti. The Middle Eastern and adjoining countries of former Soviet republics continue to experience hyperendemic anthrax.

Prevalence of Anthrax in Human Beings

The prevalence of anthrax in endemic regions of various countries could potentially act as a major source of anthrax in humans, irrespective of developed or developing countries, mainly due to international trade of animal products. It could accelerate the spill over of infection into human populations. The importance of the disease lies in its ability to affect large number of livestock at one time, and carcasses posing hazard to humans and other animals both in the vicinity and at a distance through meat, hides, hair, wool or bones. Hides, hair, wool and bones may be transported to long distances for use in industries, feedstuffs or handicrafts. Nevertheless, natural environmental exposure to infectious doses in the normal course of human life and endeavour is presumed to be less. It has been noticed that human anthrax incidence is dependent on the level of exposure to affected animals in a particular country.

The actual incidence of anthrax in developing countries is not known accurately mostly due to under-reporting and improper health care system capable of diagnosing the disease. Historical analysis of epidemiological data globally revealed the following approximate ratios: (a) one human cutaneous anthrax case to ten anthrax livestock carcasses; (b) one incident of enteric human anthrax to 30-60 anthrax-infected animals eaten; (c) 100-200 cutaneous cases for each enteric case that occurs in humans. Reported animal to human case ratios in a country or region reflect the economic conditions, quality of surveillance, social traditions, dietary behaviour, etc. in that country or region. In contrast to animals, age- or sex-related bias is generally not apparent in humans, though males generally have higher occupational risk rates in many countries.

However, artificially created aggressive exposures to overwhelmingly massive doses (many millions of spores) such as an act of bio-terrorism have potential to cause substantial devastation to human as well as animal populations within selective targeted areas.

Prevalence in India and Other Developing Countries

In India, both sporadic cases and outbreaks are being reported regularly. Most cases have occurred in agricultural labourers who are involved in handling of meat or skin of infected animals (Kumar *et al.* 2000). Since the livestock population predominantly exists in the rural areas, people living in villages and depending on livestock for their livelihood face greater risk of infection (Bhatt *et al.* 1989). Similarly, a large number of people working in tanneries and leather industries have higher risk (Lalitha and Kumar 1996). The incidence of anthrax in humans in the country is not known accurately due to the fact that a large number of cases go unreported. Moreover, sporadic human cases go unnoticed or undiagnosed due to lack of medical facility in rural areas. As a result, the true incidence of anthrax in humans is likely to be higher than that reported in the literature. However, detailed information collected from three Southern Indian states, Andhra Pradesh, Karnataka and Tamil Nadu, has confirmed the endemicity of anthrax. During the last two decades, about 70 cases of human anthrax have been encountered at Vellore, of which 26 cases had cutaneous anthrax. Previously, more than 112 cases of anthrax (71 cutaneous anthrax cases) were recorded in Vellore, Tamil Nadu (Lalitha and Kumar 1996). Several cases of cutaneous anthrax in humans have been noticed especially in areas where endemic outbreaks of anthrax in cattle, goats, sheep, etc. were recorded (Thappa and Karthikeyan 2001).

Human case rates for anthrax are highest in Africa, the Middle East and central and Southern Asia. Where the disease is infrequent or rare in livestock, it is rarely seen in humans. Infectious doses, which have not been established for humans, and the severity of the resulting infection clearly depend on several factors such as route of infection, nutritional and other states of health on the part of the infected person, and probably on the relative virulence of the infecting strain.

Prevalence in Developed Countries

In developed countries, since there are large scale mechanized livestock production systems, the incidences/chances of human contact with infected animals are relatively lower. However, in such countries with advanced agriculture and animal husbandry practices, the source of infection can either be improperly treated locally produced meat and bone meals salvaged from moribund stock or imported infected bones/contaminated meat and bone meals. Cross-contamination can occur during shipping of infected dry hides, vegetables, feedstuffs, etc. Similar cross-contamination following the reuse of feed sacks has also resulted in outbreaks of anthrax. In England, Germany, Italy and Canada, pastures flooded with tannery wastewater historically have posed a long recognized hazard to grazing stock and subsequently human incidences.

Before the vaccines and antibiotics became available, and at a time when understanding of industrial hygiene was relatively basic, workers mainly involved in industrial occupations such as processing animal products were exposed to significant number of anthrax spores on a daily basis. In Britain, 354 cases of anthrax in such industries were notified during the 13-year period 1899-1912 (Anon 1918). Although the number of such exposed persons is not known, it must have been in thousands. Nevertheless, outbreaks and epidemics do occur in humans; sometimes these are sizeable, such as the epidemic in Zimbabwe which began in 1979 (Turner 1980). The largest reported outbreak of human inhalational anthrax that took place in 1979 in the city then called Sverdlovsk (now Ekaterinburg) in the former Soviet Union near a Soviet military microbiology facility. Occasionally, the case fatality rates are substantial, such as in the Sverdlovsk incident (Abramova and Grinberg 1993) and outbreak in a mill in New Hampshire, USA, in 1957 (Brachman *et al.* 1960). In the 1900s, human inhalation anthrax occurred sporadically in the United States among textile and tanning workers, but the incidence of the illness had declined dramatically. In October and November 2001, 22 cases of confirmed or suspected inhalation and cutaneous anthrax were reported associated with the intentional release of the organism in the United States. An additional case of cutaneous disease occurred in March 2002 (Sternbach 2003).

In most of the developed countries, major sources of human anthrax infection are indirect contact with infected animals, or industrial/occupational exposure to infected or contaminated animal products (Shadomy and Smith 2008). Moreover, the industrial anthrax incidence data from many countries is infrequently, erratically or incompletely reported at the local or national levels. In the developed parts of the world, where it is now seen rarely, anthrax is largely considered as biological warfare agent (Pile *et al.* 1998, Friedlander 2000) that has resulted in major public anxiety over the recent years, especially after 2001 (Jernigan *et al.* 2001).

Disease Manifestations in Animals

Following entry, anthrax spore are carried to the lymphatics where they undergo germination and multiply. Initially, during the incubation period in the susceptible herbivore, which ranges from about 36 to 72 hours, the bacteria are filtered out by spleen and other parts of the reticuloendothelial system which leads to the hyperacute systemic phase, usually without easily discernible prior symptoms. The bacteria build up rapidly in the blood (doubling time about 0.75 to 2 hours depending on host species) to levels of >10^8/ml together with massive toxaemia at the time of death. The action of the toxin on the endothelial cell lining of the blood vessels results in their breakdown, internal bleeding and the characteristic terminal haemorrhage to the exterior which is an essential part of the organism's cycle of infection (OIE 2008).

Anthrax primarily affects livestock and wildlife with clinical features of hyperacute or acute symptoms and usually with a fatal outcome. The first signs of an anthrax outbreak are one or more sudden deaths in the affected livestock, although farmers may reflect retrospectively that the animals had shown signs such as having been off feed or having produced less milk than usual. During the systemic phase, the animals become distressed, appear to have difficulty in breathing and cease eating and drinking. Swellings in the submandibular fossa may be apparent; temperature may remain normal for most of the period or may rise. During the initial phase of infection, animal may remain responsive to the treatment, but if treatment fails it lapses into coma followed by death from shock.

Apart from sudden death, bleeding from orifices, subcutaneous haemorrhage, without prior symptoms or following a brief period of fever and disorientation is observed in ruminants. In equines and some wild herbivores, transient symptoms like fever, restlessness and dyspnoea may be apparent. In case of pigs, carnivores and primates, local oedemas and swelling of face and neck or of lymph nodes, particularly mandibular and pharyngeal and/or mesenteric, are noticed.

An anthrax affected carcass is not advised to be opened as it is known to contaminate the environment by spilled body fluids with subsequent spore formation, but sometimes if carcass has been opened, the dark unclotted blood and markedly enlarged haemorrhagic spleen (splenomegaly) are immediately apparent. Thickened mesentery and excessive oedematous and peritoneal fluid may be apparent. Petechial haemorrhages may be visible on many of the organs and the intestinal mucosa may be dark red and oedematous with areas of necrosis. In horses, the intestine and parenchymatous organs may be less affected than in sheep and cattle and the subcutaneous and intramuscular tissues may be oedematous. In pigs, enlarged mandibular and suprapharyngeal lymph nodes are visible. The intestinal anthrax may only become apparent at necropsy; smears and cultures should be made from the mesenteric fluid and lymph nodes. In other animals, in addition to any haemorrhagic exudates, severe inflammation and oedematous swelling of the lips, tongue, gums, jowls and throat may be diagnostic indicators.

Disease Manifestations in Humans

Human beings almost invariably contract anthrax directly or indirectly from infected animals. Records of person-to-person spread or laboratory-acquired anthrax are rare (Heyworth *et al.* 1975, Collins 1988, Quinn and Turnbull 1998). Based on the mode of human exposure to spores and subsequent anthrax infections, it is broadly divided in two types:

a) Non-industrial Anthrax

It can also be called an occupational anthrax, since it mainly affects individuals such as farmers, veterinarians, butchers and animal handlers. Infection occurs following the accidental exposure to aerosol or contaminated materials. This kind of infection usually manifests as the cutaneous form of anthrax. The condition may be seasonal as it is known to occur during the season of occurrence of the disease in animals, which act as source of infection for humans.

b) Industrial Anthrax

It occurs mainly in individuals who are employed in the industrial processing of bones, hides, wool and other animal products. The contaminated animal products act as source of infection.

Based on the route by which the infection is acquired, the disease is categorized into mainly three forms:

i) Cutaneous Anthrax

It is generally believed that *B. anthracis* is non-invasive. However, cutaneous form of infection is mainly acquired through skin following an insect bite or a small cut, abrasion or other lesion (bite, ulcer, etc.). Lesions are generally seen on exposed regions of the body, mostly on the face, neck, hands and wrists. The incubation period ranges from 9 hours to 2 weeks, mostly 2 to 6 or 7 days. Generally, following the entry of spores through skin lesions, a small pimple or papule appears in 1-2 days. A ring of vesicles develops around the papule around 3-4 days. Then, vesicular fluid may be exuded. Unless there is secondary infection, there is no pus and the lesion is not painful, although painful lymphadenitis may occur in the regional lymph nodes. By 5-7 days, the original papule ulcerates to form the characteristic eschar. Clinical symptoms may be more severe if the lesion is located in the face, neck or chest. The eschar begins to resolve by 10th day; resolution takes almost six weeks and is not hastened by treatment. A small proportion of cases, if untreated, develop systemic anthrax with hyperacute symptoms and finally lead to high fever, toxaemia, regional painful adenomegaly and extensive oedema; shock and death may ensue. This form of anthrax is known to account for 95 per cent or more of human cases globally. In comparison to other two lethal forms (inhalation/gastroinstestinal), cutaneous form is often considered as self-limiting following treatment with antibiotics. Studies have indicated approximately 10-20 per cent fatality in untreated humans, whereas with treatment fatality could be less than 1 per cent in the affected individuals. Cutaneous anthrax is greatly reduced in the high risk occupations by appropriate clothing and gloves, dressing of wounds and other hygienic practices.

ii) Gastrointestinal Tract Anthrax

The infection here is contracted through ingestion of either contaminated food, especially meat of an infected/dead animal, or contaminated water. There are two clinical forms of gastrointestinal anthrax. a) *Intestinal anthrax*: Symptoms include nausea, vomiting, fever, abdominal pain, haematemesis, bloody diarrhoea and massive ascites. Unless treatment commences early enough, toxaemia and shock develop, followed by death. There is evidence that mild, undiagnosed cases with recovery occur. b) *Oropharyngeal anthrax*: The main clinical features are sore throat, dysphagia, fever, regional lymphadenopathy in the neck and toxaemia. Even with treatment, the mortality is about 50 per cent (Doganay *et al.* 1986).

iii) Inhalation (Pulmonary) Anthrax

In this case, infection occurs following inhalation of airborne anthrax spores or spore-laden dust. Recorded inhalation LD_{50} in non-human primates ranges from 2,500 to 760,000 spores (Meselson *et al.* 1994, Watson and Keir 1994). The US Department of Defence bases its strategies on an estimate that the LD_{50} for humans is

8,000 to 10,000 spores (Meselson *et al.* 1994). In any event, substantial exposure is evidently necessary before the risk of inhalation anthrax becomes significant. The likelihood of inhaled spores penetrating far enough to induce inhalation anthrax depends greatly on the size of the particles. Inhalation anthrax was the cause of death in the deliberate release of anthrax spores in the United States in October 2002.

Symptoms prior to the onset of the final hyperacute phase are nonspecific and flu-like with mild upper respiratory tract signs. Earlier case reports of inhalation anthrax indicate that the illness begins insidiously with mild fever, fatigue and malaise lasting one to several days (Plotkin *et al.* 1960). Headache, muscle aches, chills and fever are noticed with development of a cough and mild pain in the chest in some cases. This mild initial phase is followed by sudden development of dyspnoea, cyanosis, disorientation with coma and deaths especially in persons who do not respond to treatment. In pulmonary anthrax, the X-ray picture of the lung is very characteristic, with extremely enlarged mediastinal lymph nodes. Death occurs within 24 hours following the onset of hyperacute phase.

Among the above three forms of the disease, gastrointestinal tract and pulmonary cases are more often fatal, largely because they go unrecognized until it is too late for effective treatment. However, serological and epidemiological evidence suggests that undiagnosed low grade gastrointestinal tract or pulmonary anthrax with recovery can also occur, and may not be infrequent among the exposed groups (Brachman *et al.* 1960, Van den Bosch 1996).

Besides the above common modes of disease transmission, mechanical transmission by biting insects is believed to be at least an occasional route of infection in humans in some countries (Rao and Mohiyudeen 1958, Davies 1983). This mechanism of infection has also been demonstrated experimentally (Turell and Knudson 1987).

iv) Anthrax Meningitis

Development of haemorrhagic meningitis is a dangerous possibility in all the above-mentioned forms of anthrax. In such events, the clinical signs of meningitis with intense inflammation of the meninges, markedly elevated CSF pressure and appearance of blood in CSF are followed rapidly by loss of consciousness and death. The case fatality rate is almost 100 per cent (Kanungo *et al.* 1990, Lalitha *et al.* 1996). However, a few instances of survival following treatment have been noticed (Khanne *et al.* 1989).

v) Anthrax Sepsis

Generally, it develops after the lymphohematogenous spread of *B. anthracis* organisms from a primary cutaneous, gastrointestinal or pulmonary lesion. Clinical features include high fever, toxaemia and shock, leading to death in a short time.

Diagnostic Procedures

A sudden death of animals with oozing of unclotted blood from natural orifices is generally suspected for anthrax. Post-mortem examination of such carcasses is discouraged to avoid environmental contamination (OIE 2008). However, the dead animals may show any number of lesions, none of which is pathognomonic or entirely consistent. Lesions most commonly seen are those of a generalized septicaemia often accompanied by an enlarged spleen having a 'blackberry jam' consistency and poorly clotted blood. Haemorrhage from the nose, mouth, vagina and/or anus at death is not a common sign.

For differential diagnosis, other causes of sudden death should also be considered. Amongst these are botulism, blackleg, peracute babesiosis, chemical poisoning (heavy metals, other poisons), ingestion of toxic plants, snake bite, lightning strike or metabolic disorders such as lactic acidosis, magnesium deficiency or bloat. Different conventional and molecular methods are currently in use for confirmatory as well as differential diagnosis.

Conventional Approaches

Generally, blood from the affected animals is primarily collected for examination. Swabs of blood, body fluids, or those taken from incisions in tissues or organs can also be used. Smears of blood or tissues from fresh anthrax-infected carcasses are used to demonstrate encapsulated *B. anthracis*. Specimens can be examined by cultural examination on blood agar and isolation of the organism in bacteriology laboratories.

a) Isolation

Smear of blood or tissues from fresh anthrax-infected carcasses are used for isolation using 5-7 per cent of sheep/horse blood agar which is a medium of choice. Further, colonies are identified by their characteristic 'medusa' colony. Isolation or demonstration of organisms is difficult, especially in terminal stage of bacteraemia or in animals that had received antibiotics before death. Similarly, recovery of organisms from old decomposed carcasses, processed specimens (bone meal, hides), or environmental samples (contaminated soil) is also often difficult which requires labour intensive procedures. However, selective medium such as PLET (polymyxin B-lysozyme-EDTA-thallous acetate agar) has been recommended for selective growth of the organisms (Knisely 1966).

b) Bacteriphage Lysis Method

The susceptibility of *B. anthracis* organisms to gamma bacteriophage is used for confirmation similar to disc diffusion method wherein zone of lysis is observed on agar plate (Brown and Cherry 1955). However, in recent times, some resistant strains have also been identified.

c) Demonstration of Capsule

Virulent encapsulated organisms are demonstrated in fresh blood or smears of tissues collected from affected animals following staining with polychrome methylene blue (M'Fadyean reaction). The capsule stains pink, whereas the bacillus cells stain dark blue. The cells are found in pairs or short chains and are often square-ended. The encapsulated virulent *B. anthracis* forms mucoid colonies on agar plates which can be visualised by capsule staining. Gram's and Giemsa stains do not stain the capsule of bacterial cell grown on nutrient agar or broth, however, they could be used as differential stains.

d) Mice or Guinea Pig Inoculation

This method is employed when recovery of organisms is less efficient, especially from clinical specimens of animals that have received antibiotic therapy. Mice or guinea pigs inoculated either subcutaneously or intramuscularly die in 48-72 hours post-inoculation. Further confirmation can be carried out as described above.

e) Ascoli's Test

This test is used to demonstrate the presence of thermostable anthrax antigen in animal tissues (Ascoli 1911). A homogenised boiled tissue lysate is allowed to react with hyperimmune serum to produce a precipitation reaction that indicates positive reaction. A simple, rapid and highly sensitive and specific chromatographic device that uses a monoclonal capture antibody detecting the anthrax specific protective antigen, has now been designed as alternative to Ascoli's test (Burans *et al.* 1996).

f) Immunofluorescence

It is used to demonstrate capsule in virulent organisms using specific hyperimmune sera (Ezzell and Abshire 1996). However, this method has not been used for routine diagnosis of anthrax.

g) Serological Methods

The best currently accepted serological procedure is the enzyme-linked immunosorbent assay (ELISA) in microtitre plates coated with PA component of the anthrax toxin, which appears to be truly specific for *B. anthracis*. This test is commonly used as a retrospective test or to evaluate the immune response in animals or humans following vaccination. However, this test is currently confined to a few specialist laboratories. At present, various versions of the ELISA exist and standard procedures are available (Turnbull *et al.* 1992, Redmond *et al.* 1997). Unfortunately, current ELISAs for detection of anthrax PA and LF are able to achieve sensitivity levels of only ≤1 to 20 ng/ml.

Molecular Approaches

The use of anthrax spores in letters in a bio-terrorist event in 2001 has led to rapid advances in development of novel detection technologies, especially from all kinds of potential environmental samples. In view of the restricted use of conventional methods for identification of causative agent, coupled with misidentification of *Bacillus* species from environmental samples, polymerase chain reaction (PCR) assays have been developed, which can be directly applied on clinical samples for rapid and specific identification of *B. anthracis* within 3-4 hours (Hutson *et al.* 1993, Beyer *et al.* 1996). Several molecular tools have been in use for more than a decade for rapid detection and epidemiological investigation of infectious diseases in animals as well as in humans (Olive and Bean 1999, Shivachandra *et al.* 2003).

Confirmation of virulent anthrax isolates can be done by amplifying toxin genes using PCR. Template DNA for PCR assay can be prepared either from a fresh colony of *B. anthracis* on nutrient agar or by suspension of growth in sterile deionised water. Following preparation of colony or culture lysate, PCR reaction can be set up using published primers (Hutson *et al.* 1993, Beyer *et al.* 1996) for confirming the presence of pX01 and pX02 plasmids, which are known to encode for the protective antigen and capsule, respectively. The primers that can be used to detect the virulent isolates are shown in Table 2.1.

Table 2.1: Primers for detecting virulent isolates of *B. anthracis*

Target Gene	Primer ID	Primer Sequence	Amplicon Size
Protective antigen (PA) (pX01 plasmid)	PA 5 3048-3029 PA 8 2452-2471	5'-TCCTAACACTAACGAAGTCG-3' 5'-GAGGTAGAAGGATATACGGT-3'	~596 bp
Capsule (pX02 plasmid)	1234 1411-1430 1301 2257-2238	5'-CTGAGCCATTAATCGATATG-3' 5'-TCCCACTTACGTAATCTGAG-3'	~846 bp

Alternative to the above-mentioned primers in PCR assays, the method described by Jackson *et al.* (1998) and Ramisse *et al.* (1996) could be used in multiplex PCR assays directly employed on tissue samples. However, PCR assays may not be applied to detect the released exotoxins in the blood/tissue samples. Multiplex real-time PCRs have been developed to specifically detect and differentiate the organism from other zoonotic agents (Janse *et al.* 2010). This ensures highly reliable detection, while template consumption and laboratory effort are kept at a minimum. Since, *B. anthracis* is closely related to the endospore forming bacteria *B. cereus* and *B. thuringiensis*, a novel multiplex real-time PCR for simultaneous specific identification of *B. anthracis* and discrimination of different *B. anthracis* virulence types has been developed (Wielinga *et al.* 2011). Although, *B. anthracis* is widely described as a genetically monomorphic species, multiple-locus variable-number tandem-repeat analysis (MLVA) or VNTR and *pagA* gene sequencing could be effectively used to determine the genetic diversity of various anthrax isolates for rapid epidemiological studies (Sue *et al.* 2007).

Tang *et al.* (2009) developed a europium nanoparticle-based immunoassay (ENIA) for sensitive detection of anthrax PA. The ENIA exhibited a linear dose-dependent pattern within the detection range 0.01-100 ng/ml and was approximately 100-fold more sensitive than ELISA. These results indicate that the universal labeling technology based on europium nanoparticles and its application may provide a rapid and sensitive testing platform for clinical diagnosis and laboratory research.

Prevention and Control Strategies

In order to effectively prevent and control anthrax, both in animals as well as in humans, several international organizations have streamlined guidelines for continued surveillance of anthrax and implementation of appropriate strategies by the respective local administrations (Turnbull *et al.* 1998, OIE 2008).

Preventive Measures

The only means of eradicating anthrax in animals, and thus in humans, is through animal vaccination (Turnbull 1998). As anthrax is almost invariably fatal in domestic animals, a preventive strategy should be adopted involving regular vaccination of all susceptible animals (cattle, sheep, goats) which are at high risk in endemic areas. Vaccination programmes should be strategically implemented in endemic areas by administering at least a month prior to the presumed anthrax outbreak period. This maximizes the likelihood that animals will develop protective immunity against the bacterium before the highest-risk period. Based on the seasonal occurrence report, it is advisable to vaccinate the susceptible animals during the latter half of May or early June.

Anthrax Vaccines for Animals

In 1937, Sterne developed anthrax spore vaccine which is used most widely for prevention of anthrax in animals (Sterne 1937). He derived a rough variant (34F2) of virulent *B. anthracis* from culture on serum agar in an elevated CO_2 atmosphere, which was incapable of forming a capsule and was subsequently found to have lost the pX02 plasmid. It has become the most widely used strain worldwide for animal anthrax vaccine production. In Central and Eastern Europe, an equivalent pX02-derivative, Strain 55, is the active ingredient of the current livestock vaccine. The Sterne strain is very effective in preventing disease in animals. Its excellent performance over several decades has not let any motivation to develop a new successor or alternative strategy. However, the residual virulence it retains for certain animal species, the limited duration of protection conferred, and the fact that it must be administered by injection make it less than ideal in certain situations, especially in developing countries (Turnbull 1991). Hence, there is a renewed interest in new generation anthrax vaccines for animals, particularly with respect to the development of a fully avirulent formulation or genetically engineered invasive organism that might carry and express protective antigen to elicit long lasting protective immune response.

Anthrax Vaccines for Humans

Initially, the vaccine was introduced for workers at high risk occupations (Brachman *et al.* 1962). In early period, although live spore vaccines were prepared for human use in China and USSR, these vaccines were not licensed for use in humans in most other countries. In UK and USA, non-living human vaccines developed in the 1950s and 1960s, respectively, are currently available (Turnbull 1991). However, protection tests in animals have indicated that currently licensed human non-living vaccines may have limited efficacy and the protection they confer is of uncertain duration. Prior to the advent of antibiotics, anthrax was treated by passive immunization with animal antisera. This practice indicated an important role of antibodies in protecting against the disease. Today there is overwhelming evidence that antibodies are key players in conferring immunity to anthrax.

Antibodies to *B. anthracis* proteins, in particular PA, confer immunity to inhalation anthrax and can be induced by vaccination. In fact, the protective effect of the anthrax vaccines licensed in USA (anthrax vaccine adsorbed [AVA], also known as BioThrax®), and in UK (anthrax vaccine precipitated) is based on induction of an antibody response to *B. anthracis* proteins, primarily PA. The vaccine is derived from a *B. anthracis* culture supernatant, whose major component is PA with trace amounts of other bacterial components, including EF and LF, which are adsorbed to aluminum hydroxide gel (Grabenstein 2003). Numerous studies have confirmed that an antibody response to PA is sufficient to provide protection (Ramirez *et al.* 2002, Leppla *et al.* 2002). However, vaccination is unlikely to provide protection when implemented after an individual has been exposed to aerosolized *B. anthracis* spores because of short incubation period and rapid progression of the disease. In this situation, prophylactic treatment with antibiotics is effective when initiated prior to the onset of symptoms and if maintained over a period of at least 60 days. A major drawback of the AVA vaccine is its lot to lot variation, ill-defined general composition, and lengthy course of administration, coupled with side effects. Six injections over a course of 18 months are considered necessary to induce protection with subsequent annual boosters recommended to maintain immunity.

These drawbacks have led to increased efforts in recent years to develop next generation vaccines based on recombinant DNA/protein technology that are more rigorously defined and confer more rapid protection. The most developed vaccine candidate is based on recombinant PA expressed and purified from *Escherichia coli* or from an asporogenic, nontoxigenic, nonencapsulated strain of *B. anthracis* (Leppla *et al.* 2002, Ramirez *et al.* 2002). Moreover, several delivery systems have also been recently developed such as anthrax toxin antigens displayed on bacteriophage T4, needle free-skin patch, toxin-hybrids, domains of toxin, fusion antigens, etc. (Matyas *et al.* 2002, Shivachandra *et al.* 2006, 2007, Li *et al.* 2006, 2007, Peachman *et al.* 2012).

Alternate or Supplemental Prophylaxis Strategies

Several strategies have been formulated to produce novel vaccines and delivery systems (Friedlander and Little 2009). Some of them are as below:

☆ Considering the need of alternate or supplemental treatment modalities, particularly those that neutralize anthrax toxins, the focus has been on passive immunization with monoclonal and polyclonal antibodies to *B. anthracis* toxin components, primarily PA and to a lesser extent LF (Little *et al.* 1997).

☆ Humanized and fully human monoclonal antibodies (mAbs) with high affinities for PA and LF have been generated using a variety of approaches. These mAbs afford significant protection from inhalation anthrax in small animal models and non-human primates in a prophylactic and therapeutic setting (*e.g.* Abthrax™, an anti-PA mAb). A potential disadvantage of mAbs is their monospecific nature, which makes it possible, in principle, to develop *B. anthracis* strains that resist their action (Albrecht *et al.* 2007). Hence, a cocktail of mAbs that target multiple epitopes of one protein or more than one *B. anthracis* toxin component would be preferable.

☆ Polyclonal immunoglobulin, 'Anthrax Immune Globulin', from the plasma of human AVA-vaccinated volunteers, is being manufactured as an alternative to mAb preparations. Its advantage is that it reflects the breadth of the human immune response to *B. anthracis*; however, there is limited availability, lot to lot variation, and it carries the risk of infectious disease transmission (Turnbull *et al.* 1988, Grabenstein 2008). Hence, there is a need to develop recombinant human polyclonal antibody.

☆ Other anthrax anti-toxins are in research phase with the main focus on soluble or conjugated forms of CMG2 as a lethal toxin antidote and a dominant-negative form of PA that interferes with heptamer formation

(Cryan and Rogers 2011). LFn-PA fusion fragments have also been used as potential vaccine candidates (Li *et al.* 2009).

☆ Small-molecule inhibitors of the enzymatic activities of LF and Oedema Factor are also considered potential candidates which are highly desirable given their potential oral availability and low cost production (Sellman *et al.* 2001, Aulinger *et al.* 2005).

☆ Antibodies that neutralize anthrax toxins have been developed and shown to provide protection from inhalation anthrax in a prophylactic and therapeutic setting. These antibodies are likely to be used in combination with antibiotics (Fowler *et al.* 2005).

☆ DNA vaccine formulations, either plasmid encoding for mutant toxins or vector delivered vaccine with prime boost strategies, have also been attempted (Gu *et al.* 1999, Hahn *et al.* 2006)

The development of new medical countermeasures against inhalation anthrax and other biological warfare agents has been spearheaded by several researchers and biotech companies in recent times. However, an ideal anthrax vaccine design would be the one which targets spore, bacillus, capsule as well as toxins, especially the Protective Antigen/Lethal Factor (Wang and Roehrl 2005).

Control Strategies

Controlling anthrax in humans depends on controlling the infection in animals. Surveillance, vaccination and proper disposal of carcasses are the most efficient ways of preventing and controlling anthrax infection in domestic herds, and thus in limiting its transmission to humans. Periodical assessment of epidemiological features of anthrax would be helpful to chalk out the strategy for precise control and prevention. The control strategies can be grouped according to whether these are applicable before an anthrax outbreak or during the outbreak or afterwards.

The strategies include timely livestock vaccination and correct management of infected carcasses. A national as well as global responsibility should oblige local governments to guarantee sufficient budget provisions for anthrax control and eradication programmes and not just for emergencies, because the livestock farmers incur severe economic losses, especially in the endemic areas. An important step in the implementation of anthrax control is the acquisition of data or information about the disease. Field data related to the characteristics of the pathogen, its ecology and determinants of its natural occurrence are useful initial tools for livestock producers and veterinary services dealing with anthrax.

Surveillance can be used to predict areas where natural livestock cases of anthrax are likely to occur. These areas should have an effective mandatory reporting system in place so that all unexpected livestock deaths during the anthrax period are reported to the veterinary authorities for immediate investigation. Field veterinarians should have the facility to make on-site diagnosis or good liaison with laboratory services to ensure diagnosis without delay.

Apart from these measures, appropriate chemotherapy should be in place to provide immediate response or treatment to the affected animals or humans. Since vaccination may not confer protection after the exposure to aerosolized spores having taken place, antibiotics (ciprofloxacin or doxycycline) administered soon after the exposure and prior to the onset of symptoms are the most effective means of preventing disease. Since spores can remain dormant in the lungs for an extended period of time, a course of oral antibiotics is recommended. Additional measures of protection need to be considered in the event of a mass exposure during intentional release/bio-terrorism. According to the commonly agreed recommendations of the Centers for Disease Control (CDC) and other organizations, a 60-days course of oral antibiotics combined with a 3-dose series of anthrax vaccine at 2-week intervals is given, following a potential exposure to aerosolized *B. anthracis* spores.

Future Directions

The advances during the past one decade, both in basic and applied research of medical countermeasures against anthrax have been remarkable. The provision of funds by various developed nations under bio-defence programme resulted in development of novel products for prevention and treatment of human anthrax as well as other biowarfare agents. However, there is a need to validate such technologies/strategies thoroughly under various phases of clinical trials before they are considered for use for public or livestock. Meanwhile, all countries need to be more vigilant by having effective surveillance, disease monitoring and reporting system, coupled with stockpile of vaccines/medicines to tackle any kind of eventualities either natural or intentional in the days ahead. A comprehensive review and redesign of the health care system for both humans and livestock is needed urgently to ensure equity and quality in healthcare. Recently, international organizations such as WHO and OIE have provided several recommendations for effective control and prevention of zoonoses, including anthrax emphasizing the One Health concept, which needs to be strictly implemented by the regional authorities.

References

1. Abramova AA and Grinberg LM. 1993. Pathology of anthrax sepsis according to materials of the infectious outbreak in 1979 in Sverdlovsk (microscopic changes). *Arkh Patol* 55: 18-23.

2. Albrecht MT, Li H, Williamson ED, LeButt CS, Flick-Smith HC, Quinn CP, Westra H, Galloway D, Mateczun A, Goldman S, Groen H and Baillie LW. 2007. Human monoclonal antibodies against anthrax lethal factor and protective antigen act independently to protect against *Bacillus anthracis* infection and enhance endogenous immunity to anthrax. *Infect Immun* 75: 5425-5433.

3. Ascoli A. 1911. Die Präzipitindiagnose bei Milzbrand. *Centralbl Bakt Parasit Infeckt* 58: 63-70.

4. Aulinger BA, Roehrl MH, Mekalanos JJ, Collier RJ and Wang JY. 2005. Combining anthrax vaccine and therapy: a dominant-negative inhibitor of anthrax toxin is also a potent and safe immunogen for vaccines. *Infect Immun* 73: 3408-3414.

5.   Beyer W and Turnbull PC. 2009. Anthrax in animals. *Mol Aspects Med* 30: 481-489.

6.   Beyer W, Glockner P, Otto J and Bohm R. 1996. A nested PCR and DNA-amplification-fingerprinting method for detection and identification of *Bacillus anthracis* in soil samples from former tanneries. *Salisbury Med Bull* 87 Special Suppl: 47-49.

7.   Bhatt P, Mohan DN and Lalitha MK. 1989. Current incidence of anthrax in animals and man in India. In: Proceedings of the International Workshop on Anthrax at Winchester, England. *Salisbury Med Bull* 68 Special Suppl: 8.

8.   Brachman PS, Gold H, Plotkin SA, Fekety FR, Werrin M and Ingraham NR. 1962. Field evaluation of a human anthrax vaccine. *Am J Public Health* 52: 632-645.

9.   Brown ER and Cherrry WB. 1955. Specific identification of *Bacillus anthracis* by means of a variant bacteriophage. *J Infect Dis* 96: 34-39.

10. Cryan LM and Rogers MS. 2011. Targeting the anthrax receptors, TEM-8 and CMG-2, for antiangiogenic therapy. *Front Biosci* 16: 1574-1588.

11. Ezzell JW and Abshire TG. 1996. Encapsulation of *Bacillus anthracis* spores and spore identification. *Salisbury Med Bull* 87 Special Suppl: 42.

12. Fowler RA, Sanders GD, Bravata DM, Nouri B, Gastwirth JM, Peterson D, Broker AG, Garber AM and Owens DK. 2005. Cost-effectiveness of defending against bio-terrorism: a comparison of vaccination and antibiotic prophylaxis against anthrax. *Ann Intern Med* 142: 601-610.

13. Friedlander AM. 2000. Anthrax: clinical features, pathogenesis, and potential biological warfare threat. *Curr Clin Top Infect Dis* 20: 335-349.

14. Friedlander AM and Little SF. 2009. Advances in the development of next generation anthrax vaccines. *Vaccine* 27(Suppl. 4): D28-D32.

15. Gajendragad MR and Uma S. 2012. Epidemiology of anthrax in India. Technical Bulletin 10, PDADMAS, Bangalore. p. 1-6.

16. Grabenstein JD 2003. Anthrax vaccine-a review. *Immunol Allergy Clin N Am* 23: 713-730.

17. Grabenstein JD. 2008. Vaccines: countering anthrax: vaccines and immunoglobulins. *Clin Infect Dis* 46: 129-136.

18. Gu ML, Leppla SH and Klinman DM. 1999. Protection against anthrax toxin by vaccination with a DNA plasmid encoding anthrax protective antigen. *Vaccine* 17: 340-344.

19. Hahn UK, Aichler M, Boehm R and Beyer W. 2006. Comparison of the immunological memory after DNA vaccination and protein vaccination against anthrax in sheep. *Vaccine* 24: 4595-4597.

20. Hutson RA, Duggleby CJ, Lowe JR, Manchee RJ and Turnbull PCB. 1993. The development and assessment of DNA and oligonucleotide probes for the specific detection of *Bacillus anthracis*. *J Appl Bacteriol* 75: 463-472.

21. Jackson PJ, Hugh-Jones ME, Adair DM, Green G, Hill KK, Kuske CR, Grinberg LM, Abrramova FA and Keim P. 1998. PCR analysis of tissue samples from the 1979 Sverdlovsk anthrax victims: The presence of multiple *Bacillus anthracis* strains in different victims. *Proc Natl Acad Sci USA* 95: 1224-1229.

22. Janse I, Hamidjaja RA, Bok JM and van Rotterdam BJ. 2010. Reliable detection of *Bacillus anthracis, Francisella tularensis* and *Yersinia pestis* by using multiplex qPCR including internal controls for nucleic acid extraction and amplification. *BMC Microbiol* 10: 314-326.

23. Jernigan JA, Stephens DS, Ashford DA, Omenaca C, Topiel MS, Galbraith M, Tapper M, Fisk TL, Zaki S, Popovic T, Meyer RF, Quinn CP, Harper SA, Fridkin SK, Sejvar JJ, Shepard CW, McConnell M, Guarner J, Shieh WJ, Malecki JM, Gerberding JL, Hughes JM and Perkins BA. 2001. Bio-terrorism-related inhalational anthrax: the first 10 cases reported in the United States. *Emerg Infect Dis* 7: 933-944.

24. Kanungo R, Sujatha S, Das AK and Rao RS. 1990. Anthrax meningitis - a clinical enigma. *Ind J Med Microbiol* 6: 149-151.

25. Knisely RF. 1966. Selective medium for *Bacillus anthracis*. *J Bacteriol* 92: 784-786.

26. Kumar A, Kanungo R, Bhattacharya S, Badrinath S, Dutta TK and Swaminathan RP. 2000. Human anthrax in India: urgent need for effective prevention. *J Commun Dis* 32: 240-246.

27. Lalitha MK and Kumar A. 1996. Anthrax-a continuing problem in southern India. *Indian J Med Microbial* 14: 63-72.

28. Leppla SH. 1999. *Bacillus anthracis* toxins. In: Alouf JE and Freer JH. (Eds.) *The comprehensive source book of bacterial protein toxins*. Academic Press, London. p. 243- 263.

29. Leppla SH, Robbins JB, Schneerson R and Shiloach J. 2002. Development of an improved vaccine for anthrax. *J Clin Invest* 110: 141-144.

30. Li Q, Shivachandra SB, Leppla SH and Rao VB. 2006. Bacteriophage T4 capsid: A unique platform for efficient surface assembly of macromolecular complexes. *J Mol Biol* 363: 577-588.

31. Li Q, Shivachandra SB, Zhang Z and Rao VB. 2007. Assembly of the small outer capsid protein, SOC, on bacteriophage T4: a novel system for high density display of multiple large anthrax toxins and foreign proteins on phage capsid. *J Mol Biol* 370: 1006-1019.

32. Li Q, Peachman K, Sower L, Leppla SH, Shivachandra SB, Matyas GR, Peterson JW, Alving CR, Rao M and Rao VB. 2009. Anthrax LFn-PA hybrid antigens: Biochemistry, immunogenicity, and protection against lethal Ames spore challenge in rabbits. *The Open Vaccine J* 2: 92-99.

33. Little SF, Ivins BE, Fellows PF and Friedlander AM. 1997. Passive protection by polyclonal antibodies against *Bacillus anthracis* infection in guinea pigs. *Infect Immun* 65: 5171-5175.

34. Matyas GR, Friedlander AM, Glenn GM, Little S, You J and Alving CR. 2004. Needle-free skin patch vaccination method for anthrax. *Infect Immun* 72: 1181-1183.

35. Meselson M, Guillemin J, Hugh-Jones M, Langmuir A, Popova I, Shelokov A and Yampolskaya O. 1994. The Sverdlovsk anthrax outbreak of 1979. *Science* 266(5188): 1202-1208.

36. OIE. 2008. Manual of diagnostic tests and vaccines for terrestrial animals. Chapter 2.1.1. *Anthrax*. World Organisation for Animal Health, Paris, France. p. 135-144.

37. Okinaka RT, Cloud K, Hampton O, Hoffmaster AR, Hill KK, Keim P, Koehler TM, Lamke G, Kumano S, Mahillon J, Manter D, Martinez Y, Ricke D, Svensson R and Jackson PJ. 1999. Sequence and organization of pXO1, the large *Bacillus anthracis* plasmid harboring the anthrax toxin genes. *J Bacteriol* 181: 6509-6515.

38. Olive DM and Bean P. 1999. Principles and applications of methods for DNA based typing of microbial organisms: a mini review. *J Clin Microbiol* 37: 1661-1669.

39. Peachman K, Li Q, Matyas G, Shivachandra SB, Lovchik J, Lyons CR, Alving C, Rao VB and Rao M. 2012. Anthrax vaccine antigen-adjuvant formulations completely protect New Zealand White rabbits against challenge with *Bacillus anthracis* Ames strain spores. *Clin Vaccine Immunol* 19: 11-16.

40. Petosa C, Collier RJ, Klimpel KR, Leppla SH and Liddington RC. 1997. Crystal structure of the anthrax toxin protective antigen. *Nature* 385: 833-838.

41. Pile JC, Malone JD, Eitzen EM and Friedlander AM. 1998. Anthrax as a potential biological warfare agent. *Arch Intern Med* 158: 429-434.

42. Plotkin SA, Brachman PS, Utell M, Bumford FH and Atchison MM. 1960. An epidemic of inhalation anthrax, the first in the twentieth century. I. Clinical features. *Am J Med* 29: 992-1001.

43. Quinn CP and Turnbull PCB. 1998. Anthrax. In: Collier L, Balows A and Sussman M (eds.) *Topley and Wilson's Microbiology and Microbial Infections*, Vol. 3, Ninth Edition. Arnold, London, UK. p. 799-818.

44. Ramirez DM, Leppla SH, Schneerson R and Shiloach J. 2002. Production, recovery and immunogenicity of the protective antigen from a recombinant strain of *Bacillus anthracis*. *J Ind Microbiol Biotechnol* 28: 232-238.

45. Ramisse V, Patra G, Garrigue H, Guesdon JL and Mock M. 1996. Identification and characterization of *Bacillus anthracis* by multiplex PCR analysis of sequences on plasmids pX01 and pX02 and chromosal DNA. *FEMS Microbiol* Lett 145: 9-16.

46. Rao M, Peachman KK, Li Q, Matyas GR, Shivachandra SB, Borschel R, Morthole, VI, Fernandez-Prada C, Alving CR and Rao VB. 2011. Highly effective generic adjuvant systems for orphan or poverty-related vaccines. *Vaccine* 29: 873-877.

47. Sellman BR, Mourez M and Collier RJ. 2001. Dominant-negative mutants of a toxin subunit: an approach to therapy of anthrax. *Science* 292: 695-697.

48. Shadomy SV and Smith TL. 2008. Zoonosis update - Anthrax. *J Am Vet Med Assoc* 233: 63-72.

49. Shivachandra SB, Kumar AA, Srivastava SK and Chaudhuri P. 2003. Nucleic acid based typing techniques: emerging tools for epidemiological studies. *Livestock International* 7(2): 20-23.

50. Shivachandra SB, Rao M, Janosi L, Sathaliyawala T, Matyas GR, Alving CR, Leppla SH and Rao VB. 2006. *In vitro* binding of anthrax protective antigen on bacteriophage T4 capsid surface through Hoc-capsid interactions: A strategy for efficient display of large full length proteins. *Virology* 345: 190-198.

51. Shivachandra SB, Li Q, Peachman K, Matyas GR, Leppla SH, Alving CR, Rao M and Rao VB. 2007. Multi-component anthrax toxin display and delivery using bacteriophage T4. *Vaccine* 25: 1225-1235.

52. Sternbach G. 2003. The history of anthrax. *J Emerg Med* 24: 463-467.

53. Sterne M. 1937. The effect of different carbon dioxide concentrations on the growth of virulent anthrax strains. *Onderstepoort J Vet Sci Anim Ind* 9: 49-67.

54. Sterne M. 1959. Anthrax. In: Stableforth AW and Galloway IA (eds.) *Infectious diseases of animals. Vol. 1. Diseases due to bacteria*. Butterworths, London. p. 16-52.

55. Sue D, Marston CK, Hoffmaster AR and Wilkins PP. 2007. Genetic diversity in a *Bacillus anthracis* historical collection (1954 to 1988). *J Clin Microbiol* 45: 1777-1782.

56. Tang S, Moayeri M, Chen Z, Harma H, Zhao J, Hu J, Purcell RH, Leppla SH and Hewlett IK. 2009. Detection of anthrax toxin by an ultrasensitive immunoassay using europium nanoparticles. *Clin Vaccine Immunol* 16: 408-413.

57. Thappa DM and Karthikeyan K. 2001. Anthrax: An overview with the Indian subcontinent. *Int J Dermatol* 40: 216-222.

58. Turell MJ and Knudson GB. 1987. Mechanical transmission of *Bacillus anthracis* by stable flies (*Stomoxys calcitrans*) and mosquitoes (*Aedes aegypti* and *Aedes taeniorhynchus*). *Infect Immun* 55: 1859-1861.

59. Turnbull PCB. 1991. Anthrax vaccines: past, present and future. *Vaccine* 9: 533-539.

60. Turnbull PCB. 1998. Anthrax. In: Palmer SR, Soulsby EJL and Simpson DIH (eds.) *Zoonoses. Biology, clinical practice and public health control*. Oxford University Press, Oxford, UK. p. 3-16.

61. Turnbull PC, Leppla SH, Broster MG, Quinn CP and Melling J. 1988. Antibodies to anthrax toxin in humans and guinea pigs and their relevance to protective immunity. *Med Microbiol Immunol* 177: 293-303.

62. Turnbull PCB, Boehm R, Cosivi O, Doganay M, Hugh-Jones ME, Lalitha MK and De Vos V. 1998. *Guidelines for the surveillance and control of anthrax in humans and animals*. WHO/EMC/ZDI/98.6. World Health Organization, Geneva, Switzerland.

63. Wang JY and Roehrl MH. 2005. Anthrax vaccine design: strategies to achieve comprehensive protection against spore, bacillus, and toxin. *Med Immunol* 4: 4.

64. Wielinga PR, Hamidjaja RA, Agren J, Knutsson R, Segerman B, Fricker M, Ehling-Schulz M, de Groot A, Burton J, Brooks T, Janse I and van Rotterdam B. 2011. A multiplex real-time PCR for identifying and differentiating *B. anthracis* virulent types. *Int J Food Microbiol* 145 Suppl 1: S137-S144.

2014, Zoonoses: Bacterial Diseases　　　　　　　　　　　　　　　　*Pages* **38–51**
Editor: **Sudhi Ranjan Garg**
Published by: **DAYA PUBLISHING HOUSE, NEW DELHI**

3

Brucellosis

H. Rahman, G. B. Manjunatha Reddy and Rajeswari Shome
Project Directorate on Animal Disease Monitoring and Surveillance,
Bengaluru – 560 024

Brucellosis is a zoonotic infection transmitted from animals to humans by the ingestion of infected food products, direct contact with an infected animal or inhalation of aerosols. Gastric intermittent fever, Malta fever, Mediterranean fever, Bang's disease, Rock fever, Gibraltar fever, Cyprus fever, undulant fever are the synonyms for brucellosis. Organisms resembling *Brucella* have been detected in carbonized cheese from the time of the Roman era. Brucellosis was predominant in the Mediterranean region and its history is associated with military campaigns. This disease was fully elucidated by Sir David Bruce, Hughes, and Zammit working in Malta (Cultuer *et al.* 2005). Bang discovered *Brucella abortus*, the causative agent of abortion in cattle and of brucellosis (undulant fever) in human beings (Bang 1897). The disease has been reported from all parts of the world and is endemic in Asia, Sub-Saharan Africa, some countries of Latin America, the Middle East and the Mediterranean, and South Eastern Europe. In India, the presence of brucellosis was first established early in the previous century and since then reported from almost all states (Renukaradhya *et al.* 2002). The disease has been reported in cattle, buffaloes, sheep, goats, pigs, dogs and humans. Despite the advances made in the diagnosis and therapy, brucellosis is still widespread and its prevalence in many developing countries is increasing. The disease has considerable impacts on human and animal health as well as socio-economics. The economic losses due to brucellosis in cattle and buffaloes have been estimated to the tune of rupees 240 million per year in India. This amount is more than half per cent of total value of all meat and milk products produced in the country. Annual loss due to human brucellosis is estimated to be 30 million man-days. Economic losses associated with brucellosis in animals are due to abortions, premature births, decreased milk production and repeat breeding that may lead to temporary or permanent infertility in infected livestock.

In human beings, the disease is rarely fatal, but can lead to severe debilitation and disability. Nevertheless, it is reported that approximately 2 per cent of the untreated patients die of brucellosis (Madkour 2001). The disease has the tendency towards chronicity and persistence, becoming a granulomatous disease capable of affecting any organ system. The disease remains the world's most common bacterial zoonosis, with over half a million new cases annually, while the prevalence rate in some countries exceeds ten cases per 100,000 population (Mantur and Amarnath 2008), being higher in people working in organized farms (Umapathy 1984). Despite being endemic in many developing countries, brucellosis is under-diagnosed and under-reported. With all domestic animals including some wild animals acting as definitive and reservoir host, human is an accidental host.

Geographical Distribution

The disease exists worldwide, particularly in the Mediterranean basin, the Arabian Peninsula, the Indian subcontinent, and in parts of Mexico, Central America and South America (Pappas *et al.* 2006) as shown in Table 3.1. Seroprevalence of brucellosis in animals in India has been depicted in Figure 3.1.

Table 3.1: Geographical distribution and animal reservoirs of *Brucella* organisms

Organism	Animal Reservoir	Geographic Distribution
B. melitensis	Goat, sheep, camel	Mediterranean, Asia, Latin America, parts of Africa and some Southern European countries
B. abortus	Cow, buffalo, camel, yak	Worldwide
B. suis	Pig (biotype 1-3)	South America, Southeast Asia, United States
B. canis	Canines	Cosmopolitan
B. ovis	Sheep	No known human cases
B. neotomae	Rodents	Not known to cause human disease
B. pinnipediae, B. cetaceae	Marine animals, mink whale, dolphin, seal	Some human cases, mainly neurobrucellosis

Etiological Agent, Vectors and Host Range

Brucellosis is a bacterial disease caused by a Gram negative, non-spore forming, facultative intracellular bacteria belonging to genus *Brucella* of family Brucellaceae (family III) with *Mycoplana* and *Ochrobactrum* of the order Rhizobiales in the class Alphaproteobacteria of the phylum Proteobacteria. The genus *Brucella* consists of seven species according to antigenic variation and primary host: *Brucella melitensis* (sheep and goats), *B. suis* (pigs), *B. abortus* (cattle), *B. ovis* (sheep), *B. canis* (dogs), *B. neotomae* (wood rats) and *B. microti, B. ceti, B. pinnipedialis* and *B. inopinata* (marine mammals).

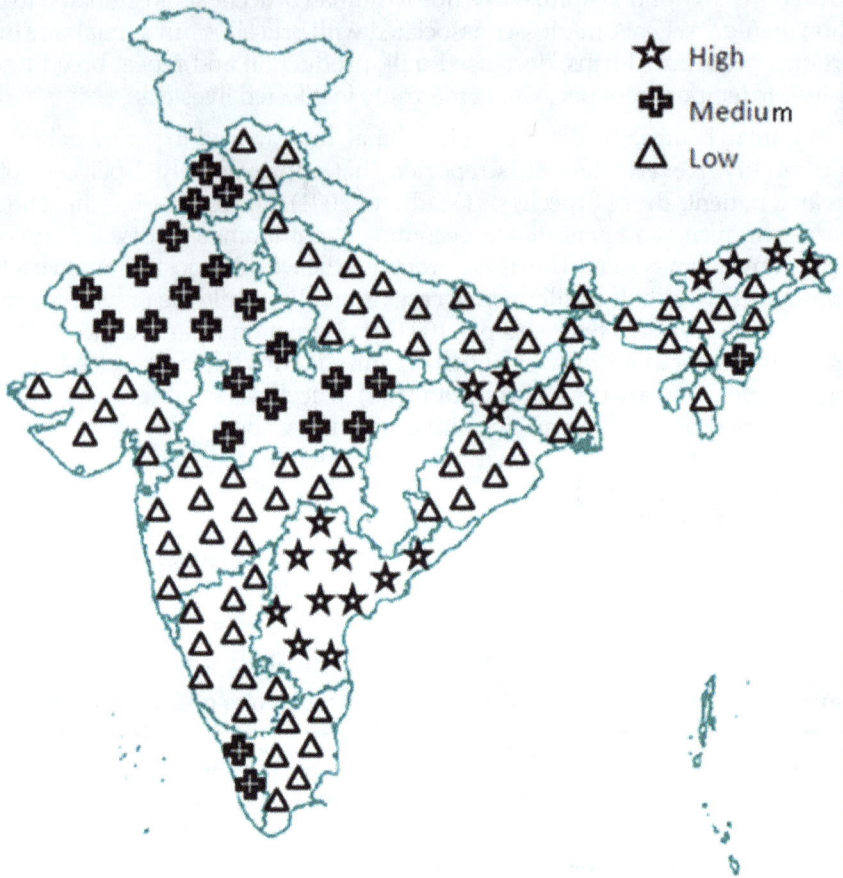

Figure 3.1: Seroprevalence of brucellosis in animals in India (2005-2010)

Brucellosis is primarily a disease of animals, especially domestic animals (cattle, buffalo, sheep and goat, pig, dogs and horses). The disease has also been reported in wild animals (ruminants), cetaceans, pinnipeds and marine mammals (Table 3.2).

Disease Transmission

Brucellosis is transmitted from animal to animal by contact, following an abortion. Most frequently, it is acquired by ingestion, but inhalation, conjunctival inoculation, skin contamination and udder inoculation from infected milking cups are other possibilities. The sexual transmission is more important in sheep, goats and pigs than in cattle and buffalo.

In humans, the possible means of acquisition of brucellosis include person-to-person transmission, infection from a contaminated environment, occupational exposure usually resulting from direct contact with infected animals, and foodborne infection.

Table 3.2: *Brucella* species and biovars, preferential hosts and pathogenicity for humans

Species	Biovars	Colony Morphology	Preferential Host	Pathogenicity in Humans
B. melitensis	1-3	Smooth	Sheep, goat	High
B. abortus	1-6, 9	Smooth	Cattle	High
B. suis	1, 3	Smooth	Pig	High
	2	Smooth	Wild boar, hare	Low
	4	Smooth	Reindeer, caribou	High
	5	Smooth	Rodent	No
B. neotomae	–	Smooth	Desert rat	Moderate
B. ovis	–	Rough	Ram	No
B. canis	–	Rough	Dog	Moderate
B. pinnipedialis	–	Smooth	Seal	? *
B. ceti	–	Smooth	Cetacean	?
B. microti	–	Smooth	Soil, vole, fox	?
B. inopinata	–	Smooth	Human	

* Although some human cases have been described, the actual pathogenicity remains unknown (Godfroid *et al.* 2010)

Environmental and Other Predisposing Factors

In cattle, sheep, goats and swine, susceptibility to brucellosis is greatest in sexually mature animals. Young animals are often resistant, although it should be noted that latent infections can occur and such animals may present a hazard when mature. The management practices like breeding, preventive healthcare, feeding and hygiene are far more important in determining the risk of infection. Breed may also affect the susceptibility, particularly in sheep. The milking breeds seem to be the most susceptible to *B. melitensis*. Polymorphism of the natural resistance associated monocyte protein (NRAMP) gene has been shown to influence substantially the susceptibility to brucellosis in cattle and pigs. The transmission of disease in sheep and goats is facilitated by commingling of flocks and herds belonging to different owners, purchasing animals from unscreened sources, sharing of male breeding stock, mingling of animals during summer grazing and in markets and in cold due to close space housing.

In human beings, the people like farmers, farm labourers, animal attendants, stockmen, shepherds, sheep shearers, goatherds, pig keepers, veterinarians, inseminators, slaughtermen, butchers, meat packers, collectors of foetal calf serum, processors of hides, skins and wool, renderers and dairy workers, who work with or are in close contact with animals and animal products, are at high risk of contracting the disease through direct contact with infected animals or through heavily contaminated environment. Laboratory staff involved in culturing *Brucella* is at particular risk and the staff employed in the maintenance of farm premises, factories or plants used for processing animal products are also under risk of exposure to infection. In case of urban population, tourists or travellers are at the main risk of exposure to disease thorough consumption of unpasteurized milk and milk products. Brucellosis is more during spring and summer season, which coincides with the period of peak abortions in animals. Since the disease is largely occupational, the majority of cases are reported in males between the ages of 20 and 45 years, although women and children are also susceptible.

Prevalence of Disease in Animals and Humans

The presence of brucellosis in India was first established as early as in 1897. Since then, it has been reported from almost all states (Renukaradhya *et al.* 2002), but the situation at different places varies widely. In different studies the seroprevalence of brucellosis in sheep and goats has been reported to be from 2.2 per cent to 26.99 per cent (Renukaradhya *et al.* 2002, Chandra *et al.* 2005, Shome *et al.* 2006, Gupta *et al.* 2007, Gupta *et al.* 2009, Maninder *et al.* 2010, Sonawane *et al.* 2011, Gupta *et al.* 2012). The seroprevalence in large animals is shown in Table 3.3.

Table 3.3: Seroprevalence of *Brucella* in large animals

Animal Species	Seroprevalence (per cent)				Reference
	RBPT	STAT	CFT	ELISA	
Buffalo	7.09	2.70	11.14	8.10	Prahlad *et al.* (1999)
Buffalo	11.50	8.75	–	16.25	Rao *et al.* (1999)
Crossbred cow	16.25	15.00	–	31.25	
Bovine	33.33	50.35	–	56.02	Chakraborty *et al.* (2000)
Cows	–	9.60	–	–	Mehra *et al.* (2000)
Heifers	–	12.60	–	–	
Buffalo	–	11.40	–	–	
Cattle					Renukaradhya *et al.* (2002)
Buffalo					
Cattle	1.25	1.95	–	–	Rajesh *et al.* (2003)
Cattle	37.38	36.45	–	40.18	Barbuddhe *et al.* (2004).
Cattle	20.47	18.89	–	26.50	Chand and Sharma (2004)
Cows		43.28	–	47.76	Mahato *et al.* (2004)
Heifers		14.89	–	17.02	
Cattle	14.70	7.19	–	–	Nasir *et al.* (2004)
Buffalo	15.38	2.91	–	–	
Cattle	18.53	9.00	–	–	
Buffalo	35.40	23.70	–	–	
Bovine	11.94	10.28	–	8.05	Bhattacharya *et al.* (2005)
Cattle	41.79	–	–	–	Kachhawaha *et al.* (2005)
Buffalo	25.56	–	–	–	
Cattle	–	1.55	–	3.11	Mishra *et al.* (2005)

Contd...

Table 3.3–*Contd...*

Animal Species	Seroprevalence (per cent)				Reference
	RBPT	STAT	CFT	ELISA	
Buffalo	–	1.97	–	4.18	
Cattle	13.83	–	–	10.40	Jai Sunder *et al.* (2005)
Bovine	11.11	–	–	13.59	Ganesan and Anuradha (2006)
Bovine	–	–	–	–	Dhand *et al.* (2005)
Mithun	11.00	20.00	–	34.00	Rajkhowa *et al.* (2005)
Bulls		–	–	–	Kanani (2007).
Bovine	–	18.07	–	–	Sharma *et al.* (2007)
Yaks				16.00	Rahman *et al.* (2007)

RBPT: Rose Bengal plate test; STAT: Standard tube agglutination test; CFT: Complement fixation test; ELISA: Enzyme-linked immunosorbent assay.

Several published reports, including the recent ones, indicate that human brucellosis is quiet common in India. The seroprevalence of human brucellosis varied from 0.8 to 8.5 per cent in several reports published before the year 2000 (Mathur 1964, Randhawa *et al.* 1974, Panjarathinam and Jhala 1986, Handa *et al.* 1998, Kadri *et al.* 2000). In subsequent reports, it has been estimated from 1.6 to 6.8 per cent (Thakur and Thapliyal 2002, Sen *et al.* 2002, Mantur *et al.* 2004, Mantur *et al.* 2006, Agasthya *et al.* 2012).

Disease Manifestations in Animals

In large ruminants the disease is caused by *B. abortus* and in small ruminants by *B. melitensis*. The disease occurs in sexually mature animals and it spreads primarily by ingestion, skin penetration or udder contamination. Semen from infected bulls is a good source of infection during artificial insemination. After gaining entry, organisms are localized in adjoining lymph nodes and multiply, and spread to other lymphoid tissues including spleen and mammary and iliac lymph nodes. They are localized in the pregnant uterus, udder, testis, accessory male sex glands and joint capsules to produce symptoms. Clinical signs are primarily based on the immune status of animal and highly susceptible pregnant cows or buffaloes suffer from hygroma and reveal abortion after six months, retained placenta and metritis. In bulls, epididymitis and orchitis involving one or both scrotal sacs may occur. The testicles are enlarged, have painful swellings and subsequently reveal liquifactive necrosis and degeneration. In mild cases sinovitis and painful swelling of affected joint is noticed. Though mortality rate is low, but on post-mortem examination, oedematous placenta, leathery plaques on chorion and necrosis of cotyledons and swelling of foetal membranes are observed. Brucellosis can be diagnosed on the basis of history of abortions and clinical signs and can be confirmed by isolation of causal organisms from visceral organs and lymph nodes of aborted foetus.

Disease Manifestations in Humans

The signs and symptoms of brucellosis are extensive and they can be similar to many other febrile illnesses. Symptoms may resemble a nonspecific flu-like illness, weakness, fever, sweating, headache, malaise, lethargy, anorexia and joint pain. In some individuals, a chronic or long lasting illness can develop with recurring attacks. In chronic disease, the organisms are found in bone marrow, spleen, liver and lymph nodes. Complications can occur including damage to bones and joints, blood clots in veins and psychological changes. The disease in peracute or acute phase may progress to a chronic one with relapse, development of persistent localized infection or a non-specific syndrome resembling the chronic fatigue syndrome. Since there are no specific clinical signs peculiar to brucellosis, the diagnosis must be supported by laboratory tests which indicate the presence of the organism or a specific immune response to its antigens.

Diagnostic Procedures

Unequivocal diagnosis of *Brucella* infections can be made only by the isolation and identification of *Brucella*, but in situations where bacteriological examination is not practicable, diagnosis must be based on serological methods. There is no single test by which a bacterium can be identified as *Brucella*. A combination of growth characteristics, serological, bacteriological and/or molecular methods is usually needed.

Collection of Samples

The specimen materials of choice from animals include blood, serum, stomach contents, spleen and lung from aborted foetus, placentomes, foetal membranes, vaginal swabs, milk, semen and arthritis or hygroma fluids from adult animals. From animal carcasses, the preferred tissues for culture are the mammary, medial and internal iliac, retropharyngeal, parotid and prescapular lymph nodes and spleen. All specimens must be packed separately and transported immediately to the laboratory cooled or preferably frozen in leak proof containers.

Bacteriological Methods

Staining of Smears

Smears from the placental cotyledon, vaginal discharge, foetal stomach content, semen etc. may be stained by the modified Ziehl-Neelsen or Kosters' method. The presence of large aggregates of intracellular, weakly acid fast organisms with *Brucella* morphology is presumptive evidence of brucellosis. Care must be taken to differentiate other acid fast organisms such as *Coxiella burnetii* and *Clamydophila abortus* which may superficially resemble *Brucella*.

Culture

For the isolation of *Brucella*, the most commonly used medium is the Farrell medium, which contains antibiotics to inhibit the growth of other bacteria present in the clinical samples. Some *Brucella* species, like *B. abortus wildtype* (biovars 1-4), need 10 per cent CO_2 for growth, while others, like *B. abortus wildtype* (biovars 5, 6, 9), *B. abortus* S19 vaccine strain, *B. melitensis*, and *B. suis*, do not need it. For liquid samples (milk or blood), sensitivity is increased by the use of a biphasic medium like the Castaneda medium. Growth may appear after 2-3 days, but cultures are usually considered negative after 2-3 weeks of incubation. Most *Brucella* strains, particularly *B. abortus biovar* 2 and *B. ovis*, grow better in media containing 5-10 per cent of sterile (equine or bovine) serum free from *Brucella* antibodies. The identification of *Brucella* spp. is based on morphology, staining and metabolic profile (catalase, oxidase and urease).

Biotyping

Biotyping of *Brucella* spp. is performed using different tests, the most important being agglutination tests with antibodies against rough or smooth LPS, lysis by phages, dependence on CO_2 for growth, production of H_2S, growth in the presence of basal fuchsine or thionine and the crystal violet or acriflavine tests. These tests will give good results when carried out using standardized procedures by experienced laboratory personnel.

Molecular Methods

The use of polymerase chain reaction (PCR) to identify *Brucella* DNA at genus, species and even biovar levels has been extended to improve diagnostic tests and a diversity of methods have been developed. Applications for PCR methods range from the diagnosis of the disease to characterization of field isolates for epidemiological purposes including taxonomic studies. Among the several PCR-based methods, the best validated methods are based on the detection of specific sequences of *Brucella* spp., such as the 16S, 23S genes, the IS711 insertion sequence or the bcsp31 gene. Nevertheless, as a general rule, brucellosis PCR techniques show a lower diagnostic sensitivity than culture methods, although their specificity is close to 100 per cent. The best results have so far been obtained by combining culture and PCR detection on clinical samples. Random amplification of polymorphic DNA (RAPD), multiplex PCR, real time PCR and real time PCR with high resolution melting (HRM) have also been used for diagnosis, and molecular characterization and typing.

Serological Tests

Detection of specific antibody in serum or milk remains the most practical means of diagnosis of brucellosis. The most efficient and cost-effective procedure usually involves screening of all samples by a cheap and rapid test which is sensitive enough to detect a high proportion of infected animals. Samples positive on screening are then tested with more sophisticated, specific, confirmatory tests for final diagnosis.

Rose Bengal Plate Test (RBPT)

RBPT is a simple spot agglutination test where drops of stained antigen and serum are mixed on a plate and any resulting agglutination signifies a positive reaction. The test is an excellent screening test for IgM antibodies.

Serum Agglutination Test (SAT)

SAT has been used extensively for brucellosis diagnosis. Although simple and cheap to perform, its lack of sensitivity and specificity restricts its use in the absence of other alternative techniques.

Complement Fixation Test (CFT)

The sensitivity and specificity of CFT is good, but it is a complex method to perform which requires good laboratory facilities and trained staff. If these are available and the test is carried out regularly with proper attention to quality assurance, CFT can be very satisfactory.

Enzyme-linked Immunosorbent Assay (ELISA)

There are different types of ELISA tests which are highly sensitive and specific, simple to perform with a minimum of equipment, and readily available from a number of commercial sources in kit form. More suitable than CFT for use in smaller laboratories, ELISA technology is now used for diagnosis of a wide range of animal and human diseases. Although, in principle, ELISAs can be used for testing serum of all species of animals and man, results may vary between laboratories depending on the exact methodology used. Not all standardization issues have yet been fully addressed. For screening, the test is generally carried out at a single dilution. It should be noted, however, that although ELISAs are more sensitive than RBT, sometimes these may not detect infected animals which are RBT positive. It is also important to note that ELISAs are only marginally more specific than RBT or CFT.

Supplementary Tests

Milk Ring Test

Milk ring test (MRT) is a simple and effective method, but can only be used with cow's milk. A drop of haematoxylin stained antigen is mixed with a small volume of milk in a glass or plastic tube. If specific antibody is present in the milk, it will bind to the antigen and rise with the cream to form a blue ring at the top of the column of milk.

Milk ELISA

ELISA may be used to test bulk milk and is extremely sensitive and specific, enabling the detection of individual infected animals in large herds in most circumstances.

Fluorescence Polarization Assay

This technique requiring special reagents and reading equipment is claimed to have advantages in sensitivity and specificity over the other methods. Evaluation has been limited however, and the procedure is not widely available.

Intradermal Test

This procedure, using a standardized antigen preparation such as Brucellin INRA or Brucellergene OCB, can be used for monitoring the status of herds in brucellosis-free areas. It is sensitive and specific but false positive reactions can occur in vaccinated animals.

Prevention and Control

It is nearly always more economical and practical to prevent the disease than to attempt to control it. The following measures are useful in preventing *Brucella* infection in animals:

☆ Careful selection of replacement animals is important. Whether purchased or produced from existing stock, these should originate from *Brucella*-free herds or flocks. Pre-purchase tests are necessary unless the replacements are from populations in geographically circumscribed areas that are known to be free of the disease.

☆ Isolation of purchased replacements for at least 30 days is helpful. In addition, a serological test prior to commingling is necessary.

☆ Prevention of contacts and commingling with herds of flocks of unknown status or those with brucellosis should be effected.

☆ If possible, laboratory assistance should be utilized to diagnose causation of abortions, premature births, or other clinical signs. Suspected animals should be isolated until a diagnosis can be made.

☆ Herds and flocks should be included in surveillance measures such as periodic milk ring tests in cattle (at least four times per year), and testing of slaughtered animals with simple screening serological procedures such as RBT.

☆ Proper disposal (burial or burning) of placentas and nonviable foetuses, and thorough disinfection of contaminated areas should be done.

☆ Cooperate with public health authorities to investigate human cases. Animal brucellosis, especially when caused by *B. melitensis*, can often be identified through investigations of cases in humans.

The general methods of control of brucellosis in animals include: (1) test and slaughter, (2) hygienic measures, and (3) vaccination. These are most effective when they are combined. Test and slaughter of seropositive animals is usually a part of organized governmental programmes where the goal is eradication. Few countries can afford the high costs of test and slaughter. In addition, there may be a lack of skilled workers, diagnostic facilities and enforceable legislation. Other negative factors may be nomadism, commingling of flocks, and reservoirs in other species. Compensation for slaughter of animals is often absent, leading to livestock owner resistance and non-cooperation, especially in developing countries like India.

It is widely agreed that vaccination is the most effective and practical method of reducing the incidence of many diseases, including brucellosis, in livestock. The live vaccines like *B. abortus* S19 and *B. melitensis* Rev 1 have proved to be the most effective agents in cattle and in sheep/goats, respectively. Strain RB51 has replaced S19 in some countries. There is some controversy about its effectiveness. S19 and Rev 1 are relatively inexpensive to produce and are highly immunogenic. They may sometimes cause abortions but this may be overcome by reducing the dose of the vaccines. It is necessary to keep the vaccine refrigerated. Post-vaccinal antibodies may interfere with the interpretation of diagnostic test results. Although immunity may not be complete in some animals, vaccination practically eliminates clinical brucellosis and, in cattle, the herd immunity exceeds 90 per cent (Nicoletti 2010).

Successful control of brucellosis will depend upon many factors: prevalence, type of animal husbandry, laboratory diagnosis, availability and quality of vaccines, available resources (money, personnel), legal authority, intersectoral cooperation, and many others. Control of human infections depends almost wholly upon control of disease in animals, which is largely a veterinary sector's responsibility.

Future Directions

Control of any disease depends on its early diagnosis. Even though there are many diagnostic tests available for brucellosis, refinement of the existing tests and development of newer techniques like lateral flow based pen side diagnostic kit and synthetic peptide antigen based tests should be developed. Despite the advancements in vaccine safety, even the current animal vaccines have certain drawbacks such as causation of abortion among pregnant vaccinates and persistent infection of the vaccine strain in them. Development of new vaccine(s) or improvements in existing vaccines, including expansion for use in more animal species, alternate route of administration (oral vaccines) and efficacy against more species of the pathogenic *Brucella* are, therefore, needed. Control of brucellosis in wildlife animals has proved more challenging. A bigger challenge for brucellosis control in wildlife and feral domestic animals, even if an effective vaccine is developed, could be proficient vaccine delivery systems for these animals. Because of higher cost involved in test and slaughter and mass vaccination programmes in developing countries, there is need for more effective treatment, whether prophylactic or in clinical cases. Investigations are required on antibiotics targeted at infected cells such as those with carriers like liposomes.

References

1. Agasthya AS, Isloor S and Krishnamsetty P. 2012. Seroprevalence study of human brucellosis by conventional tests and indigenous indirect enzyme-linked immunosorbent assay. *Scientific World J*. doi:10.1100/2012/104239.

2. Bang B. 1897. The etiology of epizootic abortion. *J Comp Pathol Therap* 10: 125.

3. Barbuddhe SB, Chakurkar EB, Bale MA, Sundaram RNS and Bansode RB. 2004. Prevalence of brucellosis in organized dairy farms in Goa region. *Indian J Anim Sci* 74: 1030-1033.

4. Bhattacharya DK, Ahmed K and Rahman H. 2005. Studies on seroprevalence of bovine brucellosis by different tests. *J Vet Pub Health* 3: 131-133.

5. Chakraborty M, Patgiri GP and Barman NN. 2000. Application of delayed-type hypersensitivity test (DTH) for the diagnosis of bovine brucellosis. *Indian Vet J* 77: 849-855.

6. Chand P and Sharma AK. 2004. Situation of brucellosis in bovines at organized cattle farms belonging to three different states. *Vet Immunol Immunopathol* 6: 11-15.

7. Chandra M, Singh BR, Shankar H, Agarwal M, Sharma G and Agrawal RK. 2005. Seroprevalence of brucellosis in chevon goats from Bareilly slaughterhouse. *Indian J Anim Sci* 75: 220-221.

8. Cultuer SJ, Whatmore AM and Commander NJ. 2005. Brucellosis- new aspects of an old disease. *J Appl Microbiol* 98: 1270-1281.

9. Dhand NK, Randhawa SS and Singh G. 2005. Carryover and persistence of maternal antibodies against brucellosis in buffalo calves. *Indian J Anim Sci* 75: 194-195.

10. Ganesan PI and Anuradha P. 2006. Rose Bengal test and dot-ELISA in diagnosis of bovine brucellosis. *Indian Vet J* 83: 907.

11. Godfroid J, Nielsen K and Claude S. 2010. Diagnosis of brucellosis in livestock and wildlife. *Croat Med J* 51: 296-305.

12. Gupta VK, Verma DK, Singh SV and Vihan VS. 2007. Serological diagnostic potential of recombinant outer membrane protein (Omp31) from *Brucella melitensis* in goat and sheep brucellosis. *Small Ruminant Res* 70: 260-266.

13. Gupta VK, Vohra Jyoti, Kumari Ranjeeta and Vihan VS. 2009. Evaluation of an ELISA using recombinant outer membrane protein 31 for diagnosis of ovine brucellosis. *Indian J Anim Sci* 79: 41-43.

14. Gupta VK, Gupta G, Pathak M, Kumar A and Vihan VS. 2012. Milk-enzyme linked immunosorbent assay for serological diagnosis of *Brucella melitensis* infection in goats. *Indian J Anim Sci* 82: 15-19.

15. Handa R, Singh S, Singh N and Wali JP. 1998. Brucellosis in north India: Results of a prospective study. *J Commun Dis* 30: 85-87.

16. Jai Sunder, Rai RB, Kundu A, Chatterjee RN, Senani S and Jeyakumar. 2005. Incidence and prevalence of livestock diseases of Andaman and Nicobar Islands. *Indian J Anim Sci* 75: 1041-1043.

17. Kachhawaha S, Singh K and Tanwar RK. 2005. Serological survey of brucellosis in cattle and buffaloes of Jodhpur region. *Vet Practitioner* 6: 43-44.

18. Kadri SM, Rukhsana A, Laharwal MA and Tanvir M. 2000. Seroprevalence of brucellosis in Kashmir (India) among patients with pyrexia of unknown origin. *J Indian Med Assoc* 98: 170-171.

19. Kanani AN. 2007. Serological, cultural and molecular detection of *Brucella* infection in breeding bulls. *Postgraduate thesis*, Anand Agricultural University, Anand.

20. Madkour MM. 2001. *Madkour's brucellosis*. Springer Verlag, New York.

21. Mahato G, Sharma K and Mahanta PN. 2004. Comparative evaluation of serological tests for detection of brucellosis in bovine. *Indian J Vet Med* 24: 46.

22. Maninder S, Singh DK, Shivaramu KV, Biswas R, Rawat S, Boral R, Singh S and Cheema PS. 2010. Serum as clinical specimen in PCR for diagnosis of ovine brucellosis. *Indian J Anim Sci* 80:17-18.

23. Mantur BG and Amarnath S. 2008. Brucellosis in India. *J Biosci* 33: 539-547.

24. Mantur BG, Akki AS, Mangalgi SS, Patil SV, Gobbur RH and Peerapur BV. 2004. Childhood brucellosis - A microbiological, epidemiological and clinical study. *J Trop Pediatr* 50: 153-157.

25. Mantur BG, Biradar MS, Bidri RC, Mulimani MS, Veerappa, Kariholu P, Patil SB and Mangalgi SS. 2006. Protean clinical manifestations and diagnostic challenges of human brucellosis in adults: 16 years experience in an endemic area. *J Med Microbiol* 55: 897-903.

26. Mathur TN. 1964. *Brucella* strains isolated from cows, buffaloes, goats, sheep and human beings at Karnal: Their significance with regard to the epidemiology of brucellosis. *Indian J Med Res* 52: 1231-1240.

27. Mehra KN, Dhanesar NS and Chaturvedi VK. 2000. Seroprevalence of brucellosis in bovines in Madhya Pradesh. *Indian Vet J* 77: 571-573.

28. Mishra V K, Arora S and Bist B. 2005. Seroprevalence of brucellosis among cows and buffaloes of Gorakhpur district of Uttar Pradesh. *J Vet Pub Health* 3: 67-70.

29. Nasir AA, Parveen Z, Shah MA and Rashid M. 2004. Seroprevalence of brucellosis in animals at Government and private livestock farms in Punjab. *Pakistan Vet J* 24(3): 144-146.

30. Panjarathinam R and Jhala CI. 1986. Brucellosis in Gujarat state. *Indian J Pathol Microbiol* 29: 53-60.

31. Pappas G, Papadimitriou P, Akritidis N, Christou L and Tsianos EV. 2006. The new global map of human brucellosis. *Lancet Infect Dis* 6: 91-99.

32. Prahlad K, Singh DK and Barbuddhe SB. 1999. Seroprevalence of brucellosis and comparison of serological tests to diagnose in buffaloes. *Buffalo J* 15: 361-370.

33. Rahamn N, Bhattachrya M, Rajkhowa J, Soud N, Nandankar U and Mukherjee S. 2007. Seroprevalence of brucellosis in yaks (*Poephagus grunniens*) in India. *Indian J Anim Sci* 77: 796-798.

34. Rajesh, JB, Tresamol PV and Saseendranath MR. 2003. Seroprevalence of brucellosis among cattle in Kerala. *Cheiron* 32: 41-43.

35. Rajkhowa S, Rahman H, Rajkhowa C and Bujarbaruah KM. 2005. Seroprevalence of brucellosis among mithuns (*Bos frontaiis*) in India. *Prev Vet Med* 69: 145-151.

36. Randhawa AS, Kalra DS and Kapur MP. 1974. Some seroepidemiologic observations on brucellosis in humans and animals. *Indian J Med Sci* 28: 133-138.

37. Rao TS, Devi VR, Babu RM and Rao AVN. 1999. Comparison of rapid plate agglutination, standard tube agglutination and dot-ELISA tests for the detection of antibodies to *Brucella* in bovines. *Indian Vet J* 76: 255-256.

38. Renukaradhya GJ, Isloor S and Rajasekhar M. 2002. Epidemiology, zoonotic aspects, vaccination and control/eradication of brucellosis in India. *Vet Microbiol* 90: 183-195.

39. Sen MR, Shukla BN and Goyal RK. 2002. Seroprevalence of brucellosis in and around Varanasi. *J Commun Dis* 34: 226-227.

40. Sharma S, Mahajan AV, Kaur K, Verma S, Meenakshi and Kumar H. 2007. Screening of dairy farms for brucellosis and paratuberculosis. *Indian Vet J* 84: 315-316.

41. Shome R, Shome BR, Deivanai M, Desai GS, Patil SS, Bhure SK and Prabhudas K. 2006. Seroprevalence of brucellosis in small ruminants. *Indian J Comp Microbiol Immunol Infect Dis* 27: 13-15.

42. Sonawane GG, Tripathi S and Dubey SC. 2011. Sero-incidence of brucellosis in small ruminants of semiarid Rajasthan. *Indian J Anim Sci* 81: 327-329.

43. Thakur SD and Thapliyal DC. 2002. Seroprevalence of brucellosis in man. *J Commun Dis* 34: 106-109.

44. Umapathy BL, Nagamani M, Bhat P and Keshavamurthy BS. 1984. Seroepidemiological studies on human brucellosis in and around Bangalore. *Indian J Comp Microbiol Immunol Infect Dis* 5: 83-87.

2014, Zoonoses: Bacterial Diseases
Editor: Sudhi Ranjan Garg
Published by: DAYA PUBLISHING HOUSE, NEW DELHI

Pages 52–87

4

Mycobacterial Zoonoses

Debabrata Mahapatra
Texas A&M Veterinary Medical Diagnostic Laboratory, Amarillo,
TX 79106, USA

Tuberculosis (TB) is a term originating from the word 'tubercle', meaning a protuberance or a nodule that encompasses various diseases caused by bacteria belonging to the *Mycobacterium tuberculosis* (MTB) complex. Apart from causing substantial economic losses in cattle herds, mycobacterial infection constitutes a major public health concern with a third of the world's population presently infected with MTB complex organisms. The growing human immunodeficiency virus (HIV) epidemic, the HIV and TB syndemic, rapid propagation of multi-drug resistant (MDR) *M. tuberculosis* isolates and the re-emerging threat of zoonotic tuberculosis caused by *M. bovis* even further complicates this global crisis. According to an estimate, in the year 2010, there were 8.8 million (range 8.5-9.2 million) incident cases of TB, 1.1 million (range 0.9-1.2 million) deaths from TB among HIV-negative people and an additional 0.35 million (range 0.32-0.39 million) deaths from HIV-associated TB. In the same year, there were 5.7 million notifications of new and recurrent cases of TB, equivalent to 65 per cent (range 63-68 per cent) of the estimated number of incident cases. Also, about 40 per cent of the world's notified cases of TB were accounted for by India and China, and a further 24 per cent and 82 per cent were contributed by Africa and the 22 high-TB burden countries (HBCs) respectively. The number of cases of MDR-TB enrolled for treatment rose to 46,000 and less than 5 per cent of new and previously treated patients were being tested for MDR-TB in most countries in 2010 (WHO 2011).

Animal tuberculosis caused by *M. bovis* is of high economic relevance with respect to livestock farming as it directly affects the animal productivity and influences international trade of animal products. Over 50 million cattle were estimated to be infected worldwide towards the end of the last century, which resulted in economic losses of about 3 billion dollars (Cosivi 1998). It can have potentially devastating effects on the ecosystem as a whole, because of region specific wildlife reservoirs. This holds true for countries such as US, UK, Africa and New Zealand. Additionally, the zoonotic potential bears a significant public health concern worldwide. From an earlier review of zoonotic tuberculosis studies, the proportion of human cases around the world was estimated to be 3.1 per cent for all forms of tuberculosis, of which pulmonary forms constituted about 2.1 per cent and the extrapulmonary form about 9.4 per cent. Also, *M. bovis*, which is considered as being the second most frequent causative agent, accounts for approximately 5 per cent of the global tuberculosis cases (Cassidy 1998) and an estimated 10-15 per cent of human cases of tuberculosis in countries where bovine TB is endemic and pasteurization of milk is rare (Ashford 2001).

One of the earliest evidences of pathognomonic bone lesions indicative of tuberculosis in bovids was found in the fossils of ice-age wooly mammoths and mastodon fossils who appear to have spread the disease over the Holartic region of Europe, Northern Asia and upper regions of North America. *Mycobacterium bovis* was apparently spread to musk oxen that brought it to the Americas via the Siberian Alaskan ice bridge to North America where bison and Bighorn sheep were infected. In fact, mycobacterium DNA has been isolated from 17000 years old remains of an Alaskan bison (Rothschild *et al.* 2001, Rothschild and Martin 2006). In modern history, particularly during the colonial era, movement of cattle along with emigration of European settlers could have largely facilitated the intercontinental spread of the disease. In the post-colonial times, cattle from Europe were exported to African countries in order to improve the dairy production in these countries. Data based on polymerase chain reaction (PCR) based spoligotyping and variable number tandem repeat (VNTR) analysis allowing identification of clonal complexes of *M. bovis* strains in Africa (Af1) and Europe (Eu 1) bolsters the above theory and confirms the evolution of *M. bovis* strains through clonal expansion in a restricted geographic location that subsequently were shared between geographically distinct countries but having political and economic ties (Muller *et al.* 2009).

Several synonyms have been used over the centuries to describe various forms of TB in humans. Pulmonary form of TB has been described as "consumption" or "phthisis", both indicating the severe wasting and hemoptysis associated with later stages of the disease. Spinal form, marked by spinal and bone deformities, has been called "Pott's disease" after an 18th century English physician. "Scrofula" was named for the extrapulmonary form presented with cervical lymphadenitis or swelling of the lymph nodes of the neck. It was also called "The King of the Evil" from the prevailing myth that it could be healed through a touch by a reigning monarch. The cutaneous form used to be called as "lupus vulgaris" in the middle ages (Smith 2003). Yet another name for cutaneous tuberculosis has been "Butchers or Prosector's wart" (Grange *et al.* 1988). Disease caused by *M. bovis* in cattle is also known as "pearl disease" owing to the pearl-like lesions (tubercles) in the pleura and mesentery (Acha and Szypres 2001).

Etiology

Currently, there are more than 120 recognized species of mycobacteria (Devulder and Flandrois 2005) with immense diversity with respect to growth characteristics, pathogenicity, adaptation in host, and responses to anti-tuberculous drugs. Of these, bacteria belonging to MTB complex and species belonging to non-tuberculous mycobacteria (NTM) are of significance with regards to disease in animal hosts and zoonoses. MTB complex, the causative agent of tuberculosis in distinct hosts, is composed of *M. tuberculosis, M. bovis, M. africanum, M. canetii* and *M. microtii*, along with more recently added species, *M. caprae* and *M. pinnipedi*, that are considered variants of *M. bovis* (Aranaz *et al.* 2003, Cousins *et al.* 2003).

Often known by several acronyms, mycobacterias other than MTB complex and *M. leprae* are called atypical mycobacteria, non-tuberculous mycobacteria (NTM), and mycobacteria other than tubercle bacilli (MOTT). There has been a dramatic increase in the number of clinically significant species over the years because of the improvement in molecular techniques. Of the many NTM species, the slow growing *Mycobacterium avium* complex (MAC) is the most important that comprises of two main species: *M. avium* and *M. intracellulare*. *M. avium* is subdivided into four subspecies: *M. avium* subsp. *avium, M. avium* subsp. *hominisuis, M. avium* subsp. *paratuberculosis*, and *M. avium* subsp. *silvaticum* (Turenne 2008). Others that cause disease in humans and animals include *M. kansasii, M. abscessus, M. chelonae, M. fortuitum, M. genavense, M. gordonae, M. haemophilum, M. immunogenum, M. malmoense, M. marinum, M. mucogenicum, M. nonchromogenicum, M. scrofulaceum, M. simiae, M. smegmatis, M. szulgai, M. terrae complex, M. ulcerans, and M. xenopi* (Griffith *et al.* 2007).

Long before the first ever description of the tubercle bacillus by Robert Koch in 1882 and identification of human and bovine tubercle bacilli as separate organisms (Sakula 1882), the epidemiological link between bovine and human tuberculosis was recognized. The easily noticeable nodular tuberculous lesions on the mesentery and pleura of slaughter cattle in previous centuries were considered harmful to human health and later associated with an infectious agent "contagium vivium" transmitted to humans from cattle (Orland 2003). Later on, Revenel in 1902 reported the occurrence of *M. bovis* in a child with tuberculous meningitis. At about the same time, Emil von Behring, leading pediatrician, thought of bovine tuberculosis in humans as an infectious disease that could be acquired in early childhood and that could cause pulmonary disease in adults after a period of latency (Zeiss and Behling 1940). Also, a significant percentage of human tuberculosis cases was thought to have been caused by *M. bovis*, especially in children and cattle-tending people in rural areas (Schmiedel 1968). But with implementation of eradication programmes in developed countries that involved mandatory pasteurization of milk, compulsory tuberculin testing of livestock and removal of positive reactors, the incidence of bovine and zoonotic tuberculosis came down significantly (Michel 2010). Today, bovine tuberculosis is emerging as an important disease due to the increasing incidence of cases in humans, particularly in immunocompromised individuals in developing countries and also in developed countries with continuous movement of immigrants.

The evolutionary relationship between *M. tuberculosis* and *M. bovis* is rather interesting. The domestication of cattle that is believed to have occurred between 10,000 and 25,000 years ago might have allowed the passage of mycobacteria from domesticated livestock to humans and their eventual adaptation to a new host. This leads to the hypothesis that *M. bovis* that caused TB-like disease in cattle could be the evolutionary precursor of *M. tuberculosis* (Stead 1977). However, in the light of new molecular epidemiological data, this hypothesis seems to be incorrect. In fact, analysis of the existence of rare synonymous single nucleotide polymorphisms (SNPs) that allow discrimination between closely related bacteria, and the study of the distribution of deletions and insertions in the genomes of MTB complex strongly suggest that both *M. tuberculosis* and *M. bovis* evolved independently and both could have another precursor species, possibly related *to M. canetti*, also referred to as *Mycobacterium protuberculosis* (Sreevatsan *et al.* 1997, Brosch *et al.* 2002, Gutierrez *et al.* 2005).

Host Range and Reservoirs

Members of the MTB complex that are phylogenetically closely related are capable of causing disease in a vast array of host species including humans. While some members, notably *M. tuberculosis*, *M. africanum* and *M. canetii*, are predominantly human or rodent pathogens (*M. microtii*), others such as *M. bovis*, *M. caprae* have a wider host spectrum (Brosch *et al.* 2002, Prodinger *et al.* 2005).

M. tuberculosis does not have an indigenous animal host or reservoir besides humans and thus infected animals most probably represent the accidental hosts (Steel 1980, Theon and Steel 1995, Une and Mori 2007). Although considered a primary human pathogen, *M. tuberculosis* infects a variety of species including fish, reptiles, birds, domestic and wild animal species most frequently living in close proximity to human beings such as in captive setting (Aranaz *et al.* 1999, Montali *et al.* 2001, Oh *et al.* 2002, Sternberg *et al.* 2002, Alexander *et al.* 2002, Alfonso *et al.* 2004, Une and Mori 2007).

Mycobacterium bovis, having the broadest host range, affects feral and domesticated cattle, sheep, goat, horse, pig, dog, cat, deer, buffalo, bison, fennec fox, ferret, badger, possum, hare, Arabian Oryx, alpaca, camel, llama, humans and non-human primates. It has also been reported in hyena, leopard, lion, kudu, baboon, cheetah, warthog, bush pig, elk, coyotes, meerkats, black rhinoceros, Lynx and Aoudad (O' Reilly and Daborn 1995, Perez *et al.* 2001, O' Brien *et al.* 2008, VerCautern *et al.* 2008, Drewe *et al.* 2009, Esple *et al.* 2009, Candela *et al.* 2009, Michel *et al.* 2009).

Susceptibility of different host species to MTB complex organisms varies depending on the route of exposure, dose of organisms and virulence of the strain. Humans, non-human primates and guinea pigs are highly susceptible to *M. tuberculosis*. Cattle, cats and rabbits are susceptible to *M. bovis*, but are quite resistant to *M. tuberculosis*. Wild hoof stock is usually susceptible to *M. bovis*, but the reports on the isolation of *M. tuberculosis* are scarce. Dogs and pigs are susceptible to both

M. bovis and *M. tuberculosis* (Lo Bue *et al.* 2010).

Many wild animals act as potential reservoirs of *M. bovis* and are capable of transmitting infection to domestic and other valuable wildlife species. Important among them are the brushtail possum (*Trichosurus vulpecula*) in New Zealand, European badger (*Meles meles*) in the United Kingdom and Republic of Ireland, bison (*Bison bison*) and white-tail deer (*Odocoileus virginianus*) in North America, African buffalo (*Syncerus caffer*) and Kafue lechwe antelopes (*Kobus leche*) in Africa, all of which can act as "maintenance hosts" capable of maintaining infection without further transmission from another species. Additionally, some species can also act as "spillover" hosts that are able to maintain the infection for a time but require periodic input from another source or "dead end" hosts unable to maintain the infection without an external source (De Lisle *et al.* 2002, Rhyan and Spraker 2010).

Disease Caused by *M. tuberculosis*

Transmission of *M. tuberculosis* can occur between humans, from humans to animals and back again from animals to humans.

Human-to-Human Infection

Tuberculosis (TB) is an infectious disease caused by the bacillus *M. tuberculosis* that typically affects the lungs (pulmonary TB) but can affect other sites as well (extrapulmonary TB). Human-to-human transmission of TB occurs when individuals with active infection cough, sneeze, spit, vocalize resulting in release and spread of aerosol clouds of infected saliva or mucus droplets containing *M. tuberculosis* to the surroundings (Figure 4.1). Exposure to these highly infectious aerosols either for a

Figure 4.1: A sneeze in progress, revealing the plume of salivary droplets expelled from mouth (Courtesy: CDC, Brian Judd and James Gathany)

brief period or even exposure to a single bacterial cell can lead to an infection (Nicas *et al.* 2005, Kamper-Jorgensen *et al.* 2009).

The mean incubation period, according to a recent report in the Netherlands, was 1.3 years with shorter time in secondary cases in young individuals and men, those with extrapulmonary TB and those not reporting previous TB or previous preventive therapy (Borgdorff *et al.* 2011). Relatively, a small proportion of people (<10 per cent) infected with *M. tuberculosis* actually go on to develop TB, most develop a latent infection. The risk of development of infection into an active disease is the highest in the first year after infection but reactivation can take place even after several decades (Lillebaek *et al.* 2002). The probability of developing TB, however, is much higher among the people infected with HIV. The disease is more common among men than women, and it affects mostly adults in the economically productive age-groups. Around two-thirds of the cases are estimated to occur in the 15-59 years age group (WHO 2011).

Mortality rates are high without treatment. Among the sputum smear-positive and HIV-negative cases of pulmonary TB, around 70 per cent have been reported to die within 10 years while among the culture-positive (but smear-negative) cases, the fatality has been reported as 20 per cent within 10 years (Tiemersma *et al.* 2011). According to Marais *et al.* (2006), children make up an estimated 10-15 per cent of the total global tuberculosis burden with high morbidity and mortality rates, primarily because the cases remain under-recognized and under-reported. The disease is rapidly progressive with frequent dissemination to extrapulmonary sites.

Human-to-Animal Infection

The human-to-human infection rotates in a cyclic fashion. Humans being the only reservoir naturally infect cattle with *M. tuberculosis* that spread it to the other cattle, eventually becoming the source of infection in humans (Yumi and Tooru 2007). Among domestic animals, *M. tuberculosis* infection has been most frequently identified in cattle (Chandrasekharan and Ramakrishnan 1969, Steel 1980, Theon 1981). Reports of TB in cattle caused by *M. tuberculosis* are usually from countries with high incidence of human TB burden and close human and cattle interactions (Berg *et al.* 2009). For years, despite lack of direct evidence, humans suffering from active TB were strongly believed to be the main source of *M. tuberculosis* infection in animals including cattle, spread via sputum and rarely urine or faeces (Theon and Steel 1995). It was not until the last decade that Ocepek *et al.* (2005) reported the first transmission of *M. tuberculosis* from a farm worker to cattle, which was unequivocally confirmed by molecular typing using the IS*6110*-based restriction fragment length polymorphism (RFLP) analysis. Very recently, Ameni *et al.* (2011) identified the transmission of *M. tuberculosis* from humans in approximately 27 per cent of cattle from two different management systems in central Ethiopia by molecular analysis. Infection of affected cattle occurred through the unique habit of farmers in the region of chewing ground, baked tobacco and discharging the juice directly into the oral cavity of cattle. Supportive evidence of this oral route of infection was the finding of tuberculous lesions in the retropharyngeal and mesenteric lymph nodes of such cattle on post-mortem examinations. Transmission of *M. tuberculosis* from humans to other animals such as canines (Erwin *et al.* 2004), non-human primates (Michel and Huchzermeyer 1998) and elephants

(Michalak *et al.* 1998) is not uncommon.

Disease Caused by *M. bovis*

Transmission of *M. bovis* can occur between animals, from animals-to-humans, less commonly from humans to animals and rarely from humans to humans.

Animal-to-Animal Infection

Bovine tuberculosis caused by *M. bovis* is a chronic progressive infectious disease characterized by the formation of typical granulomatous lesions with varying degrees of necrosis, encapsulation, and calcification (Michel *et al.* 2010). The time course of infection is usually weeks to several months if not years, although animals appear clinically normal. Clinical sign of weight loss in advanced stages is most common, besides coughing in tuberculous pneumonia, swollen lymph nodes especially of the head, and discharging abscesses of lymph nodes (de Lisle *et al.* 2002). In general, extra-pulmonary forms of bovine TB presents typically as cervical lymphadenitis in animals and humans (Figure 4.2). Behavioural change such as unusual diurnal activity in nocturnal brushtail possums has also been reported (Sauter and Morris 1995). According to an experimental study, the minimum infective dose required to stimulate specific immune response and general pathology in cattle is 1 CFU containing about 6-10 viable bacilli (Gillian 2005). A distinctive feature of *M. bovis*

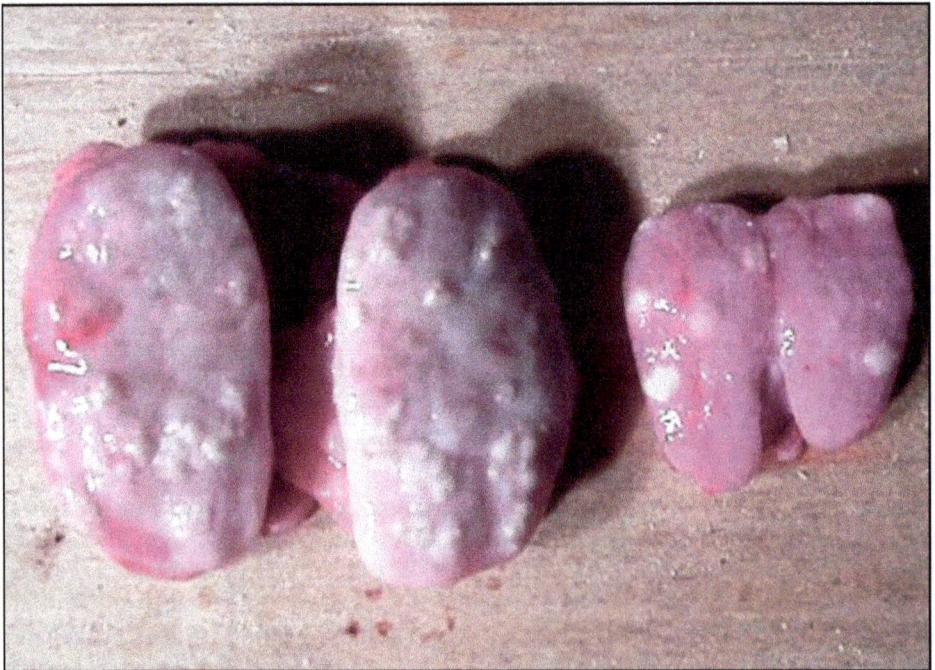

Figure 4.2: Cut sections of lymph nodes from a captive adult Asian Elephant showing multifocal to coalescing raised white foci of abscesses (granulomatous lymphadenitis) caused by infection with *Mycobacterium bovis*. (Courtesy: Dr. Scott P. Terrell, Walt Disney World, Orlando, Florida, USA).

infection is the variation in the appearance and distribution of lesions in the different host species (de Lisle *et al.* 2002).

The distribution of lesions observed in cattle and other animals on post-mortem indicates the route of transmission. Gross pathology in animals infected by inhalation of aerosols is typically restricted to the thoracic cavity with majority of lesions in the upper and lower respiratory tracts and their associated lymph nodes. This warrants inhalation not only as the primary route of *M. bovis* infection but also the most suitable means of active excretion and aerosol dissemination to other animals. In naturally infected cattle, however, lesions are most frequently located in the dorso-caudal apex regions of the lungs (Francis 1947, McIlroy *et al.* 1986) which leads to the proposition that the size of inhaled infectious droplets and the topographical orientation of lungs may also influence the airflow dynamics and lesion distribution (Cassidy 2006). The alternative routes of aerogenous infection include contaminated water droplets, eructation during rumination of infected pastures or inhalation of contaminated dust particles, the latter implicated to be the most likely way cattle get infected by badger excretions in a contaminated environment.

Animals having lesions in the mesenteric lymph nodes usually acquire infection following ingestion of bacteria directly from animals or from contaminated pastures; hence, water or fomites are considered most important routes of infection secondary to inhalation (Neill *et al.* 1994, Menzies and Neill 2000, Pollock and Neill 2002, Phillips *et al.* 2003). In animals with head lymph node lesions, particularly the retropharyngeal nodes, determination of the actual route of infection remains difficult (de Lisle *et al.* 2002). The location of lesions also provides clues about the spread of infection as in the case of badgers, kidney lesions indicate infection from contaminated urine. Rarity of lesions other than the head nodes and thorax such as mammary glands and urogenital tracts suggests these to be uncommon routes of transmission (O' Brien *et al.* 2001). Transmission of infection between cattle depends on several factors, including frequency of excretion, route of infection, infective dose, period of communicability and host susceptibility (Griffin and Dolan 1995).

Histopathologic lesions in camelids, cattle and other wild bovids such as Oryx, Kudu, Nilgai and Sable antelopes bear close resemblance (Theon 2009) and are characterized by granulomatous response with areas of central necrosis and mineralization surrounded by macrophages, giants cells and lymphocytes that are walled-off by fibrous connective tissue. Exception to this classic pattern has been reported in badgers and brushtail possums in which neutrophils are predominant but have similarity with those found in lions and leopards in which giant cell or fibrous capsule formations are present (Gallagher *et al.* 1976, Cooke *et al.* 1995, Keet *et al.* 1996). Tubercles in non-human primates do not show significant differences to distinguish these from *M. tuberculosis* infections (Theon 2009).

Animal-to-Human Infection

Zoonotic tuberculosis is most commonly caused by *M. bovis* and to a lesser extent by *M. caprae* that is prevalent in certain regions of Europe (Prodinger *et al.* 2005). Clinically, radiologically and histopathologically, tuberculosis caused by *M. bovis* is indistinguishable from that caused by *M. tuberculosis* (Cosivi *et al.* 1998,

Grange 2001). Humans can get infected through ingestion, inhalation or direct contact with mucous membranes or breached skin (prosecuter's wart). Ingestion of unpasteurized milk is still regarded as the principal vector for transmission in countries and regions where bovine tuberculosis is not controlled and commonly results in cervical lymphadenopathy, intestinal lesions, chronic skin TB (lupus vulgaris) and other extrapulmonary forms, particularly in young people (Cosivi 1998). Also, young children infected with *M. bovis* typically develop abdominal infections and older patients suffer from swollen and often ulcerated lymph nodes in the neck (Bolognesi 2007). Regardless of the route, the infectious dose in humans is unknown but estimated to be in the range of tens to hundreds of organisms by the respiratory route and millions by the gastrointestinal route. The infective dose is influenced by host species (higher for human than cattle), and host immune status (Grange 1988, O' Reilly and Daborn 1995).

M. bovis has been isolated from HIV infected persons and the disease resembles that caused by *M. tuberculosis*, hence, may manifest as pulmonary, extrapulmonary, and in severely immunocompromised individuals, a disseminated disease. However, *M. bovis* in people has been considered less virulent compared to its human counterpart and therefore less likely to lead to an overt post-primary disease, although more than half of the reactivation cases involve the lungs (Shitaye *et al.* 2006) and raises the concern of human-to-human transmission of *M. bovis* (Cosivi 1998).

Human-to-Human Infection

Unlike the transmission of *M. bovis* from cattle to humans or transmission of *M. tuberculosis* between humans, the role of human-to-human airborne transmission of *M. bovis* has been controversial with the prevailing view that transmission between humans is a rare event and only likely in HIV infected individuals. A series of reports addressing this issue obviously speak otherwise. Evans *et al.* (2007) documented transmission of *M. bovis* from a smear positive pulmonary tuberculosis patient (primary source) to five other susceptible but not necessarily immunocompetent persons who became smear positive and showed symptoms for months before being diagnosed. This pattern of transmission and evolution of pulmonary disease was similar to that of *M. tuberculosis*. Other reports include three nosocomial outbreaks of MDR-TB caused by *M. bovis* in HIV infected patients in Spain (Cobo *et al.* 2001), and report of non-HIV father and daughter in Argentina who tested positive for identical *M. bovis* isolates from their pulmonary lesions after an interval of 5 years between infections (Etchechoury *et al.* 2010). Hence, it would be prudent to realize that human-to-human transmission of *M. bovis* does exist and is possible in cases with or without severe immunosuppression. Strains are capable of mutating to MDR genotypes that maintain the pathogenicity for humans. Cattle and other animals in TB endemic areas might eventually get re-infected with MDR strains from humans (Theon *et al.* 2009).

Non-Tuberculous Mycobacterial Diseases

Non-tuberculous mycobacteria (NTM) presently comprises more than 130 known species, the majority of which are nonpathogenic free-living saprophytic organisms but can behave as opportunistic pathogens in the presence of predisposing conditions. Environment being the major reservoir, NTM organisms can be found in water, soil, dust and plants. Transmission of infection to humans and animals is mainly from aerosol, ingestion and via contact with contaminated environments without evidence of human-to-human transmission (Falkinham 1996, Tortoli 2009).

Mycobacterium avium subsp *paratuberculosis* Associated Mycobacterial Zoonosis

Mycobacterium avium subsp *paratuberculosis* (MAP), a member of MAC, is the causative agent of Johne's disease in cattle and recent evidences link this organism to Crohn's disease (CD) in humans, a chronic inflammatory disease of the GI tract histologically characterized by granulomatous reaction in the infected tissues (Falkinham 1996). MAP is highly persistent and capable of survival within livestock environment (water, faeces, soil) for long periods, posing risk to humans. However, studies indicate contaminated milk to be the main source of transmission from animals to humans other than animal derived food stuff. A study on pasteurized milk samples in UK yielded a 1.8 per cent positive rate by culture and 11.8 per cent by PCR for MAP organisms (Grant 2002, Whittington 2004).

The outcome of exposure is influenced by the microbial phenotype. While the extracellular classical ZN-positive phenotype of MAP excreted from heavily infected animals actually confers some natural resistance to the disease, exposure to the intracellular forms which have passaged through bovine macrophages in milk and cheese or through environmental protists have enhanced virulence for humans (Cirillo 1997, Patel 2006). This explains why dairy farmers, veterinarians and children exposed to MAP exposed cattle have lower incidence of CD.

Following its emergence in Europe and North America in the mid 20th century, today Crohn's has become widespread across the globe and increasingly becoming a major healthcare problem. The city of Winnipeg is a hot spot for CD due to exposure of the populace to high levels of waterborne MAP brought down from agricultural river catchments of US mid-West together with Canadian provinces (Taylor 2009). In animals and primates, MAP causes a local and systemic immune dysregulation and affects non-myelinated intestinal neurons causing chronic enteric neuropathy, whereas in humans, inflammation in Crohn's disease is caused by a two tier cooperative pathogenic mechanism. Primary intracellular infection with MAP organisms widely distributed in the gut causes immune dysregulation and a specific chronic enteric neuropathy with loss of mucosal integrity and impairment of critical functions. Grossly visible inflammatory segments result from perturbed neuroimmune response to penetration of secondary pathogens from the lumen into the gut wall (Scanu 2007, Taylor 2009). The benchmark diagnostic test for MAP infection in humans has been a nested PCR applied to single ~20mg fresh endoscopic mucosal biopsies (Bull 2003).

Recent genomic studies provide ample evidence of MAP zoonotic transmission from domestic animals to humans where MAP plays a role in the disease progression of CD patients with pathogen specific susceptibility genes and/or regions of duplications within genes (Wynne *et al.* 2009).

Aquatic Mycobacterial Zoonosis

Fish mycobacteriosis or piscine tuberculosis is often a lethal disease affecting a wide variety of fish globally, both in wild and aquarium setting with serious impacts on the aquaculture and aquaria trade and lack of a proper cure other than depopulation and facility disinfection (Jacobs *et al.* 2009). Of several species that affect the fresh water, brackish and marine fish, *Mycobacterium marinum* is the most common piscine-related NTM that infects humans (Parent *et al.* 2005). The affected humans usually have a history of a localized skin trauma followed by subsequent exposure to aqueous environment (Bartralot *et al.* 2005). Fish handlers, fish tank hobbyists, etc. are, therefore, at a high risk of developing infections. Lesions typically develop either as ulcerative or raised local granulomas and are usually located at the extremities such as hands and fingers (Parent *et al.* 2005). Less often, deep subcutaneous infection can cause tenosynovitis, osteomyelitis and arthritis (Zeegelaar 2008). Such deep-seated lesions can cause lymphocutaneous infections that result in nodular lymphangitis, which frequently ulcerate, suppurate and spread along proximal lymphatics (Kostma and DiNubile 1993, Easton 2009). The extent of cutaneous involvement, number of lesions and the systemic involvement depend on the immune status of the subject. Immunocompromised patients tend to have a more severe presentation and course of illness (Parent *et al.* 2005, Bartralot *et al.* 2005).

Environmental and Other Predisposing Risk Factors

A number of factors influence the occurrence of tuberculosis and heighten the chances of disease in certain categories of people. The ability to identify individuals at greater risk of developing tuberculosis has important implications in evolving and implementing the public health policies and patient care.

Risk Factors in Human Populations

Risk factors for mycobacterial infections in humans are well described especially for *M. tuberculosis*. The likelihood of individuals contracting primary disease and its progression to active disease is dramatically influenced by host factors. The high susceptibility individuals include those with compromised immune system either due to HIV infections, hematologic or reticuloendothelial malignances or immuno-suppressive medications such as corticosteroids, tumour necrosis factor (TNF)-alpha inhibitors, calcineurin inhibitors and cytotoxic chemotherapeutic agents. Additionally, patients with chronic diseases such as diabetes, renal disease, silicosis are at elevated risk. Independent risk factors for the disease arise when individuals are younger than four years of age, have history of long-term malnutrition or substance abuse (Jensen *et al.* 2005).

Risk factors for acquiring bovine TB exist across all age groups of susceptible human hosts (Shitaye *et al.* 2007), particularly those with immuno-suppressive or debilitating disorders. High risk individuals are the people in contact with potentially infected animals such as veterinarians, abattoir workers, meat inspectors, autopsy personnel, farmers, milkmen, animal keepers, animal dealers, laboratory personnel and the persons having potential TB infected pets (O' Donahue *et al.* 1985, Yumi and Tooru 2007). The existing age-old culture of eating raw meat and drinking raw milk, close contact between humans and animals, inadequate meat inspection, low standards of food hygiene and personal hygiene practices prevailing in certain countries and regions have been identified as significant risk factors for contracting and spread of zoonotic TB in humans. Additionally, development and spread of MDR strains and increased migration of infected population to disease-free zones have been reported to increase the risk of infection (Cosivi *et al.* 1998, Shitaye *et al.* 2007).

Risk Factors in Animal Populations

Risk factors for *M. bovis* in animals may relate to individual animals, herd or region/country (Humblet *et al.* 2009). Animal level risk factors include age, gender, breed, body condition, immune status, genetic resistance and susceptibility to bovine TB, vertical and pseudo-vertical transmissions and auto-contamination. Herd related risk factors include the history of bovine TB outbreak in the herd and human antecedent of tuberculosis in the household, herd size, type of cattle farming, management practices, intensity of the farming system and housing of cattle, and the sensitivity and frequency of diagnostic tests in use. Access to veterinary services, introduction of new animals in a herd, movements of animals, contact between animals, culling practices, contact of livestock with other domestic species and wildlife, environmental persistence of *M. bovis* and influence of climate are some other herd level factors. The region or country level factors that influence the bovine TB risk include the prevalence of the disease in the region, contiguity with other regions, international trade, trans-border animal movements, migration of animals, and movement of wildlife in the region (Humblet *et al.* 2009). Depending upon the prevailing specific epidemiological status and control programmes, there are differences in the importance of these risk factors in different cattle systems in the developed and developing countries.

Epidemiological Features and Prevalence

Although, most of the available epidemiologic data on bovine tuberculosis are obtained from studies conducted in the industrialized nations, developing countries around the world are the ones that are most profoundly affected by the disease (Michel 2010). The data available on the worldwide animal health information database of the OIE (2012) shows that 128 out of 155 countries reported the prevalence of *M. bovis* in their cattle population during the period between 2005 and 2008. Although in lesser proportion, there are reports of cases of bovine tuberculosis in elderly people following reactivation of dormant infections (Wedlock *et al.* 2002). According to WHO (2009), the populace to a large extent are increasingly vulnerable to infection due to poverty, HIV and reduced access to health care systems. Accurate assessment of the incidence of human tuberculosis is difficult usually because the diagnosis of TB still depends solely on sputum smears, but cases might account for up to 10-15 per cent of all new cases of human tuberculosis registered yearly (Cosivi *et al.* 1998, Ashford *et al.* 2001, Theon *et al.* 2006, de la Rua-Domenech 2006).

In most developed countries, with the exception of a few countries having wildlife reservoirs of *M. bovis*, the prevalence of bovine tuberculosis is at very low levels (Eurosurveillance Editorial Team 2005). Infection from *M. bovis*, according to one report has accounted for approximately 0.5 per cent to 7.2 per cent of all bacteriologically confirmed tuberculosis patients in industrialized countries. Infection in most cases is usually associated with consumption of unpasteurized dairy products imported from countries where *M. bovis* is prevalent (CDC 2005, de la Rua-Domenech 2006).

Asian Countries

Asia is home to about a third of the world's cattle population with 94 per cent of the 460 million cattle herds and 99 per cent of buffalo population located in areas where partial or no tuberculosis control programmes exist (Cosivi 1998).

India

In India, there is obvious lack of any nationwide survey on the prevalence of *M. bovis* in animals but Prasad *et al*. (2005) reported 28.8 per cent incidence of *M. bovis* in cattle based on nested PCR analysis targeting the *H*up B gene. Gokhalesh *et al*. (2006) screened Indian buffaloes for IS *1081* gene of MTB complex obtained from a variety of specimens (milk, scapular nodes, blood, pharyngeal swabs, rectal pinches, and faecal material) and reported 50 per cent of these to be positive for *M. bovis*. There is little epidemiological data about zoonotic tuberculosis caused by *M. bovis* in the country, but Prasad *et al*. (2005) reported a high (10.3 per cent) incidence of bovine TB positive cases in 2005. A more recent report by Jain (2011) sighted a 9.52 per cent incidence. The percentages of co-infection with *M. tuberculosis* in these studies were 8.7 and 4.76 per cent, respectively, along with a high (15-28 per cent) incidence of *M. tuberculosis* in cattle in the report by Prasad *et al*. (2005). Gokhalesh *et al*. (2006) have also pointed to an increasing incidence of bovine TB in humans in recent years while reporting a 10 per cent incidence of *M. tuberculosis* in Indian buffaloes. Shah *et al*. (2006) reported detection of *M. bovis* (17 per cent), *M. tuberculosis* (2.8 per cent) and both of these organisms in combination (13.4 per cent) in human cerebrospinal fluid samples. This scenario clearly indicates the transmission of tuberculous infection from human to cattle and vice versa posing a grave health hazard.

Despite these alarming figures, the concept of zoonotic tuberculosis in India has been underestimated for decades. In fact, a monograph entitled "*Zoonotic Diseases of Public Health Importance*" clearly failed to recognize bovine tuberculosis to be of zoonotic concern (NICD 2005). Where government spending on health in 2006 was about 5 per cent of GDP and a budget for human tuberculosis increased to 100 million USD (WHO 2010), it would be reasonable to think that a portion of that budget would be allocated to bovine tuberculosis in particular. However, the recent "Road Map to Combat Zoonoses in India Initiative" (RCZI) with multi-institutional partnership launched in 2008 recognizes bovine tuberculosis as a neglected zoonosis. A newsletter in 2009 reported a 15 per cent rate of *M. bovis* isolation from milk in a central military laboratory and isolation of *M. bovis* or *M. tuberculosis* from apparently healthy cattle indicative of subclinical infection. This organization also aims to advocate communication and awareness, and foster collaborative research to effectively combat zoonosis in India (RCZI 2009).

Unlike many other countries, where wild animals have been known for decades to act as reservoirs of *M. bovis* and thus posing a huge obstacle against eradication of the disease, the existence of such a maintenance host in the wild is unknown and has never been explored in India (Bose 2008). It is also unknown whether the wildlife-domestic animal interface would cause actual disease transmission, especially in the regions where the wildlife suffers massive encroachment from humans who live and forage for food or graze their cattle in the forest. The possibility of transmission of disease in such a scenario cannot be completely ruled out.

There have been controversies about successful implementation of test and slaughter policy in India. However, a study comparing two dairy herds with and without segregation and culling policies being practised clearly shows the former with a low (0.65 per cent -1.85 per cent) level of prevalence and the latter with a high (15.76 per cent) incidence. Furthermore, avoiding periodic introduction of animals into herds for breeding purposes could result in farm free of bovine TB (Mukherjee 2006).

Abattoir surveillance with lesion detection during commercial slaughter has traditionally served as a cost-effective method for passive surveillance and an important source of epidemiological data for bovine TB both in Officially TB free (OTF) and non-OTF countries. Unlike most other countries, including some African countries, a government regulated abattoir surveillance system for collection of bovine TB incidence data has not been strictly employed in India. Abattoirs in the country are mostly service-oriented; cater to the growing meat industry, and under the control of the exiting market forces and not the government. Furthermore, religious and political constraints over large animal slaughter, especially bullocks, render inadequate ante-mortem and post-mortem inspection leading to large scale illegal slaughtering in local markets without veterinary supervision (Government of India 2010). No wonder, zoonotic TB in India is still underestimated and underexplored. Illegal immigration and cattle trafficking from neighboring countries and the sheer ignorance about the disease among all strata of the populace add to the socio-economic burden of the nation.

Pakistan

Tuberculosis in Pakistan is present in epidemic proportions in animals, humans and perhaps in wildlife as well. An earlier study in 2002 identified bovine TB incidence in 6.91 per cent buffaloes and 8.64 per cent cows by tuberculin test. In 28.07 per cent and 12.18 per cent of the tuberculin positive buffaloes, the organisms were isolated from milk and nasal secretion, respectively. Similarly, isolation was done from 25 per cent and 12.5 per cent of milk and nasal secretion, respectively, in tuberculin positive cows. Subsequent study, based on comparative intradermal tuberculin test (CIDT), showed bovine TB prevalence in Nili Ravi buffaloes in Punjab province to be 10.06 per cent where old age and high milk yield played a significant role in the disease prevalence. Besides, *M. bovis* was isolated from 44.44 per cent milk samples, *M. tuberculosis* from milk and faecal samples (33.33 per cent) and atypical mycobacteria from milk samples as well. Overall the incidence of bovine TB was found to be rising with time. Yet another study using PCR methods showed 29 per cent milk samples positive for *M. bovis* and 35 per cent positive for MTB complex, again stressing the significance of possible transmission to humans. Although, accurate epidemiologic data is lacking, according to rough estimates, 20 per cent of tuberculosis patients visiting hospitals in Lahore suffer from extrapulmonary tuberculosis (Suleiman *et al.* 2002, Mumtaz *et al.* 2008, Khan *et al.* 2008).

Iran

Few studies on *M. bovis* prevalence have been conducted in Iran. An unpublished report indicates its decline from 10 per cent before the test and slaughter programme initiated in 1970 to 0.4 per cent in the West Azerbaijan province alone, and a 0.12 per cent overall incidence in 2006. Comparative data on the disease in buffaloes during the same time period is unavailable, however, a recent study suggests a 0.7 per cent incidence of *M. bovis* in buffaloes (Mosavari 2007). Due to the rapid expansion of cattle herds over the last four decades at an annual rate of 1.8 per cent, intensification in farming practices, and less stringent test slaughter programme, *M. bovis* in Iran has emerged as a major cause of cattle illness and economic loss (Tadayon 2006). Data from a recent study employing a combination of spoligotyping and VNTR analysis suggest an unprecedented genomic homogeneity of *M. bovis* isolates (perhaps genetic bottleneck) resulting from ever increasing susceptible Holstein Friesian cattle herd numbers (Tadayon 2008).

China

In China, bovine tuberculosis in cattle and humans has become a matter of great concern owing to the rapid expansion of cattle population due to increase in the number of self-employed cattle breeders in recent years, and due to rapid development of dairy-cattle industry in Northern and Western regions where animal attendants are in close contact with cattle and the practice of drinking raw milk prevails till date. Two national surveys in cows undertaken between 1985 and 1987 showed the prevalence rates of bovine TB to be 5.83 per cent and 5.43 per cent, respectively. Since then, no national surveys have been done. A case report in 2004 by Xiaomei showed a positive purified protein derivative (PPD) reactor rate of 15.6 per cent in a herd with five of the eight animal attendants also positive for *M. bovis* (Xing 1997, Xu *et al.* 1998, Fuller 2006, Yingyu 2009).

Japan

With the establishment of Bovine Tuberculosis Eradication Project in 1901, followed by the enactment of Animal Infectious Diseases Control Law in 1951, by which positive cattle are culled, the number of cases of cattle infected with *M. bovis* has reduced to 0-2 per cent per annum since 2000 (Yumi and Tooru 2007). Considering the situation in neighboring Asian countries, Japan is one of the few nations that have been able to control bovine tuberculosis with no reports of *M. bovis* isolation from wild animals since the 1970s (Chiba 1993, Yoshikawa 2006).

African Countries

Ethiopia

Ethiopia is one of the African countries where tuberculosis is endemic, highly prevalent in both cattle and human population resulting in high morbidity and mortality and where the disease is detected by tuberculin skin testing, abattoir meat inspection and rarely on bacteriological methods. According to recent estimates, bovine TB ranges between 3.4 per cent in small holding production system to about 50 per cent in intensive dairy production units. In slaughterhouses, it ranges from 3.5 to 5.2 per cent that eventually becomes a great concern for human health. Infection in humans is acquired through raw milk consumption and the extrapulmonary form causing cervical lymphadenitis is more common (Shitaye *et al.* 2007).

Nigeria

Nigeria is yet another nation where bovine tuberculosis is widespread both in humans and animals. Although the actual prevalence rate at the national level is unknown, data available from a limited survey undertaken over the past 30 years indicate that *M. bovis* prevalence is on the rise and ranges from 2.5 per cent in 1976 to 14 per cent in 2007 (Abubaker 2011). Several factors responsible for the persistence of *M. bovis* have been identified. Important among them include the failure of bovine TB control programme, lack of adequate veterinary services, eating habits and living standards of families, poor availability of diagnostics, the associated problem of AIDS/HIV/bovine TB syndemic, and illiteracy (Abubaker 2011). Moreover, a major setback is the inability of the national TB control programme to recognize the significance of *M. bovis* as a major public health problem. General lack of collaboration between human and animal veterinary medicine makes the situation worse (Abubaker 2007).

Uganda

Despite zoonotic TB being more often reported as an extrapulmonary disease, it has been shown that 0.4 to 10 per cent of sputum isolates from patients are *M. bovis* positive in African countries (Cosivi 1998). In 2002, the average number of extrapulmonary TB cases in humans in Uganda and its main milk basin located in the Mbarara district was 7.5 and 6 per cent, respectively (Ugandan Ministry of Health 2003). In a recent survey of 658 patients admitted in the same district hospital, the prevalence of *M. bovis* was less than 4.2 per cent while 8.6 per cent of the samples were from extrapulmonary sites. In addition, 42 per cent of the surveyed patients had a history of raw milk consumption and 91 per cent admitted eating raw or undercooked meat (Byarugaba 2009).

North American Countries

Canada

Bovine TB has been consistently present in a relatively small geographic location in and around the North Western part of Riding Mt. National Park since 1978 with significant yearly variations in the residing elk and deer population since 1997. Latest surveillance data in free-ranging elk and cervid in south western Manitoba carried out from 1997-2010 reveals a six-time higher prevalence of *M. bovis* in elk, suggesting it as a significant reservoir host of the deadly organism. However, infected deer are also required to maintain a true infection reservoir in the system. Lack of young age culture positive elks since 2003 indicates a declining trend in the area.

USA and the Caribbean

In the US, bovine tuberculosis, although widespread, is limited to 0.01 per cent of dairy herds compared to Johne's disease that afflicts 68 per cent of the US dairy herds. Bovine TB causes a conservative annual loss of 3 billion USD of which 3.5-4 million USD is attributed to current test and slaughter eradication programme. Bovine TB is historically endemic in white tail deer acting as a reservoir host, and with elk and a variety of carnivores and omnivores serving as spillover hosts in the state of Michigan. The incidence rate of *M. bovis* in white tail deer outside of the 12 endemic county areas decreased from 4.9 to ~1.7 per cent while the rate in elk population ranges from 0-0.6 per cent. Although, it has been prevalent in the deer population, surveillance data indicate no further increase in the incidence in Michigan residents since the first major outbreak began in 1994, and has been within or ~ 1 new case per year for a total of 13 till date.

Despite the aggressive USDA test and slaughter protocol, bovine TB, remains a significant economic and zoonotic threat accounting for approximately 1 per cent of all TB cases. With the changing epidemiologic pattern of TB in the US, incidence of disease is becoming more concentrated in certain geographic regions of the country populated with foreign-born people. A retrospective analysis of TB incidence from 1994 to 2005 in San Diego, California region revealed a 45 per cent of all culture-positive cases in children (<15 years of age) and an increased 6 per cent incidence in adults, and almost all cases from 2001-2005 were people of Hispanic origin with extrapulmonary TB (Timothy 2008). Later investigations using national TB surveillance data conducted in New York city revealed similar epidemiological profiles. Of particular interest was the fact that 83 per cent of the people interviewed reported consuming cheese from Mexico during their stay in the US, while 78 per cent were not aware whether the milk products they consumed were pasteurized. Yet, another study in San Diego children identified ingestion of unpasteurized dairy products from Mexico to be the sole risk factor for acquiring bovine TB, indicating dairy herds from Mexico to be the source of infection (Besser 2001, LoBue 2003, CDC 2005, Theon 2006, Wilkins *et al.* 2006, O' Brien *et al.* 2008, Seth *et al.* 2009).

In decreasing order of prevalence of *M. bovis* in the Caribbean, Haiti with a relatively high is followed by Dominican Republic with intermediate prevalence and slew of countries including Bermuda, Virgin Islands, Bahamas, Grenada, Antigua, Trinidad-Tobago, Jamaica, and Cuba among others (Kantor and Ritacco 2006).

Latin America
In a continent of substantial heterogeneity in geographic size, human and cattle population, existing breeding programmes and dairy industries, plus, with two of the world's main meat exporting countries, Brazil and Argentina, the real incidence of *M. bovis* is humans, like in many other developing nations, is underestimated and perhaps ignored in most Latin American countries. A recent study on the prevalence of *M. bovis* in humans in 10 Latin American countries reveals a 0-2.5 per cent incidence in relation to those caused by *M. tuberculosis* with majority of the cases diagnosed in Argentina, besides Equador, Brazil and Venezuela. Countries where *M. bovis* in humans has never been isolated were Chile, Colombia, Costa Rica, Dominican Republic, Peru and Uruguay. These were the countries, however, where negative data might underestimate the infection because of limited use of pyruvate containing medium for growth of *M. bovis*, yet Uruguay and Costa Rica are among those where

food protection and safety policies have been consistently reinforced. The bovine TB prevalence data in cattle in most Latin American countries come from tuberculin surveys of variable coverage or from systematic national records of condemnation at abattoirs (Kantor *et al.* 2008, 2010).

European Countries

The test-and-slaughter policy implemented in many developed countries has in fact eradicated *M. bovis* from 11 states of the European Union with the exception of Ireland and its neighbor United Kingdom where the disease is still prevalent (Reiriego Gordejo 2006, de la Rua-Domenech 2006). While, in the 1980s, 4-6 per cent of all cases of laboratory confirmed cases of tuberculosis in southwest Ireland were caused by *M. bovis*, in a survey during 1998-2006, 3 per cent of all *M. tuberculosis* complex isolates were *M. bovis* positive (Cotter *et al.* 1996, Ojo *et al.* 2008).

A resurgence of *M. bovis* infections in cattle in the UK since the 1990s forbids an Official TB free (OTF) status with an average annual increase in new TB herd breakdown rate of 14 per cent. Despite a more than five-fold increase in cattle herd infections during the 1990s, primarily in the southwest England and Wales, a similar trend has not occurred in the human population. A cross-sectional study linking demographic, clinical and DNA fingerprinting data on cases reported between 2005-2008 shows a decrease in annual incidence from 0.065 to 0.047 cases (<1 per 100,000 persons) with an epidemiologic trend predominantly that of reactivated disease, similar to that of other industrialized nations. Most patients (73 per cent) were natives of white ethnicity born pre-1960 before widespread introduction of pasteurization. Subsequent study carried out in London and southeast England carried out over the period 2005 to 2009 also shows similar low incidence of disease in humans. This outcome is a true reflection of a combination of strict measures taken and implemented over the years towards control and eradication of the disease (dela Rua-Domenech 2006, Mandal *et al.* 2011, Stone *et al.* 2012).

France reveals a relatively high incidence, 2 per cent of TB cases in humans caused by *M. bovis* (Mignard *et al.* 2006) compared to earlier reports of 1.5 per cent in the 1960s and 0.5 per cent in 1995 (Haddad *et al.* 2001). Interestingly in a study, 70 per cent of TB due to *M. bovis* occurred in French-born patients as opposed to traditional foreign-born patients who equaled about 55 per cent of cases with bovine TB. Reactivation of infection acquired before milk pasteurization practices rather than a primary disease was cited as the reason, because, these patients were above 50 years of age and sought treatment for a torpid infection (Mignard *et al.* 2006).

In the Netherlands, only 10 cases of *M. bovis* infection, all in imported livestock, have occurred since 1993. Of these, most cases primarily had extrapulmonary manifestations owing to infection contracted through the oral route. As far as the human infection goes, 1.4 per cent of tuberculosis cases during 1993-2007 were caused by *M. bovis*, of which the majority was in elderly people. During 1998-2007, the infections were divided equally between native Dutch and immigrants. The natives were elderly, and likely had the disease from endogenous reactivation of latent TB contracted before milk pasteurization, while the immigrant population was mostly foreign-born or first and second generation who may have travelled back and forth to their country of origin and contracted the disease which mainly manifested as extrapulmonary lesions affecting cervical and abdominal lymph nodes. The geographic distribution showed a proportional increase in *M. bovis* infections in major cities where most immigrants are settled (Majoor *et al.* 2011).

Switzerland is one among the 11 European Union states where bovine TB is very rare and the disease-free status of herds was achieved in the years between 1934 and 1959 by voluntary and few mandatory measures. Since then there has been no report of TB as per World Organization for Animal Health (OIE) International Zoo Sanitary Code (>99.8 per cent of all herds are officially acknowledged as TB free). Bovine TB occurs mainly in elderly patients who have been exposed before the eradication of bovine tuberculosis programme. Since the report of 11 *M. bovis* positive cases among 615 confirmed notifications for TB in 1999, there has been another report of a patient who had been exposed and contaminated during childhood and who apparently transmitted tuberculosis back to cattle (Fritsche 2004).

Australia

With the successful implementation of Brucellosis and Tuberculosis Eradication Campaign (BTEC) in 1970 following loss of productivity, threat to trade restrictions and risk of human disease, Australia has been declared free with respect to bovine tuberculosis since 1997 (Roberts *et al.* 1998). Presently, the incidence of *M. bovis* infection in humans is low: only about 1-2 case notifications per year have been encountered since the turn of the century. This amounts to 0.2 per cent of all cases of human tuberculosis in Australia that frequently represent latent disease often acquired decades ago (Ingram *et al.* 2010). Following the advent of pasteurization in 1950, the incidence of *M. bovis* infection declined with a shift in the epidemiology towards pulmonary disease in adults having occupational exposure to cattle or in immigrants. In a study of 236 positive cases in Australia between 1970 and 1994, 70 per cent were males with a mean age of 51-60 years, and were reported to be working in the meat or livestock industry (Cousins 1999).

New Zealand

New Zealand is host to large reservoir of *M. bovis* infection in wild and farmed animals. Disease in domestic cattle and deer is primarily attributed to the ongoing transmission from its wildlife reservoir, the brushtail possums. After its introduction into New Zealand via cattle in the 1830s-1840s, the first possum cases of tuberculosis were recorded in 1967. By 2000, tuberculosis infected possums occupied 23.6 per cent of New Zealand's 6.24 million hectares which also includes 75 per cent of *M. bovis* infected cattle herds. Being very susceptible to *M. bovis* infection, as few as 20 colony forming units lead to disease. Spread is primarily through animal-animal contact via inhalation. Infection is independent of age but more prevalent in males than females. Prevalence in possum populations averages 5 per cent but can vary from 1 per cent to 60 per cent with the higher prevalence tending to occur in possums foraging on pasture adjacent to forest. Transmission to livestock occurs when healthy cattle and deer come in close contact with infected possums and/or infected fomites. Elimination of the disease would require eliminating majority of possums. A 10-year study report (1996-2007) indicates a reduced spillover transmission of *M. bovis* to feral pigs following selective depopulation of possums. Before possum control, the overall prevalence of culture confirmed *M. bovis* infection in feral pigs ranged from 16.7 per cent to 94.4 per cent, which fell to near zero within 2-3 years (AWHN 2010, Nugent *et al.* 2011).

Given the high disease prevalence in animal hosts, incidence in human population is not uncommon. According to an estimate by Baker (2006), the incidence of *M. bovis* in humans in the period between 1995-2002 was comparatively higher, accounting for about 2.7 per cent of laboratory confirmed human tuberculosis cases that amounts to a rate of 0.2/100,000 people. Restriction endonuclease analysis of these isolates revealed a 61 per cent similarity with patterns for isolates from cattle, deer, possums, ferrets, pigs and occasionally cats, suggesting an ongoing low level transmission of infection between cattle and wildlife species including the reservoir host brushtail possums.

Diagnostic Procedures

Rapid, accurate and early diagnosis of tuberculosis is critical for timely initiation of treatment, minimizing transmission and reducing morbidity and mortality both in humans and animals. Diagnosis of extrapulmonary forms of bovine tuberculosis in humans, mainly caused by *M. bovis*, still poses a formidable challenge to the physicians. Important among the difficulties include non-specific symptoms, insidious onset, diverse presentations that bring patients to the physicians at a later stage of the disease (Jain 2011). Besides, diagnosis of tuberculosis still depends on tools such as direct sputum microscopy and chest radiographs, especially in resource-poor countries. The paucibacillary nature of samples often contributes to the delay in diagnosis. Fortunately, global research in the past decade has been focused on the identification of improved diagnostics for tuberculosis resulting in development and validation of novel tools aimed mainly at improving detection of the organism or a specific host-immune response.

Diagnosis of bovine tuberculosis by clinical examination is of very limited value given that most infected animals do not show pathognomonic clinical signs of the disease. Passive abattoir surveillance alone has been employed as a cost-effective method as a supplement to live cattle testing. Although this method alone has been found to have a low level of sensitivity (28.5 per cent) owing to inconsistencies in regional animal identification regulations and record keeping procedures (Anon 2009), a major factor for the successful eradication of bovine TB in countries such as Australia has been largely due to meticulous animal (lesion) identification at slaughter (Cousins and Roberts 2001). Diagnostic tests of bovine tuberculosis can be broadly classified into direct or indirect tests.

Direct Test

In general, diagnosing *M. bovis* in cattle involves detection of organisms through direct culture and/or by molecular methods. Direct diagnosis of TB in humans, in contrast, is largely based on sputum bacteriology and microscopy with live imaging such as radiography and magnetic resonance imaging (MRI) where available.

Culture

Isolation of *M. bovis* from tissues harvested from suspect/slaughtered animals or sputum/biopsy samples in humans by culture, followed by staining and microscopic observation still, remains the "gold standard" despite several drawbacks. Major limitations of this technique include a low sensitivity (50-70 per cent) for sputum samples due to the requirement of high bacterial load for microscopy that is most often not available from HIV infected humans or in extrapulmonary tuberculous lesions; a low yield of culturable quantities of bacteria from blood, urine, lavage and CSF specimens; a highly time consuming procedure; and the absence of a distinct species differentiation. The combined use of solid and liquid culture media in the recent times has been reported to significantly increase the overall sensitivity (Michel 2010). Lack of speciation, on the other hand, as Etchechoury *et al.* (2010) point out, is one of the main reasons why information regarding *M. bovis* contribution to global TB burden is scarce and unclear. The use of Lowenstein-Jensen media (LJ) supplemented with glycerol (that does not promote *M. bovis* growth, which usually grows faster on egg medium supplemented with pyruvate) in most laboratories is another reason. Additionally, for many low-income countries, a mycobacterial culture always becomes an expensive option over the cheaper and quicker acid-fast staining method.

Molecular Tests

Most promising molecular diagnostic tools include PCR-based tests and DNA strain typing. While PCR tests are rapid and accurate, DNA strain typing offers answers to epidemiological queries. PCR assays are vital for identification of species within the *M. tuberculosis* complex, especially in countries where zoonotic tuberculous is of grave concern. The assays typically target genes such as the *hup* B gene with simplex, nested and multiplex formats in conventional and real time chemistries (Misra *et al.* 2005, Reddington *et al.* 2011, Thacker *et al.* 2011, Jain 2011). Although they come with obvious limitations in terms of low sensitivity when applied to samples with low numbers of organisms, the overwhelming advantages outweigh the limitations.

The DNA approach for *M. bovis* strain typing has over the years employed several useful and credible techniques. Starting with restriction endonuclease analysis (REA) in the eighties, followed by restriction fragment length polymorphisms (RFLP) analysis of the insertion sequence *IS6110* and of the polymorphic GC rich sequences (PGRS), it progressed to the PCR amplification of the direct repeat (DR) region or spoligotyping, finally to the most recent and highly popular Variable Number Tandem Repeats (VNTR) or mycobacterial interspersed repetitive units analysis. In essence, typing is indispensible in providing information about the source of infection in the herd and in delineating the most likely route of transmission. It is useful to investigate the phylogeny of strains to eventually track the evolutionary strings to ascertain their historical movement between countries plus their current and global distribution. All these techniques offer advantages along with some shortfalls as well (Collins 2011).

Indirect Tests

Majority of indirect tests currently available are based on the measurement of various parameters of the immune response to *M. bovis*; cellular immune and antibody responses

Cell-Mediated Based Tests

Tuberculin Skin Test (TST)

This test is perhaps the best-known tuberculous diagnostic test ever and, although not totally perfect, it has yet to be replaced by a more accurate and satisfactory method. It constitutes a primary screening of live cattle, is based on a delayed hypersensitivity response and represents the OIE prescribed test for international trade. The *M. bovis* strain AN5 is used worldwide for bovine tuberculin production. The test measures dermal swelling 3 days after PPD is injected into the skin of the caudal fold (CFT), or the mid cervical intradermal region/neck (CIT). The skin of neck is considered more sensitive; therefore higher doses of PPD are used on the tail fold to compensate for the difference. The comparative cervical test (CCT) is used to differentiate animals infected with *M. bovis* from those sensitized to PPD-B following exposure to other mycobacteria. While there are advantages of TST in the form of low costs, high availability, long history of use because of lack of alternatives, there are some obvious limitations as well. These include difficulties in administration and result interpretation, low degree of standardization and imperfect test accuracy (De la Rua-Domenech 2006). Attempts to increase the sensitivity and specificity of TST by use of defined antigens and recombinant proteins (ESAT-6, CFP-10, Rv3615c), either alone or in combination, cocktails are being investigated currently (Schiller *et al.* 2010).

Interferon Gamma Assay

This *in-vitro* assay (Bovigam® Prionics, Switzerland) measures specific cell-mediated immune responses by circulating lymphocytes and is used either to confirm the reactors of the CFT or CIT testing or run alongside in parallel to increase the diagnostic sensitivity. In brief, heparinized whole blood is incubated with PPD specific antigens for 18-24 h during which time the antigenic stimulation induces production and release of IFN-gamma from T lymphocytes. IFN-gamma present in plasma supernatant is then quantified by sandwich ELISA. A positive result is seen where there is a preferential release of IFN-g to constituents of *M. bovis* compared to other mycobacteria. Advantages of the IFN-gamma test are its increased sensitivity, more rapid repeat testing, no requirement for a second farm visit, more objective procedures and interpretations compared to TST. Limitations among others comprise reduced specificity due to a high degree of cross reactivity between PPDa, *b* and *j* tuberculins and also differences in the kinetics of cell-mediated immune responses to *M. avium* subsp. *avium* and *M. avium* subsp. *paratuberculosis* organisms (Barry *et al.* 2011), high logistical demands, increased likelihood of a non-specific response in young animals and high costs (De la Rua-Domenech 2006). As with TST, improvements in the form of replacements by defined antigens for stimulation, use of in-tube/in-plate stimulation device, in combination with modified interpretation/cut-off and cost reduction measures may allow for useful developments for the IFN-gamma assay (Schiller *et al.* 2010).

Prevention, Management, Control and Eradication Strategies

Tuberculosis remains a major public health threat largely due to a sizeable reservoir of latently infected individuals who may relapse into active disease decades after first acquiring the infection. This holds true for all cases of TB caused by *M. tuberculosis* complex, particularly those caused by *M. tuberculosis* and *M. bovis*. The control and eradication of tuberculosis in humans caused by *M. bovis* warrants early recognition of pre-clinical infection in animals and prompt removal of any infected animals in order to eliminate a future source of infection for other animals and for humans (FSAI 2008). Basic strategies for effective prevention control and elimination of bovine TB is well-known and well-defined, although actual implementation, adhering to and follow up of these regulations is what vary greatly with respect to region/country in question worldwide. Elaborate benchmarks and guidelines have been provided by all leading international organizations such as WHO, OIE, CDC, USDA-APHIS to name a few. Based on the information available from these and other published sources (Lobue and Moser 2005, Good and Duignan 2011), some basic strategies for developed and developing countries are being described below that may be helpful in tackling the problem.

Strategies in Developed Countries

☆ Practising routine active surveillance of wildlife and livestock where wildlife reservoirs are present and performing risk assessment to include all species sharing the ecosystem with cattle to identify potential maintenance or spill over hosts and developing effective control strategies towards eradication.

☆ Practising bans on supplemental feeding of wildlife to avoid close contact with livestock and humans.

☆ Evaluating potency assays performed during tuberculin manufacturing process as it is critical to tuberculin skin test performance.

☆ Sharing experiences and technical knowhow through collaboration and partnership with scientists and personnel from other countries to consider options and incorporation as appropriate.

☆ Carefully considering human sources of infection during epidemiologic investigation of outbreaks.

☆ Abiding strictly to recommendations provided by OIE or EU for international trade of cattle that must originate from OTF countries/regions or from an OTF herd subjected to TST with negative results within 30 days prior to shipment.

Strategies in Developing Countries

☆ Recognition of the significance of *M. bovis* by the national TB control programme as major human and animal health problem.

☆ Regional and nationwide *M. bovis* surveillance programmes to estimate the incidence of disease in the cattle population.

☆ Raising public awareness about the disease in general with emphasis on transmission routes, risk factors, at risk individuals, prognosis, prevention and control.

☆ Minimization of chances of contact between domestic livestock and wild ruminants during grazing by improving animal husbandry practices.

☆ Recognition of the importance of HIV endemic and its effects on tuberculosis susceptibility in humans, especially with respect to occurrence of drug resistance strains of *M. tuberculosis* complex.

☆ Identification of potentially undiscovered wildlife reservoirs of *M. bovis*, if any, and investigation of epidemiological impacts of such reservoirs in the maintenance and spread of disease.

☆ Application of standard public health measures on patients afflicted with *M. bovis* as would be used to manage patients with contagious *M. tuberculosis* to stop person-to-person spread of the infection.

☆ Evolving cooperation between medical and veterinary medical health professions in the wake of a disease outbreak.

☆ Implementation of mandatory pasteurization of milk supplies with routine screening of meat in slaughterhouses carried out under the supervision of a qualified inspector.

☆ Adoption of stringent rules with respect to illegal animal and human trafficking/movement of immigrants across international borders.

☆ Adoption of mandatory tuberculin test and slaughter policy wherever possible.

☆ Usage of egg medium supplemented with pyruvate to facilitate growth of *M. bovis* by culture.

☆ Speciation of MTB complex isolates upon detection to differentiate between *M. tuberculosis* and *M. bovis*, not only to facilitate tracing the source of infection to better formulate health control strategies but also to put patients on effective therapeutic regimen. Patients with *M. bovis* are treated with routine drugs for tuberculosis, except pyrazinamide (PZA) due to its inherent resistance to the drug.

Future Directions

Significant advancement in the understanding of the dynamics and control of a multi-host pathogen such as *M. bovis* would require identifying, assessing and overcoming of obstacles that junction at the crossroads of several disciplines including human and veterinary medicine, microbiology, molecular and population genetics, animal and human behaviour, community ecology, conservation biology, epidemiology, sociology, political science and other related sectors at the regional/national level. Armed with the advanced knowledge and technical knowhow and the will to achieve success, the need for collaboration on an international level on all aspects of disease prevention, control and eradication has never been more urgent.

Innovation of novel tools and refinement of existing diagnostics will be another crucial factor for bovine and zoonotic TB control and eradication programmes in the years to come. New and improved diagnostics are emerging with the potential to modify or even replace existing tools both in terms of performance and cost. Sustainable use of these improved diagnostics would complement long-term cost reductions spent on efforts towards disease eradication. Improvements of the IFN-γ assay might come in the form of using ESAT-6 and CFP-10, either alone or in combination to increase the specificity, sensitivity and predictive values and become a powerful primary screening test for bovine TB. Antibody based detection assays offer the possibility for convenient, flexible and cost-effective means for bovine TB surveillance along with increased specificity. This would be appealing to resource poor countries where large scale culling is not cost-effective because of lack of government indemnity support for condemned animals, but segregating and removal of cattle at advanced stage of disease with increased risk of bacterial shedding would be feasible. Greater availability of rapid methods for detection, speciation and PZA susceptibility testing will help better understand the magnitude of current infection, discover more cases by retrospective studies, detect drug resistant strains and identify newer modes of transmission of *M. bovis* in the population. Novel biomarker proteins would have significant impact on elucidation of pathogenesis, development of mycobacterium specific diagnostic tools for monitoring the progression of diseases for monitoring disease progression, response to therapy and vaccine-based interventions. However, many of the newly developed methods may not be feasible for many developing countries due to lack of infrastructure and well-trained personnel. Nevertheless, the future of zoonotic tuberculosis control and eradication relies heavily on the concerted efforts of lawmakers, governments, scientists, human and animal health care professionals, social workers and the general populace at regional, national and global scale to work towards a common goal of eliminating the disease.

References

1. Abubaker UB, Ameh JI, Abdulkadir AI, Salisu I, Okaiyeto SO and Kudi AC. 2011. Bovine tuberculosis in Nigeria: a review. *Vet Res* 4: 24-27.

2. Acha PN and Szypres B. 2001. *Zoonosis and communicable diseases common to man and animals*. Pan American Health Organization, Washington DC, USA.

3. Alexander KA, Pleydell E, Williams MC, Lane EP, Nyange JF and Michel AL. 2002. *Mycobacterium tuberculosis*: an emerging disease of free-ranging wildlife. *Emerg Infect Dis* 8: 598-601.

4. Alfonso R, Romero RE, Diaz A, Calderon MN, Urdaneta G, Arce J, Patarroyo ME and Patarroyo MA. 2004. Isolation and identification of mycobacteria in New World primates maintained in captivity. *Vet Microbiol* 98: 285-295.

5. Ameni G, Vordermeier M, Firdessa R, Aseffa A, Hewinson G, Gordon SV and Berg S. 2011. *Mycobacterium tuberculosis* infection in grazing cattle in central Ethiopia. *Vet J* 188: 359-361.

6. Anon. 2009. *Analysis of bovine tuberculosis surveillance in accredited free states*. USDA, Animal and Plant Health Inspection Service, Veterinary Services, January 30, 2009. p. 7-23.

7.   Aranaz A, Liebana E, Gomez-Mampaso E, Galan JC, Cousins D, Ortega A, Blazquez J, Baquero F, Mateos A, Suarez G and Dominguez L. 1999. *Mycobacterium tuberculosis* subsp. *caprae* subsp. *nov*: a taxonomic study of a new member of the *Mycobacterium tuberculosis* complex isolated from goats in Spain. *Int J Syst Bacteriol* 49: 1263-1273.

8.   Ashford DA, Whitney E, Raghunathan P and Cosivi O. 2001. Epidemiology of selected mycobacteria that infect humans and other animals. *Rev Sci Tech* 20: 325-327.

9.   AWHN. 2010. *Exotic - possum and tuberculosis fact sheet*. Australian Wildlife Health Network. http://www.wildlifehealth.org.au.

10. Baker MG, Lopez LD, Cannon MC, DeLisle GW and Collins DM. 2006. Continuing *Mycobacterium bovis* transmission from animals to humans in New Zealand. *Epidemiol Infect* 134: 1068-1073.

11. Barry C, Corbett D, Bakker D, Anderson P, McNair J and Strain S. 2011. The effect of *Mycobacterium avium* complex infections on routine *Mycobacterium bovis* diagnostic tests. Vet Med International doi:10.4061/2011/145092.

12. Batralot R, Garcia-Patos V, Sitjas D, Rodriguez-Cano L, Mollet J, Martin Casabona N and Coll P. 2005. Clinical patterns of cutaneous nontuberculous mycobacterial infections *Br J Dermatol* 152: 727-734.

13. Berg S, Firdessa R, Habtamu M, Gadisa E, Mengistu A, Yamuah L, Ameni G, Vordermeier M, Robertson BD, Smith NH, Engers H, Young D, Hewinson RG, Aseffa A and Gordon SV. 2009. The burden of mycobacterial disease in Ethiopian cattle: implications for public health. *PLoS ONE* 4: e5068.

14. Besser RE, Pakiz B, Schulte JM, Alvarado S, Zell ER, Kenyon TA and Onorato IM. 2001. Risk factors for positive Mantoux tuberculin skin tests in children in San Diego, CA: Evidence for boosting and possible food-borne transmission. *Pediatrics* 108: 305-310.

15. Bolognesi N. 2007. TB or not TB: The threat of bovine tuberculosis. http//:www.scidev.net/en/features/tb-or-not-tb-the-threat-of-bovine-tuberculosis.html.

16. Borgdorff MW, Sebek M, Geskus BR, Kremer K, Kalisvaart N and Van Soolingen D. 2011. The incubation period distribution of tuberculosis estimated with a molecular epidemiological approach. *Int J Epidemiol* 40: 964-970.

17. Bose M. 2008. Natural reservoir, zoonotic tuberculosis and interface with human tuberculosis: An unsolved question. *Indian J Med Res* 128: 4-6.

18. Brosch R, Gordon SV, Marmiesse M, Brodin P, Buchrieser C, Eiglmeier K, Garnier T, Gutierrez C, Hewinson G, Kremer K, Parsons LM, Pym AS, Samper S, van Soolingen D and Cole ST. 2002. A new evolutionary scenario for the *Mycobacterium tuberculosis* complex. *Proc Natl Aca Sc* 99: 3684-3689.

19. Bull TJ, McMinn EJ, Sidi Boumedine K, Skull A, Durkin D, Neild P, Rhodes G, Pickup R and Hermon-Taylor J. 2003. Detection and verification of *Mycobacterium*

avium subsp. *paratuberculosis* in fresh ileo-colonic mucosal biopsy specimens from individuals with and without Crohn's disease. *J Clin Microbiol* 41: 2915-2923.

20. Byarugaba F, Etter EMC, Godreuil S and Grimaud P. 2009. Pulmonary tuberculosis and *Mycobacterium bovis*, Uganda. *Emerg Inf Dis* 15: 124-125.

21. Candela MG, Serrano E, Carrasco-Martinez C, Martin-Atance P, Cubero MJ, Alonso F and Leon L. 2009. Co-infection is an important factor in epidemiological studies: the first serosurvey of the aoudad (*Ammotragus lervia*). *Euro Jour Clin Microbiol Inf Dis* 28: 481-489.

22. Cassidy JP. 2006. The pathogenesis and pathology of bovine tuberculosis with insights from study of tuberculosis in humans and laboratory animal models. *Vet Microbiol* 112:151-161.

23. CDC. 2005. Human tuberculosis caused by *Mycobacterium bovis*- New York City 2001-2004. Centers for Disease Control and Prevention MMWR *Morb Mortal Wkly Rep* 54: 605-608.

24. Chandrasekharan KP and Ramakrishnan R. 1969. Bovine tuberculosis due to the human strain of *Mycobacterium tuberculosis*. *Indian J Tuberc* 16: 103-105.

25. Chiba T. 1993. *Pathology of zoo animals*. Case report in Nagoya city Higashiyama Zoo (1950-1987), Tokyo: Kindaibunngei.

26. Cirillo JD, Falkow S, Tompkins LS and Bermudez LE. 1997. Interaction of *Mycobacterium avium* with environmental amoeba enhances virulence. *Infect Immun* 65: 3759-3767.

27. Cobo J, Asensio A, Moreno A, Navas E, Pintado V, Oliva J, Gomez-Mampaso E and Guerrero A. 2001. Risk factors for nosocomial transmission of multidrug-resistant tuberculosis due to *Mycobacterium bovis* among HIV-infected patients. *Int J Tuberc Lung* 5: 413-418.

28. Collins DM. 2011. Advances in molecular diagnostics for *Mycobacterium bovis*. *Vet Microbiol* 151: 2-7.

29. Cooke MM, Jackson R, Coleman JD and Alley RG. 1995. Naturally occurring tuberculosis caused by *Mycobacterium bovis* in brushtail possums (*Trichosaurus vulpecula*): II Pathology. *NZ Vet J* 43: 315-321.

30. Cotter TP, Sheehan S, Cryan B, O' Shaughnessey E, Cummins H and Bredin CP. 1996. Tuberculosis due to *Mycobacterium bovis* in humans in the South-west region of Ireland: Is there a relationship with infection prevalence in cattle? *Tuberc Lung Dis* 77: 545-548.

31. Cousins DV and Dawson DJ. 1999. Tuberculosis due to *Mycobacterium bovis* in the Australian population: cases reached during 1970-1994. *Int J Tuberc Lung Dis* 3: 715-721.

32. Cousins DV and Roberts JL. 2001. Australia's campaign to eradicate bovine tuberculosis: the battle for freedom and beyond. *Tuberculosis* 81: 5-15.

33. Cousins DV, Bastida R, Cataldi A, Quse V, Redrobe S, Dow S, Duigan P, Murray A, Dupont C, Ahmed N, Collins DM, Butler WR, Dawson D, Rodriguez D, Loureiro J, Romano MI, Alito A, Zumarraga M and Bernadelli A. 2003. Tuberculosis in seals caused by a novel member of the *Mycobacterium tuberculosis* complex: *Mycobacterium pinnipedi* sp. *Int J Syst Evol Bacteriol* 53: 1305-1314.

34. De la Rua-Domenech R. 2006. Human *Mycobacterium bovis* infection in the United Kingdom: Incidence, risks, control measures and review of the zoonotic aspects of bovine tuberculosis. *Tuberculosis* 86: 77-109.

35. De Lisle GW, Bengis RG, Schmitt SM and O' Brien DJ. 2002. Tuberculosis in the free ranging wildlife: detection, diagnosis and management. *Rev Sci Tech Off Int Epiz* 21: 317-334.

36. Dean SD, Rhodes SG, Coad M, Whelan AO, Cockle PJ, Clifford DJ, Hewinson RG and Vordermeier HM. 2005. Minimum infective dose of *Mycobacterium bovis* in cattle. *Infection Immunity* 73: 6467-6471.

37. deKantor IN and Ritacco V. 2006. An update on bovine tuberculosis programmes in Latin American and Caribbean countries. *Vet Microbiol* 112: 111-118

38. deKantor IN, Ambroggi M, Poggi S, Morcillo N, Maria A, Telles DS, Ribeiro MO, Maria C, Torres G, Polo LC, Ribon W, Garcia V, Kuffo D, Asencios L, Lucy M, Campos V, Rivas C and deWaard JH. 2008. Human *Mycobacterium bovis* infection in ten Latin American countries. *Tuberculosis* 88: 358-365.

39. deKantor IN, LoBue PA and Theon CO 2010. Human tuberculosis caused by *Mycobacterium bovis* in the United States, Latin America and the Caribbean. *Int J Tuberc Lung Dis* 14: 1369-1373

40. Devulder G, Perouse de Montclos M and Flandrois JP. 2005. A multigene approach to phylogenic analysis using the genus *Mycobacterium* as a model. *Int J of Sys Evol Microbiol* 55: 293-302.

41. Dolenc-Voljc M and Zolnr-Dovc M. 2010. Delayed diagnosis of *Mycobacterium marinum* infection: A case report and review of literature. *Acta Dermatoven APA* 19(2): 35-39.

42. Drewe JA, Foote AK, Sutcliffe RL and Pearce GP. 2009. Pathology of *Mycobacterium bovis* infection in wild Meerkats (*Suricata suricatta*). *J Comp Path* 140: 12-24.

43. Erwin PC, Bemis DA, Mawby DI, McCombs SB, Sheeler LL, Himelright IM, Halford SK, Diem L, Metchock B, Jones TF, Schilling MG, and Thomsen BV. 2004. *Mycobacterium tuberculosis* transmission from human to canine. *Emerg Infect Dis* 10: 2258-2260.

44. Esple IW, Hlokwe TM Van Gey Pittus NC, Lane E, Tordiffe AS, Michel AL, Muller A, Kotze A and van Helden PD. 2009. Pulmonary infection due to *Mycobacterium bovis* in a black rhinoceros (*Diceros bicomis minor*) in South Africa. *J Wildlife Dis* 45: 1187-1193.

45. Etchechoury I, Morcillo N, Sequeira M, Imperiale B, Echeverria G, Lopez M, Caimi K, Zumarraga M, Cataldi A and Romano MI. 2010. Detection and typing

of *Mycobacterium bovis* from patients with pulmonary tuberculosis in Argentina. *Zoonosis Public Health* 57: 375-381.

46. Eurosurveillance Editorial Team. 2005. Trends in zoonoses in Europe, 2003. Euro Surveill 10(21): pii=2712. http://www.eurosurveillance.org/ViewArticle.aspx?ArticleId=2712.

47. Evans JT, Sonnenberg P, Smith GE, Banerjee A, Dale J, Innes JA, Hunt D, Tweddell A, Wood A, Anderson C, Hewinson GR, Smith NH and Hawkey PM. 2007. Bovine tuberculosis: Multiple human-to human transmission in the UK. *Lancet* 14: 1168.

48. Falkinham JO III. 1996. Epidemiology of infection by nontuberculous mycobacteria. *Clin Microbiol Rev* 9: 177-215.

49. Francis J. 1947. Bovine tuberculosis. Staples Press, London.

50. Fritsche A, Engel R, Buhl D, Zellweger JP, 2004. *Mycobacterium bovis* tuberculosis: from animal to man and back. *Int J Tuberc Lung Dis* 8: 903-904.

51. FSAI. 2008. *Zoonotic tuberculosis and food safety*. 2nd edition. Food Safety Authority of Ireland, Dublin, Ireland.

52. Fuller F, Huang J, Ma H and Rozelle S. 2006. Got milk? The rapid rise of China's dairy sector and its future prospects. *Food Policy* 31: 201-205.

53. Gallagher J, Muirhead RH and Burn KJ. 1976. Tuberculosis in wild badgers (*Meles meles*) in Gloucestershire: pathology. *Vet Rec* 98: 9-14.

54. Good M and Duignan A. 2011. Perspectives on the history of bovine TB and the role of tuberculin in bovine TB eradication. *Vet Med International* doi: 10.4061/2011/410470.

55. Government of India. 2010. Solid waste management manual. Slaughter house waste and dead animals. Annexure 5. Ministry of Urban Development, Government of India, New Delhi. http://urbanindia.nic.in/publicinfo/swm/chap5.pdf.

56. Grange JM 2001. *Mycobacterium bovis* infection in human beings. *Tuberculosis* 81: 71-77.

57. Grange JM, Yates MD and Collins CH. 1988. Inoculation mycobacteriosis. *Clin Exp Dermatol* 13: 211-220.

58. Grant IR, Ball HJ and Rowe MT. 2002. Incidence of *Mycobacterium paratuberculosis* in bulk raw and commercially pasteurized cows' milk from approved dairy processing establishments in the United Kingdom. *Appl Environ Microbiol* 68: 2428-2435.

59. Griffin JM and Dolan LA. 1995. The role of cattle-to-cattle transmission of *Mycobacterium bovis* in the epidemiology of tuberculosis in cattle in the Republic of Ireland: a review. *Irish Vet J* 48: 228-234.

60. Griffith DE, Aksamit T, Barbara A, Brown E, Catanzaro A, Daley C, Gordin F, Holland SM, Horsburgh R, Huitt G, Iademarco MF, Iseman M, Olivier K, Ruoss

S, Fordham von Reyn C, Wallace RJ Jr and Winthrop K. 2007. An Official ATS/ IDSA Statement: Diagnosis, treatment, and prevention of nontuberculous mycobacterial diseases. *Am J Respir Crit Care Med* 175: 367-416.

61. Gutierrez MC, Brisse S, Brosch R, Fabre M, Omais B, Marmiesse M, Supply P and Vincent V. 2005. Ancient origin and gene mosaicism of the progenitor of *Mycobacterium tuberculosis*. *PLoS Pathog* 1: e5.

62. Haddad N, Ostyn A, Karoui C, Masselot M, Thorel MF and Hughes SL. 2001. Spoligotyping diversity of *Mycobacterium bovis* strains isolated in France from 1979 to 2000. *J Clin Microbiol* 39: 3623-3632.

63. Humblet MF, Boschiroli ML and Saegerman C. 2009. Classification of worldwide bovine tuberculosis risk factors in cattle: a stratified approach. *Vet Res* 40: 50.

64. Ingram PR, Bremner P, Inglis TJ, Murray RJ and Cousins DV. 2010. Zoonotic tuberculosis: on the decline. *Commun Dis Intell* 34: 339-344.

65. Jacobs JM, Stine CB, Baya AM and Kent ML. 2009. A review of mycobacteriosis in marine fish. *J of Fish Dis* 32: 119-130.

66. Jain A. 2011. Extrapulmonary tuberculosis: A diagnostic dilemma. *Int J Clin Biochem* 26: 269-273.

67. Jensen PA, Lambert LA, Iademarco MF and RIdzon R. 2005. Guidelines for preventing the transmission of *Mycobacterium tuberculosis* in healthcare settings. *MMWR Recomm Rep* 54(RR-17): 1-141.

68. Kamper-Jorgensen Z, Lillebaek T and Anderson AB. 2009. Occupational tuberculosis following extremely short exposure. *Clin Respir Journal* 3: 55-57.

69. Keet DF, Kriek NP, Penrith ML, Michel A and Huchzermeyer H. 1996. Tuberculosis in buffaloes (*Syncercus caffer*) in the Kruger National Park: spread of disease to other species. *Onderstepoort J Vet Res* 67: 115-122.

70. Khan IA, Khan A, Mubarak A and Ali S. 2008. Factors affecting prevalence of bovine tuberculosis in Nili Ravi buffaloes. *Pakistan Vet J* 28: 155-158.

71. Kostman JR and DiNubile MJ. 1993. Nodular lymphangitis: a distinctive but often unrecognized syndrome. *Ann Int Med* 118: 883-888

72. Lillebaek T, Dirksen A, Baess I and Strunge B. 2002. Molecular evidence of endogenous reactivation of *Mycobacterium tuberculosis* after 33 years of latent infection. *J Infect Dis* 185: 401-404.

73. LoBue PA and Moser KS. 2005. Treatment of *Mycobacterium bovis* infected tuberculosis patients: San Diego county, California United States 1994-2003. *Int J Tuberc Lung Dis* 9: 333-338.

74. LoBue PA, Betacourt W, Peter C and Moser KS. 2003. Epidemiology of *Mycobacterium bovis* disease in San Diego county, 1994-2000. *Int J Tuberc Lung Dis* 7: 180-185.

75. LoBue PA, Enarson DA and Theon CO. 2010. Tuberculosis in humans and animals: an overview. *Int J Tuberc Lung Dis* 14: 1075-1078.

76. Majoor CJ, Magis-Escurra C, vanIngen J, Boeree MJ and van Sooligan D. 2011. Epidemiology of *Mycobacterium bovis* disease in humans, the Netherlands, 1993-2007. *Emerg Inf Dis* 17: 457-463.

77. Mandal S, Bradshaw L, Anderson LF, Brown T, Evans JT, Drobniewski F, Smith G, Magee JG, Barrett A, Blatchford O, Laurenson IF, Seager AL, Ruddy M, White PL, Myers R, Hawkey P and Abubaker I. 2011. Investigation transmission of *Mycobacterium bovis* in United Kingdom in 2005 to 2008. *J Clin Microbiol* 49: 1943-1950.

78. Marais BJ, Hesseling AC, Gie RP, Schaff HS and Beyers N 2006. The burden of childhood tuberculosis and the accuracy of community-based surveillance data. *Int J Tuberc Lung Dis* 10: 259-263.

79. McIlroy SG, Neill SD and McCracken RM. 1986. Pulmonary lesions and *Mycobacterium bovis* excretion from the respiratory tract of tuberculin reacting cattle. *Vet Rec* 118: 718-721.

80. Menzies FD and Neil SD. 2000. Cattle-to-cattle transmission of bovine tuberculosis. *Vet J* 160: 92-106.

81. Michalak K, Austin C, Diesel S, Bacon MJ, Zimmerman P and Maslow JN. 1998. *Mycobacterium tuberculosis* infection as a zoonotic disease: transmission between humans and elephants. *Emerg Infect Dis* 4: 283-287.

82. Michel AL and Huchzermeyer HF. 1998. The zoonotic importance of *Mycobacterium tuberculosis*: transmission from human to monkey. *J S Afr Vet Assoc* 69: 64-65.

83. Michel AL, Coetzee ML, Keet DF, Mare L, Warren R, Cooper D, Bengis RG, Kremer K and van Helden P. 2009. Molecular epidemiology of *Mycobacterium bovis* isolates from free-ranging wildlife in South African game reserves. *Vet Microbiol* 133: 335-343.

84. Michel AL, Muller B and van Helden PD. 2010. *Mycobacterium bovis* at the animal-human interface: A problem or not? *Vet Microbiol* 140: 371-381.

85. Mignard S, Pichat C and Carret G. 2006. *Mycobacterium bovis* infection, Lyon, France. *Emerg Inf Dis* 12: 1431-1433.

86. Misra A, Singhal A, Chauhan DS, Katoch VM, Srivastava K, Thakral SS, Bharadwaj SS, Sreenivas V and Prasad HK. 2005. Direct detection and identification of *Mycobacterium tuberculosis* and *Mycobacterium bovis* in bovine samples by a novel nested PCR assay: correlation with conventional techniques. *J Clin Microbiol* 43: 5670-5678.

87. Montali RJ, Mikota SK and Cheng LI. 2001. *Mycobacterium tuberculosis* in zoo and wildlife species. *Rev Sci Tech Off Int Epizoot* 20: 291-303.

88. Mosavari N, Jamshidian M, Feizabadi MM, Shapouri S, Taheri MR, Arefpajoohi R and Tadayon K. 2007. Tuberculosis in buffalo: the first report on the phenotypic and genetic characteristics of the isolated organisms in Western Azerbaijan, Iran. *Arch Razi Institite* 62: 75-82.

89. Mukherjee F. 2006. Comparative prevalence of tuberculosis in two dairy herds in India. *Rev Sci Tech Off Int Epiz* 25: 1125-1130

90. Mumtaz N, Chaudhry JI, Mahmood N and Shakoori AR. 2008. Reliability of PCR for detection of Bovine tuberculosis in Pakistan. *Pakistan J Zool* 40: 347-351

91. Neill SD, Pollock JM, Bryson DB and Hanna J. 1994. Pathogenesis of *Mycobacterium bovis* infection in cattle. *Vet Microbiol* 40: 41-52.

92. Nicas M, Nazaroff WW and Hubbard A. 2005. Toward understanding the risk of secondary airborne infection: emission of respirable pathogens. *J Occup Environ Hyg* 2: 143-154.

93. NICD. 2005. *Zoonotic diseases of public health importance*. National Institute of Communicable Diseases, Zoonosis Division, New Delhi.

94. Nugent G, Whitford J, Yockney IJ and Cross ML. 2011. Reduced spillover transmission of *Mycobacterium bovis* to feral pigs (*Sus scrofa*) following population control of brushtail possums (*Trichosurus vulpecula*). *Epidemiol Infect* 18: 1-12.

95. O' Brien DJ, Schmitt SM, Berry DE, Fitzgerald SD, Lyon TJ, Vanneste JR, Cooley TM, Hogle SA and Fierke JS. 2008. Estimating the true prevalence of *Mycobacterium bovis* in free ranging elk in Michigan. *J Wildlife Dis* 44: 802-810.

96. O' Donahue WJ, Bedi S, Bittner MJ and Preheim LC. 1985. Short course chemotherapy for pulmonary infection due to *Mycobacterium bovis*. *Arch Int Med* 145: 703-705.

97. O' Reilly LM and Daborn CJ. 1995. The epidemiology of *Mycobacterium bovis* infections in animals and man: a review. *Tubercle Lung Dis* 76: 1-46.

98. Ocepek M, Parte M, Zolnir-Dovc M and Poljak M. 2005. Transmission of *Mycobacterium tuberculosis* from human to cattle. *J Clin Microbiol* 43: 3555-3557.

99. Oh P, Granich R, Scott J, Sun B, Joseph M, Stringfield C, Thisdell S, Staley J, Workman-Malcolm D, Borenstein L, Lehnkering E, Ryan P, Soukup J, Nitta A and Flood J. 2002. Human exposure following *Mycobacterium tuberculosis* infection of multiple animal species in a metropolitan zoo. *Emerg Infect Dis* 8:1290-1293.

100. OIE. 2012. WAHID Interface http://www.oie.int/wahis/public.php?page=_ home.

101. Ojo O, Sheehan S, Corcoran DG, Okker M, Gover K, Nikolayevsky V, Brown T, Dale J, Gordon SV, Drobniewski F and Prentice MB. 2008. *Mycobacterium bovis* strains causing smear positive human tuberculosis, Southwest Ireland. *Emerg Inf Dis* 14: 1931-1933.

102. Parent LJ, Salam MM, Appelaum PC and Dosset JH. 1995. Disseminated *Mycobacterium marinum* and bacteremia in a child with severe combined immunodeficiency. *Clin Inf Dis* 21: 1325-1327.

103. Patel D, Danelishvile L, Yamazaki Y, Alonso M, Paustian ML, Bannatine JP, Meunier-Goddik L and Bermudez LE. 2006. The ability of *Mycobacterium avium* subsp. *paratuberculosis* to enter bovine epithelial cells is influenced by

preexposure to hyperosmolar environment and intracellular passage in bovine mammary epithelial cells. *Infect Immun* 74: 2849-2855.

104. Perez J, Calzada J, Leon-Vizcaino L, Cubero MJ, Velarde J and Mozos E. 2001. Tuberculosis in an Iberian lynx (*Lynx pardina*). *Vet Rec* 148: 414-415.

105. Philips CJ, Foster CR, Morris PA and Teverson R. 2003. The transmission of *Mycobacterium bovis* infection to cattle. *Res Vet Sci* 74: 1-15.

106. Pollock JM and Neill SD. 2002. *Mycobacterium bovis* infection and tuberculosis in cattle. *Vet J* 163: 115-127.

107. Prasad HK, Singhal A, Mishra A, Shah NP, Katoch VM and Thakral SS. 2005. Bovine tuberculosis in India: potential basis of zoonosis. *Tuberculosis* 85: 421-428.

108. Prodinger WM, Brandstatter A, Naumann L, Pacciarini M, Kubica T, Boschiroli ML, Aranaz A, Nagy G, Cvetnic Z, Ocepek M, Skrypnyk A, Erler W, Niemann S, Pavlik I and Moser I. 2005. Characterization of *Mycobacterium caprae* isolated from Europe by mycobacterial interspread repetitive unit genotyping. *J of Clin Microbiol* 43: 4984-4992.

109. RCZI. 2009. Bovine tuberculosis: A neglected zoonosis revisited. Road Map to Combat Zoonoses in India Initiative Newsletter, *Zoonoses* Watch 1: 2.

110. Reiriego Gordejo JF and Vermeersch JP. 2006. Towards eradication of bovine tuberculosis in the European Union. *Vet Mircobiol* 112: 101-109.

111. Revenel MP. 1902. Intercommunicability of human and bovine tuberculosis. *J Comp Pathol Thera* 15: 112-143.

112. Rhyan JC and Spraker TJ. 2010. Emergence of diseases from wildlife reservoirs. *Vet Path* 47: 34-39.

113. Rothschild BM and Martin LD. 2006. Did ice-age bovids spread tuberculosis *Naturwissenschaften* 93: 565-569.

114. Rothschild BM, Martin LD, Lev G, Bercovier H, Bar-Gal GK, Greenblatt, Donoghue H, Spigelman and Brittain D. 2001. *Mycobacterium tuberculosis* complex DNA from an extinct bison dated 17000 years before the present. *Clin Inf Dis* 33: 305-311.

115. Sakula A. 1882. Robert Koch Centenary of the discovery of tubercle bacillus. *Thorax* 37: 246-251.

116. Sauter CM and Morris RS. 1995. Behavioral studies on the potential for direct transmission of tuberculosis from feral ferrets (*Mustela furo*) and possums (*Trichosurus vulpecula*). *NZ Vet J* 43: 294-300.

117. Scanu AM, Bull TJ, Cannas S, Sanderson JD, Sechi LA, Dettori G, Zanetti S and Hermon-Taylor J. 2007. *Mycobacterium avium* subsp. *paratuberculosis* infection in cases of irritable bowel syndrome and comparison with Crohn's disease and Johne's disease: common neural and immune pathogenicities. *J Clin Microbiol* 45: 3883-3890.

118. Schiller I, Oesch B, Vordermeier HM, Palmer MV, Harris BN, Orloski KA, Buddle BM, Thacker TC, Lyashchenko KP and Waters WR. 2010. Bovine tuberculosis: A review of current and emerging diagnostic techniques in view of their relevance for disease control and eradication. *Transbound Emerg Dis* 57: 205-220.

119. Schmiedel A. 1968. Development and present state of bovine tuberculosis in man. *Bull Int Union Tuberc* 40: 5-32.

120. Seth M, Lamont EA, Janagama HK, Widdel A, Vulchanova L, Stabel JR, Waters WR, Palmer MV and Sreevatsan S. 2009. Biomarker discovery in subclinical mycobacterial infections of cattle. *PLoS ONE* 4: e5478.

121. Shah NP, Singhal A, Jain A, Kumar P, Uppal SS, Srivastava MVP and Prasad HK. 2006. Occurrence of overlooked zoonotic tuberculosis: Detection of *Mycobacterium bovis* in human cerebrospinal fluid. *J Clin Microbiol* 44: 1352-1358.

122. Shitaye JE, Getahun B, Alemyehu T, Skoric M, Treml F, Fictum P, Vrbas V and Pavlik I. 2006. A prevalence study of bovine tuberculosis by using abattoir meat inspection and tuberculin skin testing data, histopathological and IS6110 PCR examination of tissues with tuberculous lesions in cattle in Ethiopia. *Vet Med* 51: 512-522.

123. Shitaye JE, Tsegaye W and Pavlik I. 2007. Bovine tuberculosis infection in animal and human population in Ethiopia: a review. *Veterinarni Medicina* 52: 317-332.

124. Smith I. 2003. *Mycobacterium tuberculosis* pathogenesis and molecular determinants of virulence. *Clin Microbiol Rev* 16: 463-496.

125. Sreevatsan S, Pan X, Stockbauer E, Connell ND, Kreiswirth BN, Whittam TS and Musser JM. 1997. Restricted structural gene polymorphism in the *Mycobacterium tuberculosis* complex indicates evolutionarily recent global dissemination. *Proc Natl Acad Sci* 94: 9869-9874.

126. Stead WW. 1997. The origin and erratic global spread of tuberculosis. How the past explains the present and is the key to the future. *Clin Chest Med* 18: 65-77.

127. Steele JH. 1980. Human tuberculosis in animals. In: Steele JH (ed) *CRC handbook series in zoonoses. Section A. Bacterial, rickettsial and mycotic diseases*, Vol. 2. CRC Press, Inc., Boca Raton, Fla. p. 141-159.

128. Sternberg S, Bernodt K, Holmstrom A and Roken B. 2002. Survey of tuberculin testing in Swedish zoos. *J Zoo Wildl Med* 33: 378-380.

129. Stone MJ, Brown TJ and Drobniewski A. 2012. Human *Mycobacterium bovis* infections in London and Southwest England. *J Clin Microbiol* 50: 164-165.

130. Suleiman MS, Hamid ME and Purandas M. 2002. Bovine tuberculosis in dairy animals at Lahore: threat to the public health. http.www.priory.com/vet/bovinetb.htm.

131. Tadayon K, Mosavari N, Sadeghi F and Forbes KJ. 2008. *Mycobacterium bovis* infection in Holstein Friesian cattle in Iran. *Emerg Inf Dis* 14: 1919-1921.

132. Taylor JH. 2009. *Mycobacterium avium* subsp. *paratuberculosis*, Crohn's disease and the Doomsday scenario. *Gut Pathogens* 1: 15.

133. Thacker TC, Harris B, Palmer MV and Waters WR. 2011 Improved specificity for detection of *Mycobacterium bovis* in fresh tissues using IS*6110* real-time PCR. *BMC Vet Res* 7: 50.

134. Thoen CO and Steele JH. 1995. *Mycobacterium bovis infections in animals and humans*. Iowa State University Press, Ames.

135. Thoen, CO, Karlson AG and Himes EM. 1981. Mycobacterial infections in animals. *Rev Infect Dis* 3: 960-972.

136. Theon CO, LoBue PA, Enarson DA, Kaneene JB and Kantor IN. 2009. Tuberculosis: a re-emerging disease in animals and humans. *Veterinaria Italiana* 45: 135-181.

137. Tiemersma EW, van der Werf MJ, Borgdorff MW, Williams BG and Nagekerke NJ. 2011. Natural history of tuberculosis: duration and fatality of untreated pulmonary tuberculosis in HIV- negative patients: A systematic review. *PLoS ONE* 6: e17601.

138. Timothy CR, Moore M, Moser KS, Brodine SK and Strathdee SA. 2008. Tuberculosis from *Mycobacterium bovis* in binational communities, US. *Emerg Inf Dis* 14: 909-916.

139. Tortoli E. 2009. Clinical manifestation of non tuberculous mycobacterial infections. *Clin Microbiol Infec* 15: 906-910.

140. Turenne CY, Collins DM, Alexander DC and Behr MA. 2006. *Mycobacterium avium* subsp. *paratuberculosis* and *M. avium* subsp. *avium* are independently evolved pathogenic clones of a much broader group of *M. avium* organisms. *J Bacteriol* 190: 2279-2487.

141. Ugandan Ministry of Health. 2003. STD/AIDS Control Programme, STD/AIDS surveillance report. Ministry of Health, Kampala, Uganda. http//:www.health.go.ug/docs/hiv0603.pdf.

142. VerCauteren KC, Atwood TC, DeLiberto TJ, Smith HJ, Stevenson JS, Thomsen BV, Gidlewski T and Payeur J. 2008. Surveillance of coyotes to detect bovine tuberculosis, Michigan. *Emerg Inf Dis* 14: 1862-1869.

143. Wedlock DN, Skinner MA, deLisle GW and Buddle BM. 2002. Control of *Mycobacterium bovis* infections and the risk to human populations. *Microbes Infect* 4: 471-480.

144. Whittington RJ, Marshall DJ, Nicholls PJ, Marsh AB and Reddacliff LA. 2004. Survival and dormancy of *Mycobacterium avium* subsp. *paratuberculosis* in the environment. *Appl Environ Microbiol* 70: 2989-3004.

145. WHO. 2010. *A brief history of tuberculosis control in India*. World Health Organization, Geneva, Switzerland.

146. WHO. 2011. *Global tuberculosis control: WHO report 2011*. World Health Organization, Geneva, Switzerland.

147. Wilkins MJ, Meyerson J, Bartlett PC, Speildenner SL, berry DE, Mosher LB, Kannene JB, Robinson-Dunn B, Stobierski MG and Boulton ML. 2006. Human *Mycobacterium bovis* infection and bovine tuberculosis outbreak, Michigan, 1994-2007. *Emerg Inf Dis* 14: 657-660.

148. Wynne JW, Bull TJ, Seamann T, Bulach DM, Wagner J, Kirkwood CD and Michalski WP. 2011. Exploring the zoonotic potential of *Mycobacterium avium* subsp. *paratuberculosis* through comparative genomics. *PLoS ONE* 6: e22171.

149. Xiaomei L Yaxiong H, Yongbo S and Zhifang W. 2004. One case of dairy cow tuberculosis: the treatment and reflection. *Chin J Vet Med* 40: 58-59.

150. Xing XH, Zhao DX and Qiao FS. 1997. Study on transmission between human and bovine tuberculosis. *Neimenggu Prev Med* 22: 148-149.

151. Yingyu C, Yanjie C, Quantao D, Tao L, Xiang J, Chen J, Zhou J, Zhan Z, Kuang Y, Cai H, Chen H and Guo A. 2009. Potential challenges to the Stop TB plan for humans in China: cattle maintain *M. bovis* and *M. tuberculosis*. *Tuberculosis* 89: 95-100.

152. Yoshikawa Y. 2006. Tuberculosis as a zoonosis. *Kekkaku* 81: 613-621.

153. Yumi U and Tooru M. 2007. Tuberculosis as a zoonosis from a veterinary perspective. *Comp Immun Micro Inf Dis* 30: 415-425.

154. Zeegelaar JE and Faber WR. 2008. Imported tropical infectious ulcers in travelers. *Am J Clin Dermatol* 9: 219-232.

155. Zeiss H and Bieling R. 1940. *Behring Gestalt und Werk*, Berlin-Grunewald. p. 251-410.

2014, Zoonoses: Bacterial Diseases

Editor: **Sudhi Ranjan Garg**

Published by: **DAYA PUBLISHING HOUSE, NEW DELHI**

Pages **88–119**

5

Leptospirosis

V. Balamurugan, M. Nagalingam and H. Rahman

Project Directorate on Animal Disease Monitoring and Surveillance, Hebbal, Bengaluru – 560 024

Leptospirosis is among the fastest re-emerging anthropozoonosis that is causing considerable public health problems across the world. The disease affects a variety of animals too, including cattle, buffalo, goat, sheep, horse and swine, resulting in heavy economic losses to the farming community on account of abortions, stillbirths, infertility and reduced productivity. Leptospirosis has been recognized as an important occupational hazard of agriculture field workers, sewerage system workers, animal handlers, forestry workers, butchers, etc. The worldwide annual incidence of human leptospirosis is estimated at 0.1-1 per 100,000 in temperate countries and 10-100 per 100,000 in the humid tropical areas (WHO 2003). The transmission cycle of leptospirosis involves the carrier hosts, environment and human beings. Almost every known species of rodents, marsupials and mammals can be carrier and excretor of leptospires. The incidence of leptospirosis is very strongly associated with rainfall. The organisms are adapted to the environment of the tropical region with plenty of rainfall and it is often difficult to avoid exposure of people to animals or contaminated environment. Excess rainfall events that cause massive flooding are associated with the potential for huge outbreaks, particularly in densely populated regions of the tropical world. A large variety of animal species acts as carrier for leptospirosis, which makes it difficult to eliminate and perhaps even control the disease in tropical developing countries like India (WHO 2007). Therefore, early case detection or identification and prompt treatment as well as creating awareness about the disease among the people and public health personnel are the steps that could be taken to reduce the extent of the problem.

Etiological Agent

The causative agents of the disease, *Leptospira* are flexible, helicoidal (spiral) rods with one or both ends bent or hooked. The organisms are motile with cork-screw like rotating axial filament and the motility is characterized by rotation about its longitudinal axis and flexion and extension. The cells have typical double membrane structure in common with other spirochetes. The bacterial cells are thin, finely coiled, and actively motile. The size of the cell is about 0.1 µm in thickness and 6-20 µm in length. The fine coils are seen only under high power magnification and the coils have amplitude of 0.1 to 0.15 µm and a wavelength of approximately 0.5 µm (Faine 1994). They are too thin to be seen under light microscope and are best visualized under dark field microscope. They cannot be stained readily with aniline dyes and can be stained only faintly by Giemsa stain. Leptospires are best stained by silver impregnation techniques. Broadly, leptospires are of two types: saprophytes living in freshwater and using organic material as a food source but not requiring a host, and pathogens that require a host. The conventional classification of leptospires relies mainly on growth responses in medium containing 8-azaguanine (225 µg/ml) (Johnson and Rogers 1964b) and growth at low temperature *i.e.* the minimum growth temperature ranges from 13-15°C for pathogenic leptospires and 5-10°C for saprophytes (Johnson and Harris 1967). However this criterion could be misleading as some pathogenic *Leptospira* like serovar Icterohaemorrhagiae can also grow at 10°C (Kmety *et al.* 1966).

The genus *Leptospira* belongs to the division Gracillicutes, class Cotobacteria, order Spirochaetales and family Leptospiraceae (Johnson and Faine 1984). The family has three genera: *Leptospira*, *Leptonema* and *Turneria*. Leptospires were first observed by Stimson in 1907 in silver stained tissues from a patient (Stimson 1907). At that time, the bacteria observed were identified as *Spirochaeta interrogans*. In the absence of additional information, the naming of this species did not conform to the requirements of the International Code of Nomenclature. The first valid description of saprophytic *Leptospira* was provided in 1914 (Wolbach and Binger 1914) and that of pathogenic *Leptospira* in 1915 (Inada *et al.* 1916). The strain Ictero No. 1 of serovar Icterohaemorrhagiae (Spirochaeta *Icterohaemorrhagiae japonica*) was the first isolate of *leptospira*, which was recovered from a patient suffering from Weil's disease. Genus *Leptospira* was initially divided into two groups: *L. interrogans* sensu lato (pathogenic strains) and *L. biflexa* sensu lato (saprophytic strains).

Classification

The classification of this organism is complex. Serological classification by a cross agglutination absorption test (CAAT) analysis led to the definition of serovars, which are today considered to be the basic systematic unit for *Leptospira* species (Kmety and Dikken 1993). It is based on the expression of the surface-exposed epitopes in a mosaic of the lipopolysaccharide (LPS) antigens, while the specificity of epitopes depends on their sugar composition and orientation (De la Pena-Moctezuma *et al.* 1999). Two strains are considered different if after cross-absorption with adequate amounts of heterologous antigens at least 10 per cent of the heterologous titre regularly remains associated with one of the two antisera (Cerqueira and Picardeau 2009). For decades, CAAT has been used for the classification (Kreig and Holt 1984). Serogroups comprise antigenically related serovars and the list of serovars is updated periodically and new pathogenic serovars are described. At present, *Leptospira* species are traditionally classified into 29 serogroups and over 300 serovars (Cerqueira and Picardeau 2009). Serovar identification of isolates is essential to understand the epidemiology of this disease, but few laboratories are able to perform CAAT, so most isolates are not identified to serovar level because of unavailability of the test and its tedious and time consuming nature for routine typing. Serogroups identified using microscope agglutination test (MAT) have no official taxonomic status but serve the practical purpose of grouping common antigens together . With the emergence of molecular typing methods, it has become increasingly clearer that the concept of a "serovar" is no longer fully satisfactory as it may fail to define epidemiologically important strains in an adequate manner. For example, molecular typing has been shown to give better discrimination of strains of the Grippotyphosa serogroup than the serological typing (Hartskeerl *et al.* 2004).

Leptospiral genome has two chromosomes: large chromosome (CI) and small chromosome (CII). The size of CI chromosome ranges from 4,277,185 bp to 4,332,241 bp, whereas the size of CII chromosome is in the range of 358,943 bp to 350,181 bp. The chromosome of *Leptospira* is characterized by a G+C content of 35-41 mol per cent depending on species, with a genome size of 3.9-4.6 Mb (Zuerner 1991). The published leptospiral genome sequences are *L. interrogan* serovars (Lai and Copenhageni), two strains of *L. borgpetersenii* serovar Hardjo and two strains of *L. biflexa* serovar Patoc. *L interrogans* and *L. borgpetersenii* both have two circular chromosomes and both genomes are characterized by a high degree of genome plasticity, with evidence of large-scale genomic rearrangement.

The alternative genotypic classification is based on DNA hybridization; thus, the leptospires can be assigned to the species level. DNA-DNA hybridization is a gold standard method for species determination in prokaryotes. Based on genetic homology in DNA hybridization experiments, 15 genomic species (*L. interrogans, L. kirschneri, L. borgpetersenii, L. santarosai, L. noguchii, L. weilii, L. inadai, L. biflexa, L. meyeri, L. wolbachii, Leptospira* genomospecies 1, 3, 4 and 5) have been described in the genus *Leptopspira*, whereas *Leptonema* and *Turneria* have one species each (*L. illini* and *T. parva*) (Brenner *et al.* 1999). Genomic species is a group of Leptospiraceae serovars whose DNAs show 70 per cent or more homology at the optimal reassociation temperature of 55°C, or 60 per cent or more homology at a stringent reassociation temperature of 70°C and in which the related DNAs contain 5 per cent or less unpaired bases (WHO 2007).

The Subcommittee on the Taxonomy of Leptospiraceae, held in Quito, Ecuador in 2007 decided to give the status of species to the previously described genomospecies 1, 3, 4 and 5, resulting in a family comprising of 13 pathogenic *Leptospira* species: *L. alexanderi, L. alstonii (genomospecies 1), L. borgpetersenii, L. inadai, L. interrogans, L. fainei, L. kirschneri, L. licerasiae, L. noguchi, L. santarosai, L. terpstrae* (genomospecies 3), *L. weilii, L. wolffii,* with more than 260 serovars. It is expected that additional new species exist. Saprophytic species of *Leptospira* include *L. biflexa, L. meyeri, L. yanagawae* (genomospecies 5), *L. kmetyi, L. vanthielii* (genomospecies 4), and *L. wolbachii,* and contain more than 60 serovars. Genomic DNA-DNA hybridization demonstrated that various strains of *L. interrogans* sensu lato constitute not one but at least six different species (Yasuda *et al.* 1987). Recently, hybridization studies have identified 20 *Leptospira* species to date. According to this, pathogenic leptospira comprises nine species. However, antigenically related serovars are classified into two or more different species and a serogroup is often found in several species of *Leptospira*. Strains belonging to the same serovar may belong to different *Leptospira* species and *vice versa* (Cerqueira and Picardeau 2009).

These two classification systems are not always consistent. The correlation between the serological and genotypic classification systems is poor. A given serogroup is often found in several *Leptospira* species. For instance, 14 described serovars of the Bataviae serogroup are found in *L. interrogans* sensu stricto (five serovars), *L. santarosai* (five serovars), *L. kirschneri* (one serovar), *L. noguchii* (two serovars) and *L. borgpetersenii* (one serovar) (Levett 2001). Several studies have thus shown a lack of relationship between serotyping and molecular classification, suggesting that the genes determining serotype may be transferred horizontally between species. Thus, although antigenically related serovars, Hardjobovis and Hardjoprajitno, are classified in two different species (namely *L. borgpetersenii* and *L. interrogans*), the sequences of the loci encoding the LPS biosynthesis pathway are highly similar between these two serovars (De la Pena-Moctezuma *et al.* 1999). The basis of horizontal transfer between different *Leptospira* species, presumed to be responsible for the genetic exchange of the LPS determinants, is unknown (Cerqueira and Picardeau 2009).

Geographical Distribution

Leptospirosis has wide geographical distribution and occurs in tropical, subtropical and temperate zones and is more spread in tropical regions than in temperate countries. In the developed world, the incidence of the disease has come down substantially and most cases that occur now are associated with recreational exposure to the contaminated water, including swimming, rafting, canoeing, kayaking. In contrast, the incidence appears to be increasing in developing countries. Most countries in the Southeast Asian region are endemic to leptospirosis. On an average, 10,000 severe cases requiring hospitalization occur world over annually (Faine 1994, Smythe 1999).

Epidemiology

Leptospirosis was identified commonly as an occupational disease, first described in soldiers, miners, sewer workers and rice planters, all of whom worked in wet conditions (Levett 2001). High risk categories include the people who come in direct or indirect contact with infected animals, such as farmers, veterinarians, abattoir workers, rodent control workers, miners, soldiers, sewage system workers, rice field workers, banana farmers, etc. Epidemics have also been noted among sugarcane cutters, rice harvesters, swine keepers, cattle farm workers, and milkmen (Faine 1994, Levett 2001). Accidental infection, while handling *leptospira* cultures (laboratory acquired infection) and during necropsy of animals, has also been reported.

Leptospirosis may also be considered as a disease linked with the environment. In this regard, a number of outbreaks have been related to heavy rainfalls (Salvador, Nicaragua, Philippines, Peru and Argentina), hurricane (Puerto Rico) (Sanders *et al.* 1999) and cyclone (Odisha, India) (Sehgal *et al.* 2001).

Reservoir hosts (carrier-state species) for leptospirosis are often non-migratory, and even territorial. This means that leptospiral genetic diversity in a local area can remain static, and this local cultivation is one of the reasons for such a large number of serovars and their frequent localization (Levett 2001). Rats constitute the main reservoir hosts but other rodents, marsupials and feral mammals are also known to act as reservoirs. One or more species in the environment maintain the leptospira as a viable population through urinary shedding. Reservoir hosts can cross infect each other via urine or congenitally. Important reservoir animals of leptospirosis for human transmission are rats and mice living near human habitations, domestic animals such as cattle, swine, etc., and companion animals, especially dogs. Feral and domestic mammals as well as reptiles and amphibians serve as permanent maintenance hosts or reservoirs for over 250 known serovars. However, acute infection of non-reservoir hosts that do cross territorial boundaries allows for bacterial transfer and slow migration of serovars across continents. Since birds seem unable to carry leptospirosis, the speed of spread is far smaller than that of other infections. Moreover, the leptospiral diversity can be fixed at some isolated places such as islands, or on occasion the bacteria can be completely absent there. Since leptospires reproduce via binary fission, there is no DNA combination between serovars or the host and the development of new serovars via mutative virulence selection is slow. Sexual transmission is known to occur in many species. Transmission of the infection among maintenance hosts is efficient and the incidence of infection is relatively high. Because of the wide spectrum of animal species that serve as reservoirs, leptospirosis is considered the most widely spread zoonotic disease. Humans are not capable of acquiring carrier-state and are incidental or accidental host to the leptospira life cycle. Incidental hosts are not important reservoirs of infection and the incidence of transmission is low. Transmission of the infection from one incidental host to another is relatively uncommon.

Role of Rodents in Dissemination of Infection

The incidence of leptospirosis is greatly influenced by the rodent population in a geographical region. Once acquired, the organisms colonize the kidney cortex of the rodent for more or less its lifetime without showing a disease syndrome. Rodents are prolific shedders of leptospires, voiding them in urine continuously, and of significance in epidemiology of the disease (Zavitsanou and Babatsikou 2008). In a report, it was estimated that a tundra vole harboured about 8 million leptospires in its body and one million of them were excreted in urine daily (Golubev and Litvin 1983). Similarly, in another study, a drop of rodent urine tapped directly from the urinary bladder on autopsy contained an average 100 leptospires in each of the 25 fields of dark field microscopy (DFM) (Gangadhar *et al.* 2006). Pathogenic leptospires have been isolated from rodents all over the world. *Rattus rattus wroughtoni hinton*, *Rattus rattus rufescens*, *Bandicota indica* and *Bandicota bengalenses* are the rodent species involved in spreading leptospirosis to human and livestock in India. These are the most common rodent species in India and are distributed all over the country (Gangadhar *et al.* 1999, Matthias and Levett 2002, Koutis 2007). *Rattus hinton* is the variety predominantly found in agricultural fields and *R. rufescens* inhabits human and animal dwellings. The *Bandicota* varieties are ubiquitous, and are found particularly in granaries, animal feed mixing units, storehouses, catering establishments and in the underground drainage systems. Musk rats are among the major carriers of leptospires in the Netherlands. A study revealed that *Apodemus agrarius* was the principal carrier of leptospires in China. *Rattus norvagicus* and *Mus musculus* are the principal carriers of leptospires in Rio de Janeiro (Gangadhar *et al.* 2006).

Risk Factors

The core determinants of transmission of leptospirosis include the presence of carrier animals, suitability of the environment for the survival of leptospires, and the interaction between man, animals and environment. These determinants are often influenced by various socio-cultural, occupational, behavioural and environmental factors prevailing in a community. Consequently, epidemiology of the disease could be quite complex and dynamic (Sehgal 2006). The source of infection in an area is determined by the factors such as rodent density, population size of the farm and other domestic animals, sanitation of animal habitats, availability of veterinary services for prompt detection and treatment of animal leptospirosis, control programmes for animal leptospirosis etc. Important human behavioural, socio-cultural and occupational factors that influence transmission include the level of personal hygiene, individual habits and practices such as bathing in unprotected water bodies, water associated recreational activities, agricultural and other occupational practices, animal rearing practices, level of hygiene in milking and slaughterhouses, etc.

Specific risk factors of infection vary from one epidemiological setting to another. In a typical rural agrarian community, agricultural activities, other activities in farm land or outdoor environment, forestry, freshwater fishing, frequent wading through stagnant water bodies and use of unprotected water sources for bathing and washing are often implicated as risk factors for transmission of infection (Sehgal 2006). In a typical urban community in developing countries, where the level of environmental sanitation is poor and the population density is high, exposure to sewage water, barefoot walking, contact with garbage, rat infestation of houses, etc. are found to be associated with leptospirosis. In developed countries, leptospirosis usually occurs as small common source outbreaks or occasionally among travellers who visit endemic tropical areas. In such situations, exposure to specific contaminated water source is usually the risk factor (Sehgal 2006). Additionally, the travellers visiting leptospirosis endemic places continue to bring the disease. Moreover, many sporadic cases of leptospirosis are associated with activities of daily life; specifically many cases result from barefooted walking in damp environment or gardening with bare hands. It may be mentioned that a number of disease outbreaks have been associated with drinking of urine-contaminated water.

A pattern of disease seasonality has been described with a peak incidence occurring in summer or fall in temperate regions and during rainy seasons in warm-climate regions. This is attributed mainly to longer survival of leptospires for months or years, especially in warm and humid environments so allowing reinfection of animals/humans. Survival of leptospires is favoured by moisture and warm temperatures and under these conditions the organism may persist for days to weeks outside of the animal; survival is brief in dry soil or at freezing or sweltering temperatures. Therefore, leptospirosis occurs most commonly in the spring, autumn and early winter in temperate climates and more prevalence of disease during rainy season in the urban areas of sub-tropical and tropical countries like India. High incidence has been recorded among people who are exposed to wet environments because of occupational or other activities. Leptospirosis is a known health hazard of rice farmers in countries such as Indonesia and Thailand. Outbreaks have occurred in Korea on several occasions when the fields were flooded before harvest (Park *et al.* 1989). Outbreaks have occurred among general population when people are exposed to flood waters that have high chances of leptospiral contamination.

In nutshell, almost every country in South and Southeast Asia, South and Central America and several Island nations across the world are endemic to leptospirosis. The epidemiology of leptospirosis is extremely complex and dynamic with unique characteristics-specific for each socio-cultural and ecological setting. Agricultural workers constitute a major portion of the risk group. However, a few broad categories of epidemiological patterns could be distinguished, that could form the framework for the detailed analysis of the specific epidemiology of the disease in each community and thus might aid the development of public health strategies for the control of the disease.

Disease Transmission

The mode of transmission of leptospirosis is categorized as direct or indirect depending upon the source of infection. When the source of infection is animal tissue, body fluids or urine, the transmission is termed as direct. Direct transmission occurs when leptospires from acutely infected or asymptomatic carrier animals enter the body of the new host and initiate infection. This transmission among animals can be transplacental, haematogenous, by sexual contact or by suckling milk from infected dam. Presence of leptospires in genital tracts as well as transplacental transmission has been demonstrated in animals. Direct transmission from animals to human beings is common amongst the occupational groups who handle animals and animal tissue such as butchers, veterinarians, cattle and pig farmers, rodent control workers, laboratory personnel who handle lab animals etc., are at high risk. Accidental leptospiral exposure to veterinarians and rodent control workers has also been recorded. Human-to-human transmission through breast-feeding has also been recorded (Bolin and Koellner 1988). When the source of infection is environment contaminated with the urine of carrier animals, the transmission is termed as indirect (Faine 1994). Agricultural workers, sewage workers, people walking barefoot in water-logged areas, sports persons who participate in water-related sports such as rafting, canoeing, swimming, etc. are at high risk of contracting the disease through indirect transmission. Transmission of leptospira to mammals occurs mainly through indirect contact with water or soil contaminated with the urine of excretor animals.

Pathogenesis

Leptospires are presumed to enter the body via small abrasions or other breaches of the surface integument. Pathogenic leptospira rapidly invade the bloodstream after penetrating skin or mucous membranes. They may also enter directly into the bloodstream or lymphatic system via the conjunctiva, the genital tract in some animals, the nasopharyngeal mucosa, possibly through a cribriform plate, the lungs following inhalation of aerosols or through an invasion of the placenta from the mother to the foetus at any stage of pregnancy in mammals (Faine *et al.* 2000). Drinking or inhalation of contaminated water following immersion can also cause leptospirosis.

Leptospires spread almost immediately from the site of entry, via lymphatics to the bloodstream, where they circulate to all tissues. Avirulent and non-pathogenic leptospires, which reach the bloodstream, are cleared rapidly, within minutes of entry, by reticuloendothelial phagocytosis; the leptospires are found at first mainly in the lungs, later in the liver and spleen (Faine 1964). Those leptospires which survive by evading phagocytosis, by definition the virulent organisms in the inoculum, grow exponentially, with a doubling time of approximately 8 h, in the bloodstream and tissues. Growth ceases with either the appearance of opsonic-specific immunoglobulin in the plasma and subsequent rapid-specific immune clearance, or the death of the host. Leptospires are not pyogenic bacteria; they do not cause inflammatory reactions except through secondary tissue damage.

After entering the body through cuts, abrasions or mucous membranes, the organisms get localized in the liver. Here, they multiply and gain entry into the blood stream resulting in septicaemia. Certain serovars *viz.* Ballum, Hardjo, Pomona and Tarassovi produce haemolysin (sphingomyelinase) and exhibit chemotaxis towards haemoglobin. However, phospholipase C has been reported from serovar Canicola. During early septicaemia, the haemolysin produced causes extensive intravascular haemolysis. Abortions are noticed several weeks after the septicaemia stage. The foetus is generally autolyzed. Not much of the mechanism is understood as very little work has been done on human and animal abortions due to leptospirosis, even though there are several leptospira isolates from bovine aborted foetuses.

Once within the host tissues, pathogenic strains can reproduce as they are optimised for metabolism at body temperatures. Their survival depends on the lack of an effective host immune response. Leptospires that survive entry and the immediate innate response will rapidly migrate to the bloodstream and lymphatic system, so spreading throughout the host body within a very short time. Pathogenic bacteria reproduce by binary fission, so the colony increases exponentially, and the typical doubling time is about 8 h. The growth continues unchecked until the adaptive immune response develops or the host dies. No fully-resistant strains exist, so the adaptive response is always able to select a cognate antibody. Virulence is directly related to pathogenicity and a strain is virulent if it is adapted to reproduce rapidly *in-vivo* and has resistance to the innate immune response. In virulent strains, the bacteria are resistant to attack from the innate immune system and so can develop rapidly, until the adaptive system has a change to select and replicate a cognate antibody. Saprophytes and the pathogens that are less virulent seem to be easily targeted by the innate immune system and so are eliminated. Avirulent strains are rapidly removed by the immune system, but the speed of spread is far higher than in other bacterial infections and a host can show leptospira within blood samples in a matter of minutes from exposure.

Persistence

Leptospires are able to persist in some anatomically localized and immunologically privileged sites, even after antibodies and phagocytes have cleaved leptospires from all other sites. The most significant site of persistence is the renal tubule. Leptospires appear in the kidney two to four weeks after an acute infection, attached to an interdigitated area in the brush border of proximal renal tubular epithelium (Faine 1962). The type of reaction in the tissues ranges from mild to heavy scarring; animals may excrete leptospires intermittently or regularly for periods of months or years, or for their lifetimes. However, humans do not remain carriers for long time and the urine is free of leptospires at the time of clinical recovery.

Host Cell Interactions

Pathogenic leptospires have been shown to express adhesins, haemolysins and many lipoproteins prominent in leptospires and other spirochetes that could play a role in host-cell interactions. Ability to survive and to grow in tissues is major components of virulence. Once if leptospires enter through the integument, they are immediately exposed to the effects of non-specific factors of pH, redox potential, electrolytes, fatty acids, and small organic molecules, some of which may be nutrients. There are also pharmacologically active tissue peptides and mediators, some of them are activated by trauma or healing processes at the site of entry, and innate immunoglobulins. Specific antibodies and other immune mechanisms do not exist in a novice host because they result from previous overt or subclinical infections with the same or related serovars, will also affect the ability of invading leptospires to survive and grow.

Leptospiral lipopolysaccharide (LPS) has structural, chemical and immunological properties resembling those of gram negative bacterial LPS. Nevertheless, even though it clots limulus lysate, it is relatively nontoxic to cells. Animal haemorrhages and coagulation defects are features of severe leptospirosis. Thrombocytopaenia in infected humans and animals, accompanied by independent fibrinolysis consistent with disseminated intravascular coagulopathy has also been reported. Greater cytotoxicity and aggregation of platelets, with release of ATP and serotonin, were caused by LPS from virulent leptospires (Isogai *et al.* 1989). In human patients who died of severe leptospirosis with pulmonary haemorrhages, thrombocytopaenia occurred related to platelet aggregation and adhesion to vascular endothelium, in which there was often evidence of leptospiral antigen (Myers and Coltorti 1978).

Paradoxically, the most obvious adhesion of leptospires to cell surfaces in the renal tubules does not appear to damage the cells, or to result in inflammation around the affected tubules in the absence of repair or scarring from the acute infection in most animals. Similarly, the aggregates of leptospires found in acute leptospirosis are not necessarily associated with tissue damage. Peptidoglycan from leptospires induced tumour necrosis factor (TNF) alpha production in monocytes, consistent with murine tumor necrosis induced by leptospires. In human leptospirosis, there were elevated concentrations of TNF, correlated with severity of disease. Leptospiral phospholipases act on the erythrocytes and other cell membranes containing substrate phospholipids, leading to cytolysis. A clear role in pathogenesis has not been established, but they may be responsible, as may glycoprotein (GLP) toxin, for the holes in erythrocytes seen in electron micrographs of erythrocytes of calves infected with serovar Pomona (Thompson 1986). Lipases have been found as selective for erythrocytes of different animal groups depending on the predominant phospholipids in their membranes (Kasarov 1970). Likewise, haemolytic activities of cultures of serovar Pomona on erythrocytes, attributed to heat-stable long-chained fatty acids has been found to be specific for ruminants among a variety of zoo animals (Dózsa *et al.* 1960). These findings have been reported to be consistent with the actions of a phospholipase toxin (haemotoxin)-specific for cell membranes containing its substrate phopholipids. However, the haemotoxin has been reported lethal to a deer, but not to a rabbit (Dózsa *et al.* 1960). All the haemolysins described in leptospires are phospholipases. Conversely, although the sphingomyelinase and phospholipases have been characterized and studied extensively, it is not clear whether they have a significant role in pathogenesis.

Leptospira immunoglubolin-like (Lig) proteins are surface localized proteins in pathogenic *Leptospira* strains with properties that could facilitate the infection of damaged skin. Their expression is rapidly induced by the increase in osmolarity encountered by leptospires upon transition from water to host (Choy *et al.* 2007). In addition, the immunoglobulin-like repeats of the Lig proteins bind to the proteins that mediate the attachment to host tissue, such as fibronectin, fibrinogen, collagens, laminin, and elastin, plays an important role in cutaneous wound healing and repair (Choy *et al.* 2011). Hemostasis is critical in a fresh injury, where fibrinogen from damaged vasculature mediates coagulation. Fibrinogen binding by recombinant LigB inhibits fibrin formation, which could aid leptospiral entry into the circulation, dissemination, and further infection by impairing healing. LigB also binds fibroblast fibronectin and type III collagen, the two proteins prevalent in wound repair, thus potentially enhancing leptospiral adhesion to skin openings (Choy *et al.* 2011).

Public Health Significance

The significance of the disease in public health aspects acquires more importance, especially in countries like India because of large livestock, rodent and wildlife populations, poor sanitary conditions and animal health practices, and close association between man and animals, providing a congenial environment for the spread of the disease. Various factors influencing animal activity, suitability of the environment for the survival of the organism and behavioural and occupational habits of human beings can be the determinants of incidence and prevalence of the disease (Gangadhar *et al.* 2008). The disease was considered inconsequential till recently, but it is emerging as an important public health problem during the last decade or so due to sudden upsurge in the number of reported cases and outbreaks.

Leptospirosis has a major role as a public health problem in human with sporadic incidence reported from all over India throughout the year. Leptospirosis is also known to occur as a concurrent infection with other serious ailments like HIV, brucellosis, dengue and malaria, which have direct immunosuppressive consequences. Therefore, it is essential to establish the real dimension of this infection in livestock and humans in the developing countries. Considering the epidemic potential of the diseases and the nature of the environment and lifestyle of the people, leptospirosis is a constant threat to the health of people (Vijayachari *et al.* 2008). Leptospirosis is considered as an emerging global public health problem caused by different species of pathogenic *Leptospira* (Pappas *et al.* 2008). In a preliminary study, based on the partial RNA polymerase β subunit (rpoB) gene sequence analysis, the major prevalent pathogenic species of leptospira in animals and humans were *L. borgpetersenii*, *L. interrogans*, *L. kirschneri* and *Leptospira* intermediate species (Balamurugan *et al.* 2013). The latter species requires further study to determine the exact serovars or new species, as recently more new isolates have been classified in this species (Matthias *et al.* 2008). An earlier study had also demonstrated the circulation of *Leptospira inadai* in reservoir hosts in India (Gangadhar *et al.* 2000).

Prevalence of Leptospirosis in Animals and Humans

The epidemiology of leptospirosis has classically been described on the basis of serological data, using MAT, a gold standard serological technique for identifying the infecting serovar from human or animal serum samples. MAT results provide epidemiologically important data allowing the identification of the infection sources or reservoirs and have largely contributed to the current knowledge of leptospirosis epidemiology.

Leptospirosis is also considered as an emerging infectious disease in the developed world. For example, in countries like the United States, the cases of canine leptospirosis are significantly increasing. Recreational activities like water-sports are also believed to be an important factor for the emergence of this infection. In some countries, leptospirosis is endemic and infection is much more common than clinical disease. This fact is particularly true in Australia, where the literature indicates widespread serological prevalence without a significant incidence of clinical disease (Smythe *et al.* 2000). Leptospirosis is unnoticed and under-reported in developing countries of the world. Major outbreaks in South-East Asia were reported in the past due to cyclone in Orissa in India (1999) (Sehgal *et al.* 2001), flooding in Jakarta (2002) (Tangkanakul *et al.* 2005), Mumbai in India (2005) and in Sri Lanka (2008). Some major events of *leptospira* infection and outbreaks in animals and humans in the world are presented in Table 5.1 (ProMED Mail http://www.promedmail.org, Gideon e-books http://www.gideononline.com).

Indian Scenario

Leptospirosis has emerged as an important public health problem in large urban centres of India. The organism and its maintenance host appear to undergo adaptation to their environment, and these host preference and pathogenicity can change with time and geographic region. It is known that leptospirosis is widespread in almost

Table 5.1: Major events of leptospira infection and outbreaks in humans and animals in the world

Region	Year	Major Features of the Event
Germany	1883-1886	Earliest recognized account of leptospirosis of severe illness with jaundice and renal involvement in man (Landouzy 1883, Weil 1886)
Germany	1915	Specific antibodies detected in blood of Japanese miners with infectious jaundice (Uhlenhuth and Fromme 1915)
Barbados	1971-1972	Cattle 51 per cent, pig 13 per cent, sheep 18 per cent, horses 64 per cent and goats 19 per cent with predominant seropositive reactions to Hebdomadis, Autumnalis, Ballum and Pyrogenes (Damude *et al.* 1979)
United States	1974-1998	Highest incidence of leptospirosis in Hawaii (Katz *et al.* 2002)
Ethiopia	1975	91.3 per cent, 70.7 per cent, 57.1 per cent, 47.3 per cent, 15.4 per cent and 8.3 per cent seropositivity in horses, cow, pig, goats, camels and dogs, respectively (Moch *et al.* 1975)
German Federal Republic	1987	Seroprevalence study during outbreak, 16 per cent cattle, 1.2 per cent pigs, 14.45 per cent sheep, 0.3 per cent goats, 4.5 per cent horses and 8.4 per cent dogs positive to 10 different serovars (Schoenberg *et al.* 1987)
Malaysia	1987	Seroprevalence in 40.5 per cent cattle, 3 per cent buffaloes and 16 per cent pigs with serovar Hardjo in cattle and buffaloes and Pomona in pigs (Bahaman *et al.* 1987)
Ontario, Canada	1988	Outbreak, 39 per cent of all herd showed evidence of leptospiral infection (Prescott *et al.* 1988)
New Zealand	1990-1998	Annual incidence 4.4 per 100,000 populations, emergence of serovar Ballum as more frequent cause of human infection
Bangladesh	1994	Outbreak after a flood, 38 per cent seropositivity in human indicating high risk of leptospiral infection in rural population
Thailand	1995-2003	High morbidity of leptospirosis reported from provinces situated in the lower part of the Northeast region, outbreaks of leptospirosis correspond with rainy season, most infections in agricultural workers, primarily rice field farmers (Tangkanakul *et al.* 2005)
Brazil	1996-2009	37,035 cases
California	1999	Sporadic case associated with exposure to freshwater during duck hunting
Bangladesh	2000	Outbreak, 359 dengue-negative patients, leptospirosis detected by PCR
Japan	2001	Increase in leptospirosis cases, reaching more than 200 reported deaths annually until 1960, mostly in rice-field farmers
Nepal	2001	Survey in 200 bovines with infertility problems, 8.5 per cent serum samples positive for serovar Hardjo

Contd...

Table 5.1–Contd...

Region	Year	Major Features of the Event
Indonesia	2001	Outbreak, continuous health problem
Santa Clara	2001	Sporadic case associated with exposure to freshwater duck hunting
Indonesia	2002	Outbreak, notably in Jakarta after massive flooding
Turkey	2002	257/574 cattle and 16/200 sheep samples positive for leptospirosis.
China	2002-2007	About 1,500 confirmed cases and 50 deaths in mainland China
Mexico	2004-2006	Survey in dairy cattle, 1: 100 to 1: 1600 antibody titers in 10.33 per cent animals
Guyana	2005	Six deaths reported due to leptospirosis.
United Kingdom	2006	Two clinically diagnosed cases
Tajikistan	2006	15 cases, attributed to recent works on public water system
Argentina	2007	400 cases, 5 deaths
Hong Kong, China and Sri Lanka	2007	5 cases, 6 deaths within a period of 3 months
Ukraine	2007	3 cases
Sri Lanka	2007	Annual incidence 14 per 100,000, high prevalence of leptospirosis in cattle and rodents in Kandy
Queensland, Australia	2008	Queensland accounted for 71.2 per cent of leptospirosis cases in the country, serovars Zanoni, Australis and Hardjo accounting for 58.9 per cent of these.
Jamaica	2008	3 cases of school students
Peruvian Amazon	2008	881 patients with fever, 41 per cent had antibodies for leptospirosis
Bhutan	2008	Suspected leptospirosis outbreak after floods
Fiji	2009	300 cases with at least 7 deaths after a flood
France	2011	Seroprevalence in 29.7 per cent cattle, 51 per cent pigs, 78.6 per cent horses, 10-20 per cent in wild rodents with serovars Icterohaemorrhagiae, Australis, Grippotyphosa, Sejroe, Ballum, Hebdomadis, Canicola, Bataviae and Pomona

all the species of farm and domestic animals and in several rodent species in many states and union territories, including Andaman and Nicobar Islands, Andhra Pradesh, Bihar, Haryana, Karnataka, Kerala, Madhya Pradesh, Maharashtra, Punjab, Uttar Pradesh, Tamil Nadu and West Bengal (WHO 2000). The leptospirosis situation in India is a cause of concern and it is endemic in all southern states, other coastal states and Andaman and Nicobar Islands, where high prevalence has been recorded both in animals and humans (Rao *et al.* 2003). Leptospira serovars responsible for seropositivity among most of the animals and humans in India have been identified as Icterohaemorrhagiae, Hardjo, Patoc, Australis, Canicola, Grippotyphosa, Pyrogenes, Pomona, Tarassovi and Ballum. The broad spectrum of clinical presentations associated with leptospirosis also hampers case identification. The complexity in case identification and lack of diagnostic facility hinder the actual importance and losses due to leptospirosis in India, particularly in bovine leptospirosis. The study indicated a high level of the prevalence of leptospirosis in animals and humans warranting continuous investigations in order to suggest control strategies in the future. Since the isolation rate of leptospires from clinical specimens is low due to prior indiscriminate use of antibiotics, serological techniques remain the cornerstone of diagnosis. Serological study of leptospirosis in humans has been limited in India. Some major events of *leptospira* infection and outbreaks in animals and humans in India are presented in Table 5.2 (ProMED Mail http://www.promedmail.org, Gideon e-books http://www.gideononline.com).

Disease Manifestations in Humans

The incubation period of leptospirosis in humans is usually 7 to 12 days, with a range of 2 to 29 days. Signs and symptoms are highly variable (Faine 1982). Leptospirosis in humans may show a wide variety of clinical symptoms and signs including fever, severe headache, myalgias, conjunctival suffusion, jaundice, general malaise, stiff neck, chills, abdominal pain, joint pain, anorexia, nausea, vomiting, diarrhoea, oliguria/anuria, haemorrhages, skin rash, photophobia, cough, cardiac arrhythmia, hypotension, mental confusion, psychosis. Human infections vary from asymptomatic to severe. Many cases are mild or asymptomatic, and go unrecognized. Some serovars tend to be associated more often with some syndromes (*e.g.*, severe disease is often associated with serovar Icterohaemorrhagiae). However, any serovar can cause any syndrome or any clinical signs.

Severe manifestations of leptospirosis include any combination of jaundice, renal failure, pulmonary haemorrhage, myocarditis, and hypotension refractory to fluid resuscitation. Other complications include aseptic meningitis and ocular involvement including uveitis. As originally described in the 19th century, Weil's disease is characterized by a triad of fever, jaundice, and splenomegaly. Current usage of the term "Weil's disease" refers to fever, jaundice, and renal failure and is often considered synonymous with severe leptospirosis.

In humans, leptospirosis is usually a biphasic illness (Ferrar 1990). Now-a-days, atypical form of leptospirosis also frequently reported from many parts of India and other countries. The first phase, called the acute or septicaemia phase, usually begins abruptly and lasts approximately a week. This phase is characterized by non-

Table 5.2: Major events of leptospira infection and outbreaks in humans and animals in India

Region	Year	Major Features of the Event
Andaman and Nicobar Islands	1931	Extensive survey of disease outbreak, *Leptospira* isolated (Taylor and Goyle 1931)
Madras, Tamil Nadu	1983	Outbreak in cattle, 35 of 75 (47 per cent) human serum samples seropositive (Ratnam *et al.* 1983)
Madras, Tamil Nadu	1984–1985	Acute renal failure in 19 human patients due to leptospirosis
Andaman and Nicobar Islands	1986	Outbreak in a food-fed commune
Madras, Tamil Nadu	1988	Outbreak after monsoon, 33/40 (82.5 per cent) suspected patients with specific leptospiral antibodies.
Andaman, Andaman and Nicobar Islands	1988	Outbreaks of Andaman haemorrhagic fever, identified as leptospirosis in 1993
Madras, Tamil Nadu	1992	Outbreak (48 cases) (Venkataraman and Nedunchelliyan 1992)
Diglipur, North Andaman, Andaman and Nicobar Islands	1993	Outbreak of acute febrile illness (18 cases) with haemorrhagic manifestations and pulmonary involvement, 66.7 per cent victims with significant titres of antibodies against leptospira, first report of pulmonary leptospirosis in India (Sehgal *et al.* 1995)
Madurai, Tamil Nadu	1994	Outbreak after severe flooding, increase in individuals with uveitis
Madras, Tamil Nadu	1997	Outbreak (1,127 confirmed cases), involvement of serovar Icterohaemorrhagiae (40.2 per cent), Canicola (20.0 per cent) and Automnalis (14.9 per cent) (Natarajaseenivasan and Ratnam 1997)
Valsad and Surat, Gujarat	1997	Outbreaks (562 cases) with fatalities
Orissa	1999	Outbreak of febrile illness with haemorrhagic manifestations, particularly pulmonary haemorrhage following a cyclone and flooding, 143 suspect cases, 28 confirmed cases, 11 fatalities, serovars Pomona, Hebdomadis and Canicola identified (WHO 2000, Sehgal *et al.* 2001)
Mumbai, Maharashtra	2000	Outbreak followed by local flooding, serological results indicated 18/53 children with acute leptospirosis (Karande *et al.*, 2003)
Gujarat	2002	Outbreak, 16 fatal cases (Clerke *et al.* 2002)

Contd...

Table 5.2–Contd...

Region	Year	Major Features of the Event
Maharashtra	2002	Concurrent outbreaks of leptospirosis and dengue, outbreak (74 cases) involving *Leptospira interrogans* serovar Copenhageni (Karande *et al.* 2005)
Gujarat	2003	Outbreak (27 fatal cases, 177 under treatment), involving 131 villages in south Gujarat
Madras (Chennai), Tamil Nadu	2003	Outbreak in a nurses' hostel
Kerala	2003	616 individuals developed symptoms, 115 died
Karnataka	2003	55.95 per cent cases of leptospirosis gave history of contact with animals, study conducted at KMCH, Manipal.
Gujarat	2004	Outbreak (550 cases, 75 fatal) of suspected leptospirosis
Mumbai, Maharashtra	2005	Outbreak (100 or more fatal cases) of suspected leptospirosis following local flooding
Chittoor district, Andhra Pradesh	2005	Outbreak (49 cases) caused by contact with stagnant water (Sohan *et al.* 2008)
Kolkata, West Bengal	2005	Outbreak involving 10 persons (Mathur *et al.* 2009).
Karnataka	2006	Outbreak, 258 cases
Kerala	2006	Outbreak, 11 cases, 1 fatal
Maharashtra	2006	Outbreak, 150 cases or more, at least 60 fatal
Karnataka	2007	Outbreak, 1,516 cases
Chennai, Tamil Nadu	2011	Seropositivity 54.47 per cent among domestic animals, 72.23 per cent in wild animals in captivity, 37.03 per cent in rodents, overall seropositivity 56.68 per cent for *Leptospira interrogans* serovar Australis.

specific signs including fever, chills, headache and conjunctival suffusion. Myalgia, which typically affects the back, thighs or calves, is often severe. Occasionally, a transient skin rash occurs. Other symptoms may include weakness, photophobia, lymphadenopathy, abdominal pain, nausea, vomiting, sore throat, cough, chest pain and haemoptysis. Mental confusion, neck stiffness and other signs of aseptic meningitis have been reported in this phase. Jaundice can be seen in more severe infections. These symptoms last for approximately 4 to 9 days, typically followed by a 1 to 3 days period during which the temperature drops and the symptoms abate or disappear.

The second phase of leptospirosis, called the immune phase, is characterized by the development of anti-leptospira antibodies, and the excretion of the organisms in the urine. This phase can last up to 30 days or more, but it does not develop in all patients. Some patients also have pulmonary symptoms, with clinical signs ranging from cough, dyspnoea, chest pain, and mild to severe haemoptysis, to adult respiratory distress syndrome.

Differential Diagnosis

The following diseases should be considered in the differential diagnosis of leptospirosis in human, influenza; dengue and dengue haemorrhagic fever; hantavirus infection, including hantavirus pulmonary syndrome or other respiratory distress syndromes; yellow fever and other viral haemorrhagic fevers; rickettsiosis; borreliosis; brucellosis; malaria; pyelonephritis; aseptic meningitis; chemical poisoning; food poisoning; typhoid fever and other enteric fevers; viral hepatitis; pyrexia of unknown origin (PUO); primary HIV seroconversion; legionnaire's disease; toxoplasmosis; infectious mononucleosis and pharyngitis.

Disease Manifestations in Animals

Clinical signs in leptospirosis depend on the immunity of the animals and virulence and quantity of the serovar to which they are exposed. All mammals appear to be susceptible to at least one species of *Leptospira* (Merck Veterinary Manual 2012). Disease is rare in cats, and less common in sheep than cattle. The primary reservoir hosts for most leptospira serovars are wildlife, particularly rodents. Reservoir hosts among domestic animals include cattle, pigs, sheep and dogs. The specific reservoir host (s) varies with the serovar and the geographic region. Disease in reservoir hosts including rodents is more likely to be asymptomatic, mild or chronic. The serovars associated with disease in different animal species are given below (Merck Veterinary Manual 2012).

- ☆ *Cattle*: Hardjo, Pomona, Grippotyphosa, Canicola, Icterohaemorrhagiae
- ☆ *Sheep and goats*: Hardjo, Pomona, Grippotyphosa, Ballum
- ☆ *Pigs*: Pomona, Grippotyphosa, Bratislava, Canicola, Icterohaemorrhagiae, Tarassovi, Muenchen
- ☆ *Horses*: Hardjo, Pomona, Canicola, Icterohaemorrhagiae
- ☆ *Dogs*: Pomona, Grippotyphosa, Canicola, Icterohaemorrhagiae, Pyrogenes, Paidjan, Tarassovi, Ballum, Bratislava

The reservoir hosts associated with different serogroups/serovars of *leptospira* are as follows.

- ☆ *Rats*: Serogroups Icterohaemorrhagiae and Ballum
- ☆ *Mice*: Serogroup Ballum
- ☆ *Cattle*: Serovars Hardjo, Grippotyphosa and Pomona
- ☆ *Sheep*: Serovars Hardjo and Pomona
- ☆ *Pigs*: Serovars Pomona, Tarassovi and Bratislava
- ☆ *Dogs*: Serovars Canicola and Bataviae

Leptospira infections may be asymptomatic, mild or severe, and acute or chronic. The clinical signs are often related to kidney disease, liver disease or reproductive dysfunction. Chronically infected animals are often asymptomatic. The incubation period varies from 4 to 12 days in dogs. Abortions usually occur 3 to 10 weeks after infection in cattle, and 15 to 30 days after infection in pigs. In seals and sea lions, the symptoms reported include depression, polydipsia, fever, abortions and neonatal deaths (Merck Veterinary Manual 2012).

Acute leptospirosis occurs mainly in calves. The symptoms may include fever, anorexia, conjunctivitis and diarrhoea. Severely affected animals may also develop jaundice, haemoglobinuria, anaemia, pneumonia, or signs of meningitis such as incoordination, salivation and muscle rigidity. Leptospirosis in sheep and goats is similar to the disease in cattle. In swine, subclinical infections are common; however, clinical leptospirosis is most often characterized by reproductive signs including late term abortions, infertility, stillbirths, mummified or macerated foetuses, and increased neonatal mortality (OIE 2005). The clinical signs and severity of disease are highly variable in dogs. Some infections are asymptomatic or mild, while others are severe or fatal. Leptospirosis in horses is most commonly associated with uveitis or abortions. The disease is typically seen as a self-limiting mild fever with anorexia, although in severe forms haemolysis and vasculitis can result in petechial haemorrhages on mucosal surfaces, haemoglobinuria, anaemia, icterus, conjunctival suffusion, depression, and weakness. Leptospiral uveitis is common in human and horses. The organisms get attached to the neutrophils and form agglutinins. This complex gets attached to the cells of cornea leading to opacity of cornea and the lens.

Pathology

The primary or central characteristic lesion in leptospirosis is disruption of the integrity of the cell membrane of the endothelial cells lining small blood vessels in all parts of the body, which leads to capillary leakage and extravasation of cells, including haemorrhages. Other lesions follow as secondary effects. These effects can be attributed to the action of a GLP toxin of leptospires. Widespread petechial haemorrhages are apparent in all organs and tissues, particularly the lungs, omentum and pericardium. Ischaemia from damaged blood vessels in the renal cortex leads to renal tubular necrosis, particularly of the proximal convoluted tubules. The resulting anatomical damage causes renal failure that can be fatal. Liver cell necrosis caused by ischaemia and destruction of hepatic architecture leads to the characteristic jaundice of the severe type of leptospirosis (Arean 1962). Blood clotting mechanisms are affected by liver failure, aggravating the haemorrhagic tendencies and there may also be thrombocytopenia. Leptospires enter the cerebrospinal fluid (CSF) in the early septicemia phase of the illness, but there is little evidence of inflammatory response in the CSF (Wesley 1995). The anterior chamber of the eye is invaded by leptospires during acute infection, but they are trapped there and cannot move out after the local vasodilation and inflammation subsides. Antibodies from circulation can enter and cause an acute hypersensitivity uveitis (Lucchesi and Parma 1999).

Diagnostic Procedures

An array of laboratory tests is available for diagnosing leptospirosis. However selection of the right specimens and tests and correct interpretation of test results are important. Laboratory diagnosis is broadly divided into two categories. The direct evidence includes either demonstration of leptospires or its DNA and isolation of organism from clinical specimens. Detection of specific antibodies to leptospires (serological diagnosis) is indirect evidence.

Bacteriological Methods

Isolation of leptospires from clinical specimens is the strongest evidence for confirmatory diagnosis. Leptospiremia occurs during the first stage of the disease, beginning before the onset of symptoms, and ends by the first week of the illness. Thus blood cultures should be taken as soon as possible after the patient's presentation and before antibiotics treatment. CSF and dialysate fluid can also be cultured during the first week of illness. Urine can be cultured from the second week of symptomatic illness. Even from aborted foetus's heart blood samples can be used for isolation and confirmation due to abortion in animals (Figure 5.1). Collection of urine following treatment of the animals with a diuretic enhances the chances of detecting the organism. The duration of urinary excretion varies but may last for several weeks. In important cases involving individual animals (*e.g.* clearing an infected stallion to return to breeding), negative tests on three consecutive weekly urine samples has been considered to be a good evidence that an animal is not shedding leptospires in urine.

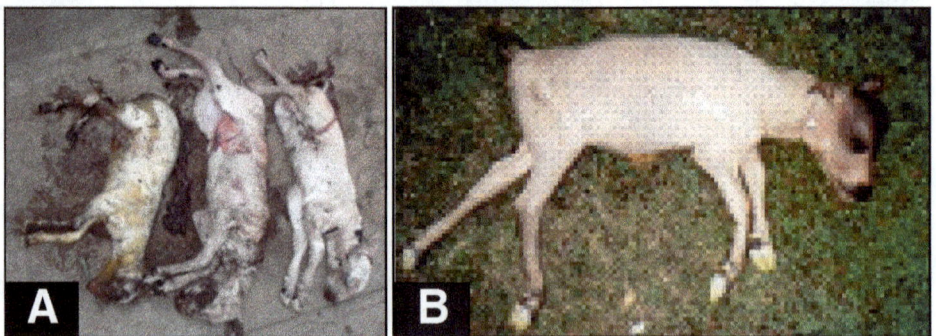

Figure 5.1: Aborted foetuses of goats (A) and cattle (B)
that yielded pathogenic leptospirae

A wide variety of culture media can be used for cultivation of leptospires. Media that contain rabbit serum include Korthof's medium, Fletcher's medium and Stuart's medium. Rabbit serum contains nutrients including high concentrations of bound vitamin B_{12} which helps in the growth of leptospires. All these media can be used for isolation of leptospires from the clinical specimens and for the maintenance of leptospires but not for preparation of antigens for MAT. In protein-free medium, long chain fatty acids are treated with charcoal to detoxify the free fatty acids which are highly toxic to leptospires. The antigenicity and other characters of the cells grown in this medium are similar to those media described above. In fatty acid albumin medium, long chain fatty acid is used as a nutritional source and serum albumin as detoxicant. This medium is popularly known as Ellinghausen-McCullough-Johnson-Harris (EMJH) medium and widely used for isolation, maintenance and preparation of antigens for MAT and for growing leptospires in bulk. Culture media can be enriched by addition of 1 per cent foetal calf serum or rabbit serum to cultivate the fastidious leptospiral serovars. Selective culture media, a modified EMJH medium (Johnson and Harris 1967) supplemented with 200-300 µg/ml of 5-fluorouracil, is ideally suited for isolation and maintenance of pathogenic leptospires (Figure 5.2A) (Johnson and Rogers 1964a). Liquid medium with 1 per cent bovine serum albumin (BSA) solution containing 5-fluorouracil at 100-200 µg/ml should be used as transport medium for the submission of samples. Liquid medium can be converted into semisolid and solid by adding agar or agarose. Semisolid media contain 0.1-0.2 per cent agar whereas solid media contain 0.8-1 per cent agar. Semisolid medium is commonly used for isolation of leptospires and for maintaining the cultures. Colonies are subsurface and visible at 7-21 days of incubation (Figure 5.2A). Solid medium is not ideal for isolation or maintenance of leptospires and mainly used for research purpose to clone the leptospires from mixed leptospiral cultures. Addition of 0.4-1 per cent rabbit serum to semisolid culture medium enhances the chances of isolating fastidious leptospiral serovars (Dinger 1932). Cultures should be incubated at 29±1°C for at least 16 weeks, and preferably for 26 weeks. Cultures should be examined by dark-field microscopy (DFM) every 1-2 weeks for identification. The DFM can be used as a rapid diagnostic procedure in leptospirosis (Vijayachari *et al.* 2001a).

Isolation and identification is the method of choice to identify circulating serovars in a particular geographical region. In addition, locally isolated and identified strains will be more useful to be used as antigens in MAT, as local strains were found to be more sensitive and strongly reactive than reference strains. However, the test has the following demerits:

☆ Culture methods are very tedious, complicated, expensive, technically demanding, time consuming, requiring prolonged incubation (minimum 45 days before declaring a sample negative) and may not be successful (low sensitivity) always.

☆ Leptospirae are highly infectious organisms requiring 'Biosafety level II' facilities.

☆ Prior administration of antibiotics to the patients animals before sample collection greatly reduces the chances of successful isolation.

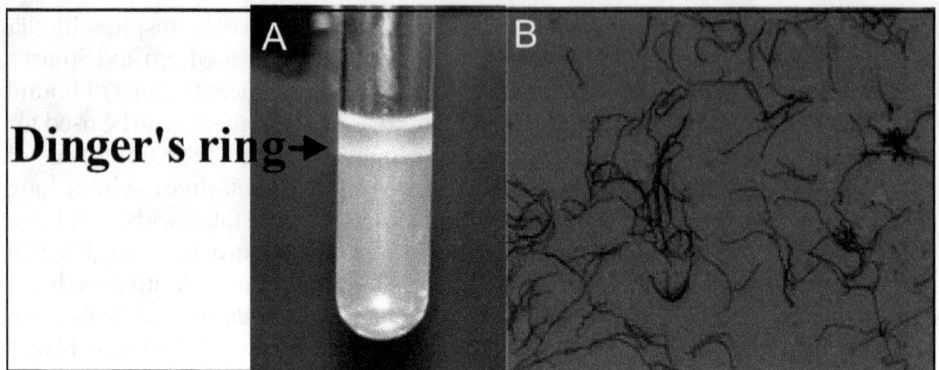

Figure 5.2: Leptospira isolated in EMJH semisolid medium showing positive Dinger's Ring (A) and culture stained by ADMAS silver staining technique x400 (B)

For experimental animal inoculation, blood or other clinical material is injected into the animal's abdominal cavity; golden hamsters are being used most often for this purpose. A specimen of fluid from the abdominal cavity is collected with a sharp pipette at regular intervals and examined by DFM for the presence of leptospirae. In addition, the animal is observed for clinical signs of disease. This method now seems to be rarely used, probably because culturing *in vitro* yields comparable results and avoids animal suffering.

Microscopic Methods

Leptospires may be visualized in clinical specimens by DFM or by immunofluorescence or light microscopy after appropriate staining like silver impregnation technique (Gangadhar and Rajasekhar 1998) (Figure 5.2B). Approximately 10^4 leptospires/ml are necessary for one cell per field to be visible by DFM (Turner 1970). The quantitative buffy coat method has a sensitivity of approximately 10^3 leptospires/ml. Microscopy of blood is of value only during the first 7-10 days of the acute illness during leptospiremia. DFM examination of body fluids such as blood, urine, CSF, and dialysate fluid has been used but is both insensitive and lacks specificity. Immunofluorescence staining of bovine urine, water and soil and immunoperoxidase staining of blood and urine has been applied to increase the sensitivity of direct microscopic examination. Histopathological stains and immunohistochemical methods have been applied for the detection of leptospires in tissues. This method is simple and rapid, but, it has low sensitivity and specificity. Further, it requires technical expertise and is necessary to differentiate serum proteins and fibrin strands in blood from actual leptospirae.

Immunological Methods

MAT using live culture antigen (approximately 2×10^8 leptospires per ml) is the most widely used serological test. For optimum sensitivity, it should use antigens representative of all the serogroups known to exist in a particular region. The specificity of MAT is good; antibodies against other bacteria usually do not cross-react with *leptospira* to a significant extent (OIE 2008). However, there is significant serological cross-reactivity between serovars and serogorups of leptospira and an animal infected with one serovar is likely to have antibodies against the infecting serovar that cross-react with other serovars usually at a lower level in MAT. Many laboratories perform a screening test at a final serum dilution of 1 in 100 and then retest sera with titres of ≥100 to determine an endpoint using doubling dilutions of serum samples beginning at 1 in 100 through to 1 in 12,800 or higher.

As an individual test, MAT is very useful in diagnosing acute infection (Vijayachari *et al.* 2001b); the demonstration of a four-fold change in antibody titres in paired acute and convalescent serum samples is of diagnostic value (OIE 2008). In addition, a diagnosis of leptospirosis is likely based on the finding of very high titres with a consistent clinical picture in case of single serum test. MAT is relatively serovar/serogroup specific and test of choice for sero-epidemiologic studies. However, it has some limitations like: (1) Representative serovars of all serogroups have to be maintained; (2) The procedure is complex and time consuming; (3) Reading results require experienced personnel; (4) It is not possible to distinguish between IgM and IgG antibodies; (5) Second serum sample essential; (6) False negativity in the early course of the disease; (7) Selection of battery of serovars is not easy; (8) The age and density of antigens; and (9) Reproducibility.

Enzyme linked immunosorbent assay (ELISA) has been developed for serodiagnosis using a wide variety of antigen preparations, from leptospiral sonicates to recombinant outer membrane proteins such as LipL32, LipL41, LipL48, OmpL1, OmpL37, OmpL47, LigA, LigB, etc. The assay obviates the need for maintenance of live cultures and is amenable to automation. However, sensitivity and specificity do not match those of the MAT, and reliance on ELISA alone is not recommended. ELISA can detect IgM antibodies earlier than MAT using single antigenic preparation. Heat stable antigens, which are stable at room temperature for long periods, can be used in ELISA allowing rapid mass screening of samples. However, test has some demerits like infecting serovar cannot be assessed, calibration of cut off value and significant titre are required and comparatively less specific.

Rapid screening tests for leptospiral antibodies in acute infection have also been developed using broad reacting genus-specific antigens to detect the immune response to the infecting leptospires. Since the prevalence of leptospiral serogroups varies geographically, antigenic characteristics of the pathogen causing infection may vary from one place or location to another. The sensitivity of the tests in any given setting, therefore, depends on the ability of test antigens to detect antibodies produced against the site-specific leptospira serovars. Microcapsule agglutination test (MCAT), LEPTO Dipstick, Macroscopic slide agglutination test (MSAT), LEPTO Lateral flow, Indirect haemagglutination assay (IHA) and LEPTO dry dot tests are some of the rapid diagnostic tests used for screening. Generally, the sensitivity of these tests is low during the first week of illness, then increases to a peak by 10-12 days or during the second week of the disease. IgM antibodies become detectable during the first week of illness allowing the diagnosis to be confirmed and treatment can be initiated (Chernukha *et al.* 1976). IgM detection has been shown to be more sensitive than MAT when the first specimen is taken early in the acute phase of the illness. All these rapid tests are costly and not a confirmatory test. It is easy to perform and read and results can be obtained within 30 seconds to 15 minutes. It does not require any special expertise or equipment.

Molecular Techniques

Identification of isolates to serovar level is an essential step to understand the epidemiology of the leptospirosis in both humans and animals in any geographic region. However, serovar identification remains a relatively blunt tool with which to investigate details of epidemiology. Because of the difficulties associated with serological identification of leptospiral isolates, there has been great interest in molecular methods for identification and subtyping (Herrman 1993). Polymerase chain reaction (PCR)-based assays are now being used in many diagnostic and reference laboratories for the detection of leptospires in clinical samples (Vijayachari and Sehgal 2006). A variety of primer sets for PCR assays targeting different genes of the leptospira genome have been described. However, only two PCR assays have been extensively evaluated and gained widespread use for diagnosis purpose. One is the genus-specific assay which amplifies DNA from both pathogenic and non-pathogenic serovars (Merien *et al.* 1992), and the other is the approach described by Gravekamp *et al.* (1993) and evaluated by Brown *et al.* (1995), which requires two sets of primers in order to detect all pathogenic *leptospira* species. Improved sensitivity has been achieved by quantitative PCR either using *Taq*Man probes or SYBR green fluorescence (Levett *et al.* 2005). Quality control of PCR assays requires careful attention to laboratory design and workflow to prevent contamination of reagents, and use of appropriate control samples. This assay gives relatively quick results in the early stage of the disease when antibodies are not yet developed at detectable level, but it has some demerits like need of sophisticated equipment and skilled/technical expertise, complicated and expensive, etc. Molecular tools employed for the classification of *Leptospira* species include pulse field gel electrophoresis (PFGE), restriction fragment length polymorphism (RFLP), arbitrarily primed PCR, fluorescent amplified fragment length polymorphism (FAFLP), variable number tandem repeat (VNTR) etc. However, these techniques lack reproducibility or have low sensitivity or specificity (Levett 2005, Vijayachari and Sehgal 2006.).

Treatment

Antibiotic therapy should be started as soon as the diagnosis of leptospirosis is suspected regardless of the phase of the disease or duration of symptoms. Treatment is effective within 7 to 10 days of infection and it should be given immediately on diagnosis or suspicion. The antibiotic of choice is benzyl penicillin by injection in doses of 5 million units per day for five days (Guidugli *et al.* 2000). Patients who are hypersensitive to penicillin may be given erythromycin 250 mg 4 times daily for 5 days. Doxycycline 100 mg twice daily for 10 days is also recommended (Takafugi *et al.* 1984). Tetracyclines are also effective but contraindicated in patients with renal insufficiency, in children and in pregnant women (OIE 2005, 2008).

For mild leptospirosis, doxycycline (hydrochloride, hyclate) is the drug of choice (Takafugi *et al.* 1984). Alternative drugs include amoxicillin and azithromycin dihydrate. For moderate-severe leptospirosis, penicillin G remains the drug of choice.

Severe cases usually treated with high doses of intravenous penicillin (IV) (benzylpenicillin 30 mg/kg up to 1.2 g IV 6-hourly for 5-7 days). Less severe cases can be treated with oral antibiotics such as amoxycillin, ampicillin, doxycycline (2mg/ kg up to 100 mg 12 hourly for 5-7 days) or erythromycin. Third-generation cephalosporins such as ceftriaxone and cefotaxime, and quinolone antibiotics also appear to be effective (Hospenthal and Murray 2003). In general, antibiotic therapy should be completed for 7 days, except for azithromycin dihydrate which could be given for 3 days. Jarisch-Herxheimer reactions (resembles bacteria sepsis can occur after initiation of antibacterials) have also been reported in patients with leptospirosis treated with penicillin. It is postulated that the inflammatory process results from activation of the cytokine cascade during the degeneration of spirochetes (Faine 1994). Patients receiving penicillin should be monitored for these reactions. Step-down therapy can be instituted once if patient is clinically stable and is able to tolerate oral medication. For animal infections, tetracycline and oxytetracycline, erythromycin, enrofloxacin, tiamulin, and tylosin have been reported to be successful in acute cases. Oxytetracycline, amoxicillin, and enrofloxacin may be useful to treat chronic infections.

Prevention

The risk to humans can be minimized by avoiding contamination of water with animal urine, control of rodents and ensuring the proper vaccination of pets. Annual vaccinations, confinement rearing, and chemoprophylaxis are to be employed for animals. Selecting replacement stock from herds that are seronegative for leptospirosis, chemoprophylaxis and vaccination of replacement stock are important. Human leptospirosis can be controlled by reducing its prevalence in wildlife, rodents and domestic animals. Although difficult in controlling the disease in wild animals, leptospirosis in domestic animals can be controlled through recommended vaccination. The most effective preventive measure is avoidance of high-risk exposure (*i.e.* wading in floods and contaminated water, contact with animal's body fluid). If high risk exposure is unavoidable, appropriate personal protective measures including wearing boots, goggles, overalls and rubber gloves are to be continued or followed. Pre-exposure antibiotic prophylaxis is not routinely recommended. However, in those individuals who intend to visit highly endemic areas and are likely to get exposed (*e.g.* travellers, soldiers, those engaged in water-related recreational and occupational activities), pre-exposure prophylaxis may be considered for short-term exposures. The recommended regimen for pre-exposure prophylaxis for non-pregnant, non-lactating adults is doxycycline (hydrochloride and hyclate) 200 mg once weekly, to begin 1 to 2 days before exposure and continued throughout the period of exposure. Currently, there is no recommended pre-exposure prophylaxis that is safe for pregnant and lactating women. Doxycycline is the recommended post-exposure chemoprophylactic agent for leptospirosis (Faine 1998).The duration of prophylaxis depends on the degree of exposure and the presence of wounds. Individuals should continue to monitor themselves for fever and other flu-like symptoms and should continue to wear personal protective measures since antibiotic prophylaxis is not 100 per cent effective. The decision to give prophylaxis depends on the risk exposure assessment.

Immunization of domesticated animals is carried out to protect them from leptospirosis so that productivity is maximized and to protect humans in contact with these animals (Rao *et al.* 2003). Immunization has been widely used for many years as a means of inducing immunity in animals and humans, with limited success. Immunity to leptospirosis is largely humoral and is relatively serovar specific. Thus, immunization protects against disease caused by the homologous serovar or antigenically similar serovars only. Therefore, vaccines must contain serovars representative of those present in the population to be immunized. Early vaccines were composed of suspensions of killed leptospires cultured in serum-containing medium, and side effects were common. Modern vaccines prepared using protein-free medium are generally without such adverse effects (Levett 2001). Human leptospirosis can be controlled by reducing its prevalence in wild and domestic animals. Although little can be done in wild animals, leptospirosis in domestic animals can be controlled through vaccination with inactivated whole cells or an outer membrane preparation (Palaniappan *et al.* 2007). Human vaccines were very scanty. In France, a monovalent vaccine containing only serovar Icterohaemorrhagiae is licensed for human use. Immunization with polyvalent vaccines has been practised in the Far East, where large number of cases occur in rice field workers, such as in China, Japan and Thailand. A vaccine containing serovars Canicola, Icterohaemorrhagiae and Pomona has also been developed in Cuba (Levett 2001).

Dogs are immunized to protect them and human companions. Commercial vaccines containing suspensions of killed *L. interrogans* serovars Icterohaemorrhagiae, Pomona, Grippotyphosa and Canicola are widely available. The vaccines are given either subcutaneously or intramuscularly in two initial doses, one month apart, followed by annual boosters (Faine 1998). Vaccines protect against disease and renal shedding under experimental conditions, but transmission of serovar Icterohaemorrhagiae from immunized dogs to humans has also been reported (Feigin *et al.* 1973). Moreover, immunized dogs may be infected with serovars other than those contained in commercial vaccines. Hence, it is always imperative to include major prevalent serovars in addition to the traditional vaccine strains, based on the incidence of canine infection to combat the increasing serovars, in the population. In developed countries, pigs and cattle are widely immunized, as are domestic dogs, but in most developing countries, vaccines which contain the locally relevant serovars are not available. Most vaccines require booster doses at yearly intervals (Levett 2001). Most bovine and porcine vaccines contain serovars Hardjo and Pomona. In North America, commercial vaccines also contain serovars Canicola, Grippotyphosa and Icterohaemorrhagiae.

Future Perspectives

Various factors influencing animal activity, suitability of the environment for the survival of the leptospirae and behavioural and occupational habits of human beings can be the determinants of incidence and prevalence of the disease. The disease was considered inconsequential till recently, but it is emerging as an important public health problem during the last decade or so due to sudden upsurge in the number of reported cases and outbreaks. Effective prevention and control measures can be achieved through proper diagnostics and prophylactic aids to curtail further spread in most of the zoonotic diseases. Improved sanitary conditions such as proper treatment and disposal of human waste, higher standards for public water supplies, improved personal hygiene procedures and sanitary food preparation are vital to strengthen the control measures. A clear understanding of epidemiology of the disease with reservoir host, especially the virulence and transmissibility of leptospirosis, could help in understanding the severity and thereby to take appropriate control measures.

Research should focus on molecular biology of the pathogen so as to develop diagnostics and prophylactics in a modern way to combat the infection in short time. To safeguard the public health from pathogens of zoonotic infections, application of skills, knowledge and resources of veterinary public health is essential. Further, the control measures for emerging and re-emerging zoonotic pathogens are demanding, as there is population explosion. Novel, highly sensitive and specific techniques comprising genomics and proteomics along with conventional methods would be useful in the identification of these organisms, thereby allowing the application of therapeutic, prophylactic and preventive measures on time. The first line of control measures for any disease is the surveillance. Control and prevention strategies should be designed based on transmission pattern and characteristics of microbes; involvement of carrier, reservoir and incidental hosts; environmental factors; and epidemiology of the disease.

References

1. Arean VM. 1962. The pathologic anatomy and pathogenesis of fatal human leptospirosis (Weil's disease). *Am J Pathol* 40: 393-423.

2. Bahaman AR, Ibrahim A and Adaus H. 1987. Serological prevalence of leptospiral infection in domestic animals in west Malayisa. *Epidermiol Inf* 99(2): 379-392.

3. Balamurugan, V, Gangadhar, NL, Mohandoss, N *et al.* 2013. Characterization of *leptospira* isolates from animals and humans: Phylogenetic analysis identifies the prevalence of intermediate species in India. *Springer Plus* 2: 362.

4. Bolin CA and Koellner P. 1988. Human-to-human transmission of *Leptospira interrogans* by milk. *J Infect Dis* 158: 246-247.

5. Brenner DJ, Kaufmann AF, Sulzer KR, Steigerwalt AG, Rogers FC and Weyant RS. 1999. Further determination of DNA relatedness between serogroups and serovars in the family Leptospiraceae with a proposal for *Leptospira alexanderi* sp. nov. and four new *Leptospira* genomospecies. *Int J Syst Bacteriol* 49 (2): 839-858.

6. Brown PD, Gravekamp C, Carrington DG, Van de Kemp H, Hartskeerl RA, Edwards CN, Everard COR, Terpstra WJ and Levett PN. 1995. Evaluation of the polymerase chain reaction for early diagnosis of leptospirosis. *J Med Microbiol* 43: 110-114.

7. Cerqueira GM and Picardeau M. 2009. A century of *Leptospira* strain typing. *Infect Genet Evol* 9(5): 760-768.

8. Chernukha YG, Shishkina ZS, Baryshev PM and Kokovin IL. 1976. The dynamics of IgM- and IgG-antibodies in leptospiral infection in man. *Zentbl Bakteriol* 236: 336-343.

9. Choy HA, Kelley MM, Chen TL, Moller AK, Matsunaga J and Haake DA. 2007. Physiological osmotic induction of *Leptospira interrogans* adhesion: LigA and LigB bind extracellular matrix proteins and fibrinogen. *Infect Immun* 75(5): 2441-2450.

10. Choy HA, Kelley MM, Croda J, Matsunaga J, Babbitt JT, Picardeau AK and Haake DA. 2011. The multifunctional LigB adhesin binds homeostatic proteins with potential roles in cutaneous infection by pathogenic *Leptospira interrogans*. *PLoS ONE* 6 (2): e16879.

11. Clerke AM, Leuva AC, Joshi C and Trivedi SV. 2002. Clinical profile of leptospirosis in south Gujarat. *J Postgrad Med* 48(2): 117-118.

12. Damude DF, Jones CJ and Myers DM. 1979. A study of leptospirosis among animals in Barbados. *WI Trans Med Hyg* 73(2): 161-168.

13. De la Pena-Moctezuma AD, Bulach M, Kalambaheti T and Adler B. 1999. Comparative analysis of the LPS biosynthetic loci of the geneticsubtypes of serovar Hardjo: *Leptospira interrogans* subtype Hardjoprajitnoand *Leptospira borgpetersenii* subtype Hardjobovis. *FEMS Microbiol Lett* 77: 319-326.

14. Dinger JE. 1932. Duurzaamheid der smetkracht van leptospirenkweeken. *Ned Tijdschr Geneeskd* 72: 1511-1519.

15. Dózsa, L, Kemenes F and Szent-Iványi T. 1960. Susceptibility of the red blood cells of ruminants with a four-chambered stomach (Pecora) to the haemotoxin of pathogenic leptospires. *Acta Vet Acad Sci Hung*. 10: 35-44.

16. Faine S. 1962. Factors affecting the development of the carrier state in leptospirosis. *J Hyg* 60: 427-434.

17. Faine S. 1964. Reticuloendothelial phagocytosis of virulent leptospires. *Am J Vet Res* 25: 830-835.

18. Faine S. 1982. *Guidelines for the control of leptospirosis*. World Health Organization, offset publication, Geneva, Switzerland. 67: 29.

19. Faine S. 1994. *Leptospira and leptospirosis*. CRC Press, Boca Raton, Florida.

20. Faine S. 1998. *Leptospirosis. Topley and Wilson's Microbiology and Microbial infections*. 9th Edition Vol. 3. p. 849-869.

21. Faine S, Adler B, Bolin C and Perolat P. 2000. *Leptospira and leptospirosis*. 2nd edn. Medical Science Press, Melbourne, Australia.

22. Feigin RD, Lobes LA, Anderson D and Pickering L. 1973. Human leptospirosis from immunized dogs. *Ann Intern Med* 79: 777-785.

23. Ferrar WE. 1990. *Leptospira* species (Leptospirosis). In: Mandel GL, Douglas RG and Bennett JE (ed) *Principles and practice of infectious diseases*. Churchill Livingstone, New York. p. 1813-1827.

24. Gangadhar NL. 1999. Rodents and leptospirosis: A global perspective. In: *Proc. 1ˢᵗ Natl Leptospirosis Conf*: Bangalore. p. 11-13.

25. Gangadhar NL and Rajasekhar M. 1998. A modified silver impregnation staining for leptospirosis *Indian Vet Journal* 7: 349-351.

26. Gangadhar NL, Rajasekhar M, Smythe LD, Norris MA, Symonds ML and Dohnt MF. 2000. Reservoir hosts of *Leptospira inadai* in India. *Rev Sci Tech* 19 (3): 793-799.

27. Gangadhar NL, Prabhudas K, Gajendragad MR, Shashibhushan J and Kakoli Ahmed. 2006. Leptospirosis: An enigma of zoonosis for the developing world. *Infect Dis J Pak* 15(10): 20-24.

28. Gangadhar NL, Prabhudas K, Bhushan S, Sulthana M, Barbuddhe SB and Rahman H. 2008. Leptospira infection in animals and humans: a potential public health risk in India. *Rev Sci Tech* 27(3): 885-892.

29. Golubev MV and Litvin Vlu. 1983. Population ecology of leptospires, an experience with evaluation of leptospira count in a carrier and intensity of elimination in the urine. *Zh Microbiol Epidemiol Immunobiol* 6: 60-63.

30. Gravekamp C, van de Kemp H, Franzen M, Carrington D, Schoone GJ, van Eys GJJM, Everard COR, Hartskeerl RA and Terpstra WJ. 1993. Detection of seven species of pathogenic leptospires by PCR using two sets of primers. *J Gen Microbiol* 139: 1691-1700.

31. Guidugli F, Castro AA and Atallah AN. 2000. *Antibiotics for treating leptospirosis* (Cochrane Review), Cochrane Library, Issue 2. Update Software, Oxford, UK.

32. Hartskeerl RA, Goris MG, Brem S, Meyer P, Kopp H, Gerhards H and Wollanke B. 2004. Classification of leptospira from the eyes of horses suffering from recurrent uveitis. *J Vet Med B Infect Dis Vet Public Health* 51(3): 110-115.

33. Herrmann JL. 1993. Genomic techniques for identification of *Leptospira* strains. *Pathol Biol* 41: 943-950.

34. Hospenthal DR and Murray CK. 2003. *In vitro* susceptibilities of seven leptospira species to traditional and newer antibiotics. *Antimicrob Agents Chemothero* 2003: 47: 2646-2648.

35. Inada R, Ido Y, Hoki R, Kaneko R and Ito H. 1916. The etiology, mode of infection, and specific therapy of Weil's disease (spirochaetosis icterohaemorrhagica). *J Exp Med* 23(3): 377-402.

36. Isogai E, Kitagawa H, Isogai H, Matsuzawa T, Shimizu T, Yanagihara Y and Katami K. 1989. Effects of leptospiral lipopolysaccharide on rabbit platelets. *Zentralbl Bakteriol* 271: 186-196.

37. Johnson RC and Faine S. 1984. Leptospira. In: Krieg NR and Holt JG (ed.). *Bergey's manual of systematic bacteriology*, Vol. 1. Williams and Wilkins, Baltimore, Md. p. 62-67.

38. Johnson RC and Harris VG. 1967. Differentiation of pathogenic and saprophytic leptospires. 1. Growth at low temperatures. *J Bacteriol* 94: 27-31.

39. Johnson RC and Rogers P. 1964a. 5-Fluorouracil as a selective agent for growth of leptospirae. *J Bacteriol* 87: 422-426.

40. Johnson RC and Rogers P. 1964b. Differentiation of pathogenic and saprophytic leptospires with 8-Azaguanine. *J Bacteriol* 88: 1618-1623.

41. Karande S, Bhatt M, Kelkar A, Kulkarni M, De A and Varaiya A. 2003. An observational study to detect leptospirosis in Mumbai, India, 2000. *Arch Dis Child* 88(12): 1070-1075.

42. Karande S, Gandhi D, Kulkarni M, Bharadwaj R, Pol S, Thakare J and De A. 2005. Concurrent outbreak of leptospirosis and dengue in Mumbai, India, 2002. *J Trop Pediatr* 51(3): 174-181.

43. Kasarov LB. 1970. Degradation of the erythrocyte phospholipids and haemolysis of the erythrocytes of different animal species by Leptospirae. *J Med Microbiol* 3: 29-37.

44. Katz AR, Ansdell VE, Effler PV, Middleton CR and Sasaki DM. 2002. Leptospirosis in Hawaii, 1974-1998: Epidemiologic analysis of 353 laboratory confirmed cases. *Am J Trop Med Hyg* 66(1): 61-70.

45. Kmety E and Dikken H. 1993. *Classification of the species Leptospira interrogans and history of its serovars*. University Press, Groningen, Netherlands.

46. Kmety E, Plesko I, Bakoss P and Chorvath B. 1966. Evaluation of methods for differentiating pathogenic and saprophytic leptospira strains. *Ann Soc Belges Med Trop Parasitol Mycol* 46(1): 111-122.

47. Koutis CH. 2007. *Special epidemiology editions*, Technological Educational Institute of Athens, Athens, Greece.

48. Kreig NR and Holt JG. 1984. The spirochetes. In: *Bergey's Manual of Systematic Bacteriology*. Williams and Wilkins, London. p. 39-70.

49. Landouzy LTJ. 1883. Fievre bilieuse ou hepatique. *Gaz. Hopital* 56: 809.

50. Levett PN. 2001. Leptospirosis. *Clin Microbial Rev* 14: 296-326.

51. Levett PN, Morey RE, Galloway RL, Turner DE, Steigerwalt AG and Mayer LW. 2005. Detection of pathogenic leptospires by real-time quantitative PCR. *J Med Microbiol* 54: 45-49.

52. Lucchesi PM and Parma AE. 1999. A DNA fragment of *Leptospira interrogans* encodes a protein which shares epitopes with equine cornea. *Vet Immunol Immunopathol* 71: 173-179.

53. Mathur M, De A and Turbadkar D. 2009. Leptospirosis outbreak in 2005: LTMG hospital experience. *Indian J Med Microbiol* 27(2): 153-155.

54. Matthias MA and Levett PN. 2002. Leptospiral carriage by mice and mongooses on the island of Barbados. *West Indian Med J* 51: 10-13.

55. Matthias MA, Ricaldi JN, Cespedes M, Diaz MM, Galloway RL, Saito M, Steigerwalt AG, Patra KP, Ore CV, Gotuzzo E, Gilman RH, Levett PN and Vinetz JM. 2008. Human leptospirosis caused by a new, antigenically unique *Leptospira*

associated with a Rattus species reservoir in the Peruvian Amazon. *PLoS Negl Trop Dis* 2; 2(4): e213.

56. *Merck Veterinary Manual*. 2012. www.merckvetmanual.com.

57. Merien F, Amouriauz P, Perolat P, Baranton G and Saint Girons I. 1992. Polymerase chain reaction for detection of *Leptospira* spp. in clinical samples. *J Clin Microbiol* 30: 2219-2224.

58. Moch RW, Ebner EE, Barsoum LS and Bortros BA. 1975. Leptospirosis in Ethopia. A serological survey in domestic animal and wild animals. *J Trop Med Hyg* 78(2): 38-42

59. Myers DM and Coltorti EA. 1978. Broadly reacting precipitating and agglutinating antigen of leptospirae. *J Clin Microbiol* 8: 580-590.

60. Natarajaseenivasan K and Ratnam S. 1997. Seroprevelence of leptospiral infection in an agriculture based village in Tamil Nadu. *Cherion* 26(5-6): 80-83.

61. OIE. 2005. Leptospirosis. In: Manual of diagnostic tests and vaccines for terrestrial animals. *OIE terrestrial manual*.

62. OIE. 2008. Chapter 2.1.9. Leptospirosis. In: Manual of diagnostic tests and vaccines for terrestrial animals. *OIE terrestrial manual*.

63. Palaniappan R, Ramanujam S and Chang YF. 2007. Leptospirosis: pathogenesis, immunity and diagnosis. *Curr Opin Infect Dis* 20: 284-292.

64. Pappas G, Papadimitriou P, Siozopoulou V, Christou L and Akritidis N. 2008. The globalization of leptospirosis: worldwide incidence trends. *Int J Infect Dis* 12(4): 351-357.

65. Park SK, Lee SH, Rhee YK, Kang SK, Kim KJ, Kim MC, Kim KW and Chang WH. 1989. Leptospirosis in Chonbuk Province of Korea in 1987: a study of 93 patients. *Am J Trop Med Hyg* 41(3): 345-351.

66. Prescott JF, Miller RB, Nicolson VM, Martin SW and Lesnick T. 1998. Seroprevelence and association with the abortion of Leptospirosis in cattle in Ontario. *Can J Vet Rev* 52(2): 210-215.

67. Rao RS, Gupta N, Bhalla P and Agarwal SK. 2003. Leptospirosis: A review. Bra *J Infect Dis BJID* 7: 178-193.

68. Ratnam S. Sundararaj T and Subramanian S. 1983. Serological evidence of leptospirosis in a human population following an outbreak of the disease in cattle. *Trans R Soc Trop Med Hyg* 77(1): 94-98.

69. Sanders EJ, Rigau-Perez JG, Smits HL, Deseda CC, VorndamVA, Aye T, Spiegel RA, Weyant RS and Bragg SL. 1999. Increase of leptospirosis in dengue-negative patients after a hurricane in Puerto Rico in 1996. *Am J Trop Med Hyg* 61: 399-404.

70. Schoenberg AN, Staak C and Kampe U. 1987. Leptospirosis in the German Federal Republic. Results from a programme of investigation into leptospirosis in animals in 1984. *J Vet Med* 34(2): 98-108.

71. Sehgal SC. 2006. Epidemiological patterns of leptospirosis. *Indian J Med Microbiol* 24(4): 310-311.

72. Sehgal SC, Murhekar MV, Sugunan AP. 1995. Outbreak of leptospirosis with pulmonary involvement in North Andaman. *Indian J Med Res* 102: 9-12.

73. Sehgal SC, Sugunan AP and Vijayachari P. 2001. Outbreak of leptospirosis after cyclone in Orissa. *National Med J India* 15 (1): 22-23.

74. Smythe LD. 1999. Leptospirosis worldwide, 1999. *Wkly Epidemiol Rec* 74: 237-242.

75. Smythe L, Dohnt M, Symonds M, Barnett L, Moore M, Brookes D and Vallanjon M. 2000. Review of leptospirosis notifications in Queensland and Australia: January 1998–June 1999. *Commun Dis Intell* 24: 153-157.

76. Sohan L, Shyamal B, Kumar TS, Malini M, Ravi K, Venkatesh V, Veena M and Lal S. 2008. Studies on leptospirosis outbreaks in Peddamandem Mandal of Chittoor district, Andhra Pradesh. *J Commun Dis* 40(2): 127-132.

77. Stimson AM. 1907. Note on an organism found in yellow-fever tissue. *Public Health Rep* 22: 541.

78. Takafuji, ET, Kirkpatrick JW, Miller RN, Karwacki JJ, Kelley PW, Gray MR, McNeill KM, Timboe HL, Kane RE and Sanchez JL. 1984. An efficacy trial of doxycycline chemoprophylaxis against leptospirosis. *N Engl J Med* 310: 497-500.

79. Tangkanakul W, Smits HL, Jatanasen S and Ashford DA. 2005. Leptospirosis: an emerging health problem in Thailand. *Southeast Asian J Trop Med Public Health* 36(2): 281-288.

80. Taylor J and Goyle AN. 1931. *Leptosirosis in Andamans*. Indian Med Res Memoirs 20. Thacker, Spink and Co., Calcutta, India.

81. Thompson JC. 1986. Morphological changes in red blood cells of calves caused by *Leptospira interrogans* serovar *pomona*. *J Comp Pathol* 96: 512-527.

82. Turner LH. 1970. Leptospirosis III. Maintenance, isolation and demonstration of leptospires. *Trans R Soc Trop Med Hyg* 64: 623-646.

83. Uhlenhuth P and Fromme W. 1915. Experimentelle untersuchungen u¨ber die sogenannte Weilsche Krankheit (ansteckende Gelbsucht). *Med Klin* 44: 1202-1203.

84. Venkataraman KS and Nedunchelliyan S. 1992. Epidemiology of an outbreak of leptospirosis in man and dog. *Com Imm Microbiol and Infect Dis* 15(4): 243-247.

85. Vijayachari P and Sehgal SC. 2006. Recent advances in the laboratory diagnosis of leptospirosis and characterisation of leptospires. *Indian J Med Microbiol* 24(4): 310-311.

86. Vijayachari P, Sugunan AP, Umapathi T and Sehgal SC. 2001a. Evaluation of darkground microscopy as a rapid diagnostic procedure in leptospirosis. *Indian J Med Res* 114: 54-58.

87. Vijayachari P, Sugunan AP and Sehgal SC. 2001b. Evaluation of microscopic agglutination test as a diagnostic tool during acute stage of leptospirosis in high and low endemic areas. *Indian J Med Res* 114: 99-106.

88. Vijayachari P, Sugunan AP and Shriram AN. 2008. Leptospirosis: an emerging global public health problem. *J Biosci Rev* 33(4): 557-569.

89. Weil A. 1886. Ueber eine eigentu¨mliche, mit Milztumor, Icterus und Nephritis einhergehende akute Infektionskrankheit. Dtsche. *Arch Klin Med* 39: 209-232.

90. Wesley FR. 1995. Leptospirosis. *Clin Infect Dis* 21: 1-8.

91. WHO. 2000. Leptospirosis, India: Report of the investigation of a post-cyclone outbreak in Orissa, November 1999. *Wkly Epidemiol Rep* 75: 217-223.

92. WHO. 2003. *Human Leptospirosis: Guidance for diagnosis, surveillance, and control*. World Health Organization, Malta. p. 1-122.

93. WHO. 2007. *Leptospirosis Laboratory Manual*, RMRC, Port Blair and World Health Organization (country office for India). p. 1-69.

94. Wolbach SB and Binger CA. 1914. Notes on a filterable Spirochete from fresh water. *Spirocheta biflexa* (new Species). *J Med Res* 30(1): 23-26.

95. Yasuda PH, Steigerwalt AG, Sulzer KR, Kaufmann AF, Rogers F and Brenner DJ. 1987. Deoxyribonucleic acid relatedness between serogroups and serovars in the family Leptospiraceae with proposals for seven new *Leptospira* species. *Int J Syst Bacteriol* 37: 407-415.

96. Zavitsanou A and Babatsikou F. 2008. Leptospirosis: epidemiology and preventive measures. *Health Sci J* 2 (2): 75-82.

97. Zuerner RL. 1991. Physical map of the chromosomal and plasmid DNA comprising the genome of *Leptospira interrogans*. *Nucleic Acids Res* 19: 4857-4860.

2014, Zoonoses: Bacterial Diseases   Pages **120–142**
Editor: **Sudhi Ranjan Garg**
Published by: **DAYA PUBLISHING HOUSE, NEW DELHI**

6

Plague

R.K. Vaid and B.C. Bera
Veterinary Type Culture Collection,
National Research Centre on Equines, Hisar – 125 001

Plague is one of the oldest recorded infectious diseases the world over, which had a profound effect on political history, art and religion of human race (Simpson *et al.* 1903, 1905, Bray 1996). It is an exceptionally virulent, vector-borne metazoonosis transmitted from rodents, especially rats, through the bites of infected fleas. Plague, an anthropozoonosis with cosmopolitan distribution all over the planet, exhibits an impressive ability to overcome mammalian host defenses.

The causative agent of plague, *Yersinia pestis*, presents itself in septicaemic, bubonic and pneumonic forms. The rat flea, *Xenopsylla cheopsis*, is most often implicated in transmission of the infection. Pharyngeal/enteric plague can arise as a result of exposure to infectious aerosols or by ingestion of infected meat (Christie *et al.* 1980). Many different species of mammals, including rats, squirrels, mice, prairie dogs and gerbils are considered to be animal reservoirs for the agent of plague worldwide. *Y. pestis* persists in the environment as the result of a stable and constant rodent-flea infection cycle. The organism can spread to the circulatory system and cause a septicaemic form, thereby afflicting various organs via hematogenous route. The infected patients often develop necrosis in cutaneous blood vessels, as a result of which skin colour turns blackish. This led to the origin of common name "the black death" for the disease. Primary pneumonic plague occurs on inhalation of infected organism causing a primary infection in lungs, which is highly transmissible via aerosols. Bubonic plague is caused as a result of an infected flea bite. The bacteria spread to the nearest lymph node and are ingested by macrophages. Bacterial multiplication results in swelling of the inguinal and other lymph nodes forming often hot, red and painful buboes, giving rise to the name bubonic plague.

The first historically recorded major epidemic of plague occurred in China in 224 BC. Plague infection is believed to have caused greater than 150 epidemics, however, three major pandemics with a time gap of about half a century each stand out prominently in history (Butler and Dennis 2005). In the last 2000 years, there have been three separate pandemics of plague that have led to decimations of large human populations: the Justinian plague between the 5th and 7th centuries, the black death in Europe between the 13th and 15th centuries, and the modern plague from the latter half of the 1800s to the present. A figure of 200 million has been suggested as a credible number for the plague death toll throughout the recorded history (Duplaix 1988). The first Justinian episode began in Egypt which subsequently spread to Middle East and to Europe in 6th century. The Black Death episode began in the area of Black Sea and spread to Europe, resulting in death of more than a quarter of population in Europe. The modern pandemic is considered to have originated in Yunnan, China in 1855 which then spread to India, where it resulted in death of more than 10 million people. It also spread to Egypt and eastern Africa, Portugal and Scotland in Europe, Paraguay in Latin America, and to as far as the Philippines (Manila), Australia (Sydney) and United States (San Francisco) via steamships. The outbreaks in the form of fatal infections are continuing in the modern age (Mittal *et al.* 2004, Yin *et al.* 2007).

The occurrence of two plague epidemics in western parts of India in 1994 in which close to 600,000 residents deserted the bustling textile town of Surat demonstrates that plague remains one of the most dreaded infectious diseases. The third pandemic spread to India via sea route, or through *fakirs/sadhus* coming from Garhwal (Hankin 1905) to Mumbai in 1898, where by 1903, it was killing a million people per year. A total of 12.5 million Indians are estimated to have died of plague between 1898 and 1918. The death and spread of sporadic plague outbreaks has greatly reduced in 21st century due to the advent of effective public health measures and since about 1950, antibiotics. However, the 3rd pandemic, which is on the decline, is believed to have established stable enzootic foci on every major inhabited continent except Australia (Barnes and Quan 1992, WHO 1995). In spite of improved sanitation and public health surveillance, plague is still a significant health problem in Africa, Asia and South America, with 2000 cases every year and a global case fatality rate of 5 to 15 per cent. A multi-drug resistant strain of *Y. pestis* has also been reported from Madagascar (Galimand *et al.* 1997).

Plague bacilli are considered Category "A" critical agents with significant threat to public health and national security due to their easy dissemination and person-to-person transmission (CDC 2000). The potential of *Y. pestis* as a bioweapon is a huge threat to human health (Ingelsby *et al.* 2000). A WHO model predicts that release of 50 kg of *Y. pestis* over a city of 5 million may result in 500,000 cases with 100,000 deaths directly and indirectly (WHO 1970). Heavy financial cost of emergency response systems implemented internationally during the Surat plague epidemic and the resulting losses to the country's tourism and other industries underline the panic quotient of plague.

Geographical Distribution

Despite the worldwide decline in the incidence of plague, the number of countries affected by plague remains substantial (Figure 6.1). During 1954-1997, human plague was recorded in 38 countries, 7 of which (Brazil, Democratic Republic of Congo, Madagascar, Myanmar, Peru, United States, Vietnam) were affected virtually every year. In the remaining countries, outbreaks of human plague occurred during the years when plague resurged globally. The reasons for this apparent worldwide cycle are not fully understood. The disease is endemic in Africa, India, China, Mongolia, Myanmar, Vietnam, Indonesia, South Africa, USA, and South America. Isolated outbreaks continue to this day in many regions of the world (Perry and Fetherston 1997, Prentice and Rahalison 2007). Enzootic Northern American focus is the largest in the world, however, the former USSR has the second largest enzootic plague area in the world with 10 separate foci. Australia and Europe have not reported cases after the World War II (Poland and Barnes 1979, Velimirovic 1990). In Africa, mainly the Democratic Republic of Congo and Madagascar account for 96 per cent of the world cases since 1990.

Brazil, Peru and the United States notified the disease in humans nearly every year in the 18 year period (1980-1997). In North America, natural foci of plague occur in 15 western states of the United States, in south-western Canada on the border with the United States, and in northern Mexico. In South America, foci have been recorded in Andean Alpine region and Brazil, with the Andean foci represented by Bolivia, Ecuador and Peru. Plague is also reported from Argentina and Venezuela (Dennis *et al.* 1999).

In India, large plague epidemics occurred during the first half of this century and resulted in millions of deaths. However, until the recent epidemic in Beed (Karnataka), which is a known endemic region, the last laboratory-confirmed cases

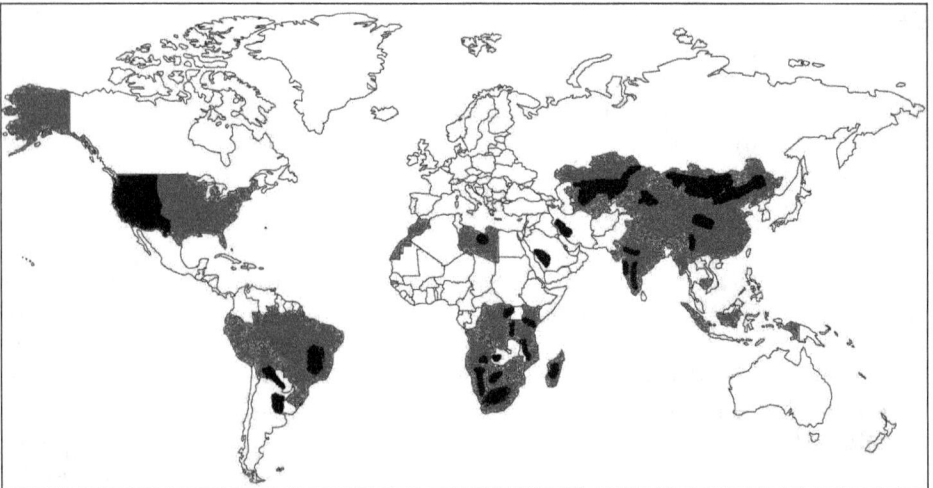

Figure 6.1: Distribution of plague foci (in dark shade) and plague endemic areas (in grey shades) in the world (Illustration by Saurabh Vaid)

of human plague in the country were reported in 1966 (Sant *et al.* 1972, Akiev 1982, WHO 1990, 1994). In recent times, plague outbreak of primary pneumonic form took place in Shimla district in the state of Himachal Pradesh in February 2002. Sixteen cases of plague were reported with a 25 per cent case fatality rate (4/16). The infection was confirmed to the molecular level with PCR and gene sequencing (Anonymous 2002). A previous outbreak in this region during 1983 was suggestive of pneumonic plague (22 cases, 17 deaths) but was not confirmed. A localized outbreak of bubonic plague occurred in village Dangud (population 332) in district Uttarkashi in the second week of October 2004 (Mittal *et al.* 2004). Modern distribution of plague foci and disease outbreak locations in India are depicted in Figure 6.2.

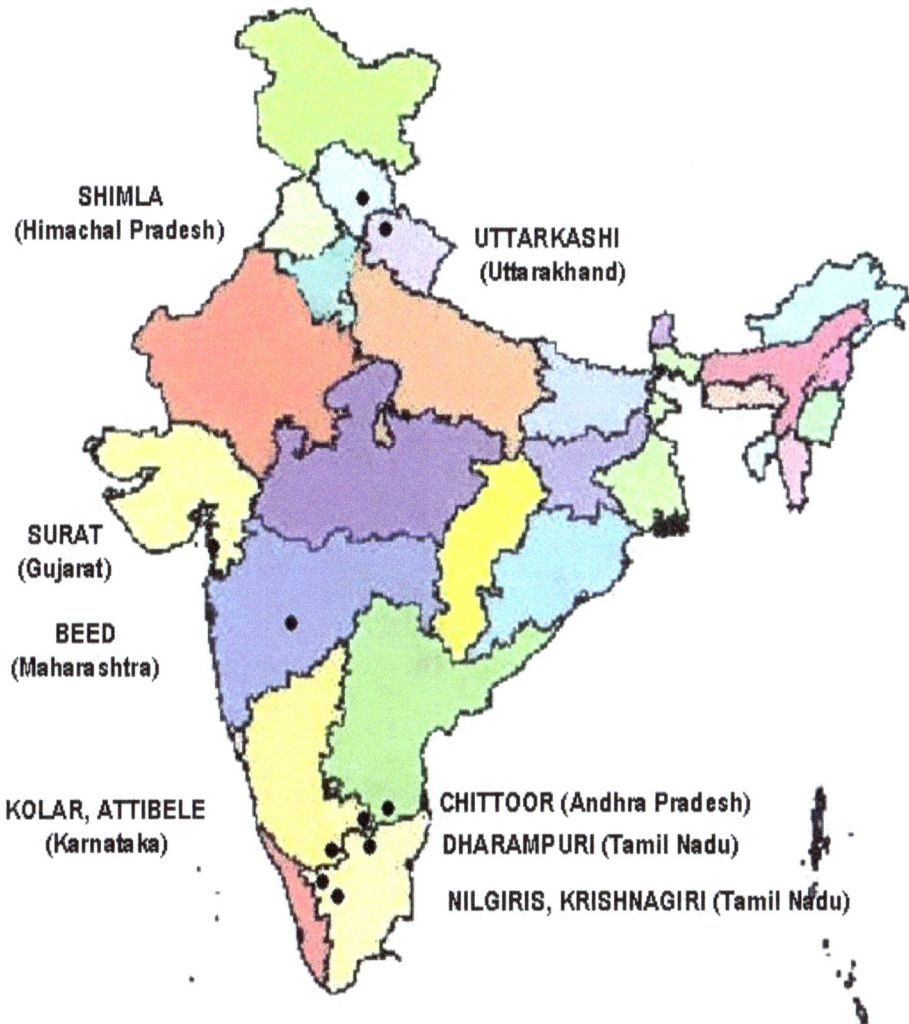

Figure 6.2: Distribution of plague in India

In China, there are ten geographical foci of plague (Xu 1997). According to a report, there are 12 natural plague foci, covering greater than 291 counties in 19 provinces (Ji *et al.* 1990). The Yunnan Province includes two plague foci, the *Rattus flavipectus* plague focus and the *Apodemus chevrieri* and *Eothenomys miletus* plague focus, respectively. *Yersinia pestis* isolated from these foci belong to biovar *Orientalis* and *Antiqua*, respectively (Zhang and Liu 2007).

In Africa, plague foci are in South Africa, Lesotho, Namibia, Uganda and Zimbabwe. Although the number of plague outbreaks in this sub-region has declined considerably in recent years, the infection persists in many areas where human plague has not been apparent for years. Severe outbreaks have occurred in recent years in Kenya, Tanzania, Zaire, Mozambique and Botswana, with smaller outbreaks in other east African countries. Plague also has been reported in scattered foci in western and northern Africa.

Etiological Agent

The genus *Yersinia of* family Enterobactericeae contains 11 species, of which 3 are human pathogens: *Y. pestis*, *Y. pseudotuberculosis*, and *Y. enterocolitica* (Perry and Fetherston 1997). Plague is caused by *Y. pestis* (Yersin 1894). Other species in the genus include *Y. frederiksenii*, *Y. kristensenii*, *Y. bercovieri*, *Y. aldovae*, *Y. molleretti*, *Y. intermedia*, *Y. rohdei* (Winn *et al.* 2006).

Yersinia pestis is a non-motile, Gram-negative rod, occasionally a cocco-bacilli with a bipolar (safety-pin like) staining with Wright, Giemsa or Wayson stain (Figure 6.3) (Aleksic and Bockemuhl 1999). It is a lactose non-fermenter, urease and indole

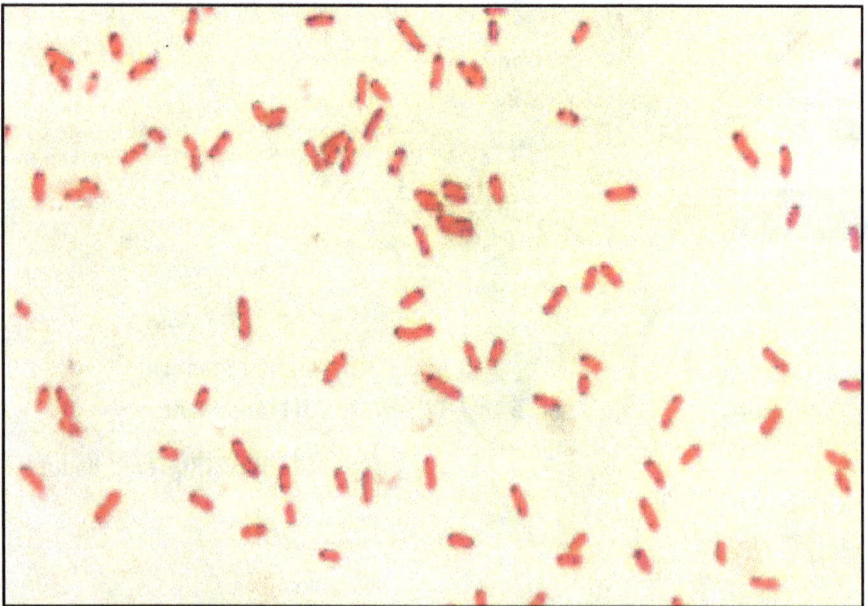

**Figure 6.3: Plague bacilli *Yersinia pestis* showing bipolarity
(Photo courtesy: CDC, Larry Stauffer)**

negative bacterium, which grows aerobically at 28°C on blood agar or MacConkey agar, typically requiring 48 hours for observable growth. The colonies at 24 hours are smaller than other members of Enterobacteriaceae and may be missed. *Yersinia* easily grows on nonselective media such as tryptic soy agar (Figure 6.4), fortified with sheep blood and on Mueller Hinton Agar. The colonies are small (1-2 mm) and like other members of family, these are grey in colour on blood agar. The colonies look non-mucoid and have a characteristic hammered copper appearance (Perry and Fetherston 1997).

Initially called as *Pasteurella*, these microorganisms were named *Yersinia* in 1894 after the French bacteriologist, Alexandre Emile Jean Yersin. As with other classical pathogens, *Y. pestis* has undergone many changes in its nomenclature from initial *Bacterium pestis* to *Bacillus pestis* until 1923, and *Pasteurella pestis* after Louis Pasteur, who was Yersin's mentor at Pasteur Institute, and, finally *Y. pestis* in 1970 (Butler 1983). Comparative genomics and DNA microarray analyses indicate that *Y. pestis* is a clonal variant of *Y. pseudotuberculosis* that diverged only within the last 1,500 to 20,000 years (Achtman *et al.* 1999, Hinchcliffe *et al.* 2003), however, the reclassification proposal of *Y. pestis* as a subspecies of *Y. pseudotuberculosis* has not been implemented due to plague's long history and laboratory and public health safety considerations associated with the disease (Bercovier *et al.* 1980, Wayne 1986).

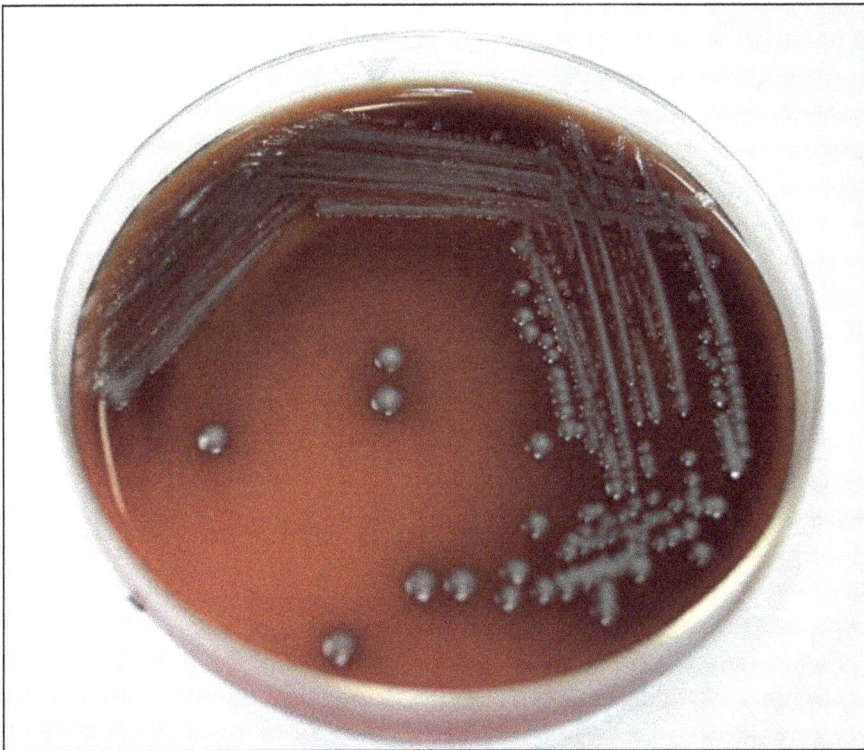

Figure 6.4: *Yersinia pestis* colonies on blood agar
(Photo courtesy: CDC, Megan Mathias, J. Todd Parker)

Based on the ability of *Y. pestis* strains to ferment glycerol and reduce nitrate, these are divided into three biovars: Antigua, Mediavalis and Orientalis. Biovar Antigua strains can ferment glycerol and reduce nitrate; biovar Mediavalis strains can only ferment glycerol but cannot reduce nitrate; and biovar Orientalis strains cannot ferment glycerol but can reduce nitrate. Moreover, each biovar has been linked to one of the three pandemics. Biovar Antigua with the Justinian plague; biovar Mediavalis with the black death; and biovar Orientalis, which is widespread, is associated with modern plague (Devignat 1951). Now-a-days, biovar Antigua strains are isolated in Africa, and might be descended from the bacteria that caused the first pandemic. Many researchers now favour the inclusion of a fourth biovar, Pestoides (Anisimov *et al.* 2004).

The clonal divergence of *Y. pestis* from *Y. pseudotubeculosis* took the former to form a niche which has few equivalents in Kingdom Prokarya. This niche is the gut of an arthropod, called oriental rat flea, *X. cheopis*, which is considered to be the classical vector of plague. The species is thought to have originated in Egypt, but during the 19th century spread to all parts of the world.

Vectors

The understanding of the epidemiology and transmission of plague infection from rodent reservoirs to human hosts depends essentially on determination of the flea species involved in plague transmission in a given area. Most human plague is

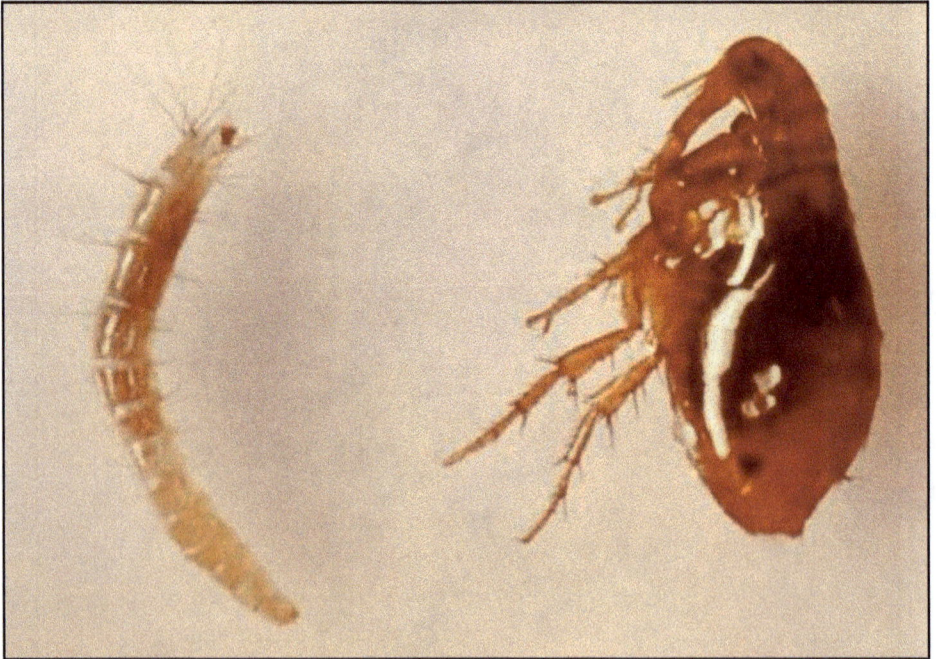

Figure 6.5: *Xenopsylla cheopis* **larva (left) and an adult female (right)**
(Photo courtesy: CDC/WHO)

the bubonic form, transmitted by the bites of infected fleas. *Y. pestis* appears to be inherently resistant to the flea immune response (Hinnebusch *et al.* 1996). There are an estimated 2,500 species and subspecies of fleas that constitute 220 genera and 15 families in the insect order Siphonaptera (Lewis 1998). Of these, approximately 80 species associated with some 200 species of wild rodents have been found to be infected with *Y. pestis* in nature, or to be susceptible to experimental infection (Pollitzer 1954). About 12 species of flea with a worldwide distribution are implicated in the transmission of domiciliary plague (Gratz and Brown 1983). However, many more species of the fleas of the order *Siphonaptera* have been implicated in the transmission of sylvatic plague (Gratz 1980). Dozens of flea species in western North America have been found to be naturally infected with plague (Hubbard 1968, Gage and Kosoy 2004).

In India, as of 1973, 76 species of fleas have been recorded (Iyengar 1973). However, the role of *X. astia*, *X. brasielensis* and *X. cheopis* is considered to be cardinal in plague transmission in the country. The village area of Mamla in Beed experienced a heavy nuisance of flea activity during August 1994 (Saxena and Verghese 1996). The most important rat flea vector of *Y. pestis* in urban or domestic situations (found on wild rodents) is *X. cheopis* (Figure 6.5), while *X. astia* predominates on wild rodents. *X. brasiliensis* is also frequently found on rodents.

Nosopsyllus fasciatus has also been found infected with *Y. pestis*. During a small outbreak of plague in Nepal, *Pulex irritans* was reported to be the vector (Laforce *et al.* 1971). *Xenopsylla astia* is a parasite of both gerbils and rats. It ranges from the Arabian peninsula through Iran to southeast Asia and to Korea (Hass and Walton 1973) and has been found on the east coast of Africa. It is a less efficient vector than *X. cheopis*. *Xenopsylla brasiliensis* is native to all Africa, south of the Sahara, where it is the most common vector in some areas (Kilonzo 1985). *Nosopsyllus fasciatus*, the northern rat flea, is one of the most prevalent fleas in Europe on commensal rats (Dobec and Hrabar 1990). Its distribution is virtually global, and had been encountered from United States to China (Ye 1983).

Host Range

Plague is primarily a disease of rodents but humans become accidental victims due to flea bite or through contact with an infected host, whether animal or human. *Yersinia pestis* microorganisms are harboured mainly by rodent animal species such as rats, mice, rabbits, prairie dogs and ground squirrels that are natural reservoirs of plague. The animal hosts of plague are classified as enzootic (maintenance) hosts and epizootic (amplification) hosts (Gordon and Knies 1947). The infectious agent is believed to be maintained by one or more enzootic or maintenance host species like wild rodents that are partially resistant to infection (Gage and Kosoy 2004). The disease is possibly maintained in nature through transmission between rodents by their flea ectoparasites, temporally characterized by enzootic periods, which are broken by epizootic events when an amplifying host becomes infected resulting in rapid long distance spread of plague (Girard *et al.* 2004, Webb *et al.* 2006). For the most part, sylvatic rodent reservoirs are the species that are susceptible to infection but resistant to the disease. This metazoonoses has animal reservoirs on nearly every major continent except Australia, and the etiological agent exhibits major virulence capability in overcoming not only the host defenses of arthropods alone, but also of the mammalian hosts. Over 200 species of mammals have been reported to be able to become infected with *Y. pestis* (Oyston and Williamson 2011). At least 220 species of rodents are known to be infected with plague bacillus. Among the commensal rodents, *Rattus rattus* and *Mus musculas* are hosts for *Y. pestis*.

In Indian subcontinent, a large number of rodent species are known that include some 46 genera, 135 species and many subspecies. Many species of rodents have been reported as actual or potential reservoirs of plague. Depending on the region, more important species are *Bandicota bengalensis, Tatera indica, Rattus norvegicus, R. rattus* and *R. rattus diardii*. In India, wild rodent *Tatera indica* has been incriminated as the main reservoir. In the three decades (1960s to 1989) of serosurveillance in India, 188,025 rodent sera were examined, out of which only 12 from *Tatera indica* were found positive for *Y. pestis* antibodies. Only two *R. rattus* sera were reported serologically positive for *Y. pestis* in 1988. The gerbil *Tatera indica*, the Indian field mouse *Mus budooga,* and the squirrels *Funambulus pennanti* and *F. palmarum* have all been found positive for plague in various foci. The population densities of rats that have been shown important reservoirs of plague including *B. bangalensis, R. norvegicus* and *R. rattus* are high in most urban areas of the country (Krishnaswami *et al.* 1970, Renapurkar and Sant 1974, Renapurkar 1981, Prakash 1988, Renapurkar 1989).

Carnivores such as domestic dogs, ferrets, polecats, black bears, badgers, raccoons, skunks and coyotes are reported to be very resistant to plague; therefore, such carnivores in general have been suggested as sentinel animals for predicting plague activity (Salkeld and Stapp 2006, Bei *et al.* 2008). Serosurveillance of dogs can give an idea about the infectious origin and nidus. Dogs present with nonspecific fever and lethargy do not directly transmit plague to humans (Orloski and Eidson 1995), however, rodent fleas living on dogs can transmit the infection (Poland and Barnes 1979). Proximity of dogs to humans like sleeping on the same bed in enzootic areas is considered a significant risk factor (Hannah Gould *et al.* 2008). Compared to dogs, domestic cats become acutely ill and generally develop buboes and pneumonic lesions similar to those seen in humans (Rust *et al.* 1971). A high rate of over 33 per cent mortality has been reported in cats while 44 per cent cats became ill but recovered (Gasper *et al.* 1993, Watson *et al.* 2001). Domestic cats can also be infected with *Y. pestis* through ingestion of plague infected rodents or through the bite of infected fleas (CDC 1981, Gage *et al.* 2000). Human plague cases through contact with infected domestic cats have been reported (Gage *et al.* 2000). Serologic survey of cats would thus help determine the infected origin and range in case of an epidemic (Bei *et al.* 2008). *Ctenocephalides felis*, the cat flea, has become completely cosmopolitan in its distribution and is frequently found on other hosts too, including dogs, humans, other mammals and birds (Hallet *et al.* 1970). Both the cat flea *Ctenocephalides felis* and the dog flea *Ctenocephalides canis* are able to transmit plague to humans from pet animals. It is noteworthy that a recent occupational hunter-linked plague outbreak in India is suspected to have occurred due to hunting and skinning of a wild cat by the index case patient who later succumbed to infection after transmitting plague to his community members (Gupta and Sharma 2007).

In the United States, many rodent species are considered reservoir hosts and a single particular species of rodents has not been implicated. *Mastomys natalensis* species complex, one of Africa's most prevalent wild rodents, plays important role in the natural cycle of plague in Africa (Isaacson *et al.* 1983). In Russian grassland steppes, the reservoir species is believed to be principally the marmot, the large ground squirrel group, which has 11 species. These are found all over the world from Alps, Pyrenees to Rockies, and in India, in Ladakh region. The great gerbil *Rhombomys opimus* is a major host of *Y. pestis* in the central Asian deserts (Pollitzer 1966, Gage and Kosoy 2004).

Disease Transmission

The ecology of plague is extremely complex, involving many different rodent-flea cycles. The transmission of disease agent to humans is usually through rodent fleas which harbour *Y. pestis*. Majority of the flea species, important in plague transmission in an epidemic, are ectoparasites of commensal or peridomestic rodents, which are often found on livestock and household animals owing to their close proximity to humans and their habitations. Rodents of various types may be sylvatic or wild in nature or these may be urban or peri-urban by habitat (Figure 6.6). Environment or ecology plays an important role in the event of disease in different settings (Gottfried 1983). These settings can be divided into wild rodent cycle or sylvatic plague and domestic cycle or urban plague. Plague is transmitted between

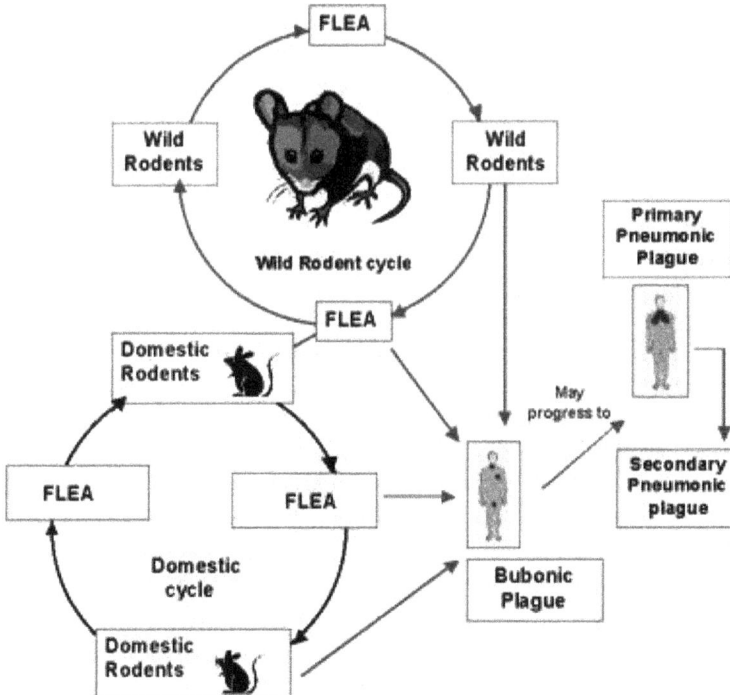

Figure 6.6: Life cycle of plague organism

rodents and to other animals via wild rodent fleas, cannibalism or (possibly) contaminated soil. Wild plague exists in its natural foci independent of human populations and their activity. Domestic plague is intimately associated with rodents living with humans and can produce epidemics in both human and animal populations.

While enzootic hosts serve to maintain plague, epizootic hosts amplify and may allow the spread of disease into new areas, where new enzootic hosts may become infected. Epizootic rodents are associated with high to moderate susceptibility and uniformly high mortality. The disease is mainly restricted to wild rodents, in which the etiological agent is maintained and transmitted among rodents and their colonies by flea-rodent cycle through flea bite. Fleas acquire *Y. pestis* from an infected blood meal. Infection in the flea is only restricted to the alimentary canal and other parts like salivary glands, reproductive system, and the hemocoele do not get affected. The etiological agent is also not transmitted transovarially, and artificially infected larvae have been shown to clear the organism in a day. Consequently, maintenance of plague in nature is absolutely dependent upon cyclic transmission between fleas and rodents and other mammals (Bibikova 1977, Hoogstraal 1980).

Occasionally, the organisms can be spread to humans through direct contact with animals. These fleas have wide distribution and varying roles as vectors of plague; however, they all readily feed on humans (Brooks *et al.* 1978). As an efficient plague vector, flea must be able to ingest the plague bacilli with its blood meal. Further, it must live long enough for the pathogen to multiply in sufficient numbers in order to transfer the pathogen to an animal or human host in sufficient concentrations to cause an infection. In addition to it, fleas must be present in sufficiently large numbers to maintain the infection in the local rodent hosts (Christie 1982).

General mechanism of *Y. pestis* transmission by fleas was described by Bacot and Martin (1914). According to this model, blockage of the proventriculus by a mass of *Y. pestis* precedes transmission, and *Y. pestis* is regurgitated to the bite site when blocked fleas attempt to feed. The model was later amended by proposing that fleas with only partial obstruction of the proventriculus were better transmitters (Bacot 1915). Although several investigators have established that biological transmission is the only reliable means of transmission (Burroughs 1947, Pollitzer 1954), there is evidence that mechanical transmission may also play role in the ecology of plague. Human-to-human transmission via *P. irritans* has been hypothesized to have contributed to the plague pandemics of medieval Europe (Beaucournu 1999). *Pulex irritans* is now worldwide in its distribution and has a wide range of hosts. It is found in the wild on foxes, badgers, squirrels and rats as well as on pigs, goats, dogs, cats and humans. Sometimes, other mammalian species have been demonstrated to be an uncommon source of plague, such as camel in North Africa.

Humans may be exposed to infection in sylvatic setting by coming in occupational contact with wild rodents, foxes, gerbils while hunting in forest, eating or flaying the game, or through flea bite during their sojourn in a wild setting. Wild carnivores such as ferrets, Siberian polecats, black bears, badgers, coyotes, raccoons, and skunks appear to be highly resistant to plague. In most of these animals, ingestion of plague infected rodents causes inapparent to mild disease and seroconversion. Infected wild rodents may spill to peri-urban areas due to environmental events like drought or earthquake. After the ingestion of infected wild rodents, domestic cats are a source of pulmonary plague as well as bubonic plague in urban plague.

After the spread of plague in an urban setting, other modes of transmission take place. Patients with pulmonary plague can be a source of infection. Also, it has been suggested that direct contacts with infected animals may be a route of transmission. Transmission from domestic cat results from direct inoculation by bite or scratch (bubonic plague) or by inhalation (pulmonary plague). Inter-human transmission by inhalation from patient with pulmonary plague is possible. However, it is an exceptional route of transmission requiring close contacts with the index case (less than 2 metres). Direct inter-human transmission by human ectoparasite (human flea, *P. irritans*; human louse, *Pediculus humanus*) has been reported.

Environmental and Other Predisposing Factors

In order to relate the occurrence of bubonic plague to climate and for the sake of forecasting, studies on seasonal prevalence of plague have been undertaken, including in India (Greenwood 1913, Latham 1900, Rogers 1928). The effects of environmental factors such as temperature, humidity, host specificity, and flea and mammalian host abundances and densities, as well as characteristics of *Y. pestis* strains are poorly understood and may drastically affect the vector efficiency in nature (Kartman 1958, Brubaker and Surgalla 1961, Bahmanyar and Cavanaugh 1976). The type species, *Y. pestis*, grows at temperatures from 4 to 40°C (optimum 28-30°C); the optimum pH for growth ranges between 7.2 and 7.6; however, the extremes of pH 5 to 9.6 are tolerated (Holt *et al.* 1994). At higher environmental temperatures (28-30°C), blockage of alimentary proventriculus of fleas decreases and clearance of the infection increases. Outbreaks in epizootic rodent hosts periodically decimate rodent populations in Africa, the Americas, Asia, and the former USSR (Bahmanyar and Cavanaugh 1976, Gage *et al.* 1992). Such epizootics and consequent human disease are favoured by sufficient populations of fleas and susceptible rodents as well as cooler temperatures with above average rainfall. Evidence of an association between climatic factors and plague incidence has been documented in Kazakhstan where temperature increase and rainfall affect the gerbil population and increase the plague risk (Davis *et al.* 2004, Kausrud *et al.* 2007).

Disease Manifestations in Animals

Yersinia pestis organism is per se found in animals. The plague bacilli, which are of importance to humans, are derived mainly from rodents (rats, mice, squirrels, etc.). At the start of 20th century, when plague struck India, a report submitted to Government of India in 1904 mentioned that the experiments undertaken demonstrated that pigs, calves, buffaloes, sheep, hens, ducks, geese, turkeys and pigeons were susceptible to plague, in addition to rats, and particularly so when fed with plague (Bannerman and Kapadia 1906). The conclusions were drawn that the ordinary farmyard animals needed to be watched and regarded as possible sources of plague, the eating of contaminated food might be a common way of acquiring plague, and therefore inspection of food stuffs assumed a new and important aspect (Bannerman and Kapadia 1906).

From an epidemiological perspective in peridomestic setting, cats and dogs are the most important animals after rodents, which get affected and play important role in disease transmission. Of the domestic livestock species common in the United States, cattle, horses, sheep, and pigs are not known to develop plague (Gage 1998). Human outbreaks of plague attributable to ingestion of *Y. pestis* infected camel meat in several Middle East countries have been reported, and goats and camels with plague develop clinical illness. Evidence of infection in wild species other than rodents has been reported in mule, deer, prong-horned antelope and non-human primates (Gage 1998). Foxes, bears, raccoons, and skunks and other wild species that appear to be resistant to plague-induced illness include wild cats such as mountain lions and bobcats. It is significant that the Indian plague outbreak in Shimla in 1992 is purported to have been resulted from killing and flaying of a sick wild cat by a hunter (Gupta and Sharma 2007).

Disease Manifestations in Human Beings

Plague has been observed in three clinical forms in humans: bubonic, septicaemic and pneumonic plague, while pharyngeal/enteric plague and meningeal plague are the other occasional forms. Bubonic plague is the classical and most common form of the disease, which results after the bite from an infected flea or by contamination of a fresh skin or bite puncture lesion. Once inside the blood stream, bacteria are phagocytosed by lymphocytes and carried to the local lymph nodes draining the site of infection where they multiply. In humans, the resulting swollen, tender, inflamed lymph node is termed a bubo. Bacteria multiply in phagocytes and reappear in the bloodstream until death, which results from endotoxic shock. Patients present with fever, chills, headache and a painful bubo. Local bacterial proliferation is sometimes evident in the form of an abscess or ulcer at the site of infection. Symptoms of fever and malaise develop 2-6 days post-infection, although the bubo may not be evident early in infection, and deeper lymph nodes may not be visible as lymphadenopathy develops. Patients may develop a significant bacteraemia. The case fatality rate is 40-60 per cent but this can be reduced to around 14 per cent by appropriate antimicrobial treatment (Craven *et al.* 1993). Secondary pneumonia arising from spread to the lungs occurs in approximately 10 per cent cases, and these patients pose a significant risk to contacts due to production of infectious aerosols, and as such quarantine is recommended.

When the buboes or lymphadenopathy is absent but *Y. pestis* infection with bacteraemia is present, this presentation is termed primary septicaemic plague that occurs in about 10-25 per cent cases (Hull *et al.* 1987). Due to septicaemic plague resembling those of most Gram-negative septicaemic infections showing fever, chills, headache, malaise, etc., there is confounding in diagnosis which delays the appropriate antibiotic therapy. This is the reason why mortality rates are higher in septicaemic form than in the bubonic plague. Untreated septicaemic plague is almost always fatal.

Primary pneumonic plague arises as a result of inhalation of plague bacilli in infectious aerosols, such as would be produced when there are pneumonia complications in bubonic plague. The epidemic spread in human settlements primarily occurs when a patient develops secondary pneumonia. Incubation period is short, about 1-3 days, after which there is sudden onset of flu-like symptoms including fever, headache, chills, bodyache, weakness and chest discomfort. A cough develops with sputum production, which may be bloody, and increasing chest pain and difficulty in breathing. Pneumonic plague is invariably fatal unless antibiotic therapy commences within 24 hours of the development of clinical symptoms (Wendte *et al.* 2011).

The current WHO definitions of plague cases in humans are as follows:

★ *Suspect case*: compatible clinical and epidemiological features; microscopic observation of Gram-negative bipolar coccobacilli on a smear from a clinical sample.

★ *Presumptive case*: direct detection of *Y. pestis* F1 antigen or a single serum sample positive for anti-F1 antibody, without notification of former exposure to the disease or vaccination or an isolate from a clinical specimen with biochemical reactions compatible with *Y. pestis* or a positive PCR test.

★ *Confirmed case*: strain isolated from a clinical sample identified as *Y. pestis* by phage lysis of cultures at 20-25 °C and at 37°C or a four-fold increase in anti-F1 antibody titre in paired sera.

The main biological approaches for diagnosing human plague are F1 antigen detection by direct fluorescence assay, ELISA, rapid diagnostic test or immunohistochemistry; serology (anti-F1 antibody detection) by ELISA, agglutination test, rapid diagnostic test or immunoblot; *Y. pestis* culture and identification by biochemistry profile compatibility or phage lysis, and antibiotic resistance screening; and microscopic observation of a smear after Gram or Wayson staining.

Prevention and Control Strategies

The number of countries notifying human cases of plague has been increasing steadily since 1954, especially due to increased reporting from Africa. After a country has notified plague cases, it remains at risk as plague has re-emerged in many countries where it was dormant for more than two decades like in India, Indonesia and Uganda. The disease can re-emerge in a country after decades of presumed absence, as in Algeria in 2003, or in a new or reactivated focus in an endemic country like Madagascar and Democratic Republic of Congo. The occurrences of outbreaks of plague in Surat (Gujarat) and Beed (Maharashtra) in 1994 and in district Shimla (Himachal Pradesh) in 2002 confirm that plague infection continues to exist in sylvatic foci in many parts of India which is transmitted to humans occasionally. It is therefore important that the states monitor the plague activity in known sylvatic foci regularly and have a system of surveillance to facilitate prompt diagnosis and treatment of cases to control the disease. If the disease can be diagnosed early and a supervised comprehensive response mounted in time, the states can prevent the disease from proceeding to pneumonic stage which is potentially dangerous from human endemic angle as it can assume epidemic dimensions.

The classical microbiological diagnostic test would be insufficient in a bioterrorism attack scenario due to these being slow and labour intensive, and on the other hand, rapid antigen detection or nucleic acid-based molecular tests are not widely available. It is, therefore, required that such tests be developed and standardized.

Recommendations for antimicrobial treatment of plague in a bio-terrorism event and a limited causality situation have been developed (Ingelsby *et al.* 2000). The main recommended antimicrobial drugs are available in USA through National Pharmaceutical Stockpile. It is useful to plan and review such a strategy for each country, where lower economic status and dependence on civil medical support makes such populace more vulnerable to sudden epidemic situations. It is important to review the drug availability and related laws on their use for treatment of plague such as gentamicin and ciprofloxacin. For effective post-exposure antimicrobial prophylaxis of selected populations, rapid identification of population at risk and administration of drugs in timely fashion would need emergency preparedness.

In case of plague epidemic, the isolation of patients and their close contacts and quarantine of exposed populations would need security forces to enforce orderliness to avoid chaos and mismanagement. Vaccines for plague have been traditionally based on killed and live attenuated preparations, which are presently unavailable and present issues of safety and efficacy. Recombinant protein subunit vaccines against F1 and V antigens in combination or alone have been developed (Titball and Williamson 2001).

Future Directions

The incidents of plague in India and worldwide demonstrate that such ancient diseases which have been deemed to be buried in the past by control measures could return back anytime and may result not only in direct death and pestilence of the nosoarea population, but additionally cause a distressing disturbance to the economic well being of states and the cities, apart from being detrimental to the political set up of a country. In today's world of high speed national and international travel and transportation as well as high population density urban conglomeration, the disease transmission and resultant secondary outbreaks can cripple the confidence and morale of even a well oiled up disease control management team. It is thus imperative for the governments to take a holistic view of infectious disease control scenario in general and diseases like plague in particular. The mechanism of prevention, control and management mitigation of such an epidemic should be vested with a well-oiled surveillance system which should be working round the clock to take prompt action in case of detection of an index case. Such a system needs to imbibe and practise modern methods of system managements in which, apart from the availability and usage of man and machines of high efficiency, there are clear-cut mandates and mechanism of procedures such as Standard Operative Procedures of various clinical procedures, definitions of cases and clinical cut-off values for making quick and unambiguous decisions. The communication networks should be fast and command hierarchy should be clear in taking and sticking to investigatory, procedural, diagnostic and control measure decisions. Data recording, processing and retrieval system should be impeccable. There should be clear-cut and well-rehearsed mechanisms of interaction between the central and state government authorities so that unnecessary time wastages can be avoided in the event of plague outbreaks.

We have learnt many lessons from the experience of the recent epidemic in Beed and Surat in India. Unlike the ignoble past, now and in future, we should be capable to immediately investigate and be able to find the cause and nature of the disease. This is important to understand the epidemiology of disease in order to take rapid public health action for control of disease and prevent its spread to other areas. Such investigation must identify the causative organism beyond reasonable doubt by both microbiological and genomic methods. Research institutions that carrying out entomological and other epidemiological studies on vector borne and zoonotic diseases and the nodal agencies for plague control activities should forge linkages with other institutions and medical colleges.

Despite that research on plague vector biology, immunology and epidemiology has increased our appreciation about *Y. pestis* and its pestilence, much remains to be done. The question of mechanism of long-term survival of plague foci being a function of host or microbial virulence remains to be elucidated. Determining DNA sequence is the most comprehensive and fruitful way of obtaining and understanding information about the genome of any living organism. The Indian *Y. pestis* isolates which are hopefully preserved at pristine locations need to be sequenced completely and compared across time and locations to find useful information related to their origin and biology like elsewhere (Dacheaux *et al.* 2004, Gibbons *et al.* 2012). We need to develop a thorough comprehension of genomics and gene expression studies in order to understand the virulence factors in different strains. The researchers need to pool their knowledge and resources in order to develop a new vaccine that induces long-lived immunity against bubonic and pneumonic plague and to develop rapid diagnostic methods usable in remote areas.

The peculiar eco-epidemiological niche of plague bacilli in a robust class of animals like wild rodents, which thrive in various types of landscapes, poses a great challenge in eradication of the disease. It thus becomes important from prevention and control perspective that the prevalence of disease be adequately understood in domestic animals. Further, quick recognition and treatment of plague in domestic animals is important to safeguard and protect the health of companion animals, their owners, and the surrounding community.

As the disease agent has the capability of high communicability, the measurement of risk involved in the population near the endemic foci, disease dynamics in relation to change in the climate and vector populations, and area-wise action plan for its rapid management need to be evolved. The plague outbreak which occurred in Surat in 1994 caused widespread panic, leading to economic losses of estimated cost of US$ 600 million (Ganapati 1995, Fritz *et al.* 1996). One cannot afford to see a repeat of such an event.

References

1. Achtman M, Zurth K, Morelli G, Torrea G, Guiyoule A and Carniel E. 1999. *Yersinia pestis*, the cause of plague, is a recently emerged clone of *Yersinia pseudotuberculosis*. *Proc Natl Acad Sci USA* 96: 1403-1404.

2. Akiev AK. 1982. Epidemiology and incidence of plague in the world, 1958-79. *Bull WHO* 60: 165-169.

3. Aleksic S and Bockemuhl J. 1999. *Yersinia* and other enterobacteriaceae. In: Murray P (ed) *Manual of Clinical Microbiology*. American Society for Microbiology, Washington DC.

4. Anisimov AP, Lindler LE and Pier GB. 2004. Intraspecific diversity of *Yersinia pestis*. *Clin Microbiol Rev* 17: 434-464.

5. Anonymous. 2002. Outbreak of pneumonic plague in village Hatkoti, district Shimla, Himachal Pradesh, India, February 2002. Directorate General of Health Services, Ministry of Health and Family Welfare, New Delhi.

6. Bacot AW. 1915. Further notes on the mechanism of the transmission of plague by fleas. *J Hygiene Plague Suppl*. 4, 14: 774-776.

7. Bacot AW and Martin CJ. 1914. Observations on the mechanism of the transmission of plague by fleas. *J Hyg (Plague Suppl.)* 3, 13: 423-439.

8. Bahmanyar M and Cavanaugh DC. 1976. *Plague Manual*. World Health Organization, Geneva, Switzerland.

9. Bannerman WB and Kapadia RJ. 1906. Domestic animals and plague. XXVII report on experiments undertaken to discover whether the common domestic animals of India are affected by plague. *J Hyg (Lond)* 6(2): 179-211.

10. Barnes AM and Quan TJ. 1992. Plague. In: Gorbach SL, Bartlett JG and Blacklow NR (ed) *Infectious diseases*. The WB Saunders Co, Philadelphia. p. 1285-1291.

11. Beaucournu JC. 1999. Diversity of flea vectors as a function of plague foci. *Bull Soc Pathol Exot* 5: 419-421.

12. Bei L, Ying G, Zhaobiao G, Yun L, Ziwen Z, Qing Z, Yanfeng Y, Zhizhong S and Ruifu Y. 2008. Serologic survey of the sentinel animals for plague surveillance and screening for complementary diagnostic markers to F1 antigen by protein microarray. *Am J Trop Med Hyg* 79: 799-802.

13. Bercovier H, Mollaret HH, Alsonso JM, Brault J, Fanning GR, Steigerwalt AG and Brenner DJ. 1980. Intra- and interspecies relatedness of *Yersinia pestis* by DNA hybridization and its relationship to *Yersinia pseudotuberculosis*. *Curr Microbiol* 4: 225-229.

14. Bibikova VA. 1977. Contemporary views on the interrelationships between fleas and the pathogens of human and animal diseases. *Annual Review Entomol* 22: 23-32.

15. Bray RS. 1996. *Armies of pestilence: The impact of disease in history*. Barnes and Noble Books, New York.

16. Brooks JE, Walton DW, Tun UMM, Naing UH and Htun UPT. 1978. Field trials of insecticidal dusts for the control of fleas on small mammals in Rangoon, Burma. Unpublished document No.WHO/VBC/78.697, World Health Organization, Geneva.

17. Brubaker RR and Surgalla MJ. 1961. Pesticins. I. Pesticin-bacterium interrelationships and environmental factors influencing activity. *J Bacteriol* 82: 940-949.

18. Burroughs AL. 1947. Sylvatic plague studies. The vector efficiency of nine species of fleas compared with *Xenopsylla cheopis*. *J Hygiene* 45: 371-396.

19. Butler T. 1983. *Plague and other Yersinia infections*. Plenum Press, New York.

20. Butler T and Dennis DT. 2005. *Yersinia* species including plague. In: Mandell GL, Bennet JE and Dolin R (ed) *Principles and practice of infectious diseases*, 6th edn. Harcourt Health Sciences, St. Louis.

21. CDC. 1981. Human plague associated with domestic cats - California, Colorado. Centers for Disease Control and Prevention. *MMWR Morb Mortal Wkly Rep* 30: 265-266.

22. CDC. 2000. Biological and Chemical Terrorism. Strategic plan for preparedness and response: Recommendations of the CDC Strategic Planning Workgroup. Centers for Disease Control and Prevention. *MMWR Morb Mortal Wkly Rep* 49 (RR4): 1-14.

23. Christie AB, Chen TH and Elberg SS. 1980. Plague in camels and goats: Their role in human epidemics. *J Inf Dis* 141: 724-726.

24. Craven RB, Maupin GO, Beard ML, Quan TJ, Barnes AM. 1993. Reported cases of human plague infections in the United States, 1970-1991. *J Med Entomol* 30: 758-761.

25. Dacheaux D, Elliot JM, Derbise A, Hauser LJ and Garcia E. 2004. Insights into the evolution of *Yersinia pestis* through whole-genome comparison with *Yersinia pseudotuberculosis*. *Proc Natl Acad Sci USA* 101: 13826-13831.

26. Davis S, Begon M, De Bruyn L, Ageyev V, Viljugrein H, Stenseth N and Leirs H. 2004. Predictive thresholds for plague in Kazakhstan. *Science* 304: 736-738.

27. Dennis DT, Gage KL, Gratz N, Poland JD and Tikhomirov E. 1999. *Plague manual: epidemiology, distribution, surveillance and control*. World Health Organization, Geneva.

28. Devignat R. 1951. Variétés de l'espèce. *Pasteurella pestis. Bulletin WHO* 4: 247-263.

29. Dobec M and Hrabar A. 1990. Murine typhus on the northern Damatian islands, Yugoslavia. *J Hyg Epidem Microbiol Immunol* 34: 175-181.

30. Duplaix N. 1988. Fleas - the lethal leapers. *Natl Geogr* 173: 672-694.

31. Fritz CL, Dennis DT, Tipple MA, Campbell GL, McCance CR *et al.* 1996. Surveillance for pneumonic plague in the United States during an international

emergency: A model for control of imported emerging diseases. *Emerg Infect Dis* 2: 30-36.

32. Gage KL. Plague. 1998. In: Collier L (ed) *Topley and Wilson's microbiology and microbial infections*. Arnold, London, p. 885-903.

33. Gage KL and Kosoy MY. 2004. Natural history of plague: perspectives from more than a century of research. *Annual Rev Entomol* 50: 505-528.

34. Gage KL, Lance SE, Dennis DT and Montenieri JA. 1992. Human plague in the United States: a review of cases from 1988-1992 with comments on the likelihood of increased plague activity. *Border Epidemiology Bull* 19: 1-10.

35. Gage KL, Dennis DT, Orloski KA, Ettestad P, Brown TL, Reynolds PJ, Pape WJ, Fritz CL, Carter LG and Stein JD. 2000. Cases of cat-associated human plague in the Western US, 1977-1998. *Clin Infect Dis* 30: 893-900.

36. Galimand M, Guiyoule A, Gerbaud G, Rasoamanana B, Chanteau S, Carniel E and Courvalin P. 1997. Multidrug resistance in *Yersinia pestis* mediated by a transferable plasmid. *New Engl J Med* 337: 677-680.

37. Ganapati M. 1995. India's pneumonic plague outbreak continues to baffle. *BMJ* 311: 706.

38. Gasper PW, Barnes AM, Quan TJ, Benziger JP, Carter LG, Beard ML and Maupin GO. 1993. Plague (*Yersinia pestis*) in cats: description of experimentally induced disease. *J Med Entomol* 30: 20-26.

39. Gibbons HS, Krepps MD, Ouellette G, Karavis M, Onischuk L, Pascale L, Broomall S, Sickler T, Betters JL, McGregor P, Donarum G, Liem A, Fochler E, McNew L, Rosenzwieg CN and Skowronski E. 2012. Comparative genomics of 2009 seasonal plague (*Yersinia pestis*) in New Mexico. *PLoS ONE* 7(2): e31604 doi:10.1371/journal.pone.0031604.

40. Girard JM, Wagner DM, Vogler AJ, Keys C, Allender CJ *et al.* 2004. Differential plague transmission dynamics determine *Yersinia pestis* population genetic structure on local, regional, and global scales. *Proc Nat Acad Sci* 101: 8408-8413.

41. Gordon JE and Knies PT. 1947. Flea versus rat control in human plague. *Am J Med Sci* 213: 362-376.

42. Gottfried RS. 1983. *The black death. Natural and human disaster in medieval Europe*. The Free Press, New York.

43. Gratz NG. 1980. Problems and developments in the control of flea vectors of disease. In: Traub R and Starcke H (eds) *Fleas*. AA Balkema, Rotterdam.

44. Gratz NG and Brown AWA. 1983. *Fleas: biology and control*. Unpublished document No. WHO/VBC/83.874. World Health Organization, Geneva.

45. Greenwood M. 1913. The factors that determine the rise, spread and degree of severity of epidemic diseases. *International Congress of Medicine*, London 18: 49-72.

46. Gupta ML and Sharma A. 2007. Pneumonic plague, northern India, 2002 [letter]. *Emerg Infect Dis* 13: 4.

47. Hallett AF, McNeil D and Meyer KF. 1970. A serological survey of the small mammals for plague in Southern Africa. *South African Med J* 44: 831-837.

48. Hankin EH. 1905. On the epidemiology of plague. *J Hyg (Lond)* 5(1):48-83.

49. Hannah Gould L, Pape J, Ettestad P, Griffith KS and Mead PS. 2008. Dog-associated risk factors for human plague. *Zoonoses Public Health* 55: 448-454.

50. Hass GE and Walton DW. 1973. Fleas (*Siphonoptera*) infesting small mammals from the western Oriental region. *Korean J Parasitol* 2: 102-107.

51. Hinchcliffe SJ, Isherwood KE, Stabler RA, Prentice MB, Rakin A, Nichols RA, Oyston PCF, Hinds J, Titball RW and Wren BW. 2003. Application of DNA microarrays to study the evolutionary genomics of *Yersinia pestis* and *Yersinia pseudotuberculosis*. *Genome Res* 13: 2018-2029.

52. Hinnebusch BJ, Perry RD and Schwan TG. 1996. Role of the *Yersinia pestis* hemin storage (*hms*) locus in the transmission of plague by fleas. *Science* 273: 367-370.

53. Holt JG, Krieg NR, Sneath PHA, Staley JT and Williams ST (eds). 1994. *Bergey's manual of determinative bacteriology*, 9th edn. Williams and Wilkins Co., Baltimore, Madison. p. 175-289.

54. Hoogstraal H. 1980. The roles of fleas and ticks in the epidemiology of human diseases. In: Traub R and Starcke H (ed) *Fleas*. AA Balkema, Rotterdam, Netherlands.

55. Hubbard CA. 1968. *Fleas of western North America*. Iowa State College Press, Ames, Iowa.

56. Hull HF, Montes JM and Mann JM. 1987. Septicemic plague in New Mexico. *J Inf Dis* 155: 113-118.

57. Ingelsby TV, Dennis DT, Henderson DA, Bartlett JG, Ascher MS, Eitzen E, Fine AD, Friedlander AM, Hauer J, Koerner JF, Layton M, McDade J, Osterholm MT, O'Toole T, Parker G, Perl TM, Russel PK, Schoch-Spana M and Tonat K. 2000. *Plague as a biological weapon: medical and public health management*. Working Group on Civilian Biodefense. JAMA 283: 2281-2290.

58. Isaacson M, Taylor P and Arntzen L. 1983. Ecology of plague in Africa: response of indigenous wild rodents to experimental plague infection. *Bulletin WHO* 61: 339-344.

59. Iyengar R. 1973. The *Siphonoptera* of the Indian subregion. *Oriental Insects* 3 (Suppl): 1-102.

60. Ji S, He J, Teng Y, Zhan X, Lei C and Wang W. 1990. The discovery and research of plague nature foci in China. *Chin J Epidemiol* 11: 1-14.

61. Kartman L, Prince FM, Quan SF and Stark HE. 1958. New knowledge on the ecology of sylvatic plague. *Ann NY Acad Sci* 70: 668-711.

62. Kausrud KL, Viljugrein H, Frigessi A, Begon M, Davis S, Leirs H, Dubyanskiy V and Stenseth NC. 2007. Climatically driven synchrony of gerbil populations allows large scale plague outbreaks. *Proc Royal Soc B* 274 (1621): 1963-1969.

63. Kilonzo BS. 1985. DDT resistance in *Xenopsylla brasiliensis*, the common plague vector in Tanzania. *Insect Science Application* 6: 111-114.

64. Krishnaswami AK, Ray SN and Chandrahas RK. 1970. Serological survey of small mammals in the south Indian plague focus. *Indian J Med Res* 58: 1407-1412.

65. Laforce FM, Achatya IL, Stott G, Brachman PS, Kaufman AF, Clapp RF and Shah NK. 1971. Clinical and epidemiological observations on an outbreak of plague in Nepal. *Bulletin WHO* 45: 693-706.

66. Latham M. 1900. The climactic conditions necessary for propagation and spread of plague. *J Royal Meterological Soc London* 26: 37-94.

67. Lewis RE. 1998. Résumé of the Siphonaptera (Insecta) of the world. *J Med Entomol* 35: 377-389.

68. Mittal V, Rana UV, Jain SK, Kumar K, Pal IS, Arya RC, Ichhpujani RL, Lal S and Agrawal SPJ. 2004. Quick control of bubonic plague outbreak in Uttarkashi, India. *J Commun Dis* 36: 233-239.

69. Orloski KA and Eidson M. 1995. *Yersinia pestis* infection in three dogs. *J Am Vet Med Assoc* 207: 316-318.

70. Oyston PCF and Williamson D. 2011. Plague: Infections of companion animals and opportunities for intervention. *Animals* 1(2): 242-255.

71. Perry RD and Fetherston JD. 1997. *Yersinia pestis* - Etiologic agent of plague. *Clin Microbiol Rev* 10: 35-66.

72. Poland JD and Barnes AM. 1979. Plague. In: Steele JH (ed) CRC *handbook series in zoonoses. Section A. Bacterial, rickettsial, and mycotic diseases*, vol. I. CRC Press, Inc, Boca Raton, Florida. p. 515-559.

73. Pollitzer R. 1954. *Plague*. World Health Organization, Geneva.

74. Pollitzer R. 1966. *Plague and plague control in the Soviet Union*. Fordham University, Bronx, NY.

75. Prakash I. 1988. Changing patterns of rodent populations in India. In: Prakash I (ed) *Rodent pest management*. CRC Press, Boca Raton. p. 179-190.

76. Prentice MB and Rahalison L. 2007. Plague. *Lancet* 369(9568): 1196-1207.

77. Renapurkar DM. 1981. Serological investigations of urban and rural commensal rodent plague in Maharashtra. *J Commun Dis* 13: 110-114.

78. Renapurkar DM. 1989. Plague surveillance studies: Summary of serological investigations. *Pestology* 13(10): 4-6.

79. Renapurkar DM and Sant MV. 1974. *Changing ecology and plague*. Bulletin Hafkine Institute 2(1): 40-43.

80. Rogers L. 1928. The yearly variations in plague in India in relation to climate, forecasting epidemics. *Proc Royal Soc Ser B* 103: 42-72.

81. Rust JH, Cavanaugh DC, O'Shita R and Marshall JD. 1971. The role of domestic animals in the epidemiology of plague. I. Experimental infection of dogs and cats. *J Inf Dis* 124: 522-526.

82. Salkeld DJ and Stapp P. 2006. Seroprevalence rates and transmission of plague (*Yersinia pestis*) in mammalian carnivores. *Vector Borne Zoonot* 6: 231-239.

83. Sant MV, Nimbkar YS, Renapurkar DM. 1972. Is plague lurking in Maharashtra? A survey in Beed district. *Indian J Med Sci* 26: 480-484.

84. Saxena VK and Verghese T. 1996. Ecology of flea-transmitted zoonotic infection in village Mamla district Beed. *Current Science* 71(10): 800.

85. Simpson WJ. 1903. Report on tie causes and continuance of plague in Hong Kong and suggestions as to remedial measures. Waterlow and Sons, London.

86. Simpson WJ, Aberd MD, Lond FRCP and Camb DPH. 1905. *A treatise on plague: dealing with the historical, epidemiological, clinical, therapeutic, and preventive aspects of the disease*. Cambridge University Press, London.

87. Titball RW and Willamson ED. 2001. Vaccination against bubonic and pneumonic plague. *Vaccine* 19: 4175-4184.

88. Velimirovic B. 1990. Plague and glasnost. First information about human cases in the USSR in 1989 and 1990. *Infection* 18: 388-393.

89. Watson RP, Blanchard TW, Mense MG and Gasper PW. 2001. Histopathology of experimental plague in cats. Vet Pathol 38: 165-172.

90. Wayne LG. 1986. Actions of the Judicial Commission of the International Committee on Systematic Bacteriology on requests for opinions published in 1983 and 1984. *Int J Syst Bacteriol* 36: 357-358.

91. Webb CT, Brooks CP, Gage KL and Antolin MF. 2006. Classic flea-borne transmission does not drive plague epizootics in prairie dogs. *Proc Nat Acad Sci* 103: 6236-6241.

92. Wendte JM, Ponnusamy D, Reiber D, Blair JL and Clinkenbeard KD. 2011. *In vitro* efficacy of antibiotics commonly used to treat human plague against intracellular *Yersinia pestis. Antimicrob Agents Chemother* doi:10.1128/AAC.01481-10.

93. WHO. 1970. *Health aspects of chemical and biological weapons*. World Health Organization, Geneva. p. 98-109.

94. WHO. 1990. *WHO consultation on plague*. New Delhi, India, 11 to 15 September 1989. World Health Organization, Geneva.

95. WHO. 1994. Human plague in 1992. *Weekly Epidemiol Rec* 2: 8-10.

96. WHO. 1995. Human plague in 1993. *Weekly Epidemiol Rec* 70: 45-48.

97. Winn Jr W, Allen S, Janda W, Koneman E, Procop G, Schreckenberger P and Woods G. (Eds). 2006. *The Enterobactericeae. Koneman's color atlas and textbook of diagnostic microbiology*, 6ᵗʰ edn. p. 239-270.

98. Xu R. 1997. Plague: Geographical foci situation in China. *Vector Ecology Newsletter* 28: 4-5.

99. Ye SM. 1983. Investigation on rat fleas in the Luda district. *Acta Entomologica Sinica* 26: 403-408.

100. Yersin A. 1894. La peste bubonique a Hong Kong. *Ann Inst Pasteur* 8: 662-667.

101. Yin JX, Dong XQ, Liang Y, Wang P, Siriarayaporn P and Thaikruea L. 2007. Human plague outbreak in two villages, Yunnan Province, China, 2005. *Southeast Asian J Trop Med Public Health* 38: 1115-1119.

102. Zhang G and Liu Z. 2007. The results of plague surveillance in China, 2006. *Chin J Control Endem Dis* 22: 1-6.

2014, Zoonoses: Bacterial Diseases
Editor: **Sudhi Ranjan Garg**
Published by: **DAYA PUBLISHING HOUSE, NEW DELHI**

Pages **143–172**

7

Salmonellosis

Bhoj Raj Singh, Chandan Prakash, Younis Farooq and Sandeep Kumar
Centre for Animal Disease Research and Diagnosis,
Indian Veterinary Research Institute, Izatnagar – 243 122

Salmonellosis has acquired immense public health importance worldwide. In recent years, zoonotic salmonellosis has become important due to the change in the agriculture and animal husbandry practices, increased global trade of animal products, change in food habits, increased consumption of raw or undercooked animal products and emergence of multi-drug resistant (MDR) *Salmonella* strains. Increasing incidence of salmonellosis and substantial mortality in patients with underlying HIV infection has led to greater interest in its epidemiology, pathogenesis and immune mechanism.

Genus *Salmonella* comprises of two species, *Salmonella enterica* and *Salmonella bongori. Salmonella enterica* is divided into six subspecies, namely, *S. enterica* subspecies *enterica, S. enterica* ssp. *salamae, S. enterica* ssp. *arizonae, S. enterica* ssp. *diarizonae, S. enterica* ssp. *houtenae* and *S. enterica* ssp. *indica* (Reeves *et al.* 1989). Speciation in the genus can be done on the basis of biochemical characteristics (Table 7.1) but is mainly based on 16s rRNA sequence while sub-speciation within *enterica* subspecies is based on 23s rRNA sequence analysis. Genomic strains showing more than 70 per cent homology in DNA hybridization experiment and difference in melting temperature of DNA (δTm) below 5°C are considered as members of one geno-species (Wayne *et al.* 1987). Besides, five more gene (proline permease, glyceraldehyde-3-phosphate dehydrogenase, malate dehydrogenase, 6-phosphate-gluconate dehydrogenase, isocitrate dehydrogenase) sequences are in use to construct phylogenic tree of the bacteria.

Table 7.1: Differentiation of some special *Salmonella* species and subspecies

Species/Subspecies	Dulcitol	Lactose	Sorbitol	Mucate	D tartarate	Citrate	Malonate	ONPG
S. bongori	+	–	+	+	–	+	–	+
S. enterica ssp. enterica	+	–	+	+	+	+	–	–
S. enterica ssp. arizonae	–	(–)	+	+	–	+	+	+
S. enterica ssp. diarizonae	–	(+)	+	D	(–)	+	+	+
S. enterica ssp. houtenae	–	–	+	–	D	+	–	–
S. enterica ssp. indica	D	(–)	–	+	+	(+)	–	D
S. enterica ssp. salamae	+	+	+	+	D	+	+	(–)

D: Delayed; ( ): Variable.

Subspecies can be further divided into different serovars on the basis of the presence of somatic 'O' and flagellar 'H' antigen (Kauffmann 1961). More than 2500 serotypes have been reported so far. Most of the serotypes belong to *enterica* species. Some serovars are host adapted, such as Typhi and Paratyphi A in humans, Abortusovis in sheep, Choleraesuis in pigs, Abortusequi in horses and Gallinarum in fowls. Although, the host-specific salmonellae contain themselves to their host species, but they may sometimes also cause infection in other non-specific host species. Thus, S. Typhi has been isolated from sick camels and S. Abortusequi and S. Choleraesuis from humans.

Salmonellosis may be caused by any of the salmonellae, irrespective of serovars, but few serovars are more common. Prevalence of a particular serovar varies according to the geographic area, time, host population affected, and the interactions among the causative agent, host and environment. The most predominant and virulent serotype causing zoonotic salmonellosis includes *S. enterica* ssp. *enterica* serovar Enteritidis (*S. enteritidis*) and *S. entrica* ssp. *enterica* serovar Typhimurium. Although more than 80 serovars of *Salmonella* are reported in India, the common zoonotic serovars isolated in India in order of decreasing frequency of occurrence are: Typhimurium, Virchow, Enteritidis, Weltevreden, 3,10:r:-, Anatum, Bareilly, Stanley and Saintpaul (Verma *et al.* 2000, 2001, Singh 2003, 2005, Singh *et al.* 2009). Some important serovars are further divided into biovars (biotypes, Table 7.2), toxinotypes, resistotypes (antibiogram based), haemolysin types, replicon types (on the basis of plasmid profiles) and phage types to understand the epidemiology of salmonellosis.

Characteristics of *Salmonella*

Salmonella is a Gram negative, facultative anaerobe, flagellated rod-shaped bacterium measuring 2-3×0.4-0.6 μm in size (Yousef and Carlstrom 2003, Montville and Matthews 2008). Salmonellae are non-fastidious organisms and can multiply under different environmental conditions inside and outside the living hosts. These can tolerate 0.4 to 4 per cent sodium chloride and can grow at 7°C to 47°C but optimally at 35°C to 37°C (Gray and Fedorka-Cray 2002). They are heat-sensitive and get killed with pasteurization. Salmonellae grow in a pH range of 3.8 to 9 with the optimum between 6.5 and 7.5. They require high water activity (a_w) for growth (0.99 to 0.94) but can survive at less than 0.2 a_w in dried and semidried foods. Complete inhibition of growth occurs at temperatures <7°C, pH <3.8 and a_w <0.94 (Hanes 2003, Singh *et al.* 2005b).

Salmonellae resist dehydration for a very long time, both in faeces and in foods. In addition, they can survive for several months in brine with 20 per cent salinity, particularly in products with a high protein or fat content, such as salted sausage and cheddar cheese (El-Gazzar and Marth 1992).

Epidemiology of Salmonellosis

Incubation Period

The incubation period is usually within the range of 12-72 h but it may extend up to a week. The incubation periods may be as short as 2.5 h if the animal product is contaminated with heavy bacterial load (Stevens *et al.* 1989).

Table 7.2: Differentiation of some special *Salmonella* serovars *S. enterica* ssp. *enterica*

Serovars	H₂S	Lysine	Ornithine	D-tartrate	Gas from glucose	Dulcitol	Maltose	Rhamnose	Sorbitol
Choleraesuis	D	+	+	(+)	+	–	+	+	(+)
Paratyphi A	–	–	+	–	+	+	+	+	+
Typhi	+	+	+	+	–	–	+	–	+
Gallinarum	+	+	–	+	–	+	+	–	–
Pullorum	+	+	+	–	+	–	–	+	(–)
Common Sal	+	+	+	+	+	+	+	+	+

Infective Dose

Usually at least 100,000 *Salmonella* cells are required to initiate infection such as in case of *S*. Bareilly and *S*. Newport (McCullough and Eisele 1951) but the infective dose may be as low as few organisms if the food product has high fat content and good buffering capacity to protect *Salmonella* from gastric acidity. Infective dose may be significantly low in young and aged persons, the persons taking antacids (for lowering gastric acidity) or oral antimicrobial drugs (that causes destruction of gut flora) and in those with any other underlying disease that causes immunosuppression, thus making them more susceptible. In animals, infective dose also depend on species and even strain (breed) of animals; comparatively resistant strains of poultry and other birds have been reported (Singh *et al.* 2002, 2003, Verma *et al.* 2004)

Seasonal Pattern

In temperate climate, mostly in developed countries with good surveillance system, the incidence of salmonellosis increases in summer months particularly in late summer. In tropical countries like India, salmonellosis is recorded more in the monsoon and post-monsoon seasons (Suresh *et al.* 2006). Survival of *Salmonella* is enhanced by high relative humidity and low temperature (Lancaster and Crabb 1953, Baker 1990).

Host Range

Salmonella organism may be broadly classified into three groups considering their host adaptability.

Group A

Human adapted *Salmonella* serovars which usually infect human beings only and are rarely recovered from animals. The examples include *Salmonella* Typhi, S. Paratyphi, S. Schottmuelleri, S. Hireschfeldii and S. Sendai. Among these, S. Schottmuelleri (S. Paratyphi B) is less adapted human serotype; it has been isolated from a wide variety of animal species in India such as from cattle, buffalo, pig, poultry, dog, fish, cockroach, etc.

Group B

It comprises mainly of animal pathogens that are animal adapted serovars and usually restricted to single host species only such as to cattle (S. Dublin), swine (S. Chloraesuis, S. Typhisuis), poultry (S. Gallinarum, S. Pullorum), sheep (S. Abortusovis) and horse (S. Abortusequi). Although Group B serovars cause infection in animals, but several report indicate their isolation from human cases of causing serious infection (Chowdry *et al.* 1993).

Group C

This group comprises of *Salmonella* which infect a wide range of animal species and humans too. Most of the zoonotic *Salmonella* serovars come under this group, for example, S. Typhimurium and S. Enteritidis. Clinical manifestation and disease severity of different serovars may vary in different host species.

Role of Pet/Companion Animals

Pet animals such as cat and dog harbour *Salmonella* and excrete these in faeces and contaminate their own body surfaces and the environment. Humans acquire infection through contaminated food, fomites, contact with contaminated pet body surface such as hair and fur, and handling of infected animals. Many salmonellosis outbreaks have been reported in young children due to close association with pet animals. In India, the prevalence of salmonellosis in dogs has been reported to be high (58 per cent) and several zoonotic serovars have been reported to cause infection in dogs. Isolation of multidrug resistant *Salmonella* serovars from pet and companion animals presents a serious public health concern.

Other Hosts and Reservoirs of *Salmonella*

Plants used as fodder for animals or as vegetables and salads for human beings may be important source of *salmonella* infection. Plants can serve as host and reservoir for the microorganism. Sometimes vegetables contaminated through water also act as reservoir of salmonella for long period. Even the beetle leaves chewed by people all over Asian continent have been identified as vectors of zoonotic salmonellae (Singh *et al.* 2006b)

Rodents and reptiles (snake, turtle, lizards, wall lizard), insects and pests are also important reservoirs of salmonella and help in spreading salmonellosis. These may act as mechanical vehicle, biological vehicle or reservoir. Rodents such as rat and mice, which have unlimited access to kitchen in rural areas, contaminate food, food vessels or kitchen surfaces. Cockroaches have also been shown to play a vital role in transmitting *Salmonella* paratyphi B var Java to human beings (Singh *et al.* 1995). Several wild animals may also carry and may suffer with salmonellosis, particularly carnivores (Chandra *et al.* 2008).

Transmission

Salmonellosis is a foodborne disease, caused by consumption of contaminated food. In US, more than 95 per cent salmonellosis cases are reported as foodborne infection (Mead *et al.* 1999). Faeco-oral route is the most common mode of transmission. Humans usually get infected after consumption of animal products either originated from infected animals or contaminated during harvesting and processing. Raw or undercooked meat and meat products (beef, pork, chicken and fish), raw milk (unpasteurized) and faulty pasteurized milk have been reported as important vehicles transmitting salmonella from animals to human beings (Gomez *et al.* 1997). In India, buffalo meat and milk are also identified as important sources of *Salmonella*. In addition to animal products, vegetables and salads sprinkled with *Salmonella* contaminated water have also been reported as major sources of *Salmonella*. Salmonellosis may also be contracted from direct contact with infected animals (Olsvik *et al.* 1985) or through fomites. Aerosol route has also caused some salmonellosis outbreaks. Sprouts, melon, chewing beetles, salad vegetables and spices eaten raw (coriander and mint) have been identified as important sources of *Salmonella*. Salmonellae have been shown to persist for long period in fodder, food plants and soil (Singh *et al.* 2007a, c, d)

Pathogenesis

The outcome of disease depends upon the interaction of the host, organism and environment. Various factors such as impairment in natural defence system of the host, greater virulence of *Salmonella* serovar, heavy bacterial contamination in the food product, young age of a person, underlying disease conditions and immunodeficient status of a person facilitate survival and multiplication of *Salmonella* organisms in the host and precipitate the disease.

Salmonella enter the body through oral route. Initially non-specific defence system of the host (gastric acidity, peristalsis, intestinal mucus, lysozyme secretions, lactoferrin in GIT, normal intestinal flora) attempts to destroy the organisms and eliminate them from the host system. In initial phase of infection, bacteria adhere to unidentified receptors on enterocytes to cause changes in the metabolic reactions in host cells. Expression of various virulence genes in bacteria lead to uptake and internalization of bacteria inside the enterocyte. Presence of various virulence factors such as adhesions, invasins and toxins of the organism determine its virulence. Enterotoxigenicity and virulence in salmonellae have been shown positively correlated (Singh *et al.* 1992, Singh and Kulshreshtha 1993).

Initially, organisms enter the non-phagocytic enterocytes to escape from the immune system where these can multiply freely with optimum supply of nutrients available in cytoplasm. After multiplication in enterocytes, the organisms invade the gut associated lymphoid tissue (GALT), where these penetrate into macrophages that line the lymphatic sinuses and form the first effective barrier to prevent further spread. Macrophages try to kill the organisms using reactive oxygen intermediate (ROI), reactive nitrogen intermediate (RNI), low pH, and defensins. *Salmonella* organisms have an arsenal of virulent factors such as cell surface charges (Singh and Sharma 2003), lipopolysaccharides (LPS) on their cell wall, inhibition of phagosome lysosome fusion, evasion mechanism from lytic effect of phagolysosome, and siderophoroes (for iron scavenging) production. The outcome of the interaction decides the intracellular survival and multiplication of the micro-organism within macrophage (Singh *et al.* 1999, Alam *et al.* 2009). If the host defence mechanism successfully limits the bacterial expansion, the infection remains localized to the intestine and GALT and usually causes mild localized enteritis. However, if the macrophages fail to resist the infection, it can cause a systemic disease. During systemic infection, the pathogens spread from GALT via efferent lymphatics and thoracic duct into the vena cava. The organisms finally reach liver and spleen where capillary system or sinusoid filters the blood and trap any pathogenic organisms. These organs are usually enlarged during systemic infection (spleenomegaly and hepatomegaly). Enteric fever in systemic disease is due to bacteremia, toxemia and partly due to host response as a protective mechanism against the pathogenic organisms.

The micro-organism's ability to cause both localized and systemic disease relies on a repertoire of elaborate virulence determinants for survival and multiplication within macrophage. *Salmonella* organisms sense hostile micro-environment and modulate expression of various virulent genes (about 35 proteins), including stress protein GroEL, while some are repressed (Buchmeier and Heffron 1990). Effecter protein expression is modulated by several regulators, including phoP/phoQ (Fields *et al.* 1989) and ompR/envZ (Lindgren *et al.* 1996). *phoP/phoQ* responds to magnesium ions which are apparently low in phagocytic vesicle (Vescovi *et al.* 1996) and ompR/ envZ probably respond to both osmolarity and pH. Mutation in genes encoding for effecter protein (as *htr*A) or genes modulating their expression (*phoP/phoQ*) have been shown to mar the virulence of salmonellae effectively (Singh *et al.* 1999). Apart from virulence gene expression, several other strategies such as absence of mannose-6-phosphate receptor (targeting signal to the late endosome) on phagocytic vesicle containing *Salmonella* and direct cytotoxic killing of macrophage ensure survival of *Salmonella* organism within macrophage.

Besides, exotoxins of *Salmonella*, specifically an enterotoxin, and cytotoxins are important in pathogenesis of salmonellosis and systemic toxicity of the organism leading to post-recovery complications. Chronic salmonellosis has been shown definitively associated with infertility (Singh *et al.* 2002, 2007b) and hair loss (Singh *et al.* 2005a); both the symptoms have been reproduced with *Salmonella* cytotoxins (Singh *et al.* 1998). All *Salmonella* cytotoxins are not uniformly present in all zoonotic salmonellae, but type I cytotoxin (CT-I) is the most important which is present in almost all *S. enterica* ssp. *enterica* serovars and may induce symptoms of systemic salmonellosis even in absence of salmonellae (Singh and Sharma 1999).

Clinical Signs

Clinical picture in salmonellosis depends upon the *Salmonella* serovar involved and multiple host factors. The disease usually occurs in two forms:

Localized Enteric Form

The enteric form of the disease is usually restricted to GIT and associated with abdominal pain, diarrhoea, nausea, vomiting, headache, etc. It is usually uncomplicated and can be easily managed with supportive fluid and electrolyte therapy. However, in terminal cases of illness when fluid accumulation starts in stomach, little hope is left for cure. Fluid accumulation in stomach is associated with organ failure induced by CT-I (Singh *et al.* 1998, 1999).

Systemic Enteric Fever

It occurs due to invasion of *Salmonella* and dissemination to visceral organs such as liver and spleen causing recurrent fever, headache, nausea, vomiting, and dirrhoea. GIT involvement is not a constant feature in systemic salmonellosis (Verma *et al.* 2003, 2004). In extreme invasive cases, this may cause metastatic abscesses in bones, aorta wall and kidney. It may also cause other complication such as nonspecific reactive arthritis and ankylosing spondylitis (Humphrey 2000). There are several types of specific signs and symptoms associated with salmonellosis including alopecia, abortions and organ-specific pathology (Singh *et al.* 1998, 2002, 2005a). Chronic salmonellosis has been shown associated with alopecia and sterility in experimental animals, though in clinical cases its significance as a cause of infertility is not yet very clear (Singh *et al.* 2002, 2007b).

Immunity to *Salmonella*

Immunity has been classified broadly as innate immunity and adaptive immunity or specific immunity. Innate immunity is present in host at its birth without prior exposure to pathogen or antigen. This is nonspecific, does not posses any memory and short lived. It is considered as a first line of defence. In contrast, host develops adaptive or specific immunity in response to antigenic determinant of a pathogen. Therefore, this is epitope specific, has memory and is long lasting. Nonspecific or innate immunity comprises serum complement, nonspecific defence cells (leucocytes) and natural killer (NK) cells. In addition to direct killing of organisms, innate immunity components regulate or modulate specific immunity mechanism by secreting various cytokines like IL-4 and IL-12 (IL: Interlukin). Cytokine expression profile will determine predominance of humoral immune response or cell mediated immunity (CMI). IL-12 mediates a helper T-cell (Th1) response, predominantly to modulate cell mediated immune response (Manetti *et al.* 1993) while IL-4 mediates a Th2 response, predominantly to modulate humoral immune response (Swain *et al.* 1990).

Being a facultative intracellular pathogen, *Salmonella* usually infects and successfully survives in macrophages (Alam *et al.* 2009). Therefore, both humoral immune response and CMI play an important role to destruct the pathogen in extracellular and intracellular phase of infection, respectively (Mastroeni *et al.* 1993). Antibodies produced by B-lymphocytes provide effector function in humoral immune response, while CMI mediated by secretion of various cytokines like IL-2, interferon gamma (INFg). Antibodies against *Salmonella* epitopes are effective against extracellular phase of salmonellosis; antibody mediates protective function in several ways:

☆ Binding of antibody with surface antigen preventing attachment of organism with host enterocytes causing failure of invasion (McGhee *et al.* 1992, Michetti *et al.* 1992).

☆ Opsonization of organism (Johnson *et al.* 1985).

☆ Activating complement mediated killing (Morgan 1990).

Cellular immunity provides protection by either destruction of infected host cell by cytolytic T cell or modulating B cell response or other T cell by T helper or T suppressor cell. However, infections with several salmonellae and other similar organisms may lead to antigenic competition as well as production of non-protective antibodies (Singh *et al.* 2006a).

Diagnosis

Prevention and control of salmonellosis can be possible only when we can detect *Salmonella* serovar infection precisely. Detection and identification of *Salmonella* organism is required to identify the source of infection, transmission pattern and transmission dynamics in disease outbreak, serovar prevalence in particular geographic area and temporal disease distribution, and determination of antibiotic sensitivity for effective treatment. Diagnosis is important in formulation of effective preventive and control strategy (Singh 2000a, b, c, 2004).

Traditional (Conventional) Techniques

Antigen Identification

Conventional diagnosis relies mainly on isolation of organisms from clinical specimens and identification by biochemical tests and serotyping. *Salmonella* isolation consists of mainly four steps: a) Pre-enrichment; b) Selective enrichment; c) Plating on selective media, and d) Biochemical identification.

Pre-enrichment step is necessary to resuscitate the injured bacteria present in dried, heated, irradiated or processed food samples. For pre-enrichment, suspected sample is inoculated in buffered peptone water and incubated at 37°C for 18-24 h. Selective enrichment follows the pre-enrichment step, which allows selective multiplication of *Salmonella* organisms by inhibiting growth of other bacteria. Tetrathionate broth, selenite F broth and Rappaport-Vassiladis (RV) broth are widely used for enrichment. For it, an aliquot of pre-enrichment broth is transferred into tetrathionate broth and selenite F broth in 1:10 ratio and in RV broth 1:100 ratio, respectively, and incubated at 40-43°C for 24-48 h. The enriched sample is then streaked onto selective plating agar media (BGA, bismuth sulphite agar, Hektoen enteric agar, xylose lysine deoxycholate agar) and incubated at 37°C for 24 h. These culture media allow selective multiplication of *Salmonella* organism as well as their differentiation from other related organisms.

Suspected *Salmonella* colony on selective agar plate is picked and inoculated into composite medium such as triple sugar iron (TSI) medium and lysine agar medium. Common tests used for identification of *salmonella* are indole (–), MR (+), VP (–), urease (–), nitrate reduction (+), phenylalanine (–), glucose (+), salicin (–), adonitol (–), inositol D (-), raffinose (–), erythritol (-), aesculin (–), sucrose (–), melibiose (-) oxidase (–), H_2S (+/-), gelatinase (–), lysine (+), ornithine (+), arginine (+), KCN (–) and alginate (–). Common confusion occurs with *Citrobacter*, which are lysine negative and melibiose fermenter. Few *Enterobacter* organisms may also interfere in salmonella identification, they are usually lysine negative (*E. gregoviae* and *E. aerogenes* are lysine positive but are positive for melibiose too) and negative for H_2S production but positive in ONPG test.

Most of the pathogenic *Salmonella* belong to *S. enterica* ssp. *enterica*. These can be easily identified and confirmed by biochemical tests such as fermentation of glucose, mannitol and dulcitol; inability to ferment sucrose, salicin and lactose; inability to hydrolyse urea; β-galactosidase positive and H_2S production. *Proteus* and *Providencia* organisms show similar reactions but these can be differentiated by urease test. *Salmonella* is urease negative while *Proteus* and *Providencia* are urease positive. Therefore, TSI and urease tests should be done simultaneously. *S. enterica* species can also be differentiated into different subspecies on the basis of various differential biochemical characters. Apart from isolation and biochemical identification of the organism, biotyping, serotyping, phage typing, antibiogram, colicin typing etc. are also important.

Antibody Based Tests

The methods of antibody identification in serum and other body fluids include Widal test for human salmonellosis and whole blood agglutination test for salmonellosis in birds. Besides Widal's test for diagnosis of typhoid in human patients, rapid plate agglutination test (RPAT), standard tube agglutination test (STAT), passive haemogglutination test (PHAT), antiglobulin test (AGT) and enzyme immunoassays are some other tests that are used in salmonellosis diagnosis. Among the recently developed immunoglobulin-based tests, enzyme immunoassay (EIA), radioimmunoassay (RIA), counter immunoelectrophoresis (CIE) and ELISA are important. Many commercially available diagnostic kits and *Salmonella* identification kits are in market in different countries and utilize multivalent antibodies raised either against specific antigens or antibodies raised against some common antigens. Some of these include TEK-ELISA (Organon Teknika, Cambridge, UK), IFR-ELISA, Report-EIA and TECRA *Salmonella* Visual A, Dulcitol-1-phosphate dehydrogenase (DPD) *mab* based kit, Cytotoxin-1-antibodies based *Salmonella* detection protocol (Singh and Sharma 2000),Chekit-S-enteritidis (ELISA) kit, Polymyxin cloth enzyme immunoassay, 1-2 Test (Bio-control, Bothel, USA), Single step *Salmonella* (SSS) by Ampcor, Camden USA Iso-Grid® of Dynal, Oslo Vi-TEK and Vi ELISA, Meat Juice ELISA kit (Singh 2000a, b, c, 2004), etc.

Most antibody-based methods are designed to identify *Salmonella* (antigen) isolated in laboratories or in clinical or food samples, but many of them can be used to identify antibodies in serum of suspected patients. Most of the tests are based on 'O' or 'H' antigens of *Salmonella* and have many chances to fail if *Salmonella* is of rare serovar. However, a few tests, either DPD-based or cytotoxin I-based ELISA, are real broad range and exploit common antigens specific to *Salmonella* genus.

Molecular Methods

DNA methods for detection and identification of pathogens have been commonly used in many research laboratories, majority of which are not applied in routine diagnostic laboratories. It is mainly due to the dependence of most methods on traditional culture techniques; hence these are not truly rapid methods. Furthermore, detection of pathogens directly from clinical samples/food samples is theoretically possible with these methods, but rarely 100 per cent successful. Major problems are the detection of dead cells, the presence of enzyme inhibitors and inaccessibility of target organisms in clinical samples. However, PCR and DNA based methods for detection of pathogens have many advantages over the classical techniques as these are quite more sensitive and specific. Rapidity can be obtained by including some modifications such as complimenting with direct chemical extraction of nucleic acids or immuno magnetic separation of bacteria or short cultivation of pathogens from food and clinical samples. The best use of DNA and PCR-based techniques has been made in characterization of strains isolated and purified by traditional methods *e.g.* for demonstration of toxin genes and virulence associated genes, and differentiating between closely related strains of the organism. Some important DNA techniques for *Salmonella* identification and epidemiology are briefly summarized below.

Polymerase Chain Reaction (PCR)

By this method, target DNA fragments are amplified using a cyclic 3 step process namely, denaturation of target DNA at high temperature, annealing of two synthetic oligonucleotide primers and polymerization with heat stable Taq DNA polymerase.

There are many types of PCR and many methods to increase their specificity *e.g.* combining two PCR methods (nested PCR), where two sets of primers are used. Although PCR reaction is capable to detect DNA of single bacterium, the volume of sample (5-10 µl) required for reaction restricts its detection limit at about 10^3 cells per ml.

PCR techniques for diagnosis of salmonellosis are in use since 1983 and probes have been developed from either randomly cloned chromosomal fragments (Cryptic DNA fragments) or specific genes either on plasmids or in chromosomes (Babu *et al.* 2008). Probes have been designed for identifying *Salmonella* to genus level as well as its different species, serogroups and serotypes too. Some PCRs, which have been used in salmonella detection, are described below.

Common PCR primers (5' – 3' orientation) for genus *Salmonella* identification (Singh 2000a, b, c, 2004)

1. Product 163bp

    F : TTATTAGGATCGCGCCAGGC

    R : AAAGAATAACCGTTGTTCAC

2. inv product 284 bp

    F : GTGAAATTATCGCCACGTTCGGGCAA

    R : TCATCGCACCGTCAAAGGAACC

3. Random genomic product 429 bp

    ST F: AGCCAACCATTGCTAAATTGGCGCA

    ST R : GGTAGAAATTCCCAGCGGGTACT

4. Hin H_2 flagellin gene (236 bp)

    Hin 1750 L : CTAGTGCAAATTGTGACCGCA

    Hin 1750 R : CCCCATCGCGCTACTGGTATC

5. H-li flagellin gene (173 bp)

    H-li 1788 : AGCCTCGGCTACTGGTCTTG

    H-li1789 R : CCGCAGCAAGAGTCACCTCA

6. GVV PQ fimbriae agf product (261 bp)

    TAF 3 : TCCGGCCCGGACTCAACG

    TAF 4 : CAGCGCGGCGTTATACCG

7. inv A and inv gene (457 bp)

    S1 : TGCTACAAGCATGAAATGG

    S2 : AAACTGGACCACGGTGACAA

8. Spv A gene based 450 bp.

    382 : CAGACCACCAGTCCGGCAC

    383 : CAGTCAATGCTCTCTCGCTG

9. hisJ gene

> Product size: 496 bp
>
>> Cohen 1, 5' ACT GGC GTT ATC CCT TTC TCT GGT G 3'
>>
>> Cohen 2, 5' ATC TTG TCC TGC CCC TGG TAA GAG A 3'

Specific PCR

MAMA-PCR

Mismatch amplification mutation assay (MAMA), developed to identify infrequent mutation, uses primers designed to have a single mismatch with mutated allele and a double mismatch with wild type, thus permitting amplification of mutated but not the wild type DNA template. Using this technique, *S. enteritidis* specific primers, based on sequence of spvA, have been developed, which produce 351 bp amplicon and its specificity is based on the change at 272 position of spv A gene.

> F: 5' GCAGACATTATCAGTCTTCAGG 3'
>
> R: 5' TCAGGTTCGTGCCATTGTCAA 3'

Serogroup Specific PCR

It is an important diagnostic tool particularly in identification of rough strains.

Oligos Used for *Salmonella* Detection by Hybridization Method

Target gene   Oligosequence (5'-3')

Random chromosomal fragment

TS11:	GTCACGGAAGAAGAGAAATCCGTACG	Tsen *et al.* (1991)
TS21:	TACATCGTAAAGCACCATCGCAATA	
TS31:	AGACCACTGACCCAGCCTAATCAA	

Random chromosomal fragment

ST15 rev:	GAGTACCCGCTGGGAATTTCTAC	Olsen *et al.* (1995)
InvA gene S3:	CTGGTTGATTTCCTGATCGC	Stone *et al.* (1994)

IS200 is not present in *S.* Agona, *S.* Arizonae, *S.* Dar-es-Salam, *S.* Panama, *S.* Infantis, *S.* virchow and S. I.9, 12 : Z. The detection limit for *S.* Typhi is 10^{3-4} (cfu/ml). In IS200, a tandem repeat of 0.3 kb is used for cross hybridization. Colorimetric single phase hybridization assay (CdorDNAH) can detect *Salmonella* by use of 16S and 23 S rRNA based probes, but it cannot detect *Salmonella enterica* ssp. *salamae*. Besides, it may give false positive results on cross reaction with *Citrobacter freundii*. Its *Salmonella* detection limit is 10^8-10^9 cfu/ml. It is produced by Genetrack The specificity and sensitivity of non-radioactive rRNA, based oligonucleotide probes are comparable with culture method and results are given usually in 48 hours.

AD-PCR (Arbitrarily Primed PCR) or RAPD-PCR

A random amplified polymorphic DNA finger printing method was developed to differentiate isolates of *Salmonella* serotypes. In this method, a small oligonucleotide is used as primer and annealing temperature is kept low to allow lot of nonspecific binding. This method has an inherent variability even in the hands of same person attempting at different times.

Commonly Used Primers for *Salmonella*

5′	TGAGCATAGACCTCA	3′
5′	GGGGGGGGG	3′
1254-5′	CCGCAGCCAA	3′
640-5′	CGTGGGGCCT	3′
650-5′	AGTATGCAG	3′

Multiplex-PCR (MP-PCR)

This is similar to normal PCR, but here more than one set of primers are used to amplify the specific product with a genus specific and another group specific set of primers. This type of PCR technique is quite common, but it has comparatively low sensitivity and specificity due to selection of a compromised PCR cycle for multiple sets of primers used in the technique.

Nested PCR

It is quite similar to multiplex PCR, but here different primer sets are added at different time; first few PCR cycles are set according to one set of primer and then second set of primers is added and PCR cycle is set accordingly. It is quite more sensitive and specific in comparison to MP-PCR.

ERIC-PCR (Enterobacterial Repetitive Intergenic Consensus PCR)

Recently a family of repetitive elements, called enterobacterial repetitive intergenic consensus (ERIC) or intergenic repeat units (IRUs), has been defined using *E. coli* and *Salmonella* Typhimurium genome. These 126 bp elements contain a highly conserved central inverted repeats are located in extragenic regions and are unrelated to REP consensus sequences. Inter-ERIC distance and pattern is specific for bacterial species and strains of different origin. This specificity can be employed to differentiate between closely related bacterial strains using single set of primers. Commonly used ERIC primers are mentioned below.

ERIC-1	5	ATGTAAGCTCCTGGGGATTCAC	3
ERIC-2	5	AAGTAAGTGACTGGGCTGAGCG	3

ERIC elements can also be used as radio labeled probe for slot blot or southern blot hybridization. The most commonly used probe is shown below.

5 GTGAATCCCCAGGAGCTTACATAAGTAAGTGACTGGGGTGAGCG 3

REP-PCR

Similar to ERIC elements of enterobacteria and Alu repeats of mammalian genome, prokaryotic genomes also has some noncoding repetitive DNA. Repetitive extragenic palindromic (REP) elements also known as palindromic units (PU) are 38 bp consensus sequence containing six totally degenerate positions including a 5 bp variable loop between each side of the conserved stem of the palindrome. Multiple functions have been thought of for these elements including role in transcription termination, mRNA stability and chromosomal domain organization *in vivo*. The distribution of REP is diverse in prokaryotic genomes and can be examined with PCR or by slot blot hybridization with radiolabelled consensus probes. These primers and probes are not only useful to type *Salmonella*, but can be used for typing of stains of any bacteria.

Primers: REP 1R-I-5   IIIICGICGICATCIGGC   3

RER 2-I-5   ICGICTTATCIGGCCTAC   3

REP ALL-I has been used as radio labeled probe.

REPALL-I 5 GCCIGATGICGICGIIIIIIIICGICTTATCIGGCCTAC   3

Ligase Chain Reaction

Although very useful technique and exploited a lot with strains of *Neisseria*, *Listeria*, *Erwinia* and *Chlamydia*, but this technique has not been exploited for *Salmonella*. This new DNA detection method uses thermostable ligase to discriminate exquisitely and amplify single base changes in the genes of interest. The enzyme specifically links two adjacent oligo nucleotides when hybridized to a complementary target only. Single base mismatch prevents ligation and amplification. As a result, this technique can distinguish closely related but distinct strains that cannot be differentiated by other techniques.

Epidemiological Tools

Zoonotic *Salmonella* serovars are major cause of foodborne diseases in human beings. These zoonotic serovars also cause systemic diseases in animals and through food chain gain entry into human host. Effective control of zoonotic salmonellosis is based on active surveillance of disease, including surveillance of the possible reservoirs and follow up on the incidents in both livestock and human populations. *Salmonella* strain typing is essential to know the changes in disease patterns, predominance of any strain and effectiveness of preventive and control strategy. Important tools in studying the epidemiology of salmonellosis in humans and animals include molecular typing, biotyping, serotyping, phage-typing, replicon typing and plasmid analysis (plasmid profiling), bacteriocin typing, pyrolysis mass spectrometry profiles, ribotyping and proteomics-protein profile. Serotyping and phage typing are two widely used methods for continuous surveillance of *Salmonella*, as these are inexpensive and definitive methods. However, their limited availability, tedious procedure, and some serovars remaining untypable with these methods are the major constraints in these methods.

Molecular typing offers a wide application in microbiology and public health such as active surveillance of disease to obtain baseline data comprising prevalence of particular serovar and its geographical and temporal distribution, locating point source infection, identification of factors contributing to persistence and spread of disease, identification of virulence subtypes, and evaluation of the effectiveness of preventive and control measures by identification of recurrent serovar infection or infection with new serovars. In molecular typing techniques, typing of any serovar does not depend on the phenotypic expression of a character. Another advantage is that most serovars can be typed with single typing scheme with little modification. Molecular typing can be done by several methods as noted below.

Restriction Enzyme Profiling

Restriction enzyme analysis comprises of direct isolation of genomic DNA from different *Salmonella* serovars. Its restriction enzyme digestion and running in agarose gel electrophoresis produces serovar specific banding pattern. This technique is easy but interpretation may be difficult because generation of so many bands makes comparison very difficult and less reproducible. This problem can be overcome by using polyacrylamide gels, which have higher sieving effect (Kapperud *et al.* 1989). This technique is now not in common use as better molecular methods are available for strain differentiation.

Ribotyping

Ribotyping is similar to restriction fragment length polymorphism (RFLP) typing. The main difference is that ribotyping highlights polymorphism in the areas surrounding the seven rRNA operons in *Salmonella*, as opposed to sequence variation within the genes. Ribotyping uses labeled hybridization probe from 16S and/or 23S ribosomal RNA sequences. This typing was first used as a means for the study of taxonomic relationships between strains of same *Salmonella* serovar (Grimont and Grimont 1986) and different workers have used this technique for studying relationship between different serovars. Ribotyping has also been used in Danish outbreak investigation with *S.* Saintpaul for identification of common source of infection. Generally, ribotyping has a relatively low discriminatory power; hence not commonly used for epidemiological purpose (Olsen 2000).

IS200 Typing

Lam and Roth (1983) identified a *Salmonella*-specific insertion sequence, termed as IS200 sequence. This insertion sequence is conserved in almost all serovars except *S.* Agona and strains of *S.* Daressalam (Olsen and Skov 1994) and *S.* Hadar (Weide-Botjes *et al.* 1998). An internal cloned fragment of this element has been used as IS200 probe universally in *Salmonella* diagnosis and epidemiology. Stanley *et al.* (1991) used this probe for *S.* Enteritidis typing and classified into possible evolutionary lines, based on the number and location of IS200 elements on *Pst*I- and *Pvu*II-digested DNA. Since then, IS200 has been applied for typing of many different serovars. IS200 typing has become a standard choice for typing in some serovars where typing schemes have not been developed.

Random Amplification of Polymorphic DNA (RAPD)

This technique uses random short oligonucleotide primers with many possible sites for hybridizations. Typing of *Salmonella* strains can be performed by comparing the size and number of amplified fragments produced (Williams *et al.* 1990). This technique has been shown to give serovar specific patterns in an investigation of a modest number of serovars (Hilton *et al.* 1996). Specificity of this PCR reaction depends mainly on the short oligonucleotides primer sequence and annealing temperature. Poor discrimination and lack of reproducibility are the main limitations of the technique (Laconcha *et al.* 1998), but the discrimination ability can be improved by careful selection of short oligonucleotide primers (Lin *et al.* 1996) and optimization of PCR reaction (Tyler *et al.* 1997).

Polymerase Chain Reaction-Restriction Fragment Length Polymorphism (PCR-RFLP)

PCR-RFLP involves PCR amplification of a specific DNA fragment and restriction enzyme digestion of the amplified DNA (Kilger and Grimont 1993). *Salmonella* organisms have been grouped on PCR-RFLP pattern, consisting of amplification of flagella gene, *fliC*, and digestion with *Taq*I and *Sca*I restriction enzyme. This pattern differentiates H-antigen groups b, i, d, j, l, v, and z10 from each other and from the group of g types (g,m, g,p or g,m,s), which could not be separated individually otherwise. Flagellar types r and e,h could be distinguished from these groups but not from each other. This technique may also be used for typing of *Salmonella* serovars using primer deduced from the *rfb* genes encoding the oligosaccharide repeating units of lipopolysaccharide (LPS) (Luk *et al.* 1993) and rRNA genes (Shah and Romick 1997). The 16S and 23S rRNA genes in *Salmonella* organism are highly conserved (Christensen *et al.* 2000), but the intergenic region between these genes (internal transcribed spacer; ITS) shows variation between operons, and a specific amplification of these regions may be used for typing purpose. When the amplicons are separated by non-denaturing polyacrylamide gel electrophoresis, serovar-specific patterns can be produced (Jensen and Hubner 1996). This typing scheme is able to discriminate the organisms up to serotype level only and this can be a viable alternative option to the conventional serotyping scheme (Christensen *et al.* 2000).

Amplified Fragment Length Polymorphism (AFLP)

This technique has successfully been applied for molecular epidemiological typing of many Gram negative and Gram positive bacteria. It involves digestion of total purified bacterial DNA using a restriction enzyme followed by ligation of the resulting fragments to a double stranded oligonucleotide adapter which is complementary to the base sequence of the restriction site. The adapters are designed such that the original restriction sites are not restored after ligation, thus preventing further restriction digestion. Selective PCR amplification of these fragments is achieved using primers corresponding to the contiguous base sequences in the adapter, restriction site plus one or more nucleotides in the original target DNA. For Hind III-based AFLP, the adapters and primers are mentioned below:

Hind III digestion

	AGCTT	A	
	A	TTCGA	
Adapter	ADHI 5	ACGGTATGCGACAG	3
	ADH2 3	GAGTGCCATACGCTGTCTCGA	5
Selective primer HI-x5		GGTATGCGACAGAGCTTX	3

X-indicates the selective base inserted into the primers. Number of Xs may also vary and X-may be either A, C, G or T.

Pulsed Field Gel Electrophoresis (PFGE)

This is a very strong technique for epidemiological purpose and in European Union (EU) countries Enter-net group is formed to make database using reference strains of *Salmonella* serovars through a standard method described hereunder.

From overnight cultures grown at 37°C, 1 ml aliquots are centrifuged (13000 rpm, 5 min) to pellet cells, cells are washed and resuspended in NSS (0.85 per cent NaCl) then heated to 40°C and mixed with equal volume of 2 per cent molten agarose and dispensed into moulds (150-200 µl/well). When cooled plugs are gently removed and lysed over two nights at 56°C in 1 ml lysis buffer (pH 9.5, 0.5 M EDTA, 1 per cent N-lauroylsarcosine containing 500 µg/ml proteinase K), then washed five times in TE buffer (pH 7.5, 10 mM Tris, 10 mM EDTA) at room temperature and can be stored in TE for months together at 4°C. Each agarose plug is digested overnight at 37°C with 30 units of X baI. PFGE is performed with (HEF DRIT system, BIORAD) in 0.5 X TBE buffer (50 mM Tris, 45 mM boric acid, 0.5 mM EDTA) at 14°C on a resolving 1 per cent agarose gels with a low molecular weight PFG ladder (DNA ladder) at 4.5 volts/cm (150 volts) for 48 hours. Pulse time ramps from five seconds to 60 seconds during the run. Gels are stained with ethidium bromide and visualized under UV light.

Restriction Fragment Length Polymorphism (RFLP)

It is a southern blot and PCR based techniques. Target genes, usually 16S or 23S rRNA genes, are amplified by using specific primers through PCR and the PCR product is digested with one or more restriction enzymes to produce a restriction profile. This profile is specific to strains of different species and origin. It is a valuable tool in the hands of epidemiologists for investigations.

rRNA Spacer Region Polymorphism (PCR-Ribotyping)

The distance variation between the conserved regions of 16S and 23S rRNA is specific according to genus, species and serovars. This specificity has been exploited for differentiation as well as identification of many bacteria including *Salmonella* (Baquar *et al.* 1994). PCR ribotyping has been used for *S.* Enteritids and *S. enterica* serogroup G. Aversions of this technique, a mapped restriction site polymorphism (MRSP), have been successfully employed to subgroup the strains of a species. Commonly used primers for MRSP are stated below:

For 16S rRNA 5' TCAAGGAGGTGATCCAGC 3'

5' CGTTTGATCCTGGCTCAG 3'

For 23S rRNA 5' TATGAACCTGCTTCCCATCGACTAC 3'

5' ATTCCGTCAGTAGCGGTGAGCGAA 3'

On 5' end any restriction site for cloning purposes can be inserted.

Biotyping

Biotyping is a method to differentiate a group of organisms based on biological parameters belonging to the same species. Biotyping may be based on substrate utilization, antibiotic susceptibility pattern, bacteriocin production and bacteriophage susceptibility. Biotyping may be used for studying the phylogenetic relationship in related bacteria, elucidation of transmission dynamics of epidemic strains, diagnosis and classification. Biotyping has been used successfully in many instances to differentiate *S.* Typhimurium and other *Salmonella* serovars into different biotypes (Duguid *et al.* 1975). Biotyping of *S.* Typhimurium is based on five major tests (xylose utilization, fermentation of meso-inositol, rhamnose, and D-tartrate and meso-tartrate resistance) and ten minor supplementary tests (fimbriation, motility, fermentation of L-tartrate, xylose, trehalose, and glycerol at 37°C, fermentation of inositol at 25°C, rhamnose utilization, nicotinamide requirement and cysteine deamination).

Serotyping

Serotyping is the method to differentiate *Salmonella* subspecies up to serotype level based on the presence of different surface antigens. Before proceeding to serotyping, it must be ensured that bacterial culture is derived from a single isolated *Salmonella* colony. *Salmonella* serovar is characterized on the basis of somatic antigen 'O' and flageller antigen 'H' according to White–Kaufmann scheme (Kauffmann 1961). Somatic antigen is present on the main body of the organism and is polysaccharide in nature while the flageller antigen is protein in nature. Most of *Salmonella* serovars express flageller antigen in two phases (called diphasic); only few serovars express flageller antigen in single phase (known as monophasic) *e.g.* *S.* Enteritidis (9,12: g,m: –). *S.* Gallinarum and *S.* Pullorum serovar are nonflagellated, nonmotile organisms; hence they do not express any flageller antigen.

The presence of somatic antigen is identified by slide agglutination test in which half loopful of *Salmonella* culture is added with antiserum and mixed with loop for 60 seconds. Formation of white granular clump characterizes positive agglutination reaction. Identification of somatic antigen starts from agglutination with polyvalent 'O' followed by agglutination with group specific antiserum and finally agglutination with individual serovar specific antiserum. Agglutination by antibodies specific for various O antigens is employed to group salmonellae into six serogroups: A, B, C1, C2, D and E. Expression of certain O antigens is controlled by the presence or absence of lysogenic phage, *e.g.* O:3,10 lysogenized by phage ε 15 will become O:3,15 and, if lysogenized by phage ε 15 and ε 34, it will become O3, 15, 34.

Identification of Flagellar Antigen

Flagellar agglutination is finer in comparison to somatic agglutination; hence this is better observed in low power magnification. Flagellar antigen is also identified by slide agglutination test similar to identification of somatic antigen. First, one drop of polyvalent 'H' antiserum is added to one loopful of *Salmonella* culture on clean glass slide. White granular clumps characterize the presence of flagellar antigen. If positive, same procedure is repeated with specific *Salmonella* 'H' antiserum. Diphasic *Salmonella* do not express both flagellar antigens at the same time. After identification of the first flagellar antigen, it is a must to induce expression of the second flagellar antigen present in the bacteria, known as phase inversion. This can be done either by Svengaurd method or Craige tube method.

In Svengaurd method, antibody against the first identified flagellar antigen is added into semisolid agar petri dish and *Salmonella* organism is inoculated in the centre of the plate and incubated at 37°C for 24 hours. Homologous antibody binds with the flagellar antigen on the surface and it will stimulate expression of the other flagellar antigen that causes bacterial motility in the media. Radiating bacterial growth from the edge of the centre is tested with different sets of flagellar antiserum by slide agglutination test. In Craige tube method, semisolid culture medium in a universal bottle contains an open-ended glass tube and the agar is melted by placing the universal bottle in a water bath at 100°C for 10-15 min. The agar is cooled to 56°C temperature and 30 μl of antiserum against the identified flagellar H antigen is added and 1 μl of *Salmonella* culture is added into the glass tube about 0.5 cm below the surface of the agar and incubated at 37°C for 24 hours. A milky growth appearing on the agar surface is used to identify the newly expressed flagellar antigen.

Phage Typing

Phage typing is a method of biotyping in which *Salmonella* serovars are classified into different groups on the basis of phage susceptibility pattern. *Salmonella* lawn culture is prepared on nutrient agar plate and phage preparation is added at their routine test dilution (RTD) concentration and incubated at 37°C overnight. Bactericidal phages infect the bacterial cells, lyse them and cause development of confluent, semi confluent lysis or small and large plaques. Plaque size and lysis pattern classify the serovars into different phage types. Phage typing scheme has been successfully used for detection and classification of *S.* Typhimurium (Guinee and Van Leeuwen 1978), *S.* Enteritidis and *S.* Bareilly (Majumdar and Singh 1973, Jayasheela *et al.* 1987). Phage typing is generally employed to determine the common source of infection in salmonellosis foodborne outbreaks. It is very reproducible when international standard sets of typing phages are used. More than 200 definitive phage types (DT) have been reported so far. For example, *S.* Typhimurium DT104 designates a particular phage type of Typhimurium isolates (Hanes 2003, Andrews and Baumler 2005).

Plasmid Profiling

Plasmid profiling was the first genotypic method used for strain differentiation in Enterobacteriacae family by Shaberg *et al.* (1981). *Salmonella* strains carry plasmids ranging from 2 to 150 kb in size. Plasmid is an extra-chromosomal element which is not necessary for survival but it carries some useful genes such as antibiotic resistance gene and bacteriocin production gene, which give selective advantage to the carrying bacteria during any selective pressure in the environment. When bacterial cell divides into daughter bacterial cells, the parental plasmid also copies itself and one copy transfers to daughter cell. Same plasmid profiling of *Salmonella* strains from a narrow geographical region indicates that these might be derived from the same bacterial clone. In plasmid profiling, plasmid are extracted and run in agarose gel electrophoresis. Dissimilar plasmid profiles clearly define the divergence among *Salmonella* strains but similar plasmid profiles should be interpreted with caution because the plasmids of same size having different genetic composition may give similar profile. This problem may be overcome by restriction enzyme digestion of extracted plasmid and banding pattern will be true to plasmid profile.

Plasmid profiling has been used in studies of zoonotic salmonellosis and investigation of outbreaks associated with *Salmonella* in milk, hamburgers, precooked roast beef, chicken and direct transmission from animals to humans.

Emergence of Multidrug Resistant *Salmonella*

Typical *Salmonella* are sensitive to chloramphenicol, trimethoprim sulfamethoxazole and ampicillin, but the antibiotic sensitivity patterns may vary even in different strains of the same serovar isolated from different geographic areas. Resistance of bacteria to antimicrobial agents is an evolutionary process for adapting to the environmental conditions. The presence of any selective pressure (antimicrobial component) favours selection of the antibiotic resistance gene carrying bacterial population by enhancing their survival and multiplication. It causes the emergence of an antimicrobial resistant bacterial population from the previously susceptible population.

The emergence of antimicrobial resistance has been recognized by the World Health Organization (WHO) and national authorities as a major public health problem due to its serious social and economic effects. Antimicrobial resistant bacteria cause increased morbidity and mortality due to failure of standard treatment, higher cost of treatment with expensive drugs, risk of transmitting antimicrobial resistance to other susceptible bacteria and emergence of multi-drug resistant organisms, and financial burden on account of research and development of new effective drugs. The emergence of multidrug resistance also causes negative impact on global trade of animal products and tourism as well as healthcare industry in developing countries like India.

The major causes of emergence of antimicrobial drug resistance are:

1. Inappropriate and irrational use of antimicrobials without studying drug susceptibility pattern of the organism,
2. Indiscriminate use of antibiotics in animal husbandry at suboptimal levels to enhance production (Tollefson *et al.* 1997),
3. Increased use of antimicrobials in agriculture and aquaculture practices,
4. Non compliance of food safety rules and regulations, for example, using antibiotics in meat and meat products as preservatives, and
5. Weak surveillance and monitoring of the emergence of antimicrobial resistant microorganisms.

Multi-drug resistant phenotypes have been increasingly described among *Salmonella* species worldwide (WHO 2000). Most antimicrobial-resistant *Salmonella* infections are acquired from contaminated foods of animal origin (Angulo *et al.* 2000). There are many reports from developed countries documenting increased incidence of multi-drug resistant *S.* Typhimurium and *S.* Enteritidis. These bacteria have gained resistance mechanism for different antimicrobial drugs such as ampicillin, amoxicillin/clavulanic acid, chloramphenicol, sulfamethoxazole, tetracycline, trimethoprim/sulfamethoxazole, nalidixic acid, neomycin, and polymyxin B. The prevalence of antibiotic resistant bacteria has increased with time. The occurrence of fluoroquinolones-resistant organisms is comparatively less common. In India, multi-drug resistant *Salmonella* have been reported in a wide range of animal species including equines, goats, buffaloes, eggs and various products of vegetable origin.

Vaccination

Vaccination against salmonellosis has gained importance due to the increased incidence of zoonotic salmonellosis and emergence of multi-drug resistant *Salmonella* strains throughout the world. As zoonotic salmonellosis usually occurs by consumption of infected or contaminated animal products, the vaccination may control salmonellosis in animals which in turn reduces the incidence of *Salmonella* infection in human beings. Vaccination in food animals should be adopted and practised with great caution because the vaccine strain may enter human body through food chain and may lead to the emergence of more virulent *Salmonella* serovars and may cause pandemic salmonellosis in human beings. Currently, there are only three types of vaccines available.

1. Live Attenuated Vaccine

Live vaccine should have sufficiently attenuated organisms with reduced virulence having complete antigenicity. Live vaccine has several advantages such as inducing cellular and humoral response both, mucosal immunity, long lived protection, no need of adjuvant and easy administration. Certain disadvantages associated with such vaccines are the presence of some immune suppressive structural component, possibility of gaining virulence in the environment or in other host, and the chances of acquiring gene from other microorganisms by natural gene transfer. Now-a-days, bacterial genomic study allows us to develop some live attenuated mutant strains through knocking out the virulence or biosynthesis genes. The most widely studied metabolically attenuated strains include the mutants deficient in biosynthesis of aromatic amino acids (*e.g. aro*A or *aro*C and *aro*D) or purines. This causes hindrance in bacterial replication in the host by restricting nutrient supply which is necessary for replication, making them nonpathogenic in host. Besides, *htr*A knockout mutants are deficient in surviving intracellularly; thus have been exploited for development of a few experimental salmonellosis vaccines. The most studied virulence attenuated vaccine strains are made by knocking out of the two-component regulatory system, phoP/phoQ in *Salmonella* pathogenicity island 2 (SPI2) loci (Singh *et al.* 2005c, Singh 2009).

2. Killed Whole Cell Vaccine

In this, whole organism is killed by physical or chemical means. This vaccine is relatively safe but the disadvantages of killed vaccines are incomplete protective antibody response, lack of cellular immune response, and the failure to produce secretary immunoglobulin (sIg) for mucosal immunity (Singh 2009). The shortcoming of the killed vaccines may be reduced by culturing vaccine candidates under iron limiting microaerophilic conditions, or through use of adjuvant to induce cell mediated immunity (CMI) and mucosal immunity (sIgA) (Baljer *et al.* 1986).

3. Subunit Vaccine

They are prepared by selecting only immunodominant antigenic epitope while avoiding inclusion of immunosuppressive epitope. Subunit vaccine may be crude or purified extract of the pathogen, synthetic peptide or recombinant protein, recombinant DNA or RNA. Subunit vaccine may be prepared from outer membrane proteins (OMPs), toxins, porins and ribosomal fractions. Certain advantages of the subunit vaccines are safety, targeted delivery where immunity required (mucosal immunity can be induced) and differentiation between vaccinated and infected animal (DIVA). Among the subunit vaccines, *Salmonella* cytotoxin I-based toxoid vaccine was patented long back for control of salmonellosis in poultry birds (U.S.A. US 6,605, 285 B2 and India 189049, 96).

Although vaccination protects animals from disease, there are some limitations associated with vaccination in food animals which are enumerated below:

☆ Protection against one serovar does not cross protect getting infection from other serovar infection.

☆ Vaccination in food animals may cause emergence of more virulent *Salmonella* serovars by capture of empty niche area which was previously colonized by less virulent or avirulent serovars. This has happened in the past in developed countries. The use of killed vaccines for eradication of S. *enterica* subsp. *enterica* serotype Abortusequi from equines and S. *enterica* subsp. *enterica* serotype Pullorum/Gallinarum from birds facilitated colonization of niche with more dangerous and potentially zoonotic serovars as S. *enterica* subsp. *enterica* serotype Typhimurium and S. *enterica* subsp. *enterica* serotyope Enteritidis in horses and birds, respectively (Woodward *et al.* 2002). These zoonotic salmonella serovars caused severe pandemic outbreaks.

☆ Vaccine strain may excrete from food animals and may enter human host. It may cause emergence of more dangerous *Salmonella* serovars.

Prevention and Control

Zoonotic salmonellosis may be effectively controlled in humans by checking access to *Salmonella* contaminated food. Animal products may get contaminated at any stage during production, processing and distribution. Control strategy must be aimed to prevent *Salmonella* infection in animals, ensuring *Salmonella* free animal products, efficient processing of foods keeping high microbiological standards in practices, and checking cross-contamination during distribution or retail marketing.

Some strategies that may be helpful in controlling salmonella infection in animal farms are listed below:

1. Isolation of infected or carrier animals,

2. Provision of *Salmonella* free feed and clean and safe drinking water,

3. Personnel hygiene in animal sheds to prevent transmission of *Salmonella* infection from animal handlers, and

4. Adoption of scientific management practices such as proper cleaning and disinfection of animal houses, mixing of organic acid in animal feed as decontaminating agent, and filling of environmental niche in newly hatched poultry chicks by commensal bacteria using gut flora from other healthy birds (Teotia *et al.* 2005).

Salmonella contamination in meat products may be acquired during slaughter process or subsequently due to carrier animals, unhygienic slaughter, cross-contamination from infected carcasses, and contamination from working surface and meat handlers. This may be prevented by stress free transportation of animals to slaughterhouse, providing adequate rest, feed and water, avoiding close confinement and overcrowding in lairage, proper disinfection of slaughtering equipment and working surfaces, hygienic disposal of waste and excreta, checking cross-contamination from infected carcasses or GIT content and offal, proper refrigeration of meat after slaughtering and creating public awareness about health hazards associated with salmonellosis. Educating the people for adopting adequate heat treatment (cooking) of meat and meat products before consumption is also essential. Similarly, proper pasteurization or boiling of milk is necessary to prevent milk-borne salmonellosis in humans.

Vaccination of food animals may be a strategy for controlling infection in animals, but this strategy must be adopted with great caution considering its public health implications. Regular disease surveillance and monitoring is essential to have a watch on the incidence of salmonellosis, prevalence of different serovars of the organism, and emergence of drug resistant *salmonella.*

References

1. Alam J, Singh BR, Hansda D, Singh VP and Verma JC. 2009. Evaluation of *aro*A deletion mutant of *Salmonella enterica* subspecies *enterica* serovar Abortusequi for its vaccine candidate potential. *Indian J Exp Biol* 47: 871-879.

2. Andrews HL and Baumler AJ. 2005. *Salmonella* species. In: Fratamico PM, Bhunia AK and Smith JL (Eds) *Foodborne pathogens: microbiology and molecular biology.* Horizon Scientific Press Ltd., UK.

3. Angulo FJ, Johnson KR, Tauxe RV and Cohen ML. 2000. Origins and consequences of antimicrobial-resistant nontyphoidal *Salmonella*: implications for the use of fluoroquinolones in food animals. *Microbial Drug Resis* 6: 77-83.

4. Babu N, Singh BR, Harishankar, Agrawal RK, Chandra M, Vijo TV, Srivastava SK and Yadav MP. 2008. Prevalence of *Salmonella* in equids determined by microbiological culture, standard tube agglutination test and PCR. *Haryana Vet* 47: 58-63.

5. Baljer G, Hoerstke M, Dirksen G, Sailer J and Mayr A. 1986. Efficacy of local and parenteral vaccination against calf salmonellosis with inactivated vaccines. *J Vet Med* 33: 206-212.

6. Baquar N, Burnens A and Stanley J.1994. Comparative evaluation of molecular typing of strains of a national epidemic due to *Salmonella brandenburg* by rRNA gene and IS200 probes and pulsed-eld gel electrophoresis. *J Clin Microbiol* 32: 1876-1880.

7. Buchmeier NA and Heffron F.1990. Induction of *Salmonella* stress proteins upon infection of macrophages. *Sci* 248: 730-732.

8. Chandra M, Singh BR, Arora BM, Shankar H, Agarwal RK and Sharma A. 2008. Isolation of *Salmonella* from the faeces of captive wolves. *J Vet Pbl Health* 6: 107-110.

9. Chowdry N, Threlfall EJ, Rowe B and Stanley J. 1993. Genotype analysis of faecal and blood isolates of *Salmonella dublin* from humans in England and Wales. *Epidemiol Infect* 110: 217-225.

10. Christensen HC, Møller PL, Vogensen FQ and Olsen JE. 2000. Sequence variation of the 16S to 23S rRNA spacer region in *Salmonella enterica*. *Res Microbiol* 151: 37-42.

11. Duguid JP, Anderson ES, Alfresson GA, Barker R and Old DC. 1975. A new biotyping scheme for *Salmonella typhimurium* and its phylogenetic signicance. *J Med Microbiol* 8: 149-166.

12. El-Gazzar FE and Marth EH. 1992. *Salmonella*e, salmonellosis, and dairy foods: A review. *J Dairy Sci* 75: 2327-2343.

13. Fields PI, Groisman EG and Heffron F.1989. A *Salmonella* locus that controls resistance to microbicidal proteins from phagocytic cells. *Sci* 243: 1059-1062.

14. Gomez TM, Motarjemi Y, Miyagawa S, Kaferstein and Stohr K. 1997. Foodborne salmonellosis. *World Health Stat Quart Rep* 50: 81-89.

15. Gray JT and Fedorka-Cray PJ. 2002. *Salmonella*. In: Cliver DO and Riemann HP (Eds) *Foodborne diseases*. Academic Press, San Diego.

16. Grimont F and Grimont PAD. 1986. Ribosomal ribonucleic acid gene restriction patterns as potential taxonomic tools. *Ann Inst Pasteur/Microbiol* 137B: 165-175.

17. Guinee PAM and van-Leeuwen WJ. 1978. Phage typing of *Salmonella*. *Methods Microbiol* 11: 158-190.

18. Hanes D. 2003. Nontyphoid *Salmonella*. In Henegariu O, Heerema N A, Dloughy S R, Vance G, Hand Vogt PH (Eds.). *International handbook of foodborne pathogens*. Marcel Dekker, Inc, New York.

19. Hilton AC, Banks JG and Penn CW. 1996. Random amplication of polymorphic DNA (RAPD) of *Salmonella*: strain differentiation and characterization of amplied sequences. *J Appl Bacteriol* 81: 575-584.

20. Humphrey T. 2000. Public health aspects of *Salmonella* infection. *Salmonella in domestic animals*. CAB International, Wallingford.

21. Jayasheela M, Singh G, Sharma NC and Saxena SN. 1987. A new scheme for phage-typing *Salmonella* bareilly and characterization of typing phages. *J Appl Bacteriol* 62: 429-432.

22. Jensen MA and Hubner R J. 1996. Use of homoduplex ribosomal DNA spacer amplication products and heteroduplex cross-hybridization products in the identication of *Salmonella* serovars. *Appl Environ Microbiol* 59: 945-952.

23. Johnson EH, Hietala S and Smith BP. 1985. Chemiluminescence of bovine alveolar macrophages as in indicator of developing immunity in calves vaccinated with aromatic-dependent *Salmonella*. *Vet Microbiol* 10: 451-464.

24. Kapperud G, Lassen, J, Dommarsnes K, Kristiansen BE, Caugant, DA, Ask E and Jahkola M. 1989. Comparison of epidemiology marker methods for identication of *Salmonella* Typhimurium isolates from an outbreak caused by contaminated chocolate. *J Clin Microbiol* 27: 2019-2024.

25. Kauffmann F. 1961. The species denition in the Enterobacteriaceae. *Int Bull Bacteriol NomenTaxon* 11: 5-6.

26. Kilger G and Grimont PAD. 1993. Differentiation of *Salmonella* phase1 agellar antigen types by restriction of the amplied iC gene. *J Clin Microbiol* 31: 1108-1110.

27. Laconcha I, Löpez-Molina N, Rementeria A, Audicana, A, Perales I and Garaizar J. 1998. Phage typing combined with pulsed-eld gel electrophoresis and random amplied polymorphic DNA increases discrimination in the epidemiological analysis of *Salmonella enteritidis* strains. *Int J Food Microbiol* 40: 27-34.

28. Lam S and Roth JR. 1983. IS200, a *Salmonella*-specific insertion sequence. *Cell* 34: 951-960.

29. Lancaster JE and Crabb WE. 1953. Studies on disinfection of eggs and incubators. *British Vete J* 109: 139-148.

30. Lin AW, Usera MA, Barrett TJ and Glodsby RA.1996. Application of random amplied polymorphic DNA analysis to differentiate strains of *Salmonella enteritidis. J Clin Microbiol* 34: 870-876.

31. Lindgren SW, Stojiljkovic I and Heffron F.1996. Macrophage killing is an essential virulence mechanism of *Salmonella typhimurium. Proc Nat Acad Sci USA* 93: 4197-4201.

32. Luk JMC, Kongmuang U, Reeves PR and Lindberg AA. 1993. Selective amplication of abequose and paratose synthase genes (rfb) by polymerase chain reaction for identication of *Salmonella* major serogroups (A, B, C2, and D). *J Clin Microbiol* 31: 2118-2123.

33. Majumdar AK and Singh SP. 1973. A phage typing scheme for *S.* bareilly. *Indian Vet J* 50: 1161-1166.

34. Mastroeni P, Villarreal Ramos B and Hormaeche CE. 1993. Adoptive transfer of immunity to oral challenge with virulent *Salmonella*e in innately susceptible BALB/c mice requires both immune serum and T cells. *Infect Immun* 61: 3981-3984.

35. McCullough NB and Eisele CW. 1951. Experimental human salmonellosis; pathogenicity of strains of *Salmonella meleagridis* and *Salmonella anatum* obtained from spray-dried whole egg. *J Infect Dis* 88: 278-289.

36. McGhee JR, Mestecky J, Dertzbaugh MT, Eldridge JH, Hirasawa M and Kiyono H.1992. The mucosal immune system: from fundamental concepts to vaccine development. *Vaccine* 10: 75-88.

37. Mead PS, Slutsker L, Dietz V, McCaig LF, Bresee JS, Shapiro C, Griffin PM and Tauxe RV. 1999. Food-related illness and death in the United States. *Emerg Infect Dis* 5: 607-625.

38. Michetti P, Mahan MJ, Slauch JM, Mekalanos JJ and Neutra MR. 1992. Monoclonal secretory immunoglobulin A protects mice against oral challenge with the invasive pathogen *Salmonella* typhimurium. *Infect Immun* 60: 1786-1792.

39. Montville TJ and Matthews KR. 2008. *Food microbiology: an introduction*, 2nd edn. ASM Press, Washington.

40. Morgan PB. 1990. Complement and infectious diseases. In: Morgan PB (ed) *Complement: clinical aspects and relevance to disease*. Harcourt Brace Jovanovich, London.

41. Olsen JE and Skov MN. 1994. Genomic lineage of *Salmonella dublin*. *Vet Microbiol* 40: 271-282.

42. Olsvik Ø, Sørum H, Birkeness K, Wachsmuth K, Fjølstad M, Larsen J, Fossum K and Feeley JC. 1985. Plasmid characterization of *Salmonella* typhimurium transmitted from animals to humans. *J Clin Microbiol* 22: 336-338.

43. Reeves MW, Evins GM, Heiba AA, Plikaytis BD and Farmer JJ III. 1989. Clonal nature of *Salmonella* Typhi and its genetic relatedness to other salmonellae as shown by multilocus enzyme electrophoresis, and proposal of *Salmonella bongori* comb. nov. *J Clin Microbiol* 27: 313-320.

44. Shaberg DR, Tompkins LS and Falkow S. 1981. Use of agarose gel electrophoresis of plasmid deoxyribonucleic acid to ngerprint Gram-negative bacilli. *J Clin Microbiol* 13: 1105-1108.

45. Shah SA and Romick TL. 1997. Subspecies differentiation of *Salmonella* by PCR-RFLP of the ribosomal operon using universal primers. *Let Appl Microbiol* 25: 54-57.

46. Singh BR. 2000a. Molecular diagnosis of salmonellosis. In: Rathore BS (Ed) *Advances in diagnosis of bacterial and viral diseases of animals*. Indian Veterinary Research Institute, Izatnagar, India.

47. Singh BR. 2000b. Recent advances in diagnosis of Salmonellosis. *ICAR Summer School on Recent Advances in Diagnosis of Zoonoses*, IVRI, Izatnagar 2-23 May 2000, Bareilly.

48. Singh BR. 2000c. Advances in diagnosis and molecular epidemiology of Salmonellosis. At *ICAR School on Diagnosis of Emerging Diseases of Livestock in India*, IVRI, Izatnagar, Bareilly.

49. Singh BR. 2003. Prevalence of different *Salmonella* serovars in animals and their environment in India in last two decades. http://upgov.up.nic.in/ivri/nsc. Accessed 30 June 2005.

50. Singh BR. 2004. Advances in diagnosis of Salmonellosis. At ICAR sponsored short course on *'Emerging poultry diseases: Their diagnosis and control by using molecular biology techniques'*. IVRI, Izatnagar, 4-13 October 2004, Bareilly.

51. Singh BR. 2005. Prevalence of *Salmonella* serovars in animals in India. http://www.aclisassari.com/acli-openlearning/uploads/lectures/Methods. Acessed 30 June 2005.

52. Singh BR. 2009. *Salmonella* Vaccines for animals and birds and their future perspective. *The Open Vaccine J* 2: 100-112

53. Singh BR and Kulshreshtha SB. 1993. Occurence of enterotoxigenic *Salmonella* serotypes in seafoods. *J Food Sci Technol* 30: 438-439.

54. Singh BR and Sharma VD. 1999. Isolation and characterization of four distinct cytotoxic factors of *Salmonella* Weltevreden. *Zbl Bakt Int J Med Microbiol* 289: 457-474.

55. Singh BR and Sharma VD. 2000. Serodetection of *Salmonella* Cytotoxins and toxinotyping of *Salmonella* strains from foods of animal origin. *Indian J Comp Microbiol Immunol Infect Dis* 21:39-48.

56. Singh BR and SharmaVD. 2003. *Salmonella* cytotoxins: Role in poultry birds. *Indian J Vet Res* 121: 1-13.

57. Singh BR, Kulshreshtha SB, Kapoor KN and Paliwal OP. 1992. Occurrence of *Salmonella* from fish/seafoods and detection of their enterotoxigenicity. *Proceedings of 3rd World congress on foodborne infections and intoxications* Vol 1. Berlin.

58. Singh BR, Khurana SK and Kulshreshtha SB. 1995. Survivability of *Salmonella paratyphi B var java* on experimentally infected cockroaches. *Indian J Exp Biol* 33: 392-398

59. Singh BR, Singh KP and Sharma VD. 1998. Experimental pathology of *Salmonella* Gallinarum cytotoxin in adult White Leghorn birds. *Indian J Vet Pathol* 22: 127-130.

60. Singh BR, Mark PS and Wallis T. 1999. Characterization of strain harbouring defined mutations in *aro*A, *htr*A, and double mutant *(aro*A-*htr*A*) of S. Abortusequi*. Tombit Report. Submitted to Indian Council of Agricultaul Research, New Delhi.

61. Singh BR, Alam, J, Hansda, D, Verma JC, Singh VP and Yadav MP. 2002. Evaluation of guinea pig model for experimental *Salmonella* serovar Abortusequi infection in reference to infertility. *Indian J Exp Biol* 40: 296-303.

62. Singh BR, Alam J, Verma JC and Singh VP. 2003. Genetic resistance of guinea fowls to *Salmonella* Gallinarum Infection. *Pantnagar J Res* 1: 48-52.

63. Singh BR, Alam J and Hansda D. 2005a. Alopecia induced by salmonellosis in guinea pigs. *Vet Record* 156: 516-518.

64. Singh BR, Chandra M, Agarwal RK and Babu N. 2005b. Curing of *Salmonella enterica* serovar Typhimurium contaminated cowpea seeds and sprouts with vinegar and chlorinated water. *J Food Proces Preserv* 29: 268-277.

65. Singh BR, Singh Y, Agarwal MC, Agarwal RK and Sharma VD. 2005c. *Salmonella* Vaccines for Veterinary Use: An overview. *Haryana Vet* 44: 1-12.

66. Singh BR, Chandra M, Agrawal RK and Babu N. 2006a. Antigenic competition among different 'O' antigens of *Salmonella enterica* subspecies *enterica* serovars during hyperimmunization in pony mares. *Indian J Exp Biol* 44: 1022-1025.

67. Singh BR, Singh M, Singh P, Babu N, Chandra M and Agarwal RK. 2006b. Prevalence of multidrug-resistant *Salmonella* on ready-to-eat betel leaves (Paan) and in water used for soaking betel leaves in North Indian cities. *J Food Prot* 69: 288-292

68. Singh BR, Chandra M, Agarwal RK and Babu N. 2007a. Effect of *Salmonella* contamination on soil fertility. *Res Crops* 8: 136-140.

69. Singh BR, Chandra M, Agarwal RK and Babu N. 2007b. Infertility associated with sub-clinical salmonellosis. *Indian J Exp Biol* 45: 834-836.

70. Singh BR, Chandra M, Agrawal RK and Babu N. 2007c. Interactions between *Salmonella enterica* subspecies *enterica* serovar Typhimurium and cowpea (*Vigna unguiculata* variety Sinensis) seeds, plants, and persistence in hay. *J Food Safety* 28: 169-187.

71. Singh BR, Kumar A, Agrawal S and Verma A. 2007d. Effect of *Salmonella* Typhimurium and *Serratia fonticola* on germination of seeds of common crops. *J Infect Dev Ctr* 1: 67-71.

72. Singh BR, Verma JC, Singh KP and Singh VP. 2009. Sero-prevalence of *Salmonella* infection in animals in north India. *Haryana Vet* 48: 43-46.

73. Singh Y, Sharma VD, Sharma SN and Singh BR. 1997. Experimental toxicity of *Salmonella* Weltevreden cytotoxin in mice. *Indian J Comp Microbiol Immunol Infect Dis* 18: 136-141

74. Stanley J, Jones C and Threlfall EJ. 1991. Evolutionary lines among *Salmonella enteritidis* phage types are identied by insertions sequence IS200 distribution. *FEMS Microbiol Let* 82: 83-90.

75. Stevens A, Joseph C, Bruce J, Fenton D, O'Mahoney M, Cunningham D, O'Connor B and Rowe B.1989. A large outbreak of *Salmonella* enteritidis phage type 4 associated with eggs from overseas. *Epidemiol Infect* 103: 425-433.

76. Suresh T, Hatha AAM, Sreenivasan D, Sangeetha N and Lashmanaperumalsamy P. 2006. Prevalence and antimicrobial resistance of *Salmonella enteritidis* and other *Salmonella*s in the eggs and egg-storing trays from retails markets of Coimbatore, South India. *Food Microbiol* 23: 294-299

77. Swain SL, Weinberg AD, English M and Huston G.1990. IL-4 directs the development of Th2-like helper effectors. *J Immunol* 145: 3796-3806.

78. Teotia UVS, Agrawal RK, Pant S, Singh BR and Raj H.2005. How to control foodborne salmonellosis. http://www.nio.org/past_events/comits/hum_patho mar_syst.jsp. Accessed 30 December 2011.

79. Tollefson L, Altekruse SF and Potter ME. 1997. Therapeutic antibiotics in animal feeds and antibiotic resistance. *Rev Sci Tech* 16: 709-715.

80. Tyler KD, Wang G, Tyler SD and Johnson WM. 1997. Factors affecting reliability and reproducibility of amplication-based DNA ngerprinting of representative bacterial pathogens. *J Clin Microbiol* 35: 339-346.

81. Verma JC and Singh BR. 2000. Prevalence and distribution of *Salmonella* serotypes in animals and poultry in India. Current status. Proceedings of Indian Association for Advancement of Veterinary Research, VI Annual conference. Bareilly.

82. Verma JC, Singh VP, Singh BR and Gupta BR.2001. Occurrence of *Salmonella* serotypes in animals in India. *Indian J Comp Microbiol Immunol Infect Dis* 22: 51-55.

83. Verma M, Paliwal OP, Singh BR, Sharma AK and Agarwal M. 2003. Experimental *Salmonella* Abortusequi infection in guinea pigs: Clinico-haematological studies. *Indian J Vet Pathol* 27: 20-23.

84. Verma M, Paliwal OP, Singh BR, Sharma AK and Tripathi BN.2004. Pathology of *Salmonella* Abortusequi infection in pregnant and non-pregnant guinea pigs. *Indian J Vet Pathol* 28: 7-10.

85. Vescovi EG, Soncini FC and Groisman EA. 1996. Mg 2+ as an extracellular signal: environmental regulation of *Salmonella* virulence. *Cell* 84: 165-174.

86. Wayne LG, Brenner DJ, Colwell RR, Grimont PAD, Kandler O, Krichevsky MI, Moore H, Moore WEC, Murray RGE, Stackebrandt E, Starr MP and Trüper HG.1987. Report of the ad hoc committee on reconciliation of approaches to bacterial systematics. *Int J Syst Bacteriol* 37: 463-464.

87. Weide-Botjes M, Kobe B, Lange C and Schwarz S. 1998. Molecular typing of *Salmonella* enterica subspecies enterica serovar Hadar: evaluation and application of different typing methods. *Vet Microbiol* 61: 215-227.

88. WHO. 2000. Overcoming antimicrobial resistance. World Health Organization. http://www.who.int/infectious-disease report/2000/index.html. Accessed 12 April 2012.

89. Williams JGK, Kubelik AR, Livak KJ, Rafalski, JA and Tingey SV. 1990. DNA polymorphisms amplied by arbitrary primers useful as genetic markers. *Nucleic Acid Res* 18: 6531-6535.

90. Woodward MJ, Gettinby C, Breshlin MF, Corkish JD and Houghton S. 2002. The efficacy of salenvac, a *Salmonella enterica* subsp. enterica serotype Enteritidis iron restricted bacterin vaccine, in laying chickens. *Avian Pathol* 31: 383-392.

91. Yousef A E and Carlstrom C. 2003. *Salmonella*. In: Yousef AE and Carstrom C (Eds) *Food microbiology: a laboratory manual*. John Wiley and Sons, New Jersey.

2014, Zoonoses: Bacterial Diseases

Editor: **Sudhi Ranjan Garg**

Published by: **DAYA PUBLISHING HOUSE, NEW DELHI**

Pages **173–186**

8

Escherichia coli Infections

D.K. Sinha, Bhoj Raj Singh, Younis Farooq and Sandeep Kumar

Centre for Animal Disease Research and Diagnosis,
Indian Veterinary Research Institute, Izatnagar – 243 122

The disease investigation reports from Central Disease Diagnosis Laboratory (CDDL), Indian Veterinary Research Institute, Izatnagar, reveal that out of 400 bacterial strains isolated from sick or morbid animals in the year 2011-12, more than 20 per cent consisted of *E. coli* isolates only, more common than any other bacteria. There are several pathotypes of *E. coli* including enterotoxigenic *E. coli* (ETEC), entheropathogenic *E. coli* (EPEC), enterohaemorrhagic *E. coli* (EHEC), uropathogenic *E. coli* (UPEC), enteroaggregative *E. coli* (EAgEC) and verotoxigenic *E. coli* (ETEC) associated with infections in human beings and animals that have zoonotic potential. Often, the *E. coli* infections are self limiting and with little chance of mortality except in very young and aged individuals. However, some *E. coli* that produce verotoxin (VT), shigatoxin (ST) or shiga like toxin (SLT) have the potential to cause severe disease as bloody diarrhoea and haemolytic uremic syndrome (HUS) in humans. HUS is characterized by haemolytic anaemia (due to abnormal breakdown of RBCs) and thrombocytopenia (reduction of platelets) leading to severe bloody diarrhoea. HUS is associated with severe acute renal failure, often requiring intensive care. *E. coli* is also involved in various disorders in animals like white scour/colibacillosis in neonatal calves, gas oedema in piglets, mastitis in milch cattle, naval ill, foal diarrhoea in horses and colisepticaemia and enteritis in poultry causing heavy economic losses in the livestock and poultry sector. Apart from domestic animals, zoo and wild animals also suffer from colibacillosis, particularly during water scarcity and flood seasons (Malik *et al.* 2000).

Etiological Agent

Theodor von Escherich (1885) first isolated and identified *E. coli* from the faeces of infants. Since then it has been reported to cause a number of foodborne infections and intoxications. It is a Gram negative, rod shaped, motile and facultative anaerobe measuring about 1.1-1.5 μm × 2-6 μm. Both smooth and rough strains (due to loss of LPS in outer membrane) of *E. coli* are found. On the basis of pathogenicity, *E. coli* strains are categorized into four types:

(a) **Enteropathogenic *E. coli* (EPEC):** non-invasive, non-toxin producing and involved in infant diarrhoea.

(b) **Enterotoxigenic *E. coli* (ETEC):** non-invasive, enterotoxin producing strain associated with most severe cholera like disease; colonizes in the gut with the help of specific fimbrial antigen (adhesins) and cause enterotoxin mediated diarrhoea.

(c) **Enteroinvasive *E. coli* (EIEC):** invades enterocytes of intestinal mucosa of colon region and produces shigellosis-like dysentery.

(d) **Enterohaemmorrhagic *E. coli* (EHEC):** It is also called as verocytotoxigenic *E. coli* (VTEC) and produces Shiga-like-Toxin (SLT). This category includes the most important 0157:H7 strain of *E. coli*, which multiplies in the lamina propria. Its pathogenicity is SLT based which is of two types: SLT-I and SLT-II.

The virulence of *E. coli* is often plasmid mediated; therefore, a loss or gain of virulence associated plasmid may alter the pathogenicity of a strain. Besides toxins and invasins, the ability of *E. coli* to use digestive enzymes like trypsin might be an important attribute of the pathogen to thrive in the intestine even when the patients are on parenteral therapy, virtually with nothing in the intestine (Singh and Sharma 2001). Similarly, adhesion of *E. coli* either due to fimbriae or through other adhesion molecules and ligands is considered very important (even more than their toxigenic potential) for causation of gastrointestinal, urogenital and respiratory tract infections. Anionic charges on the bacterial surface have been shown directly associated with the pathogenic potential of several enteropathogens, including *E. coli* and *Salmonella* (Singh and Sharma 2003).

The current model of pathogenesis indicates that Stx produced by EHEC during colonization of the intestinal tract gains entry to the host through epithelial cells and acts on submucosal immune cells that release cytokines. These in turn induce inflammation and increase the expression of the Stx receptor globotriaosylceramide (Gb3) (O'Loughlin and Robins-Browne 2001). Stx then targets the endothelium of organs in which the Gb3 receptor is expressed (*e.g.* intestine, kidneys and brain; Boyd and Lingwood 1989). Because the Gb3 receptor is a glycosphingolipid, variations in the lipid moieties of its structure may influence Stx binding (Kiarash *et al.* 1994). Stx-mediated endothelial injury activates coagulation, and inhibition of fibrinolysis leads to accumulation of fibrin and thrombosis (Tarr *et al.* 2005). The combination of Stx and O157 lipopolysaccharide (LPS) induces platelet-leukocyte aggregates and tissue factor release and thus contributes to a prothrombotic state (Stahl *et al.* 2009).

Another important virulence factor of EHEC is an outer membrane protein called intimin, which is encoded by the *eae* gene in the locus of enterocyte effacement (LEE) (Jerse *et al.* 1990, Jerse and Kaper 1991, Yu and Kaper 1992). During EHEC infection, intimin assists in colonization and induces the characteristic intimate attachment to intestinal epithelial cells and effacement of microvilli (attaching and effacing lesions) by binding to its own receptor (the translocated intimin receptor or Tir), also produced by EHEC and transferred to the host's intestinal epithelial cells by a type 3 secretion system encoded in LEE (Kenny *et al.* 1997, Paton *et al.* 1998). Expression of EHEC LEE genes is regulated by quorum sensing and is induced by the host's adrenergic hormones (Sperandio *et al.* 2003).

Some LEE-negative non-O157 EHEC strains may also produce a novel and highly potent subtilase cytotoxin (SubAB) that, when injected intraperitoneally in mice, results in microvascular thrombosis and necrosis in various organs including the brain, kidneys, and liver (Paton *et al.* 2004). However, the role of SubAB in human EHEC disease remains to be elucidated. Interestingly, the SubAB receptor is generated by metabolic incorporation of an exogenous glycan derived from food (Byres *et al.* 2008).

Host Range

All warm blooded animals harbour *E. coli* in their guts. *E. coli* of zoonotic and pathogenic potential have been isolated from diseased as well as healthy human beings, animals (pet, domestic and wild), lizards, birds (domestic, domiciled and wild) and fish and other aquatic animals more frequently than any other bacteria all over the globe including India (Singh *et al.* 1994, 1996a, 1997, 2003, Malik *et al.* 2000, Singh 2009a, 2010a, b, 2012). Cattle are considered the major reservoir of O157:H7 serotype of *E. coli,* though this serotype has also been isolated from lamb, pig, poultry and buffalo meat and their products (Singh *et al.* 1996).

Epidemiology

The faecal-oral route is the main mode of transmission of *E. coli* infection. Contaminated food, especially that of animal origin, and water are main sources of *E. coli* infection. Unsanitary condition, polluted water, overcrowding, poor personal and community hygiene, unhygienic production and transportation of food of animal origin facilitate in spread of *E. coli* infection. The less developed and developing countries, due to their poor state of hygiene and nutrition, might have more prevalence of *E. coli* infection than the developed countries. Age has significant role in occurrence of *E. coli* infection. The neonates having immature immunity, immunocompromised people with underlying diseases as AIDS, and aged are the most vulnerable to *E. coli* infection. Experimentally, it has been proved that underlying diseases may be associated with antigenic competition and may affect the development of protective immunity in young age (Singh *et al.* 2002, 2006).

In an analysis of the role of cattle as a reservoir of Shiga toxin-producing *E. coli* (STEC) during 1993-1995, Mora *et al.* (2011) reported that one-third of the calves and cows of Galicia were carriers. Furthermore, 12 per cent of the calves and 22 per cent of the farms sampled were positive for highly virulent STEC serotype O157:H7. In 1998, they conducted a study in Navarra in two slaughterhouses and five feedlots. The number of STEC O157:H7 carriers detected was very high: 10 per cent at slaughterhouse A, 19 per cent at slaughterhouse B, 23 per cent at feedlot 1, 22 per cent at feedlot 2, 8 per cent at feedlot 3, but 0 per cent in feedlots 4 and 5. These data obtained in Spain are in agreement with those reported in other countries, and confirm that around 10 per cent of cattle are colonized by STEC O157:H7 (Blanco *et al.* 2003, Blanco *et al.* 2004a, b, Martin and Beutin 2011).

Humans are infected with *E. coli* by consuming or handling contaminated food or water or through contact with infected animals. Person-to-person transmission is also possible among close contacts, in families, childcare centres, nursing homes, etc. A wide variety of food has been implicated in the outbreaks caused by ETEC, including raw (unpasteurised) milk and cheese, undercooked beef and a variety of fresh produce including sprouts, spinach, lettuce and fresh juices. In India, fresh cane juice and orange juice have resulted in cholera like epidemics, particularly in summer when the fly population is more and people take juices from the roadside unhygienic vendors (Singh *et al.* 1995, 1996b). However, the main sources of VTEC strains are ruminants, particularly cattle. Meat can become contaminated by faecal matter due to poor processing methods during slaughter and dressing. Faeces from infected animals can contaminate other foods and water (Singh *et al.* 1996a).

Disease Manifestations in Animals

In animals, though diarrhoeic illness is the most common outcome of *E. coli* infection (Singh 2009a, 2012, Singh *et al.* 2012), four major clinical forms (either acute or mild) of *E. coli* infection are recognized:

(a) *Enterotoxic form*: It occurs in the first week of birth characterized by diarrhoea without fever.

(b) *Enterotoxaemic form*: In this, the organism produces a toxin in the intestine of diseased animals that acts elsewhere in the body.

(c) *Septicaemic form*: It is associated with bacteraemia and extraintestinal localization of the organism. In this form, endotoxin is produced which may lead to shock.

(d) *Local invasive form*: It is associated with fever and damage to intestinal epithelium and ulceration.

Disease Manifestations in Humans

In humans, depending upon the type of *E. coli* strain, five forms of enteritis (either acute or chronic) are recognized.

(a) Enteropathogenic *E. coli* (EPEC)

Its causes diarrhoea and occasionally vomiting. It causes characteristic attaching-and-effacing lesions (A/E), which can be observed by intestinal biopsy in human patients (Moon *et al.* 1983) and animals, both (Taylor *et al.* 1986) models. A/E is characterized by loss of microvilli, intimate adherence of bacteria between epithelial cell membranes (Ulshen and Rallo 1980, Rothbaum *et al.* 1982), and cytoskeletal changes such as actin polymerization directly beneath the adherent bacteria (Knutton *et al.* 1989). Generally, EPEC causes infantile diarrhoea in developing countries and sporadic diarrhoea in developed countries (Nataro *et al.* 1987). Makino *et al.* (1999) reported the isolation of EPEC from a mass outbreak.

An outbreak of EPEC 0127:H6 diarrhoea occurred at two nurseries for the newborn in Chongqing (China) in May 1987. Sixty-nine neonates had diarrhoea; two deaths resulted. The source of infection was the index case's mother who had suffered with watery stools. Transmission of EPEC 0127:H6 from infant to infant took place by way of the faecal-oral route, most likely via the hands of medical staff attending their care. It was confirmed by plasmid and restriction analyses and outer membrane protein determination of a neonate who acquired EPEC during delivery through ingestion of organisms residing in the maternal birth canal (Wu and Peng 1992).

Afset *et al.* (2003) investigated the relative contribution of EPEC as a cause of infectious diarrhoea in Norwegian children. During the one-year period (2001), 598 specimens from 440 patients, less than two years old were analysed. Potential enteric pathogens were identified in 124 patients (28.2 per cent). EPEC was the most frequently identified agent (44 patients). Only one of the eae$^+$ *E. coli* isolates was classified as typical EPEC (*bfpA$^+$*). Among the 43 isolates that were classified as atypical EPEC (*bfpA$^-$*), eight strains belonged to EPEC serogroups, whereas the majority of strains (n = 35) were not agglutinated by EPEC antisera. None of the EPEC isolates were genetically related. This study demonstrated high prevalence of atypical EPEC of non-EPEC serogroups among Norwegian children with diarrhoea.

An outbreak of infantile diarrhoea was investigated in 32 children, all below two years old, in the tropical north of Australia. Rotavirus (63 per cent) and EPEC (59 per cent) were the most common pathogens identified. Of the 19 EPEC isolates, 14 (74 per cent) were of serotype 0126:H12, hitherto unreported as an EPEC serotype. EPEC-related gastroenteritis is an uncommon but recognised cause of diarrhoeal outbreaks in Australia (Barlow *et al.* 1999).

(b) Enterotoxigenic *E. coli* (ETEC)

Its causes abdominal cramps with rice water like diarrhoea which resembles cholera but there is no fever. The condition lasts for 3-4 days. It is also called as traveller's diarrhoea.

A large outbreak of gastroenteritis caused by ETEC occurred among 90 (15 per cent) of 621 passengers and 15 (4 per cent) of 395 crew aboard a cruise in the year 2002. Clinical characteristics of illness were consistent with ETEC, which was identified among 9 of 20 stool samples collected from ill passengers and crew. From 1996 through 1999, laboratory-confirmed ETEC outbreaks represented 0.2 per cent of all foodborne outbreaks reported to the US Center for Disease Control and Prevention (CDC 2001). From 1996 to 2003, 16 outbreaks of ETEC infections in the United States and on cruise ships were confirmed. *E. coli* serotype O169:H41 was identified in 10 outbreaks and was the only serotype in six. The serotype was identified in 1 in 21 confirmed ETEC outbreaks before 1996 (Beatty *et al.* 2004).

(c) Enteroinvasive *E. coli* (EIEC)

Its infection is characterized by diarrhoea culminating in dysentery with manifestation like fever, chill, abdominal cramps and muscular pain.

Vieira *et al.* (2007) conducted case control studies in 22 rural communities in northwestern Ecuador between August 2003 and July 2005 and collected stool samples to assess the presence of diarrhoeagenic bacteria. Infection was assessed by PCR specific for *LT* and *STa* genes of ETEC, the *bfp* gene of EPEC, and the *ipaH* gene of both EIEC and shigellae. The most frequently identified pathogens were EIEC (3.2 cases/ 100 persons) and shigellae (1.5 cases/100 persons), followed by ETEC (1.3 cases/ 100 persons), and EPEC (0.9 case/100 persons). EIEC exhibited similar risk factor relationships with other pathotypes analyzed but different age-specific infection rates. EIEC was found as the predominant diarrhoeagenic bacteria in this study.

(d) Enterohaemorrhagic *E. coli* (EHEC)

Its infection is characterized by bloody stool and in some cases cardiac and renal complications. EHEC may produce two immunologically distinct toxins, Stx1 or Stx2, alone or in combination. Stx can inhibit protein synthesis (Ogasawara *et al.* 1988), and O157:H7 strains that produce Stx2 may be associated with an increased risk of systemic complications (Donohue-Rolfe *et al.* 2000).

A number of foodborne outbreaks involving *E. coli* O157:H7 strains were reported world over. In 1996, an outbreak of *E. coli* O157 killed seven and hospitalised several hundred people in Scotland (Dundas *et al.* 2001). In May 2000 in Washington and November 2000 in Pennsylvania, 56 children were infected (19 required hospitalization) through contact with animals during school or family visits to farms (CDC 2001). Approximately 26 people acquired O157:H7 in August 2001 in association with attendance at a county fair in Wisconsin in Ozaukee County. The converse of increased risk for O157:H7 infections among farm families are the possibility that individuals from farms with repeated, low-level exposure may actually develop resistance to infection (Dundas *et al.* 2001).

An outbreak of non-O157 STEC at a high school in Minnesota, USA, in November 2010 was reported in which consumption of undercooked venison and not washing hands after handling raw venison were associated with illness. *E. coli* O103:H2 and non-Shiga toxin producing *E. coli* O145: NM was isolated from the ill students and venison (Rounds *et al.* 2012).

Wahl *et al.* (2011) reported an outbreak of **stx**$_1$- and **eae**-positive STEC O145:H28 infection with mild symptoms among children in a daycare centre. Extensive sampling showed occurrence of the outbreak strain as well as other STEC and EPEC strains in the outbreak population. The multilocus variable-number tandem repeat analysis (MLVA-typing) of the STEC isolates strongly indicated a common source of infection. The non-O157 STEC outbreaks are less commonly reported than O157 outbreaks.

CDC reported a multistate outbreak of *E. coli* serotype O157:H7 infections linked to romaine lettuce in the year 2011(CDC 2011). According to another report, raw milk from a farm sickened at least 11 people with serious *E. coli* O157:H7 illnesses. Two children developed hemolytic uremic syndrome, or HUS (Marler 2012).

Between May 1 and July 4, 2011, Germany reported 852 HUS cases and 3469 cases with diarrhoea (and/or with haemorrhagic colitis), of which 50 patients died (including 32 HUS patients). According to the European Centre for Disease Prevention and Control (ECDC), 49 HUS cases and 76 cases with diarrhoea were reported in 13 other European countries, including eight HUS cases from the French outbreak and two sporadic cases from Spain. Additional cases related to the outbreak were reported from the USA and Canada (ECDC 2011, Frank *et al.* 2011, Anon 2011). Several groups concluded that the outbreak was caused by a STEC strain belonging to serotype O104:H4, with virulence features common to the enteroaggregative *E. coli* (EAggEC) pathotype (Bielaszewska *et al.* 2011, ECDC 2011, Scheutz *et al.* 2011). Serotype O104:H4 is very rare and has been diagnosed and reported in humans in a few cases only. It has never been reported in animals and food.

In India, several outbreaks and cases associated with VTEC/STEC have been documented. Kapoor (1988) described hundred of incidences associated with VTEC much earlier than reported in most of the developed countries. Besides *E. coli* O:157, many of the VTEC are shown to be associated with urinary tract infections (UTI) that are refractory to antimicrobial therapy and also with pyrexia of unknown origin (PUO) in India (Kapoor 1988, Singh *et al.* 1997).

(e) Enteroaggregative *E. coli* (EAggEC)

Its strains have distinctive aggregative or "stacked-brick" pattern of adherence to cultured human epithelial cells (Nataro *et al.* 1987). These strains are associated with persistent diarrhoea in young children. They resemble ETEC strains in that the bacteria adhere to the intestinal mucosa and cause non-bloody diarrhoea without invading or causing inflammation. This suggests that the organisms produce an enterotoxin of some sort. A distinctive heat-labile plasmid-encoded toxin has been isolated from these strains, called the EAST (enteroaggregative ST) toxin. They also produce a hemolysin related to the hemolysin produced by *E. coli* strains involved in urinary tract infections. The role of the toxin and the hemolysin in virulence has not been proven. The significance of EAggEC strains in human disease is controversial. EAggEC associates mainly with persistent diarrhoea in developing countries (Nataro and Kaper 1998). Only two reports in Japan have described diarrhoeal outbreaks caused by EAggEC or EPEC. Itoh *et al.* (*1997)* reported the isolation of EAggEC from the stools of patients with severe diarrhoea in elementary and junior high schools. Huppertz *et al.* (1997) reported the association of acute and chronic diarrhoea and abdominal colic in young children in Germany with EAggEC. In addition to being isolated from sporadic cases of diarrhoea in UK, some of whom had been recent travellers abroad, EAggEC have been identified in retrospective studies of outbreaks (Scotland *et al.* 1991, 1994, Brook *et al.* 1994, Smith *et al.* 1994). Smith *et al.* (1997) reported epidemiological and laboratory studies of an outbreak of gastroenteritis that seems to be associated with EAggEC.

Diagnosis

Diagnosis of *E. coli* induced enteric illness can be made on the basis of the following:

(a) Clinical examination of the cases which mostly occur in young age group with watery diarrhoea.

(b) Isolation and identification of *E. coli* from stool, suspected food or water by first enriching the sample at 44°C and then streaking on selective media like MacConkey's agar/eosin methylene blue (EMB) agar. EHEC strain does not grow at 44°C and also does not ferment sorbitol, therefore, produces colourless colonies on sorbitol MacConkey's agar.

(c) Serological methods like slide, tube or latex agglutination tests and ELISA to detect the presence of O157:H7 antigen. Mannose-resistant haemagglutination test can be used to detect fimbrial antigens of ETEC strain. There are several toxin assay methods *viz.* agglutination test, Biken's immunodiffusion test, dot-blot method (Kapoor 1988).

(d) Molecular assay like labeled DNA gene probe, PCR and hybridization technique which are useful in detecting fimbrial antigens, enterotoxins and Shiga-like toxin.

(e) Infant mouse, rabbit illeal loop test, rabbit vascular permeability assay, cell culture techniques, etc. can be used to detect the ETEC strain.

(f) Colicin and antibiotic resistance typing of *E. coli* isolates are helpful in epidemiological studies (Kapoor 1988).

Although isolation and identification of *E. coli* from normally sterile excrements and body fluids is indicative of its association with disease, mere isolation of *E. coli* from faeces or vomit is not very significant unless its pathogenic potential is demonstrated through animal inoculation or toxin detection or DNA probes and specific PCR for toxin or virulence genes (Singh and Kulshreshtha 1994). In a recent study, *st*, *cnf*-2 genes, indicator of pathogenic potential in *E. coli*, was detected in the strains isolated from healthy as well as sick animals (Singh *et al.* 2012) indicating the problem of diagnosis and attributing any illness to *E. coli* isolates.

Treatment

Many a times there is no need of any treatment, particularly in adults, as the disease is self limiting and automatically gets cured after 3-4 flushings of the bowel. In case of infants and aged people, *E. coli* infection may be serious requiring special attention; fluid therapy (oral and parenteral) is the most advocated cure. Although antibiotic, fluid and electrolyte therapy will cure the diseased in most cases of enteric *E. coli* infection, the use of antibiotics is often disputed and in several cases, the severity of symptoms increases, which may be due to release of more toxins by the pathogens on use of antibiotics (Kapoor 1988). Moreover the emergence of multiple drug resistance in *E. coli* and other pathogens commensal with *E. coli* may lead to failure of antibiotic chemotherapy (Singh 2009, 2010a, b, 2012). Moreover, therapeutic use of antibiotics in enteric infections is disputed as antimicrobial therapy may disturb the beneficial microflora in the intestine. Prebiotics and probiotics (often composed of lactic acid producing bacteria) are often recommended for restoration of normal microflora of the gut, which can cure infection even without the use of antibiotics, either through direct killing of the enteropathogen or through competitive exclusion (Singh *et al.* 2004).

Prevention and control strategies

E. coli infection can be prevented and controlled by the following measures:

☆ Maintenance of high standard of hygiene during production, processing and distribution of food, HACCP compliance of food processing units.

☆ Monitoring and surveillance of dairies, poultry farms and abattoirs to ensure rapid detection of infected cases and to enforce healthy slaughter practices.

☆ Monitoring of animal handlers, food handlers to detect cases and/or carriers of pathogenic *E. coli*. A major problem is to ascertain the role of *E. coli* isolated from the patients because similar *E. coli* could be isolated from apparently healthy individuals/animals and their environment and also show long lasting persistence of potentially pathogenic *E. coli* harbouring one or more virulence gene (Singh *et al.* 2012).

☆ Pasteurization/heat treatment of food products of animal origin. Although all enterpathogens including *E. coli* are killed through proper pasteurization, thermotolerant *E. coli* resistant to killing through batch pasteurization may be important cause of food spoilage as well as foodborne *E. coli* infections (Singh 2009a). Such thermotolerant *E. coli* have been reported to be prevalent in equine faecal samples.

☆ Disinfection of water and use of potable water in dairy, food units, etc.

☆ Vaccination against ETEC *E. coli*, use of fimbrial vaccine like Ecolan, Jencine K 99 and Scour guard, etc.

☆ Good individual hand hygiene and food handling practices.

☆ Recent, though controversial, work suggests that feeding cattle a hay ration rather than grain can dramatically reduce *E. coli* infection and shedding.

☆ Additionally, inoculation of cattle with probiotic bacteria has been shown experimentally to reduce shedding of *E. coli* by competitive exclusion. In fact, pig moratlity associated with enteric infections including colibacillosis could be successfully reduced with the use of very simple indigenous probiotic. In Nagaland, axone, a probiotic chutney made through Umami fermentation of soybean, has been found very effective probiotic in pigs as well as birds for controlling diarrhoeal diseases, reducing mortality and enhancing weight gain (Singh *et al.* 2009a, b).

☆ Farm environments can be made safer for visitors. Prevention strategies should be developed to reduce the risk of transmission of enteric pathogens at petting zoos; farms open to the public, animal exhibits, and other venues where the public has contact with farm animals. The strategies include the use of hand washing, controlled and supervised contact with animals, and clear separation of food-related activities from areas housing animals (Crump *et al.* 2002).

References

1. Anonymous. 2011. EHEC/HUS O104:H4-The outbreak is considered to be over. The press release of Robert Koch Institute (RKI) EHEC/HUS O104:H4. http://www.rki.de/cln_117/nn_217400/EN/Home/PM__EHEC.html. Accessed 28 April 2012.

2. Bielaszewska M, Mellmann A, Zhang W, Köck R, Fruth A, Bauwens A, Peters G and Karch H. 2011. Characterisation of the *Escherichia coli* strain associated with an outbreak of haemolytic uraemic syndrome in Germany, 2011: a microbiological study. *Lancet Infect Dis* 11: 671-676.

3. Blanco J, Blanco M, Blanco JE Mora A, González EA, Bernárdez MI, Alonso MP, Coira A, Rodríguez A, Rey J, Alonso JM and Usera MA. 2003. Verotoxin-producing *Escherichia coli* in Spain: prevalence, serotypes, and virulence genes of O157:H7 and non-O157 VTEC in ruminants, raw beef products, and humans. *Exp Biol Med (Maywood)* 228: 345-351.

4. Blanco M, Blanco JE, Mora A, Dahbi G, Alonso MP, González EA, Bernárdez MI and Blanco J. 2004a. Serotypes, virulence genes, and intimin types of Shiga toxin (verotoxin)-producing *Escherichia coli* isolates from cattle in Spain and identification of a new intimin variant gene (eae-xi). *J Clin Microbiol* 42:645-651

5. Blanco M, Padola NL, Krüger A, Sanz ME, Blanco JE, González EA; Dahbi G, Mora A, Bernárdez MI, Etcheverría AI, Arroyo GH, Lucchesi PM, Parma AE and Blanco J. 2004b. Virulence genes and intimin types of Shiga-toxin-producing *Escherichia coli* isolated from cattle and beef products in Argentina. *Int Microbiol* 7: 269-276.

6. Brook MG, Smith HR, Bannister BA, McConnell M, Chart H, Scotland SM, Sawyer A, Smith M and Rowe B. 1994. Prospective study of verocytotoxin-producing, enteroaggregative and diffusely adherent *Escherichia coli* in different diarrhoeal states. *Epidemiol Infect* 112: 63-67.

7. CDC. 2001. Outbreaks of *Escherichia coli* O157:H7 infections among children associated with farm visits - Pennsylvania and Washington, 2000. *MMWR Weekly* 50: 293-297.

8. CDC. 2011. Multistate Outbreak of *E. coli* O157:H7 Infections Linked to Romaine Lettuce. http://www.cdc.gov/ecoli/2011/ecoliO157/romainelettuce/120711/index.html. Accessed 28 April 2012.

9. Crump JA, Sulka AC, Langer AJ, Schaben C, Crielly AS, Gage R, Baysinger M, Moll M, Withers G, Toney DM, Hunter SB, Hoekstra RM, Wong SK, Griffin PM and van Gilder TJ. 2002. An outbreak of *Escherichia coli* O157:H7 infections among visitors to a dairy farm. *N Engl J Med* 347: 555-560.

10. Dundas S, Todd WTA, Stewart AI, Murdoch PS, Chaudhuri AKR and Hutchinson SJ. 2001. The Central Scotland *Escherichia coli* O157:H7 outbreak: risk factors for the haemolytic uraemic syndrome and death among hospitalized patients. *Clin Infect Dis* 33: 923-931

11. ECDC. 2011. Shiga toxin/verotoxin-producing *Escherichia coli* in humans, food and animals in the EU/EEA, with special reference to the German outbreak strain STEC O104. Stockholm: European Centre for Disease Prevention and Control and European Food Safety Authority. doi:10.2900/55055.

12. Frank C, Werber D, Cramer JP, Askar M, Faber M, an der Heiden M, Bernard H, Fruth A, Prager R, Spode A, Wadl M, Zoufaly A, Jordan S, Kemper MJ, Follin P, Müller L, King LA, Rosner B, Buchholz U, Stark K and Krause G. 2011 Epidemic profile of Shigatoxin- producing *Escherichia coli* O104:H4 outbreak in Germany - Preliminary Report. *N Engl J Med* 365: 1771-1780.

13. Gorbach SL, Kean BH, Evans DG and Bessudo D. 1975. Travelers' diarrhoea and toxigenic *Escherichia coli*. *N Engl J Med* 292: 933-936.

14. Huppertz H-I, Rutkowski S, Aleksic S and Karch H. 1997. Acute and chronic diarrhoea and abdominal colic associated with enteroaggregative *Escherichia coli* in young children living in Western Europe. *Lancet* 349: 1660-1662.

15. Itoh Y, Nagano I, Kunishima M and Ezaki T. 1997. *Laboratory investigation of enteroaggregative Escherichia coli O untypeable:H10 associated with a massive outbreak of gastrointestinal illness. J Clin Microbiol* 35: 2546-2550.

16. Kapoor KN. 1988. *Escherichia coli* of public health significance and their toxigenic potential. *Ph.D. Thesis*, Rholikhand University, Bareilly. 526 pp.

17. Knutton S, Williams PH and McNeish AS. 1989. *Actin accumulation at sites of bacterial adhesion to tissue culture cells: basis of a new diagnostic test for enteropathogenic* Escherichia coli. *Infect Immun* 57: 1290-1298.

18. Makino S, Asakura H, Shirahata T, Ikeda T, Takeshi K, Arai K, Nagasawa M, Abe T and Sadamoto T. 1999. *Molecular epidemiological study of a mass outbreak caused by enteropathogenic Escherichia coli O157:H45. Microbiol Immunol* 43: 381-384.

19. Malik P, Singh BR, Malik PK and Sharma VD. 2000. Pathogenic attributes of *Escherichia coli* isolates from free-ranging Sambars (*Cervus unicolor*). *Indian J Anim Sci* 70: 1108-1111.

20. Marler B. 2012. *E. coli* cases in Central Missouri announced. http://www.foodpoisonjournal.com/foodborne-illness-outbreaks/e-coli-cases-in-central-missouri-announced/Food Poison Journal. Accessed 28 April 2012.

21. Martin A and Beutin L. 2011. Characteristics of Shiga toxin-producing *Escherichia coli* from meat and milk products of different origins and association with food producing animals as main contamination sources. *Int J Food Microbiol* 146: 99-104

22. Moon HW, Whipp SC, Argenzio RA, Levine MM and Giannella RA. 1983. *Attaching and effacing activities of rabbit and human enteropathogenic Escherichia coli in pig and rabbit intestines. Infect Immun* 41: 1340-1351.

23. Mora A, Herrera A, López C, Dahbi G, Mamani R, Pita JM, Alonso MP, Llovo J, Bernárdez M, Blanco JE, Blanco M and Blanco J. 2011. Characteristics of the

Shiga-toxin-producing enteroaggregative *Escherichia coli* O104:H4 German outbreak strain and of STEC strains isolated in Spain. *International Microbiol* 14: 121-141.

24. Nataro JP and Kaper JB. 1998. *Diarrheagenic Escherichia coli. Clin Microbiol Rev* 11: 142-201.

25. Nataro JP, Kaper JB, Robins-Browne R, Prado V, Vial P and Levine MM. 1987. *Patterns of adherence of diarrheagenic Escherichia coli to HEp-2 cells. Pediatr Infect Dis J* 6: 829-831.

26. Rothbaum R, McAdams AJ, Giannella R and Partin JC. 1982. *A clinicopathological study of enterocyte-adherent Escherichia coli: a cause of protracted diarrhea in infants. Gastroenterol* 83: 441-454.

27. Rounds JM, Rigdon CE, Muhl LJ, Forstner M, Danzeisen GT, Koziol BS, Taylor C, Shaw BT, Short GL and Smith KE. 2012. Non-O157 Shiga toxin producing *Escherichia coli* associated with venison. *Emerg Infect Dis* 18: 279-282.

28. Ryder RW, Sack DA, Kapikian AZ, McLaughlin JC, Chakraborty J and Mizanur Rahman AS. 1976. Enterotoxigenic *Escherichia coli* and Reo-virus-like agent in rural Bangladesh. *Lancet* 1: 659-663.

29. Scheutz F, Moller Nielsen E, Frimodt-Moller J, Boisen N, Morabito S, Tozzoli R, Nataro J and Caprioli A. 2011. Characteristics of the enteroaggregative Shiga toxin/verotoxin-producing *Escherichia coli* O104:H4 strain causing the outbreak of haemolytic uraemic syndrome in Germany, May to June 2011. *Euro Surveill* 16(24), pii: 19889.

30. Scotland SM, Smith HR, Said B, Willshaw GA, Cheasty T and Rowe B. 1991. Identification of enteropathogenic *Escherichia coli* isolated in Britain as enteroaggregative or as belonging to a subclass of attaching and effacing *Escherichia coli* that does not hybridize with EPEC adherence factor probe. *J Med Microbiol* 35: 278-283.

31. Scotland SM, Willshaw GA, Cheasty T, Row B and Hassall JE. 1994. Association of enteroaggregative *Escherichia coli* with travellers' diarrhoea. *J Infect* 29: 115-116.

32. Singh A, Warke S and Kalorey D. 2012. Prevalence of virulence genes in bovine and environmental *E. coli*. In: Proceedings of National Seminar on 'One health initiative and addressing food safety'. 16-17 Feb. 2012, Thrissur, Kerala. Abstract OH/X/62, p. 266.

33. Singh BR. 2009a. Thermotolerance and multidrug resistance in bacteria isolated from equids and their environment: source of pasteurization resistant bacteria. *Vet Rec* 164: 746-750.

34. Singh BR. 2009b. Axone a potential probiotic for pigs. http://www.nagalandpost.com/ShowStory.aspx?npoststoryiden=UzEwMTc2MjA=-wUiTK04CgWc= Published 28 September 2009. Accessed 28 April 2012.

35. Singh BR. 2010a. Multiple drug-resistance (MDR) in faecal bacteria isolated from north-west Indian apparently healthy and sick foals of less than three months of age. *Indian J Comp Microbiol Immunol Infect Dis* 30: 85-90.

36. Singh BR. 2010b. Occurrence of multiple drug resistant (MDR) aerobic bacteria in vaginal swabs of mares and their association with infertility. *Indian J Comp Microbiol Immunol Infect Dis* 30: 105-112.

37. Singh BR. 2012. Drug resistance in enteroptahogens of zoonotic significance isolated from diarrhoeic pigs and house sparrow perching in piggery in Jharnapani, Nagaland.: NOTO-Are-Med https://www.notoare.com/index.php/index/explorer/getPDF/17378234. Accessed 28 April 2012.

38. Singh BR and Kulshreshtha SB. 1994. Incidence of *Escherichia coli* in fishes and seafoods; Isolation, serotyping, biotyping and enterotoxigenicity evaluation. *J Food Sci Technol* 31: 324-326.

39. Singh BR. and Sharma VD. 2001. Utilization of trypsin as a source of carbon and nitrogen by pathogenic bacteria. *Indian J Vet Res* 13: 57-61.

40. Singh BR and Sharma VD. 2003. Relationship between pathogenic potential and surface characteristics of *Salmonella* serovars. *Pantnagar J Res* 1: 53-58.

41. Singh BR, Kulshreshtha SB and Pawde AM. 1994. Enteropathogens of zoonotic significance on perianal area of pet dogs of Bareilly. *Indian Vet Med J* 18: 182-184.

42. Singh BR, Kulshreshtha SB and Kapoor KN. 1995. An orange juice-borne diarrhoeal outbreak due to enterotoxigenic *Escherichia coli*. *J Food Sci Technol* 32: 504-506.

43. Singh BR, Kapoor KN, Kumar A, Agarwal RK and Bhilegaonkar KN. 1996a. Prevalence of enteropathogens of zoonotic significance in meat/milk products. *J Food Sci Technol* 33: 251-254.

44. Singh BR, Kapoor KN, Rathore RS, Yadav AS, Agarwal RK, Bhilegaonkar KN and Kumar A. 1996b. Prevalence of enteropathogens in fresh juices of fruits and vegetables. *Beverage and Food World* 23: 51-54.

45. Singh BR, Verma JC and Singh VP. 2002. Age and sex has significant effect on agglutinin titres in rabbits hyperimmunized with O and h antigens of *Salmonella* and *Escherichia coli*. *Newsletter IVRI*, XXIII (002) 3.

46. Singh BR, Shrama VD, Chanda Mudit, Agrawal Meenu, Kavitha R, Sharma G and Agrawal RK. 2003. Dye and heavy metal resistance in pathogenic bacteria of animal and public health significance. *Pantnagar J Res* 1: 86-89.

47. Singh BR, Verma JC, Sharma VD and Singh VP. 2004. Inhibitory effect of *Lactobacillus acidophilus* on common enteropathogens of zoonotic significance. *J Food Sci Technol* 41: 566-570.

48. Singh BR, Chandra Mudit, Agrawal RK and Babu N. 2006. Antigenic competition among different 'O' antigens of *Salmonella enterica* subspecies *enterica* serovars during hyperimmunization in pony mares. *Indian J Exp Biol* 44: 1022-1025.

49. Singh BR, Ebibeni N and Dhali A. 2009a. Effect of Axone (fermented soybean) feeding on humoral immune response and growth and breeding performance of Swiss albino mice. In: Chandrashekharaiah N, Thulasi A, Umaya R, Suganthi and Pal DT (Eds) *Diversification of animal nutrition research in changing scenario*, Vol 2, Proceedings of 13th Biennial Animal Nutrition Conference, December 17-19, 2009, Bangalore. p. 122, Abstract NFRS-12.

50. Singh BR, Ebibeni N and Bhatt BP. 2009b. Effect of Axone (fermented soybean) feeding on growth performance, intestinal physiology and humoral response of suckling and grower pigs in Nagaland state of north eastern hill region. In: Chandrashekharaiah N, Thulasi A, Umaya R, Suganthi and Pal DT (Eds) *Diversification of animal nutrition research in changing scenario*, Vol 2, Proceedings of 13th Biennial Animal Nutrition Conference, December 17-19, 2009, Bangalore. p. 122-123, Abstract NFRS-13.

51. Smith HR, Smith SM, Willshaw GA, Rowe B, Cravioto A and Eslava C. 1994. Isolates of *Escherichia coli* O44: H18 of diverse origin are enteroaggregative. *J Infect Dis* 170: 1610-1613.

52. Taylor CJ, Hart A, Batt RM, McDougall C and McLean L. 1986. *Ultrastructural and biochemical changes in human jejunal mucosa associated with enteropathogenic* Escherichia coli *(O111) infection. J Pediatr Gastroenterol Nutr* 5: 70-73.

53. Ulshen MH and Rallo JL. 1980. Pathogenesis of *Escherichia coli* gastroenteritis in man - another mechanism. *N Engl J Med* 302: 99-101.

54. Wu SX and Peng RQ. 1992. Studies on an outbreak of neonatal diarrhoea caused by EPEC 0127:H6 with plasmid analysis restriction analysis and outer membrane protein determination. *Acta Paediatr* 81: 217-221.

2014, Zoonoses: Bacterial Diseases
Pages 187–209
Editor: Sudhi Ranjan Garg
Published by: DAYA PUBLISHING HOUSE, NEW DELHI

9

Campylobacteriosis

Kuldeep Dhama[1], S. Rajagunalan[2], Ruchi Tiwari[3] and Sanjay Kapoor[4]

[1]*Division of Pathology,* [2]*Division of Veterinary Public Health, Indian Veterinary Research Institute, Izatnagar – 243 122*
[3]*Department of Microbiology and Immunology, College of Veterinary Sciences, Pandit Deen Dayal Upadhyaya Veterinary University, Mathura – 281 001*
[4]*Department of Veterinary Microbiology, College of Veterinary Sciences, Lala Lajpat Rai University of Veterinary and Animal Sciences, Hisar – 125 004*

Campylobacter, a Gram negative bacterium, has been long associated with animal diseases. The organism gained more importance particularly during the last 35 years as it has been recognized as a major cause of human illnesses ranging from gastroenteritis to Guillain-Barre Syndrome. Amongst *Campylobacters,* mainly thermophilic *Campylobacter* species have been implicated in foodborne infections and *Campylobacter jejuni* has now become the most frequently reported organism from cases of bacterial gastroenteritis in humans in many countries. *Campylobacter* species are widely distributed in nature. Most of them are adapted to the intestinal tract of warm-blooded animals and birds. *C. jejuni* is particularly adapted to poultry, which probably forms the largest reservoirs of the species. Contaminated water and foods of animal origin, especially improperly processed chicken have been implicated as vehicles for the transmission of campylobacteriosis.

Campylobacter fetus subspecies *jejuni* is the most common pathogen of enteritis in humans. The other pathogenic *Campylobacter* species include *Campylobacter coli*, and *Campylobacter lari* and nonpathogenic *C. sputorum*, *C. faecalis*, *C. mucosalis* and *C. bubulus* are usually found as normal flora of poultry, cattle, sheep and other domesticated animals. They are rarely associated with diseases. *C. sputorum* is a strict anaerobe found in gingival crevices of about 5 per cent in human beings (Moore *et al.* 2005, Dhama *et al.* 2011a, Man 2011).

Worldwide there is concern over the increase in foodborne illnesses. *Campylobacter* is among the most common bacteria causing gastroenteritis. A report of the World Health Organization (WHO) on diarrhoeal zoonoses has indicated that direct or indirect contact of human beings with poultry and other domesticated animals gives rise to higher risk of campylobacteriosis among rural communities. Surveys of farm animals in a number of countries have revealed the incidence of this organism up to 70 per cent in sheep, cattle and pigs, and up to 100 per cent in chickens. The significance of *Campylobacter* as diarrhoegenic agent in human beings is considered with more attention in developing countries like India. High incidence of *Campylobacter* diarrhoea as well as its duration and possible sequelae make it highly important from a socio-economic perspective (WHO 2011).

Historical Background

The history of campylobacters dates back to the 19th century when Theodore Escherich, a German scientist, in 1886 found spiral shaped bacterium in the colon of children with diarrhoea but his work remained largely unrecognized. McFadyean and Stockman (1913) first cited a *Vibrio* as the cause of abortions in ewes. Smith (1918) and Smith and Taylor (1919) associated these curved bacilli with abortion in cattle and named the organism as *Vibrio fetus*. Later, Jones and Little (1931) isolated a new vibrio, *V. jejuni*, which caused 'winter dysentery' in calves. It differed antigenically from *V. fetus*. In 1944, Doyle described *V. coli* as the causative agent of 'swine dysentery' in pigs. In 1946, Levy observed spiral organism in the blood of people affected with gastroenteritis which he could not recover by culturing. In 1947, Vinzent and coworkers isolated *V. fetus* from blood of three pregnant women, two of whom went on to have spontaneous abortions. Elizabeth King in 1957 isolated an organism which differed biochemically and antigenically from *V. fetus* from the blood of a patient. She gave the name 'related vibrio' to this new organism which grew at 42°C. Sebald and Veron in 1963 proposed a new genus *Campylobacter* as *V. fetus*, *V. jejuni* and *V. coli* differed biochemically and by DNA base composition from other members of the genus *Vibrio* (Moore *et al.* 2005, Silva *et al.* 2011). Dekeyser *et al.* (1972) isolated campylobacters and demonstrated their presence in the stools of at least 5 per cent of diarrhoeic patients. Skirrow in 1977 developed a selective medium for isolation of the campylobacters which resulted in widespread isolation of campylobacters from various samples. Consequently campylobacters were recognized as one of the most common bacterial causes of diarrhoea and as the microorganisms of great public health significance throughout the world. Among thermotolerant campylobacters, *C. jejuni* alone is responsible for 80-90 per cent of foodborne *Campylobacter* infections in humans, followed by *C. coli*, and to lesser extent, *C. lari* (Ryser and Marth 1989, Butzler 2004, Fitzgerald *et al.* 2008).

Etiological Agent

Campylobacteriosis is a disease condition caused by members of the genus *Campylobacter,* consisting of Gram negative, non-spore forming, thermophilic bacteria. Campylobacters are microaerophilic, spiral-shaped or S-shaped organism with a pair of cells exhibiting gull wing morphology. They are highly motile with a characteristic cork screw or darting type motility, produced with the help of single polar flagella at one or both ends. However, *C. gracilis* is non-motile and *C. showae* has multiple flagella. In old cultures, a coccoid form of the organism can be seen, described as viable but non-culturable (VBNC) forms, which are considered as degenerative forms but it remains controversial (Ziprin *et al.* 2002, Silva *et al.* 2011).

The word *Campylobacter* is derived from the Greek word '*Kampulos*' meaning bent or curved and '*bacter*' meaning rod. The genus *Campylobacter* included under family *Campylobacteriaceae,* consist of 25 species and 10 subspecies (ICSP 2011, Man 2011). They were initially placed under the genus *Vibrio* but later a new genus *Campylobacter* was created for them (Table 9.1).

Being microaerophilic, campylobacters require an atmosphere consisting of 3-5 per cent oxygen, 10 per cent carbon dioxide and 85 per cent nitrogen (Park 2002). Some species require hydrogen in addition, for optimal growth (Debruyne *et al.* 2008). The organisms are oxidase positive, catalase positive and do not ferment carbohydrates but obtain energy by metabolizing amino acids and from TCA cycle intermediates. Thermophilic campylobacters do not exhibit true thermophily (growth at 55°C) but prefer to grow at 41-43°C, so the term thermotolerant has been considered as more appropriate (Levin 2007). They also produce good growth at 37°C (Shane 1992) but are incapable of growing below 30°C (Moore *et al.* 2005).

Except cytolethal distending toxin, no other classical virulence factors have been identified among campylobacter to be associated with their pathogenesis (Bang *et al.* 2003, Dasti *et al.* 2010). Various pathogenic mechanisms like production of cholera-like enterotoxin and cytotoxin and ability to adhere and invade epithelial cells have been proposed to play role in cases of enteritis. Several genes have now been identified as virulent, pathogenic and important for adherence and toxin production. The *cdt*A, *cdt*B and *cdt*C genes have been identified for expression of cytotoxin production. The *fla*A, *cad*F (campylobacter adhesion to fibronectin F), and *rac*R (reduce ability to colonize R) genes have been identified to be responsible for expression of adherence and colonization. The products of *vir*B11 and *pld*A (phospholipase A) genes help in invasion (Dhama *et al.* 2011a). Although, flagella have role in motility, these are also suspected to act as virulence factor helping in colonization and invasion mechanism (Fernando *et al.* 2007, Dasti *et al.* 2010). The flagella act as a type III secretion system and help in secretion of campylobacter invasion antigen (Cia) which plays role in colonization of *C. jejuni* (Guerry 2007, Biswas *et al.* 2007, Fernando *et al.* 2007).

Campylobacters are considered to be much more sensitive to hostile conditions than *Salmonella* and *E. coli* (Humphrey *et al.* 2007). They are able to survive at refrigeration temperatures for several weeks but freezing is detrimental to them as it reduces their viability. They are also susceptible to heat treatment and may survive only for few days at room temperature (Levin 2007). Campylobacters can survive and remain pathogenic for long periods in aquatic environment (Abulreesh *et al.* 2006). The optimal water activity is around 0.997 and their D value is less than one minute indicating that they are rapidly inactivated by heat. The optimal pH for growth is between 6.5 and 7.5 (Silva *et al.* 2011).

Table 9.1: Species of *Campylobacter*, their sources and human disease conditions

Species	Source	Disease Conditions in Man
C. avium	Turkey, chickens	–
C. canadensis	Whooping cranes	–
C. coli	Many animal species and birds	Gastroenteritis, bacteremia, meningitis spontaneous abortion
C. concisus	Dog and cat	Gastroenteritis, reactive arthritis, Barrett esophagitis
C. cuniculorum	Rabbit	–
C. curvus	Dog	Gastroenteritis, ulcerative colitis, bacteremia
C. fetus subsp. *fetus*	Cattle, sheep, horse, turtle, kangaroo,	Gastroenteritis, abcess in CNS, meningitis, vertebral osteomyelitis, bacteremia, endocarditis, septic abortion
C. fetus subsp. *venerealis*	Cattle	septic abortion
C. gracilis periodontits	Dog	Crohn's disease, ulcerative colitis, brain abscess,
C. hominis	Human	Crohn's disease, ulcerative colitis
C. helveticus	Cat, dog	Diarrhoea
C. hyointestinalis subsp. *hyointestinalis*	Cattle, sheep	Diarrhoea and septicemia
C. hyointestinalis subsp. *lawsonii*	Pig	Diarrhoea and septicemia
C. insulaenigrae	Seal, sea lion, porpoise	Gastroenteritis, diarrhoea, vomiting, septicemia
C. jejuni subsp. *jejuni*	Many animal species and birds	Gastroenteritis, celiac disease, inflammatory bowel disease (IBD), GBS, UTI, meninigitis, HUS
C. jejuni subsp. *Doylei*	Many animal species and birds	–
C. lanienae	Cattle, sheep, pig	–
C. lari subsp. *concheus*	Many bird species, cattle, sheep, horse, Rhesus monkey, oyster, mussel, scallop, sea water	–
C. lari subsp. *lari*	Many bird species, cattle, sheep, horse, Rhesus monkey, oyster, mussel, scallop, sea water	Gastroenteritis, bacteremia
C. mucosalis	Dog	Gastroenteritis
C. peloridis	Shell fish	–

Contd...

Table 9.1–*Contd...*

Species	Source	Disease Conditions in Man
C. rectus	Dog	Gastroenteritis, Crohn's disease, ulcerative colitis, soft tissue necrosis
C. showae	Dog	Crohn's disease, intraorbital abscess
C. sputorum subsp. bubulus	Cattle, dog, pig, sheep	Gastroenteritis, bacteremia, axillary abscess
C. sputorum subsp. sputorum	Cattle, dog, pig, sheep	Gastroenteritis, bacteremia, axillary abscess
C. subantarticus	Black-browed albatross, gentoo penguin, gray-headed albatross	
C. troglodytis	Chimpazee	–
C. upsaliensis	Cat, dog	Gastroenteritis, spontaneous abortion, breast abscess
C. ureolyticus	Horse	Gastroenteritis, Crohn's disease, ulcerative colitis, soft tissue abscess, arthritis
C. volucris	Black headed gull	–

Source: ICSP (2011), Man (2011).

Host Range

Both domestic and wild animals and birds act as major reservoirs for campylobacters. Campylobacters are commensal in the gastrointestinal tract of animals and birds. Poultry is a major source of the organism with an isolation rate as high as 90-100 per cent (Nannapaneni *et al.* 2009, Hakkinen *et al.* 2009). Higher body temperature of poultry (42°C) is supposed to favour the growth of campylobacters (Dasti *et al.* 2010). Campylobacters have also been isolated from other avian species like turkeys, quails, psittacines (parrots), waterfowls (ducks), passeriforms (finches and canaries), pigeons, etc. (Altekruse *et al.* 1994, Abulreesh *et al.* 2006, Dhama *et al.* 2011a). Among domestic animals, cattle, sheep, swine, dogs and cats act as reservoir; isolation from horse and mouse has also been reported (Zweifel *et al.* 2008). Poultry birds are considered as the major source of *C. jejuni*, while pigs act as the major source of *C. coli*. A single animal or bird may be simultaneously infected with different types or strains of campylobacters.

Disease Transmission

C. jejuni live in the small intestine and colon of normal birds and animals. They do not cause any significant disease in animals or poultry. The infected animals and birds can contaminate the water sources and spread the campylobacter infection. Contamination of water bodies may also occur by improperly treated abattoir effluents and sewage disposal. Human gets infection mainly from animals and various environmental sources. The main mode of transmission is by ingestion of contaminated food and untreated water (Figure 9.1).

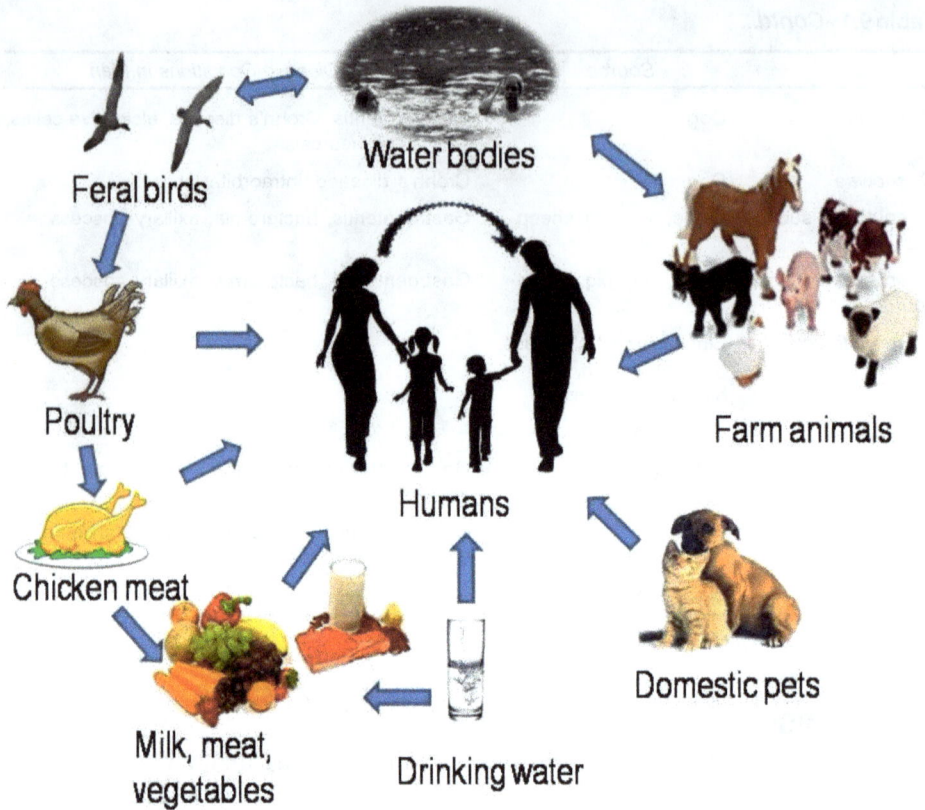

Figure 9.1: Transmission pathways of campylobacters

Epidemiological studies have revealed a major association between *Campylobacter* infection in humans and handling and consumption of raw or undercooked poultry meat. Worldwide, much of the poultry meat presented at retail shops/slaughterhouses is contaminated with *C. jejuni* and/or *C. coli* (Hariharan 2002, Dhama *et al.* 2011*a, b*). In poultry, transmission occurs by horizontal route by faecal-oral route from the environment and other contaminated feed and water sources, litter, concrete, soil, equipments and environment in the poultry house (Ridley *et al.* 2008, Dhama *et al.* 2011a). Campylobacters may also spread by the hands, footwear and clothing of farm workers. Litter can remain infective for long periods, and infected birds can continue to excrete *C. jejuni* for up to two months (Kassem *et al.* 2010). Infected poultry have been reported to carry 10^6-10^8 CFU/g campylobacters in their ceaca (Hermans *et al.* 2011). After introduction of infection into the flock, it spreads very rapidly and the whole flock may get infected within 2-3 days but there is a delay in the colonization of campylobacters during initial stages (Conlan *et al.* 2007). Transmission through drinking water, especially nipple drinkers, has been found to play an important role (Cokal *et al.* 2011).

Table eggs are not regarded as a vehicle of foodborne *Campylobacter* infection. There is no evidence of vertical transmission either on the surface of eggs or by transovarial transmission (Jacobs-Reitsma 1997, Callicott *et al.* 2006) or transmission from hatcheries horizontally (Petersen *et al.* 2001). However, meat and other edible products derived from infected animals like milk and offal may contain the organism (Kulkarni 1992, Sharma and Prasad 1997, Varma *et al.* 2000, Kirkpatrick and Tribble 2011). *The bacteria* are capable of surviving in meat, milk, food, or water maintained at 4°C for several weeks. Raw undercooked chicken is one of the major risk factor for human infection responsible for nearly 55-71 per cent of human infection and its risk factor is high due to the high level of consumption (Humphrey *et al.* 2007, French *et al.* 2008). Consumption of unpasteurized milk and dairy products is also an important source of the organism (Butzler and Oosterom 1991, Silva *et al.* 2011). Transmission by contact with water during recreational activities may also occur (Woolridge and Ketley 1997). Campylobacters have also been isolated from various vegetables (Kumar *et al.* 2001, Verhoeff-Bakkenes *et al.* 2011), which get contaminated from untreated water or by cross-contamination with raw meat or chicken (Wong *et al.* 2007). Use of contaminated kitchen utensils during processing could contaminate the food materials. Role of flies in the transmission of *Campylobacter* is also suspected as they can act as mechanical vectors (Ekdahl *et al.* 2005).

Person-to-person transmission of campylobacters is not a common feature, although the organism is excreted in huge numbers in the faeces of infected individuals; asymptomatic carriage is well known. Occupational risk is high for people who are in close contact with animals, for example, farmers, veterinarians, animal handlers and abattoir workers (Wong *et al.* 2007). Poultry workers are three times at more risk of developing campylobacteriosis compared to general population, the role of airborne transmission has been suspected in them (Wilson 2004). The infection can also be acquired through contact with infected pets, especially in children who play with them (Silva *et al.* 2011). Travel to endemic countries has been reported to be a reason for spread of campylobacters, especially drug resistant campylobacters (Vlieghe *et al.* 2008, Shah *et al.* 2009).

Environmental and Predisposing Factors

Campylobacter are ubiquitously distributed in the environment. Several predisposing factors are responsible for them. These include ingestion of contaminated food and water, lack of effective water distribution system, poor hygienic practices like not washing hands after handling of infected animals, especially during occupational exposure, contact with pets, improper handling of food material in the kitchen resulting in cross-contamination, improper heat treatment of food materials, consumption of raw or improperly pasteurized milk and milk products, stagnation of water in and around animal sheds, contamination of water bodies by wildlife, etc. Flies are also considered as a risk factor in the transmission of campylobacters (Butzler and Oosterom 1991, Wagenaar *et al.* 2006).

Prevalence in Animals

Campylobacters, being commensal organism in different warm blooded animals and bird species, are prevalent worldwide. Several reports on isolation and identification of campylobacters are available from different countries. These organisms can affect humans, animals and a variety of birds such as chickens, turkeys, psittacines (parrots), passeriforms (finches and canaries), pigeons, Chillian flamingo, common peafowl, fantail pigeons, ostrich, etc. Animals like llama, Cape hyrax, orangutan, chimpanzee, African elephant, gorilla, Japanese macaque, have also yielded campylobacter isolates (Sharma and Prasad 1997, Misawa *et al.* 2000, Hakkinen *et al.* 2009, Sippy *et al.* 2012). Parsons *et al.* (2010, 2011) reported that 46-50 per cent of dogs in boarding kennels, 68-73 per cent in rescue kennels and 38 per cent of those attending veterinary practices in UK are infected with campylobacters, especially with *C. upsaliensis.*

Suzuki and Yamamoto (2009b) based on literature survey have concluded that on an average 58.8 per cent of retail poultry meats and 60.3 per cent of poultry by-products were contaminated with *Campylobacter* sp. in Japan. Later, these workers indicated 50 per cent or more prevalence in poultry meat in most countries (Suzuki and Yamamoto 2009a). Biasi *et al.* (2011) found campylobacters in 18.9 per cent of pig carcasses in Brazil, while in France, Denis *et al.* (2011) reported 25 per cent of sows positive for *Campylobacter* (*C. coli*). Hussain *et al.* (2007) reported raw chicken (48 per cent), beef (10.9 per cent) and mutton (5.1 per cent) in Pakistan contaminated with campylobacters. Salihu *et al.* (2009) reported that campylobacters could be isolated from 12.9 per cent of cattle in Nigeria. Some reports on campylobacter prevalence in India have been summarized in Table 9.2.

Prevalence in Humans

Campylobacters are the leading cause of bacterial gastroenteritis throughout the world (WHO 2011, FoodNet 2012). Although cases occur as sporadic infection, common source infection can occur especially in large gatherings. In developed countries, monitoring and surveillance programmes and investigation of foodborne outbreaks are being carried out regularly, while in developing countries including India such programmes are not so regular. This combined with unavailability of adequate laboratory facilities and expertise in the developing world makes it quite difficult to get the actual estimates of the disease burden. Even in developed countries, not all cases are reported; so the true incidence of campylobacteriosis is underreported. It has been indicated that for every reported case of campylobacteriosis nine others were not reported (Wheeler *et al.* 1999). In US, the incidence of campylobacter showed a decreasing trend between 1996 and 2010 (FoodNet 2012), while in England and Wales, there is an increase in the proportion of outbreaks caused by campylobacters (Gormley *et al.* 2010). The EFSA (2012) has reported that campylobacteriosis is the most frequently reported foodborne illness in EU with over 1.9 lakhs cases annually but the actual numbers are expected to be around 9 lakhs per year. The notification rate in 2010 for campylobacteriosis in children less than 5 years of age was 126.8 per 100,000 populations (EFSA 2012). New Zealand reported the highest incidence rate for campylobacteriosis in 2006 with 384 cases per 100,000 population which with due intervention strategies in poultry industry decreased to 90.6 per 100,000 (Campbell *et al.* 2012). In Japan, *Campylobacter* and *Salmonella* caused more number of food poisoning next to norovirus, with 3071 campylobacteriosis cases in 2008 (IASR 2010).

Table 9.2: Prevalence reports on campylobacters in India (2005-2012)

Place	Prevalence	References
Mumbai and Pune, Maharashtra	95 per cent prevalence in 40 poultry meat samples	Bandekar *et al.* (2005)
Lucknow, Uttar Pradesh	13.5 per cent occurrence (47/348) among persons with diarrhoea and 0.6 per cent (2/351) among persons without diarrhoea	Jain *et al.* (2005)
Pune, Maharashtra	Out of 126 faecal samples (bufflao, cow, ox, sheep, goat) and 86 water samples from Mula and Pavana rivers, 16.7 per cent and 3.9 per cent isolation of *C. jejuni*, *C. coli*, and *C. lari*, respectively, from faecal samples, and 6.9 per cent, 9.3 per cent and 27.3 per cent, respectively, from river water samples	Baserisalehi *et al.* (2005)
Chandigarh	10.1 per cent (14) prevalence of *C. jejuni* among 138 individuals receiving antibiotic therapy	Vaishnavi and Kaur (2005)
Gannavaram, Andhra Pradesh	21 per cent (42/200) lamb meat carcasses and 9.5 per cent (19/200) rectal swabs of sheep positive for *C. jejuni*	Varma *et al.* (2005)
Chennai, Tamil Nadu	Isolation of *C. jejuni* from 16 out of 200 HIV infected individuals	Kownhar *et al.* (2007)
Vellore, Tamil Nadu	5.7 per cent (9/158) prevalence among children with diarrhoea and 3.1 per cent (3/99) among children without diarrhoea	Aijampur *et al.* (2008)
Bareilly, Uttar Pradesh	*C. jejuni* prevalence 22.72 per cent (65/286) in poultry faecal samples, high prevalence in rainy season (33.65 per cent), followed by summer (20.39 per cent) and winter (11.39 per cent)	Singh *et al.* (2008)
New Delhi	*C. jejuni* infection reported from 27.7 per cent patients with GBS compared to 2.3 per cent in neurological and 2.3 per cent in non-neurological controls	Kalra *et al.* (2009)
Bareilly, Uttar Pradesh	24 per cent (culture based) and 30 per cent (PCR based) prevalence among 50 cloacal samples of poultry and 6 per cent (culture based) and 8 per cent (PCR based) prevalence in 50 eggs collected from the same birds	Somvanshi *et al.* (2009)
Bareilly, Uttar Pradesh	Overall 12.79 per cent prevalence of *C. jejuni* and *C. coli* in poultry meat	Singh *et al.* (2009)
Chandigarh	*C. jejuni* infection in 30 per cent (15/50) individuals with GBS and in 8 per cent (3/40) control individuals	Sharma *et al.* (2011)
Vellore, Tamil Nadu	4.5 per cent (8/394) prevalence in children and 64 per cent (16/25) in chickens; no campylobacters isolated from 589 cows, 2 buffaloes, 11 bullocks and 25 goats	Rajendran *et al.* (2012)
Mathura, Uttar Pradesh	Campylobacter prevalence 51 per cent in dogs	Kumar *et al.* (2012)

Based on the published reports, the situation of campylobacter prevalence in humans and animal species in India is briefly summarized in Table 9.2.

Disease Manifestations in Animals

Thermophilic campylobacters are part of the intestinal flora in warm blooded animals and birds, and they seldom cause clinical disease in them (WHO 2011). Infected birds may remain as healthy carriers and excrete the organism in huge numbers in their faeces. The organism is generally nonpathogenic in mature poultry. However, some strains are invasive and/or toxic for some young birds (newly hatched chicks and poults below 2 weeks), and those colonized early in life may show signs of enteritis and death. Affected birds show signs of lethargy, loss of appetite and weight loss, intestinal distension, hepatitis and pass yellowish or watery diarrhoea and there is decrease in egg production in layers. They are suspected to be associated with vibrionic hepatitis in poultry characterized by focal grayish white lesions on the liver with areas of necrosis and haemorrhages (Jennings *et al.* 2011). *C. jejuni* could also cause sporadic cases of abortion in cattle (Wagenaar *et al.* 2006) and sheep (Sahin *et al.* 2008).

Disease Manifestations in Humans

Almost all people are susceptible to the infection but usually young children and young adults are affected more with *Campylobacter* enteritis. Campylobacteriosis is a sporadic disease characterized by a self limiting gastroenteritis with or without blood in the stools, fever (up to 40°C), abdominal cramping, myalgia, headache, malaise, nausea and vomiting, which may last for more than 24 hours. The incubation period ranges from 3-5 days and the course of the disease may last for not more than ten days (Hariharan 2002, Wong *et al.* 2007, Dhama *et al.* 2011a, b, WHO 2011). There is a clear epidemiological difference in the course of disease between developed and developing countries. In developed countries, the disease occurs mainly in young adults characterized by acute abdominal pain and bloody diarrhoea while in developing countries, it is mainly restricted to children less than five years of age with non-inflammatory, watery diarrhoea. This difference was suggested to be due to the level of immunity among people because of constant exposure to the organism in developing countries from environmental sources. Asymptomatic carriage is frequent in adults in developing countries. Also there is a typical seasonal variation in the incidence rate of campylobacteriosis in developed countries with more cases occurring in summer and autumn while such variations are not observed in developed countries due to less extreme temperature variations and continuing surveillance programmes (Coker *et al.* 2002). Mortality due to campylobacteriosis is rare, occurring in very young and elderly patients (WHO 2011). The diarrhoeic stools may be offensive smelling and contain mucous, blood, pus or leukocytes due to inflammatory changes caused by the infection. The organism can also cause bacteremia, especially in pregnant women. Debilitated and immunocompromised individuals such as AIDS patients are at a greater risk and more severe and prolonged disease (Wooldridge and Ketley 1997, Kownhar *et al.* 2007, Dhama *et al.* 2011a, b, Kirkpatrick and Tribble 2011). Diarrhoea is consistent with the idea that toxins play a role in this disease. Strains of *C. jejuni* may produce a number of toxins, mainly a cytolethal distending toxin (CDT), non-CDT cytotoxins, including Shiga toxins, and hemolysins. The organism produces a cytotoxin that has immunological similarities to cholera toxin.

In some cases, there may be complications like septicaemia, toxic megacolon, hepatitis, pancreatitis, reactive arthritis (Reiter's syndrome), meningitis, recurrent colitis, acute cholecystitis, haemolytic uraemic syndrome (HUS), thrombotic thrombocytopenic purpura and Guillain-Barré syndrome (GBS) (Vucic *et al.* 2009, Talukder *et al.* 2011, Sharma *et al.* 2011, Hardy *et al.* 2011, Dhama *et al.* 2011a). The incidence of reactive arthritis was reported to be 1-5 per cent in the campylobacter infected individuals with an annual incidence rate of 4.3 per 100,000 population (Pope *et al.* 2007). In humans, its potentially serious outcomes confirm its importance as a significant public health hazard.

Guillain-Barré Syndrome is an acute disease of the peripheral nervous system of humans where an acute neuromuscular paralysis is caused by immune mediated demyelination of nerve cells (Sharma *et al.* 2011, Hardy *et al.* 2011). First described by Guillain, Barré and Srohl in 1916, GBS is characterized by ascending paralysis, conduction block with segmental demyelination of the nerves, macrophage and lymphocytic infiltration of the nerves, and elevated protein with no cells or very few cells in the cerebrospinal fluid. It is a rare disease condition in which the affected person may recover fully or may have permanent nerve damage. It results in respiratory paralysis, severe neurological dysfunction, flaccid paralysis and death in a small but a significant proportion of the patients. Among various causes of GBS, campylobacter infection has been reported as the common antecedent infection in several cases. Investigators from different parts of the world have isolated *C. jejuni* from the stools of patients with GBS at the onset of neurologic symptoms. The incidence of GBS was reported to less than 1 per 1000 infections (Allos 2001). As campylobacter shares certain antigens with the neurons, auto antibodies are produced against neuronal gangliosides due to molecular mimicry (Young *et al.* 2007). Isolates of *C. jejuni* from GBS cases may belong to any of the serotypes, though Penner serotype O:19 is the most common one.

Diagnosis

Conventional Bacterial Culture Technique

The samples for detection of campylobacters include faeces, rectal/cloacal swab of the affected birds or patients, caecal contents and swabs, intestinal contents, food, water, milk, etc. Faecal specimens should be collected and placed in semisolid Cary-Blair or Stuart or Campylobacter thioglycolate broth transport medium. Direct demonstration of pathogen with characteristic 'S' or comma shape in fresh faeces/stool can be performed with the help of dark field/phase contrast microscopy in wet preparation or after staining. However, confirmatory diagnosis is based on the isolation and identification of the causative agent. Campylobacters being fastidious and microaerophilic organisms require selective media containing oxygen scavenging materials and antibiotics to inhibit other competing organisms present in the sample. Various selective culture media containing different supplements and selective antimicrobial agents have been developed for isolation of campylobacters. Major antimicrobial agents used in these media in different combinations are cephlosporins, trimethoprim, polymyxin, novobiocin, vancomycin, teicoplanin, bacitracin, rifampicin, cyclohexamide and amphotericin B. Supplements like blood, serum, charchoal sodium metabisulphate, sodium pyruvate, ferrous sulphate are also added to improve the isolation rate (Hurd *et al.* 2012).

Protocols for isolation of *Campylobacter* vary with the type of sample and there is no gold standard method for isolation of all campylobacters. Depending on the nature of sample, it is subjected to an enrichment procedure in a selective medium to allow recovery of the stressed campylobacters, followed by plating on to selective agar medium. For faecal samples, direct plating on to selective agars without enrichment could yield good isolation rate. A filtration technique has also been used for isolation of campylobacters from faecal samples by actively or passively passing the faecal suspension through membrane filters of 0.45 μm or 0.65 μm pore size and then plating on to selective media (Debruyne *et al.* 2008).

Major selective enrichment broths and agar media for camylobacters include Bolton broth, Preston broth, Park and Sanders broth, Charcoal cefoperazone deoxycholate broth, campylobacter enrichment broth, Perston agar, Skirrows agar, Butzlers agar, Campy-BAP, Balser Wang medium, modified charcoal cefoperazone deoxycholate agar, CAT medium, Karmali agar, campycefex agar, etc. Different media may have different efficacy. Use of two or more media increases the likelihood of detecting campylobacters compared to a single medium (Oyarzabal *et al.* 2005).

The culture plates require incubation at 42°C in a microaerobic environment. Isolation can take several days to a week. Some species may require hydrogen in addition for optimal isolation. The microaerobic environment can be provided with the help of candle jar or by use of gas generation packs or carbon dioxide incubators. The colonies typically take 2-4 days to appear and are smooth circular (usually 1 mm diameter), slightly raised grey coloured and glistening. Preliminary identification is then carried out by microscopic examination of stained smear of the organism and by observing the organism for typical motility pattern using phase contrast microscopy or hanging drop method. This is followed by biochemical testing to confirm the genus and species of isolates.

The organisms can also be propagated *in vitro* in tissue culture systems including Chinese hamster ovary cells, HeLa cells and human epithelial cell lines. Fertile chicken eggs serve as convenient system for its isolation and propagation.

Rapid Techniques

As the recovery of campylobacters by culture-based method is poor and time taking, other cost-effective and rapid diagnostic methods have been evolved for detection and characterization of campylobacters. Rapid diagnostics include PCR and multiplex PCR methods targeting various genes of campylobacter (Bang *et al.* 2001, 2003, Al-Amri *et al.* 2007), real time PCR (Rudi *et al.* 2004, Ridley *et al.* 2008), DNA microarray technique (Volokhov *et al.* 2003), rapid enzyme immuno assay (Tolcin *et al.* 2000), ELISA (Schmidt-Ott *et al.* 2005), immunochromatography (Kawatsu *et al.* 2008), LAMP assay (Yamazaki *et al.* 2008), latex agglutination assay (LAT) (Miller *et al.* 2008), complement fixation test (Gurturk *et al.* 2002), passive haemagglutination test (Oza *et al.* 2002) and others. These recent tools provide fast, sensitive and specific diagnosis. These can deliver quantitative data too and can be well-employed for molecular epidemiological surveys of *Campylobacter* infection and to identify the source of infection.

Identification of source of infection is also very important in epidemiological investigations. The typing techniques that may be used in campylobacteriosis investigation include biotyping, phage typing, serotyping, PCR RFLP, RAPD, *fla*-SVR, ribotyping, PFGE, MLST, AFLP, Fla-DGGE (denaturing gradient gel electrophoresis). These are helpful in discriminating different isolates/strains and thus in identifying the likely source of the organism.

Prevention and Control

Among other sources of infection, poultry birds carry huge number of campylobacters in their intestine which could contaminate the carcass during slaughter and processing and transmit the infection to human beings. Thus controlling the campylobacter infection in poultry and contamination of poultry meat would substantially reduce the human infection. Measures to reduce the campylobacter colonization in poultry can reduce the level of contamination of the final product (Smith and Fratamico 2010). Adoption of hygienic and biosecurity measures in poultry houses such as the use of foot bath, proper hand washing before and after entering the pens, and adopting hygienic and efficient disposal of dead birds, body discharges, and wastes, restricting the entry of visitors and their direct contact with the flock, keeping other domestic animals away from the poultry farm premises, control of rodents, wild birds, pests and insects, etc. are some of the important measures. Some other measures include regular chlorination of drinking water, addition of organic acids to drinking water, disinfection of farm equipments, effective disinfection of the premises before and after stocking, and health education of farm workers.

The control over relative humidity in the poultry houses is important in the transmission of campylobacters as it has been observed that colonization of the organisms is delayed in chickens raised under low relative humidity conditions (Line 2006). Certain measures for hygienic slaughter of poultry such as improved washing of carcasses, use of counter-flow scalding, elimination of immersion chillers, separate slaughter of infected and non-infected flocks (scheduled processing), disinfection of water used for carcass washing, disinfection of premises and equipments of abattoir etc. play vital role in reducing the risk associated with foodborne campylobacteriosis (Wagenaar *et al.* 2006, Dhama *et al.* 2011a). Some measures that can reduce the contamination level on poultry carcass include the use of chlorine and its derivatives, organic acids, trisodium phosphate, irradiation and UV treatment (Corry and Atabay 2000). Irradiation using cobalt-60 and electron beam generation are cost-effective procedures. Freezing has damaging effect on campylobacter and reduces their viability in chicken carcass or products stored under frozen conditions (Wagenaar *et al.* 2006). Gamma irradiation (cold pasteurization) is an effective method to eliminate *C. jejuni* from poultry meat and products. Over all, the hazard analysis critical control point (HACCP) approach covering the full chain of poultry production and processing is needed to combat the *Campylobacter* problem. Corrective measures, including end-product decontamination when necessary, must be introduced at defined critical points such as the points of risk of campylobacter introduction (Hariharan 2002, Dhama *et al.* 2011a).

Other intervention strategies like competitive exclusion and the use of probiotics, bacteriophages, bacteriocins, avian egg antibodies, etc. constitute the emerging potential tools for antibacterial therapy and biocontrol of bacterial contamination of foodstuffs. Selective poultry breeding to reduce colonization is also being tried (Wagenaar *et al.* 2006, Newell *et al.* 2011).

Similar practices are required to control and eliminate the contamination in other animal foodstuffs. In food establishments and kitchen, adequate hygienic measures should be taken while preparing food. These include hand washing, sanitizing the cutting boards, and adequate heat treatment of food materials while cooking (>70°C).

Care should be taken while handling animals, especially pet animals. Water sources should be protected from contamination originating from livestock, wild animals and birds (Dhama *et al.* 2011a). Treatment of animals having gastroenteritis, adequate cooking of meat and offal before feeding to animals, isolation and treatment of infected animals and regular efficient disinfection of farm premises are also important.

Surveillance and monitoring programmes for campylobacters need to be strengthened. Additionally, it is important to educate people about the risks associated with handling raw meat, and consuming undercooked or contaminated products. In case of illness, timely and appropriate medical treatment is also very important.

There is no effective vaccine against campylobacteriosis either in animals or humans (Wallis 2004, Jagusztyn-Krynicka *et al.* 2009).

Therapeutic Guidelines

The disease is a self limiting and does not require antibiotic treatment. Fluid therapy and electrolytes may be recommended depending on the dehydration status of the patients. In severe cases and in immunocompromised patients antibiotics treatment may be prescribed. Macrolides and fluoroquinolones (erythromycin, ciprofloxacin) constitute the drug of choice (Moore *et al.* 2005, Young *et al.* 2007, Kirkpatrick and Tribble 2011). If *C. jejuni* is considered a problem in companion bird aviaries or in exotic species, antibiotics such as fluoroquinolones, gentamicin, chloramphenicol, ciprofloxacin, etc. can be administered in drinking water. Erythromycin has been reported to be very effective in campylobacteriosis. However, the problem of development of antibiotic resistance and the occurrence of drug residues in animal products need to be given due consideration.

Future Directions

Control of campylobacters in the food chain requires the development of innovative and sequential intervention strategies. Several intervention strategies to reduce intestinal colonization of chickens have been tried, yet further research and supporting field data are required. Evolving techniques for rapid and sensitive detection of *C. jejuni* is also necessary for the maintenance of a safe food and water supply.

Currently there are no commercial vaccines available to control *Camplylobacter* infection. Three major challenges have been identified in the development of vaccine against campylobacter: (1) identification of protective antigens, (2) induction of a rapid, potent immune response, and (3) the need for novel adjuvants to potentiate the immunity against campylobacters (de Zoete *et al.* 2007). With the advent of molecular biology tools and availability of whole genome nucleotide sequences of campylobacters, more research is required to develop effective vaccines against campylobacteriosis for use in humans and animals.

Development of resistance to antimicrobial agents in microorganisms, including campylobacters, represents a major public health problem worldwide. While controlling the infection in animals and decontaminating the animal products, it is essential to implement strict regulations for use of antimicrobial agents in food animal production systems.

References

1. Abulreesh HH, Paget TA and Goulder R. 2006. *Campylobacter* in waterfowl and aquatic environments: incidence and methods of detection. *Environ Sci Technol* 40: 7122-7131.

2. Ajjampur SSR, Rajendran P, Ramani S, Banerjee I, Monica B, Sankaran P, Rosario V, Arumugam R, Sarkar R, Ward H and Kang G. 2008. Closing the diarrhoea diagnostic gap in Indian children by the application of molecular techniques. *J Med Microbiol* 57: 1364-1368.

3. Al-Amri A, Senok AC, Ismaeel AY, Al-Mahmeed AE and Botta GA. 2007. Multiplex PCR for direct identification of *Campylobacter* spp. in human and chicken stools. *J Med Microbiol* 56: 1350-1355.

4. Allos BM. 2001. *Campylobacter jejuni* Infections: update on emerging issues and trends. *Clin Infect Dis* 32: 1201-1206.

5. Altekruse SF, Hunt JM, Tollefson LK and Madden JM. 1994. Food and animal sources of human *Campylobacter jejuni* infection. *J Am Vet Med Assoc* 204: 57-61.

6. Bandekar JR, Raut AD and Kapadnis BP. 2005. Occurrence of *Campylobacter* in local poultry meat in Pune and Mumbai. http://www.barc.gov.in/publications/nl/2005/200510-19.pdf. Accessed 1 August 2012

7. Bang DD, Pedersen K and Madsen M. 2001. Development of a PCR assay suitable for *Campylobacter* spp. mass screening programs in broiler production. *J Rapid Meth Aut Mic* 9: 97-113.

8. Bang DD, Nielsen EM, Scheutz F, Pedersen K, Handberg K and Madsen M. 2003. PCR detection of seven virulence and toxin genes of *Campylobacter jejuni* and *Campylobacter coli* isolates from Danish pigs and cattle and cytolethal distending toxin production of the isolates. *J Appl Microbiol* 94: 1003-1014.

9. Baserisalehi M, Al-Mahdi AY and Kapadnis BP. 2005. Antimicrobial susceptibility of thermophilic *Campylobacter* spp. isolated from environmental samples. *Indian J Med Microbiol* 23: 48-51.

10. Biasi RS, de Macedo REF, Malaquias MAS and Franchin PR. 2011. Prevalence, strain identification and antimicrobial resistance of *Campylobacter* spp. isolated from slaughtered pig carcasses in Brazil. *Food Control* 22: 702-707.

11. Biswas D, Fernando UM, Reiman CD, Willson PJ, Townsend HG, Potter AA and Allan BJ. 2007. Correlation between *in vitro* secretion of virulence-associated proteins of *Campylobacter jejuni* and colonization of chickens. *Curr Microbiol* 54: 207-212.

12. Butzler JP and Oosterom J. 1991. *Campylobacter*: pathogenicity and significance in foods. *Int J Food Microbiol* 12: 1-8.

13. Butzler JP. 2004. *Campylobacter* from obscurity to celebrity. *Clin Microbiol Infect* 10: 868-876.

14. Callicott KA, Friethriksdóttir V, Reiersen J, Lowman R, Bisaillon JR, Gunnarsson E, Berndtson E, Hiett KL, Needleman DS and Stern NJ. 2006. Lack of evidence for vertical transmission of *Campylobacter* spp. in chickens. *Appl Environ Microbiol* 72: 5794-5798.

15. Campbell D, van der Logt P and Hathaway S. 2012. Surveillance for action managing foodborne *Campylobacter* in New Zealand. *WPSAR* 3: 1-3.

16. Cokal Y, Caner V, Sen A, Cetin C and Telli M. 2011. The presence of *Campylobacter jejuni* in broiler houses: Results of a longitudinal study. *Afr J Microbiol Res* 5: 389-393.

17. Coker AO, Isokpehi RD, Thomas BN, Amisu KO and Obi CL. 2002. Human campylobacteriosis in developing countries. *Emerg Infect Dis* 8: 237-244.

18. Conlan AJ, Coward C, Grant AJ, Maskell DJ and Gog JR. 2007. *Campylobacter jejuni* colonization and transmission in broiler chickens: a modelling perspective. *J R Soc Interface* 4: 819-829.

19. Corry JEL and Atabay HI. 2001. Poultry as a source of *Campylobacter* and related organisms. *J Appl Microbiol* 90: 96-114.

20. Dasti JI, TareenAM, Lugert R, Zautner AE and Gross U. 2010. *Campylobacter jejuni*: a brief overview on pathogenicity-associated factors and disease-mediating mechanisms. *Int J Med Microbiol* 300: 205-211.

21. Debruyne L, Gevers D and Vandamme P. 2008. Taxonomy of the family *Campylobacteraceae*. In: Nachamkin I, Szymanski C, Blaser M. (eds.) *Campylobacter*, 3rd ed. ASM Press. Washington, DC. p. 3-25.

22. Dekeyser P, Gossuin-Detrain M, Butzler JP and Sternon J. 1972. Acute enteritis due to related vibrio: first positive stool cultures. *J Infect Dis* 125: 390-392.

23. Denis M, Henrique E, Chidaine B, Tircot A, Bougeard S and Fravalo P. 2011. Campylobacter from sows in farrow-to-finish pig farms: risk indicators and genetic diversity. *Vet Microbiol* 154: 163-170.

24. deZoete MR, van Putten JP and Wagenaar JA. 2007. Vaccination of chickens against *Campylobacter*. *Vaccine* 25: 5548-5557.

25. Dhama K, Hansa A and Anjaneya. 2011a. *Campylobacter* infection in poultry and its zoonotic significance. *Poultry Punch* 27: 50-60.

26. Dhama K, Tiwari R and Basaraddi MS. 2011b. Avian diseases transmissible to humans. *Poultry Technology* 6: 28-32.

27. EFSA scientific report. 2012. The European Union summary report on trends and sources of zoonoses, zoonotic agents and food borne outbreaks in 2010. *EFSA Journal* 10: 2597.

28. Ekdahl K, Normann B and Andersson Y. 2005. Could flies explain the elusive epidemiology of campylobacteriosis? *BMC Infect Dis* 5: 11.

29. Fernando U, Biswas D, Allan B, Willson P and Potter AA. 2007. Influence of *Campylobacter jejuni* fliA, rpoN and flgK genes on colonization of the chicken gut. *Int J Food Microbiol* 118: 194-200.

30. Fitzgerald C, Whichard J and Nachamkin I. 2008. Diagnosis and antimicrobial susceptibility of *Campylobacter* species. In: Nachamkin I, Szymanski C and Blaser M. (eds.) *Campylobacter*, 3rd ed. ASM Press. Washington, DC. p. 227-243.

31. FoodNet. 2012. http://www.cdc.gov/foodnet/factsandfigures/2009/incidence.html Accessed 1 August 2012.

32. French NP, Carter P, Collins-Emerson J, Midwinter A, Mullner P and Wilson D. 2008. Comparing 'source attribution' models for human campylobacteriosis. http://www.svepm.org.uk/posters/2008/Comparing per cent 20'source%20attribution'%20models%20for%20human%20campy-lobacteriosis.pdf. Accessed 1 August 2012.

33. Gormley FJ, Little CL, Rawal N, Gillespie IA, Lebaigue S and Adak GK. 2010. A 17-year review of foodborne outbreaks: describing the continuing decline in England and Wales (1992-2008). *Epidemiol Infect* 9: 1-12.

34. Guerry P. 2007. *Campylobacter* flagella: not just for motility. *Trends Microbiol* 15: 456-461.

35. Gurturk K, EkinIH, Aksakal A and Solmaz H. 2002. Detection of *Campylobacter* antibodies in sheep sera by a Dot-ELISA using acid extracts from c. fetus ssp. fetus and *C. jejuni* strains and comparison with a complement fixation test. *J Vet Med B Infect Dis Vet Public Health* 49: 146.

36. Hakkinen M, Nakari UM and Siitonen A. 2009. Chickens and cattle as sources of sporadic domestically acquired *Campylobacter jejuni* infections in Finland. *Appl Environ Microbiol* 75: 5244-5249.

37. Hardy TA, Blum S, McCombe PA and Reddel SW. 2011. Guillain-barré syndrome: modern theories of etiology. *Curr Allergy Asthma Rep* 11:197-204.

38. Hariharan. 2002. *Campylobacter jejuni*: Public health hazards and potential control methods in poultry: a review. *Vet Med Czech* 49: 441-446.

39. Hermans D, Van Deun K, Martel A, Van Immerseel F, Messens W, Heyndrickx M, Haesebrouck F and Pasmans F. 2011. Colonization factors of *Campylobacter jejuni* in the chicken gut. *Vet Res* 42: 1-14.

40. Humphrey T, O'Brien S and Madsen M. 2007. Campylobacters as zoonotic pathogens: a food production perspective. *Int J Food Microbiol* 117: 237-257.

41. Hurd S, Patrick M, Hatch J, Clogher P, Wymore K, Cronquist AB, Segler S, Robinson T, Hanna S, Smith G and Fitzgerald C. 2012. Clinical laboratory practices for the isolation and identification of *Campylobacter* in Foodborne Diseases Active Surveillance Network (FoodNet) sites: baseline information for understanding changes in surveillance data. *Clin Infect Dis* 54: 440-445.

42. Hussain I, ShahidMahmood M, Akhtar M and Khan A. 2007. Prevalence of *Campylobacter* species in meat, milk and other food commodities in Pakistan. *Food Microbiol* 24: 219-222.

43. IASR (Infectious Agents Surveillance Report). 2010. *Campylobacter* enteritis in Japan, 2006-2009. http://idsc.nih.go.jp/iasr/31/359/tpc359.html Accessed 1 August 2012.

44. ICSP 2011. International Committee on Systematics of Prokaryotes. http://www.bacterio.cict.fr/c/campylobacter.html. Accessed 1 August 2012.

45. Jacobs-Reitsma WF. 1997. Aspects of epidemiology of *Campylobacter* in poultry. Vet Q 19: 113-7.

46. Jagusztyn-Krynicka EK, Łaniewski P and Wyszyñska A. 2009. Update on *Campylobacter jejuni* vaccine development for preventing human campylobacteriosis. *Expert Rev Vaccines* 8: 625-645.

47. Jain D, Sinha S, Prasad KN and Pandey CM. 2005. *Campylobacter* species and drug resistance in a North Indian rural community. *Trans R Soc Trop Med Hyg* 99: 207-214.

48. JenningsJL, Sait LC, Perrett CA, Foster C, Williams LK, Humphrey TJ and Cogan TA. 2011. *Campylobacter jejuni* is associated with, but not sufficient to cause vibrionic hepatitis in chickens. *Vet Microbiol* 149: 193-199.

49. Kalra V, Chaudhry R, Dua T, Dhawan B, SahuJK and Mridula B. 2009. Association of *Campylobacter jejuni* infection with childhood Guillain-Barré syndrome: a case-control study. *J Child Neurol* 24: 664-668.

50. Kassem II, Sanad Y, Gangaiah D, Lilburn M, Lejeune J and Rajashekara G. 2010. Use of bioluminescence imaging to monitor *Campylobacter* survival in chicken litter. *J Appl Microbiol* 109: 1988-1997.

51. Kawatsu K, Kumeda Y, Taguchi M, Yamazaki-Matsune W, Kanki M and Inoue K. 2008. Development and evaluation of immunechromatographic assay for simple and rapid detection of *Campylobacter jejuni* and *Campylobacter coli* in human stool specimens. *J Clin Microbiol* 46: 1226-1231.

52. Kirkpatrick BD and Tribble DR. 2011. Update on human *Campylobacter jejuni* infections. *Curr Opin Gastroenterol* 27: 1-7.

53. Kownhar H, Shankar EM, Rajan R, Vengatesan A and Rao UA. 2007. Prevalence of *Campylobacter jejuni* and enteric bacterial pathogens among hospitalized HIV infected versus non-HIV infected patients with diarrhoea in southern India. *Scand J Infect Dis*. S39: 862-866.

54. Kulkarni NM, Sherikar AT, Sherikar AA, Tarwate BG, Paturkar AM and Murugkar H. 1992. Incidence of *Campylobacter jejuni* in chicken and their carcasses sold at Bombay city. *Indian J Anim Sci* 62: 1024-1027.

55. Kumar A, Agarwal RK, Bhilegaonkar KN, Shome BR and Bachhil VN. 2001. Occurrence of *Campylobacter jejuni in* vegetables. *Int J Food Microbiol* 67: 153-155.

56. Kumar R, Verma AK, Kumar A, Srivastava M and Lal HP. 2012. Prevalence and antibiogram of *Campylobacter* infections in dogs of Mathura, India. *Asian J Anim Vet Adv* DOI: 10.3923/ajava.2012.

57. Levin RE. 2007. *Campylobacter jejuni*: a review of its characteristics, pathogenicity, ecology, distribution, subspecies characterization and molecular methods of detection. *Food Biotechnol* 21: 271-347.

58. Line JE. 2006. Influence of relative humidity on transmission of *Campylobacter jejuni* in broiler chickens. *Poult Sci* 85: 1145-1450.

59. Man SM. 2011. The clinical importance of emerging *Campylobacter* species. *Nat Rev Gastroenterol Hepatol* 8: 669-685.

60. Miller RS, Speegle L, Oyarzabal OA and Lastovica AJ. 2008. Evaluation of three commercial latex agglutination tests for identification of *Campylobacter* spp. *J Clin Microbiol* 46: 3546-3647.

61. Misawa N, Shinohara S, Satoh H, Itoh H, Shinohara K, Shimomura K, Kondo F and Itoh K. 2000. Isolation of *Campylobacter* species from zoo animals and polymerase chain reaction-based random amplified polymorphism DNA analysis. *Vet Microbiol* 71: 59-68.

62. Moore JE, Corcoran D, Dooley JS, Fanning S, Lucey B, Matsuda M, McDowell DA, Megraud F, Miller BC, O'Mahony R, O'Riordan L, O'Rourke M, Rao JR, Rooney PJ and Whyte AP. 2005. *Campylobacter*. *Vet Res* 36: 351-382.

63. Nannapaneni R, Chalova VI, Crandall PG, Ricke SC, Johnson MG and O'Bryan CA. 2009. *Campylobacter* and *Arcobacter* species sensitivity to commercial orange oil fractions. *Int J Food Microbiol* 129: 43-49.

64. Newell DG, ElversKT, Dopfer D, Hansson I, Jones P, James S, Gittins J, Stern NJ, Davies R, Connerton I, Pearson D, Salvat G and Allen VM. 2011. Biosecurity-based interventions and strategies to reduce *Campylobacter* spp. on poultry farms. *Appl Environ Microbiol* 77: 8605-8614.

65. Oyarzabal OA, Macklin KS, Barbaree JM and Miller RS. 2005. Evaluation of agar plates for direct enumeration of *Campylobacter* spp. from poultry carcass rinses. *Appl Environ Microbiol* 71: 3351-3354.

66. Park SF. 2002. The physiology of *Campylobacter* species and its relevance to their role as foodborne pathogens. *Int J Food Microbiol* 74: 177-188.

67. Parsons BN, Porter CJ, Ryvar R, Stavisky J, Williams NJ, Pinchbeck GL, Birtles RJ, ChristleyRM, German AJ, Radford AD, Hart CA, Gaskell RM and Dawson S. 2010. Prevalence of *Campylobacter* spp. in a cross-sectional study of dogs attending veterinary practices in the UK and risk indicators associated with shedding. *Vet J* 184: 66-70.

68. Parsons BN, Williams NJ, Pinchbeck GL, Christley RM, Hart CA, Gaskell RM and Dawson S.2011. Prevalence and shedding patterns of *Campylobacter* spp. in longitudinal studies of kennelled dogs. *Vet J* 190: 249-254.

69. Petersen L, Nielsen EM and On SL. 2001. Serotype and genotype diversity and hatchery transmission of *Campylobacter jejuni* in commercial poultry flocks. *Vet Microbiol* 82: 141-154.

70. Pope JE, Krizova A, Garg AX, Thiessen-Philbrook H and Ouimet JM. 2007 *Campylobacter* reactive arthritis: a systematic review. *Semin Arthritis Rheum* 37: 48-55.

71. Rajendran P, Babji s, George AT, Rajan DP, kang G and Ajjampur SS. 2012. Detection and identification of *Campylonacter* in stool samples of children sand animals from vellore, south india. *Indian J Med Microbiol* 30: 85-88.

72. Ridley AM, Allen VM, Sharma M, Harris JA and Newell DG. 2008. Real-time PCR approach for detection of environmental sources of *Campylobacter* strains colonizing broiler flocks. *Appl Environ Microbiol* 74: 2492-2504.

73. Rudi K, Hoidal HK, Katla T, Johansen BK, Nordal J and Jakobsen KS. 2004. Direct real-time PCR quantification of *Campylobacter jejuni* in chicken faecal and cecal samples by integrated cell concentration and DNA purification. *Appl Environ Microbiol* 70: 790-797.

74. Ryser ET and Marth EH. 1989. "New" food-borne pathogens of public health significance. *J Am Diet Assoc* 89: 948-954.

75. Sahin O, Plummer PJ, Jordan DM, Sulaj K, Pereira S, Robbe-Austerman S, Wang L, Yaeger MJ, Hoffman LJ and Zhang Q. 2008. Emergence of a tetracycline-resistant *Campylobacter jejuni* clone associated with outbreaks of ovine abortion in the United States. *J Clin Microbiol* 46: 1663-1671.

76. Salihu MD, AbdulkadirJU, Oboegbulem SI, EgwuGO, Magaji AA, Lawal M and Hassan Y. 2009. Isolation and prevalence of *Campylobacter* species in cattle from Sokoto state, Nigeria. *Vet Ital* 45: 501-505.

77. Schmidt-Ott R, Brass F, Scholz C, Werner C and Gross U. 2005. Improved serodiagnosis of *Campylobacter jejuni* infections using recombinant antigens. *J Med Microbiol* 54: 761-767.

78. Shah N, DuPont HL and Ramsey DJ. 2009. Global etiology of travelers' diarrhoea: systematic review from 1973 to the present. *Am J Trop Med Hyg* 80: 609-614.

79. Shane SM. 1992. The significance of *Campylobacter jejuni* in poultry: a review. Avian Pathol 21: 189-213.

80. Sharma A, Lal V, Modi M, Vaishnavi C and Prabhakar S. 2011. *Campylobacter jejuni* infection in Guillain-Barré syndrome: a prospective case control study in a tertiary care hospital. *Neurol India* 59: 717-721.

81. Sharma S and Prasad DN. 1997. *Campylobacter* - an emerging food-borne pathogen of concern. *Indian Dairyman* 49: 11-13.

82. Silva J, Leite D, Fernandes M, Mena C, Gibbs PA and Teixeira P. 2011. *Campylobacter* sp. as a foodborne pathogen: A Review. *Front Microbiol* 2: 200.

83. Singh R, Singh PP, Rathore RS, Dhama K and Malik SVS. 2008. Studies on effect of seasonal variations on the prevalence of *Campylobacter jejuni* in poultry faecal samples collected from Western Uttar Pradesh, India. *Indian J Comp Microbiol Immunol Infect Dis* 29: 1-2.

84. Singh R, Singh PP, Rathore RS, Dhama K and Malik SVS. 2009. Prevalence of *Campylobacter jejuni* and *Campylobacter coli* in chicken meat and carcasses collected from local poultry farms. *Indian J Comp Microbiol Immunol Infect Dis* 30: 35-38.

85. Sippy R, Sandoval-Green CMJ, Sahin O, Plummer P, Fairbanks WS, Zhang Q and Blanchong JA. 2012. Occurrence and molecular analysis of *Campylobacter* in wildlife on livestock farms. *Vet Microbiol* 157: 369-375.

86. Skirrow MB. 1977. *Campylobacter* enteritis: a "new" disease. *Br Med J* 2: 9-11.

87. Smith JL and Fratamico PM. 2010. Fluoroquinolone resistance in *Campylobacter*. *J Food Prot* 73: 1141-1152.

88. Somvanshi S, Yadav AS, Rajkumar RS and Singh RP. 2009. Detection of thermophilic *Campylobacter* from live chicken and their eggs by using conventional and molecular techniques. *Indian J Poult Sci* 44: 399-401.

89. Suzuki H and Yamamoto S. 2009a. *Campylobacter* contamination in retail poultry meats and by-products in the world: a literature survey. *J Vet Med Sci* 71: 255-261.

90. Suzuki H and Yamamoto S. 2009b. *Campylobacter* contamination in retail poultry meats and by-products in Japan: A literature survey. *Food Control* 20: 531-537.

91. Talukder RK, Sutradhar SR, Rahman KM, UddinMJ and Akhter H. 2011. Guillian-Barre syndrome. *Mymensingh Med J* 20: 748-756.

92. Tolcin R, La Salvia MM, Kirkley BA, Vetter EA, Cockerill FR and Procop GW. 2000. Evaluation of the Alexon-trend ProSpecT *Campylobacter* microplate assay. *J Clin Microbiol* 38: 3853-3855.

93. Vaishnavi C and Kaur S. 2005. Is *Campylobacter* involved in antibiotic associated diarrhoea? *Indian J Pathol Microbiol* 48: 526-529.

94. Varma KS, Jagadeesh N, Mukhopadhyay HK and Dorairajan N. 2000. Incidence of *Campylobacter jejuni* in poultry and their carcasses. *J Food Sci Technol Mysore* 37: 639-641.

95. Varma KS, Subramanyam KV, Satyanaryana A and Reddy KM. 2005. Isolation of *Campylobacter jejuni* from rectal swabs and carcasses of sheep. *Indain J Comp Microbiol Immunol Infect Dis* 26: 27-28.

96. Verhoeff-Bakkenes L, Jansen HA, in 't Veld PH, Beumer RR, Zwietering MH and van Leusden FM. 2011. Consumption of raw vegetables and fruits: a risk factor for *Campylobacter* infections. *Int J Food Microbiol* 144: 406-412.

97. Vlieghe ER, Jacobs JA, Van Esbroeck M, Koole O and Van Gompel A 2008. Trends of norfloxacin and erythromycin resistance of *Campylobacter jejuni*/*Campylobacter coli* isolates recovered from international travelers, 1994 to 2006. *J Travel Med* 15: 419-425.

98. Volokhov D, Chizhikov V, Chumakov K and Rasooly A. 2003. Microarray-based identification of thermophilic *Campylobacter jejuni*, *C. coli*, *C. lari*, and *C. upsaliensis*. *J Clin Microbiol* 41: 4071-4080.

99. Vucic S, Kiernan MC and Cornblath DR. 2009. Guillain-Barré syndrome: an update. *J Clin Neurosci* 16: 733-741.

100. Wagenaar JA, Mevius DJ and Havelaar AH. 2006. *Campylobacter* in primary animal production and control strategies to reduce the burden of human campylobacteriosis. *Rev Sci Tech* 25: 581-594.

101. Wallis TS. 2004. Vaccination against *Salmonella*, enterohaemorrhagic *E. coli* and *Campylobacter* in food-producing animals. *Dev Biol* 119: 343-350.

102. Wheeler JG, Sethi D, Cowden JM, Wall PG, Rodrigues LC, Tompkins DS, Hudson MJ and Roderick PJ. 1999. Study of infectious intestinal disease in England: rates in the community, presenting to general practice, and reported to national surveillance. The Infectious Intestinal Disease Study Executive. *BMJ* 318: 1046-50.

103. WHO 2011 *Campylobacter* fact sheet. http://www.who.int/mediacentre/factsheets/fs255/en/index.html. Accessed 1 August 2012

104. Wilson IG. 2004. Airborne *Campylobacter* infection in a poultry worker: case report and review of the literature. *Commun Dis Public Health* 7: 349-353.

105. Wong TL, Hollis L, Cornelius A, Nicol C, Cook R and Hudson JA. 2007. Prevalence, numbers, and subtypes of *Campylobacter jejuni* and *Campylobacter coli* in uncooked retail meat samples. *J Food Prot* 70: 566-573.

106. Wooldridge KG and Ketley JM. 1997. *Campylobacter*-host cell interactions. *Trends Microbiol* 5: 96-102.

107. Yamazaki W, Taguchi M, Ishibashi M, Kitazato M, Nukina M, Misawa N and Inoue K. 2008. Development and evaluation of a loop-mediated isothermal amplification assay for rapid and simple detection of *Campylobacter jejuni* and *Campylobacter coli*. *J Med Microbiol* 57: 444-451.

108. Young KT, Davis LM and DiritaVJ. 2007. *Campylobacter jejuni*: molecular biology and pathogenesis. *Nat Rev Microbiol* 5: 665-679.

109. ZiprinRL, Hume ME, Young CR and Harvey RB. 2002. Inoculation of chicks with viable non-colonizing strains of *Campylobacter jejuni*: evaluation of protection against a colonizing strain. *Curr Microbiol* 44: 221-223.

110. Zweifel C, Scheu KD, Keel M, Renggli F and Stephan R. 2008. Occurrence and genotypes of *Campylobacter* in broiler flocks, other farm animals, and the environment during several rearing periods on selected poultry farms. *Int J Food Microbiol* 125: 182-187.

2014, Zoonoses: Bacterial Diseases
Pages 210–244
Editor: Sudhi Ranjan Garg
Published by: DAYA PUBLISHING HOUSE, NEW DELHI

10

Listeriosis

S.R. Garg and V.J. Jadhav
Department of Veterinary Public Health and Epidemiology,
Lala Lajpat Rai University of Veterinary and Animal Sciences,
Hisar – 125 004

Listeriosis is caused by *Listeria monocytogenes*. The organism is found frequently in the environment and causes abortions, stillbirths and encephalitis in a wide range of animals. Due to the characteristic symptoms of meningoencephalitis, the disease is also known as 'circling disease' in animals. Epidemiological association of the occurrence of listeriosis in farm animals with feeding of poorly made silage gives it a name 'silage disease' too. Listeriasis and listerellosis are the other synonyms frequently used in scientific documents. In human beings, the disease manifests stillbirths, abortions, meningoencephalitis and septicaemia. The highest incidence of listeriosis is usually seen in women, neonates, elderly people and patients with underlying immunosuppression.

L. monocytogenes emerged as a foodborne pathogen in the late 1970s and early 1980s that caused numerous outbreaks in humans in North America. The reported incidence of human cases of foodborne listeriosis is low compared to the other foodborne bacterial pathogens like *Campylobacter* and *Salmonella*, but the mortality rate in human listeriosis is the highest among these. It is believed that listeriosis would have a greater public health impact on society in the future. It can be attributed to an expected increase in the number of susceptible individuals, including the elderly population, as well as an increase in the consumption of refrigerated foods and also those foods that have extended shelf-life due to technological interventions (Pagotto *et al.* 2006). Due to its serious consequences, the food safety programmes of many countries, including India, have adopted a policy of 'zero tolerance' to the microorganism.

Certain characteristics of *L. monocytogenes* like its ability to grow at refrigeration temperatures and to sustain wide range of pH (4.4 to 9.6) heighten the risk of its transmission through foods. Apart from tolerating low moisture environments, the organism can grow under aerobic or facultatively anaerobic conditions and even at high salt concentrations. As a result, it can pose threat to the safety of packaged foods (Thigeel *et al.* 2011).

Like other major foodborne diseases, listeriosis is responsible for heavy economic losses too. According to one estimate, the loss of productivity and mortality costs due to listeriosis in Australia alone is $ 83.1 million, in addition to the health care costs of $ 2.4 million per annum (Abelson *et al.* 2006). The annual economic losses during the year 2000 due to listeriosis have been estimated at $ 1.3 billion in USA (Todd 2007). The cost per case of listeriosis in USA was estimated as $ 1.3 million in the year 2010 (Scharff 2012). The implication of *L. monocytogenes* in disease outbreaks and the economic losses have made the organism of much scientific interest throughout the world.

Geographical Distribution

Listeriosis has worldwide distribution with sporadic occurrence. It is more common in countries situated in temperate rather than tropical zones. In 2009, human listeriosis notification rate of 0.4 per 100,000 people was recorded in the European Union with highest prevalence in Denmark, Spain and Sweden (Lahuerta *et al.* 2011). In Australia in 2003, the rate was three infections per million population for non-pregnancy listeriosis cases and 4.6 infections per 100,000 births per year for maternal-foetal infections (Anonymous 2012). Amongst the African countries, the disease has been reported in Zambia, Togo, Nigeria, Central African Republic and Botswana (Morobe 2009).

Etiological Agent

Listeria species are Gram-positive, rod-shaped, and nonspore-forming bacteria. The bacterial cells are 1-2 µm long and may exist as singles or in pairs. These organisms are motile due to the presence of peritrichous flagella and display tumbling motility (Figure 10.1). For many years after its discovery, the genus *Listeria* contained only one species, *L. monocytogenes*. Other five species, *L. innocua*, *L. ivanovii*, *L. welshimeri*, *L. seeligeri* and *L. grayi*, were added subsequently. *L. ivanovii* has two subspecies: *L. ivanovii subsp. ivanovii* and *L. ivanovii subsp. londoniensis*. Two new species have been proposed: *L. marthii* (Graves *et al.* 2010) and *L. rocourtiae* (Leclercq *et al.* 2010). Among these, *L. monocytogenes* is pathogenic to humans and animals, while *L. ivanovii* is pathogenic to animals. Other species are considered nonpathogenic. *L. seeligeri*, though considered nonpathogenic, possesses a part of the virulence gene cluster, which is present in pathogenic *L. monocytogenes* and *L. ivanovii* (Gouin *et al.* 1994).

Listeria organisms are characterized serologically on the basis of their somatic and flagellar antigens. In this classification, the letters a, b, c, etc. refer to the flagellar antigens while numericals *i.e.* 1/2, 3, 4, etc. describe the somatic antigens.

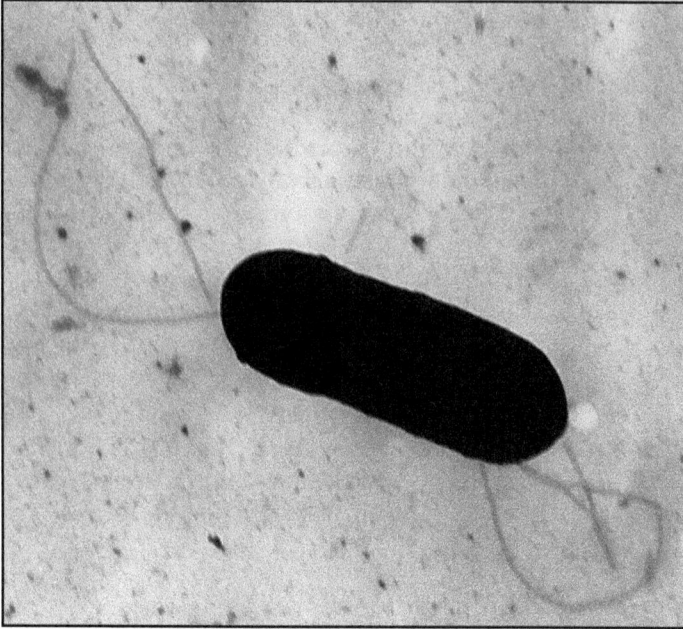

Figure 10.1: Electron micrograph of a flagellated *Listeria monocytogenes* bacterium, magnified 41,250X. (Courtesy: CDC, Dr. Balasubr Swaminathan, Peggy Hayes)

L. monocytogenes organisms have serotypes 1/2a, 1/2b, 1/2c, 3a, 3b, 3c, 4a, 4ab, 4b, 4b(x), 4c, 4d, 4e, 7. Serotypes of other *Listeria* species include 4ab, 6a and 6b of *L. innocua*; 6a and 6b of *L. welshimeri*; 1/2b, 4c, 4d and 6b of *L. seeligeri* and serotype 5 of *L. ivanovii* (Pagatto *et al.* 2006). Most of the cases of animal and human infections are caused by serotypes 1/2a, 1/2b, and 4b (Schuchat *et al.* 1991).

Listeria spp. are generally considered ubiquitously distributed in the natural environment. *L. monocytogenes* and other *Listeria* spp. have been isolated from different environmental sources, including soil, water, vegetation, sewage, animal feeds and farm environments and food processing environments. The organism is relatively resistant to NaCl, and the possibility of its growth at pH 5-9 is a root cause of some outbreaks of listeriosis in ruminants fed with poorly prepared silage that does not achieve sufficient acidity. Further, the organism can multiply over a wide range of temperatures (1 to 45°C) increasing the possibility of infection through silage in animals and through food products maintained at room or refrigeration temperatures in human beings.

Pathogenesis and Virulence Factors

L. monocytogenes causes severe illness owing to its unique ability to induce its own phagocytosis by the host cells, followed by replication within those cells and direct transfer to another cell. Inside the host cell, it remains protected from many host defenses such as antibodies and complement system. Foodborne *Listeria* are taken up by the enterocytes or M cells in the small intestinal lining and then multiply in the underlying phagocytic cells. Thereafter, bacteria are carried from the intestine in macrophage cells in blood or lymph to the liver and spleen where most of them are killed by neutrophils acting with Kupffer cells. If the T cell-mediated immune response of host is inadequate, listeriae multiply in hepatocytes and macrophages and are carried in the blood to various organs, particularly the brain and/or uterus where they penetrate the blood-brain/placental barrier. *L. monocytogenes* produces several virulence factors that facilitate this invasion process (Doyle 2001). Different virulence factors and their role in pathogenesis are depicted in Table 10.1.

Table 10.1: Virulence factors of *Listeria*

Virulence Factor	Role in the Pathogenesis
Internalis	Induces phagocytosis by nonprofessional phagocytes such as epithelial cells and thereby helps the organism penetrate in the body.
Surface protein p 104	Helps in the process of adhesion to intestinal cells.
Listeriolysin O	A bacterial pore-forming toxin, plays significant role in lysing the vacuolarmembrane and allowing *L. monocytogenes* to escape into cytoplasm of the cell.
Act A protein	Act A induces polymerization of globular actin molecules to form polarized actin filaments, bacterial cells move along these filaments to the cell membrane and cause portions of the membrane to bulge outwards, forming structures called listeriopods which are engulfed by adjacent cells allowing dissemination of *L. monocytogenes* without exposure to antibodies.
Phospholipases	Phosphatidylinositol-specific phospholipase C (PI-PLC) and phosphatidylcholine-specific phospholipase C (PC-PLC) play role in invasion, PI-PLC aids in escape from the primary vacuole while PC-PLC is active during cell-to-cell spread of bacteria.
Metalloprotease	Activation of inactive precursor of phospholipase (PC-PLC).
Clp proteases and ATPases	Group of proteins having ability to act as proteolytic enzymes aid in disruption of the vacuolar membrane and the intracellular survival of listeriae.
Protein p60	Murein hydrolase enzyme that catalyzes a reaction during the final stage of cell division of *L. monocytogenes*.

Source: Doyle (2001).

Host Range

Numerous animal species are susceptible to listeric infection. A large proportion of healthy asymptomatic animals shed *L. monocytogenes* in their faeces. Its occurrence has been reported in buffaloes, cattle, sheep, goats and less frequently in pigs, horses, dogs, turkeys, fowls and ducks (Wesley 2007). The host range also extends to laboratory animals such as rabbits, guinea pigs, mice, etc. *L. monocytogenes* has been detected in a number of minor wildlife species like canary, chinchilla, ferret, fish, gerbil and mink (Pal 2007).

Transmission of Infection

L. monocytogenes is widely prevalent in soil, water and animal excreta (manure) that may lead to contamination of vegetables and forage crops and transmission of the infection to humans as well as to the livestock. Animals can carry the bacterium without any apparent signs of infection and pass on the organisms to meat and dairy products as contaminants. The bacterium has been found in a variety of raw foods as well as in processed foods that might have become contaminated after processing. Humans and animals may also contract the infection by direct contact with infected faecal matter.

Transfer of the pathogen from mother to foetus or neonate has been well documented. Infection following inhalation of infected aerosols has also been reported in both humans and animals. Venereal spread has been documented in cattle. It is estimated that 5 per cent of the human population carries the bacterium asymptomatically in the gut (Lorber 1997). *L. monocytogenes* has been isolated from human semen (Toaff *et al.* 1962); the possibility of sexual transmission thus cannot be ignored. Larsson *et al.* (1978) reported nosocomial transmission of *L. monocytogenes* causing enterocolitis and meningitis in newborn infants.

Risk Groups

Pregnant women constitute a major risk group. The disease can affect the mother and baby both. Foetal or neonate death may follow the maternal consumption of contaminated products. In an outbreak of listeriosis in North Carolina, USA, between late 2000 and early 2001, following the ingestion of infected Mexican-style soft cheese, of 11 women reporting clinical signs, 10 were pregnant. As a result, there were five stillbirths and three premature births while two infected neonates required treatment (Anonymous 2001). Immunocompromised patients are likely to suffer serious illness with fatalities. This is also true for elderly people or infirm patients. Infection is possible in healthy persons too but they are less prone to display serious or fatal sequelae. A study in the USA has established that patients who use high quantities of antacids or who take cimetidine may be at increased risk of contracting disease due to inhibition of gastric acid (Ho *et al.* 1986). Animal handlers and veterinarians are also at greater risk due to their direct contact with contaminated body fluids.

Prevalence of Disease in Animals

Indian Perspective

In India, listeriosis for the first time was probably described by Mahajan in 1936 as "circling disease" in sheep in Hyderabad State, but he could not establish the cause of the disease (Phadke *et al.* 1979). Dhanda *et al.* (1959) reported *Listeria* infection in case of abortion in an ewe. Since then, a number of disease outbreaks have been reported from different parts of the country. Uppal *et al.* (1981) reported an outbreak of listeriosis in which 131 animals showed signs of neurological disturbances with 74 fatalities.

Abortions associated with *L. monocytogenes* and *L. ivanovii* have been reported in migratory flocks of sheep and goats in Himachal Pradesh (Sharma *et al.* 1996, Nigam *et al.* 1999). From the same region, Mahajan and Katoch (1997) isolated both of these organisms from sheep and goats with endometritis. *L. monocytogenes* was isolated from endometritis cases in buffaloes by Shah and Dholakia (1983). Dutta and Malik

(1981) isolated three strains of *L. monocytogenes* from 902 clinical specimens from different species of animals. Dash *et al.* (1998) reported an outbreak of circling syndrome due to *L. monocytogenes* in crossbred Landrance pigs at an organised farm. Samanta (2009) reported clinical case of listeriosis in a 3-month old calf in West Bengal.

The prevalence of *Listeria* in 725 clinical samples comprising of blood, milk and faecal swabs collected from mastitis affected bovines has been found to be 1.66 per cent. Of these, four (0.55 per cent) specimens were confirmed as *L. monocytogenes* and one (0.14 per cent) as *L. ivanovii* (Rawool *et al.* 2007). In buffaloes with history of reproductive disorders, Shakuntala *et al.* (2006) reported 4.4 per cent prevalence of *L. monocytogenes* and 7.4 per cent of other *Listeria* spp. Similarly, Kaur *et al.* (2010) reported isolation of *Listeria* spp. from 19.8 per cent of farm animals with a history of reproductive disorders. They also detected *L. ivanovii* from one of the 54 samples from 18 ewes.

Some other major episodes of the disease in the country are summarized in Table 10.2.

Listeria may remain in the body of healthy animals. Prevalence rates of 0.7 per cent for *L. monocytogenes* and 0.9 per cent for *L. innocua* have been reported in apparently healthy sheep in Gujarat state (Yadav and Roy 2009). From the same region, *L. monocytogenes* (1.8 per cent) and *L. innocua* (3.4 per cent) were isolated from faecal samples of apparently healthy mammals and birds at Baroda Zoo (Yadav *et al.* 2011). *L. monocytogenes* was isolated from brain of apparently healthy cattle (6.7 per cent) and buffaloes (5 per cent) slaughtered in Mumbai and Thane district of Maharashtra (Shirke *et al.* 2000). In another study, *L. ivanovii* was isolated in higher proportions from apparently healthy goats (14.52 per cent) than those suffered with abortion (7.5 per cent) and mastitis (5.56 per cent) (Elezebeth *et al.* 2007).

Isolation of pathogenic *Listeria* spp. in faecal samples of captive wild animals has also been studied. *L. monocytogenes* was isolated from eight (16 per cent) of 50 faecal samples from six different mammals and one bird. One isolate from a jackal was pathogenic (Kalorey *et al.* 2006).

A number of serological studies also reveal the prevalence of *Listeria* infection in animals. Srivastava and Khera (1980) reported the prevalence of *Listeria* antibodies in 27.4 per cent bovine samples in titres ranging from 1:50 to 1:400. Nigam *et al.* (1999) reported the association of seroprevalence of *L. monocytogenes* serovars and *L. ivanovii* with infertility and repeat breeding in animals. Barbuddhe *et al.* (2000a) demonstrated listeria seropositivity in 41.13 per cent and 33.76 per cent goats and sheep, respectively. Buffaloes have shown 4 per cent and 25.35 per cent seropositivity for phosphatidylinositol-specific phospholipase C (PIPLC) (Chaudhari *et al.* 2004) and anti listeriolysin O (ALLO), respectively (Barbuddhe *et al.* 2002). A serological survey revealed listerial antibodies in 28-31 per cent cattle and buffaloes in Gujarat state (Rathod *et al.* 2010) and 18.75 to 21.43 per cent in sheep and goats (Vasava *et al.* 2005). In another study, serum sample of 113 slaughtered goats showed the presence of ALLO antibodies in 19.16 to 21.68 per cent samples (Bhanu Rekha *et al.* 2006, Elezebeth *et al.* 2007).

Table 10.2: Some outbreaks of listeriosis in animals in India

State/Place	Year/Period	Major Features	Reference
Tamil Nadu (erstwhile Madras State)	1936-37	32 outbreaks covering many villages, organism isolated from an infected sheep suffering from circling disease	Viswanathan and Ayyer (1950)
Jammu and Kashmir	1977-78	111 abortions out of 800 lambings in sheep, *L. monocytogenes* serotype 4b isolated from stomach contents of one of 22 aborted foetuses	Vishwanathan and Uppal (1981)
Andhra Pradesh	1981	3 strains of *L. monocytogenes* serotype 4b isolated from brain of a buffalo	Uppal *et al.* (1981)
Assam	1982	Outbreak of listerial meningoencephalitis in pigs at a farm affecting 75 animals with 27 mortalities	Rahman *et al.* (1985)
Uttar Pradesh	1995	Circling movements with lateroventral deviation of the head, sudden ataxia and epileptic seizures followed by death of 27 pigs, *L. monocytogenes* isolated from brain tissues of dead animals	Dash *et al.* (1998)
Haryana	1998	Abortions caused by *L. ivanovii* in a flock of 254 crossbred sheep	Chand and Sadana (1998)
Tirupati	2000	Outbreak of disease with neurological signs involving twisting of neck in a flock of 700 broilers, 80 per cent morbidity and 40 per cent mortality, ascitic fluid in the peritoneum, and petechial haemorrhages in liver, heart, spleen, kidneys and brain, *Listeria* isolated from tissue homogenates	Vijayakrishna *et al.* (2000)
Punjab	2006	3 outbreaks of encephalitis caused by *L. monocytogenes* in migratory flocks of sheep and goats, 7.89 per cent cumulative morbidity, 7.08 per cent mortality and 89.85 per cent case fatality rate	Kumar *et al.* (2007)
Kashmir valley	2008-09	15-38 per cent prevalence of *L. monocytogenes* in 52 cases of abortions or stillbirths in sheep	Sharif *et al.* (2011)

Global Perspective

Like in India, listeriosis in animals is widespread in the other parts of the world too. Hyslop (1975) reported cases of bovine encephalitis, abortion and septicaemia due to *L. monocytogenes* in Canada. In UK, 0.05 to 0.13 per cent of about 30,000 bovine abortions occuring annually have been attributed to listeriosis (Anonymous 1994). In France, 428 outbreaks of animal listeriosis were detected in 1998-1999 (Vaissaire 2000). Most of them were related to cattle (259), followed by sheep (108). Other species included poultry, birds, horses, dogs, roe deer, rabbits, hares and wild boars. Ovine and caprine listeric encephalitis in Iraq is also reported (Yousif *et al.* 1984). Meredith and Schneider (1984) put on record an outbreak of listerial meningoencephalitis involving sheep in the Western Cape Province of South Africa that was attributed to poor management practices. Some other major episodes of listeriosis are listed in Table 10.3.

Listeria may be present even in those farms where there are no cases of listeriosis. Findings of a case control study conducted in the USA involving 24 case farms with at least one recent case of listeriosis and 28 matched control farms with no listeriosis cases showed similar overall prevalence of *L. monocytogenes* in cattle case farms (24.4 per cent) and control farms (20.2 per cent) (Nightingale *et al.* 2004). However, sheep and goat farms showed a significantly (P < 0.0001) higher prevalence in case farms (32.9 per cent) than in control farms (5.9 per cent). Faecal shedding and strain diversity of *L. monocytogenes* in healthy ruminants and swine in Northern Spain showed prevalence of the organism in 46.3 per cent of dairy cattle, 30.6 per cent beef cattle and 14.2 per cent sheep herds, but not in swine (Esteban *et al.* 2009). Serotyping of 114 isolates reveled 4b as the most prevalent serovar (84.2 per cent), followed by 1/2a (13.2 per cent).

Some studies have epidemiologically incriminated contaminated silage as the source of infection in animals. Wiedmann *et al.* (1996) reported two outbreaks of epidemic listeriosis in sheep and goats and two outbreaks in cattle due to *L. monocytogenes* infection. Strains of this organism were also detected in silage samples consumed by the animals in these three outbreaks. Similarly, Wiedmann *et al.* (1997) reported the isolation of *L. monocytogenes* from a herd of sheep having cases of listeric encephalitis. Silage and the farm equipment used to transport silage were also found to carry the same organism. An outbreak in a flock of 55 sheep caused by feeding grass silage contaminated with *L. monocytogenes* was reported in Austria (Wagner *et al.* 2005).

Incidence in Foods

Listeria spp. are known to be prevalent in a wide variety of food products. These include animal products as well as foods of plant origin.

Milk and Dairy Products

There are reports of natural infection in milk due to bovine mastitis (Donnelly and Baigent 1986). Kampelmacher (1962) reported a cow to shed *L. monocytogenes* in all four quarters at levels between 2,000 and 20,000 cells/ml milk though the udder was not inflamed and the milk appeared normal. However, Gitter *et al.* (1980)

Table 10.3: Some outbreaks of listeriosis in animals

State/Place	Year/Period	Major Features	Reference
California, USA	1988	Outbreak of disease characterized by torticollis, incoordination, and depression in a commercial broiler farm, *L. monocytogenes* isolated from brain tissue	Cooper (1989)
Southern Illinois, USA	1990	Outbreak of listeriosis in silage fed sheep, 3.1 per cent ewes (29/936) and 1.3 per cent lambs (17/1262) died	Nash *et al.* (1995)
USA	1993	Two outbreaks of epizootic listerial encephalitis, one in sheep and one in goats, *L. monocytogenes* isolated from silage and brain tissue samples	Wiedmann *et al.* (1994)
UK	1994	An outbreak of eye disease in a herd of 240 fallow deer, *L. monocytogenes* isolated from conjuctival swabs and silage	Welchman *et al.* (1997)
Saudi Arabia	2001	Outbreak of ovine septicaemic listeriosis, 7.1 per cent morbidity and 2.4 per cent mortality, adult animals and pregnant ewes primarily affected, abortions not observed	Al-Dughaym *et al.* (2001)
USA	2011	Listeriosis outbreak in dairy farm due to unusual *L. monocytogenes* serotype 4b strain, 3 animals died out of 9 affected	Bundrant *et al.* (2011)

suggested that the main source of *L. monocytogenes* in milk was probably faecal contamination and the organisms were a rare cause of mastitis. The incidence of *L. monocytogenes* in cow's raw milk has been reported as 3.8 per cent in Scotland (Fenlon and Wilson 1989), 4.4 per cent in the Netherlands (Beckers *et al.* 1987), 3 per cent in Switzerland (Breer and Schopfer 1989), 1.6 per cent in Canada (Davidson *et al.* 1989), 4.2 per cent in USA (Lovett *et al.* 1987), 9.5 per cent in Brazil (Moura *et al.* 1993) and 45.9 per cent in Spain (Dominguez *et al.* 1985). The overall incidence of *L. monocytogenes*, *L. innocua*, *L. welshmeri* and other *Listeria* species in cow milk has been estimated as 3.48 per cent, 6.16 per cent, 1.22 per cent and 0.77 per cent, respectively, by pooling the data from published surveys (Ryser and Marth 1991). In a recent study, *Listeria* species were isolated from 22 per cent raw milk and 4 per cent cottage cheese samples from Addis Ababa in Ethiopia. Among these *L. monocytogenes* was detected in 13 per cent raw milk samples and 1 per cent cottage cheese samples (Gebretsadik *et al.* 2011).

In bulk milk tank samples, the incidence of *L. monocytogenes* has been recorded as 11 per cent in Canada (Slade *et al.* 1988) and 4.4 per cent in Korea (Baek *et al.* 2000). Van Kessel *et al.* (2011) reported 7.1 per cent prevalence of *L. monocytogenes* in bulk tank milk and in-line filters from the dairies in the United States with serotype 1/2a complex being the most common serotype, followed by 1/2b and 4b. In a study, the incidence of *L. monocytogenes* in raw goat milk and sheep milk has been found to be 7.8 per cent and 1.9 per cent, respectively (Katic and Stojanovic 1992). In some instances, *L. monocytogenes* has been detected in pasteurized milk samples (Greenwood *et al.* 1991) and ready-to-eat milk products (Pochop *et al.* 2012).

In India, *L. monocytogenes* and other *Listeria* species have been detected in the samples of milk and milk products. A large survey of milk samples collected from 2060 dairy cows from central parts of India revealed the presence of *Listeria* spp. in 139 (6.75 per cent) samples, including *L. monocytogenes* (5.1 per cent), *L. innocua* (0.9 per cent), *L. welshimeri* (0.1 per cent) and *L. seeligeri* (0.1 per cent) (Kalorey *et al.* 2008). Several other reports showed the presence of *Listeria* in raw and market milk (Bhilegaonker *et al.* 1997, 1999, Pednekar *et al.* 1997, Katre *et al.* 2009), ice cream (Warke *et al.* 2000, Singh *et al.* 2012) and cheese (Bhat and Willayat, 2011).

Widespread *Listeria* contamination in cheese has drawn particular attention of the scientists. *L. monocytogenes* as high as 10^7 cfu/g have been detected in some naturally contaminated cheese (Ryser and Marth 1987). Maria *et al.* (1998) reported 10.68 per cent incidence of *L. monocytogenes* in some Brazilian cheeses. The organism has also been shown to survive in products such as cultured butter milk, butter and yoghurt (Griffith and Deibel 1990). It has been reported that *L. monocytogenes* can survive in some instances for upto 30 days after the manufacture of yoghurt at pH values as low as 4.0. *L. monocytogenes* may grow well in contaminated fluid dairy products including soymilk at temperatures ranging from 4 to 35°C (Rosenow and Marth 1987, Farber *et al.* 1990).

Meat and Meat Products

A variety of meat is reported to be contaminated with *L. monocytogenes*, though the incidences vary greatly. Lee and McClain (1987) detected *L. monocytogenes* in beef (70 per cent), pork sausage (43 per cent) and poultry (48 per cent). Skovgaard and

Morgen (1988) also reported 67 per cent incidence of *Listeria* species including 28 per cent *L. monocytogenes* in minced beef in Denmark. Wung *et al.* (1992) found *L. monocytogenes* occurring in 11 per cent meat samples in China. Among 52 different varieties of sausage, salami and paté examined by Gunasinghe *et al.* (1994) in UK, 7 (13 per cent) samples were positive for *L. monocytogenes* and 13 (25 per cent) for *Listeria* species. Johnson *et al.* (1990) detected *L. monocytogenes* in the interior muscle cores of 5 of 110 samples of beef, pork and lamb roasts. These organisms were probably present in the muscles at the time of slaughter. In India, Barbuddhe (1996) reported 6.6 per cent goat meat samples positive for *L. monocytogenes*. Bhilegaonker *et al.* (1999) reported the presence of *Listeria* species in 21 samples of goat and buffalo meat out of 46 samples, but only 2 samples of buffalo meat carried *L. monocytogenes*.

Poultry meat is also known to carry *Listeria* organisms. The presence of *L. monocytogenes* in poultry meat has been reported in 33.3 to 63 per cent samples in UK, 23 to 48 per cent in USA, 41.1 per cent in Japan, 58.8 per cent in Taiwan, 48 per cent in New Zealand and 30.2 per cent in Korea (Pini and Gilbert 1988, Genigeorgis *et al.* 1989, Johnsen *et al.* 1990, Wong *et al.* 1990, McGowan *et al.* 1994). Kwantes and Isaac (1971) reported 57 per cent incidence of *L. monocytogenes* in fresh and frozen poultry samples. Gitter (1976) isolated the organisms from 15 per cent of oven ready poultry samples. *L. monocytogenes* was isolated from 47 per cent neck skin samples of dressed poultry in Denmark (Skovgaard and Morgen 1988). Ojeniyi *et al.* (1996) isolated *L. monocytogenes* from different kinds of samples collected from seven Danish poultry abattoirs. The incidence rates of *L. monocytogenes* in spent hen samples and broiler chickens obtained from Belgian and French abattoirs were found to be 50 per cent and 10-15 per cent, respectively (Uyttendaele *et al.* 1997).

Guerini *et al.* (2007) studied the prevalence of *Listeria* in beef processing plants in the USA. The prevalence in carcasses ranged from undetectable in some warm season samples to as high as 71 per cent during cool weather. At a sheep meat packing plant in Brazil, 34.8 per cent, 4.3 per cent and 1.5 per cent samples from carcasses yielded *L. innocua, L. monocytogenes* and *L. ivanovii,* respectively (Antoniollo *et al.* 2003). More recently, *L. monocytogenes* was isolated from raw as well as cooked meat from Shandong province of China (Jia *et al.* 2011).

Listeria spp. have also been isolated from street-vended ready-to-eat foods comprising of a variety of meat and poultry products sold in Egypt (El-Shenawy *et al.* 2011), Mexico (Diaz-Lopez *et al.* 2011), Trinidad (Syne *et al.* 2011), Jordan (Awaisheh 2010) and Latvia (Berzins *et al.* 2009). Prevalence of *L. monocytogenes* was recorded as 15.8 per cent at small scale Spanish factories producing traditional fermented sausages (Martin *et al.* 2011), 60 per cent in Finnish cold-smoked pork (Berzins *et al.* 2010) and 20.5 per cent in Italian vacuum-packaged sliced salami (Di-Pinto *et al.* 2010). Consumption of processed meat products is increasing at very high rate in modern times favouring attack rate of *L. monocytogenes*. The risk assessment model has predicted that processed meats could be responsible for upto 40 per cent of the cases of listeriosis in Australia (Ross *et al.* 2009).

Fish and Other Seafoods

Freshwater fish, marine fish and other seafoods reveal widespread *Listeria* contamination. Yucel and Balci (2010) reported 30 per cent incidence of *Listeria* spp. in freshwater fish and 10.4 per cent in marine fish used for human consumption in Turkey. *L. monocytogenes* (44.5 per cent) in freshwater fish and *L. murrayi* (83.5 per cent) were the most commonly isolated species in marine fish. In an earlier study, the incidence of *L. monocytogenes* in seafoods has been reported as 10.5 per cent in Taiwan (Wong *et al.* 1990). In a study in Malaysia, 44.1 per cent prawn samples and 33 per cent oysters samples revealed the presence of *L. monocytogenes* (Arumugaswamy *et al.* 1994).

Embarek (1994) recorded 6-36 per cent prevalence of *L. monocytogenes* in ready-to-eat cold smoked salmon and cooked fish products. Heinitz and Johnson (1998) reported 14 per cent incidence of *L. monocytogenes* in smoked finfish and smoked shell fish examined over a 5-year period in the USA. Monfort *et al.* (1998) detected *Listeria* spp. in 66 samples of shell fish out of 120 samples analyzed in Brittany (Western France) that included *L. innocua* (44), *L. seeligeri* (11) and *L. monocytogenes* (11). Microbiological analysis of sushi, a traditional Japanese food, mostly consisting of rice and raw fish, sold in restaurants in Hannover city of Germany revealed the presence of *L. monocytogenes* in 1.2 per cent samples (Atanassova *et al.* 2008).

Wang *et al.* (2011) reported 4.1 per cent prevalence of *L. monocytogenes* in seafood imported in the USA. Examination of ready-to-eat foods in Japan revealed frequent (5.7 to 12.1 per cent) *L. monocytogenes* contamination in minced tuna and fish roe products (Miya *et al.* 2010). Zarei *et al.* (2012) reported 14 per cent prevalence of *L. monocytogenes* in raw/fresh fish and shrimp marketed in Iran. A study of *L. monocytogenes* in New Zealand mussel (*Perna canaliculus*) processing plants confirmed its presence in raw and processed product. The study also highlighted the importance of cross-contamination from external and internal environments (Cruz and Fletcher 2011). In India, Jeyasekaran *et al.* (1996) reported the presence of *L. monocytogenes* in 17.2 per cent finfish and 12.1 per cent shellfish. *L. innocua* was the most common species detected in finfish (65.5 per cent) and shellfish (36.1 per cent).

Fruits and Vegetables

A recent study conducted by Schwaiger *et al.* (2011) on vegetable samples collected directly from farms and supermarkets in Germany showed the presence of *Listeria* spp. in 1 per cent samples, including *L. monocytogenes*. In another study, *L. monocytogenes* were detected in 0.7 per cent of samples of fresh and minimally processed fruit and vegetables and sprouts collected from retail establishments in Spain during 2005-2006 (Abadias *et al.* 2008). Francis and O'Beirne (2006) reported 2.9 per cent incidence of *L. monocytogenes* in samples of modified atmosphere packaged fresh cut fruits and vegetables from chill cabinets of a supermarket in Ireland. *Listeria* organisms have also been detected in uncooked ready-to-eat organic vegetables on retail sale in UK (0.5 per cent), but *L. monocytogenes* was not present (Sagoo *et al.* 2001). However, the presence of *L. monocytogenes* in 0.1 per cent samples of melon and beansprouts from retail and production premises in UK at concentration of 10^2 cfu/g or more have been reported (Little and Mitchell 2004).

In India, a very high incidence of *L. monocytogenes* has been observed in tomatoes (11.1 per cent), coriander and spinach leaves (50 per cent) and cabbage (25 per cent) samples (Pingulkar *et al.* 2001). However, in another study, the organism was not present in 124 sprouted seed samples collected from various markets in Mumbai city (Saroj *et al.* 2006).

Prevalence of Disease in Human Beings

Human Listeriosis in India

Listeriosis has been recorded in human population in India, particularly in women patients with bad obstetric history. *L. monocytogenes* has been isolated from the patients with history of abortions, miscarriages, stillbirths or neonatal deaths (Krishna *et al.* 1966, Bhujwala *et al.* 1973, Bhujwala and Hingorani 1975). The organism has also been reported as a cause of meningitis and hydrocephalus in children born of infected mothers (Gogate and Deodhar 1981). The studies on association of *Listeria* with human spontaneous abortions in India have revealed the presence of *Listeria* sp. in 14.8 per cent patients and that of pathogenic *L. monocytogenes* in 3.3 per cent patients (Kaur *et al.* 2007). Many individual cases of listeriosis are on record in the country. In a case of meningitis caused by *L. monocytogenes* in a 17 years old immunocompetent patient, the organism was isolated from the cerebrospinal fluid (CSF) and subsequent treatment with ampicillin resulted in dramatic recovery (Kalyani *et al.* 2006). A similar type of case was reported in a previously healthy, 20 months old female child with no underlying predisposition who showed poor response to treatment with vancomycin and ceftriaxone, but improved dramatically after substitution with ampicillin and amikacin (Peer *et al.* 2010). A case of neonatal listeriosis from Himachal Pradesh has been reported documenting isolation of *L. monocytogenes* from CSF and blood of the baby, and also from the genital tract of the mother (Mokta *et al.* 2010). The genotypic analysis of 17 *L. monocytogenes* isolates recovered from human patients in India during 2006 to 2009 revealed 88.24 per cent of the strains belonging to serovar group 4b, 4d, 4e and 11.76 per cent to serovar group 1/2b, 3b (Kalekar *et al.* 2011).

Seropositivity of human listeriosis has been recorded as 4.38 per cent (Dutta and Malik 1978). It is estimated at 4 per cent for serotype 4b and 32 per cent for 1/2a (Korde 1992). In another study, *L. monocytogenes* agglutinins were detected in most of the 200 healthy individuals in south India (Pramod and Ravindran 1997). Barbuddhe *et al.* (1999) detected antibodies to listeriolysin O in sera samples of 49.2 per cent individuals including abattoir workers (161), dairy farm workers (22) and patients with various clinical conditions (55). Another study by Barbuddhe *et al.* (2000b) revealed 40 per cent seropositivity for listeria infections among abattoir workers.

Human Listeriosis in Other Countries

The incidence of listeriosis in developed countries is markedly higher, probably due to better diagnostic facilities and better disease monitoring system as compared to the developing countries (WHO 1988). In the United States, *L. monocytogenes* is estimated to cause nearly 1,600 illnesses each year with more than 1,400 hospitalizations and 250 deaths (Scallan *et al.* 2011). In the surveillance programme for the year 2009, 524 cases of listeriosis were reported, 19 per cent of which were associated with pregnancy. The severity of listeriosis was quite high as 13 to 19 per cent of these patients and at least 7 per cent of live born infants with listeriosis died (CDC 2011). In a 9-year period prior to 1984, Switzerland experienced a mean of 3 cases of human listeriosis per year but in a 15-month period in 1983-84, 25 cases were seen (Malinverni *et al.* 1985). There were 687 cases in France during the year 1987 (Cossart *et al.* 1989).

Several outbreaks of foodborne listeriosis are on record in different countries. An outbreak occurred in the USA where 49 patients including 7 foetuses and infants and 42 adults showed 29 per cent mortality. The persons acquired infection by consuming pasteurized milk or milk of a specific brand in Massachusetts in 1983 (Flemming *et al.* 1985). Another outbreak involving 86 cases with 33.7 per cent mortality after consuming Mexican style fresh cheese was recorded in southern California in 1985 (Linnan *et al.* 1988). Massachusetts Department of Public Health investigated a listeriosis outbreak due to *L. monocytogenes*, which implicated pasteurized, flavored and nonflavored, fluid milk produced by a local dairy as the source of the outbreak (Cumming *et al.* 2008). An outbreak of foodborne listeriosis occurred in France in the year 1992 that involved 279 cases, including 22 abortions and 63 deaths. The major vehicle was pork tongue (Jacquet *et al.* 1995). Another outbreak was reported in France involving 39 cases in 1993 due to consumption of rillettes, and yet another in 1995 (20 cases) where raw milk cheese was the source of infection (Goulet *et al.* 1995, 1998). Winter *et al.* (2009) reported listeriosis outbreak in patients admitted in one of the hospitals in Germany due to consumption of contaminated ready-to-eat scalded sausages. During 2000-2001, three listeriosis outbreaks causing febrile gastroenteritis were observed in Italy sourced to corned beef and ham, delicatessen meat and cheese as epidemiologically implicated food (Sim *et al.* 2002, Frye *et al.* 2002, Carrique-Mass *et al.* 2003). Some other outbreaks are listed in Table 10.4.

Disease Manifestations in Animals

There are four common manifestations of listeriosis in animals: CNS infection in ruminants causing menigoencephalitis in adults and meningitis in the young; septicaemia in monogastrics and neonatal ruminants involving liver and other organs; abortion and perinatal deaths in all species; and mastitis in ruminants. Affected animals usually show only one of these clinical pictures. The route of infection for encephalitis appears to be due to small wounds in the buccal mucosa, while ingestion and inhalation of *Listeria* leads to septicaemia and abortions (Stone 1997).

Encephalitis is most prevalent in the seasons when animals are confined and silage feeding is the greatest. Wilesmith and Gitter (1986) reported 2.5 per cent mean attack rate for *Listeria* encephalitis in adult sheep. The disease occurs in sheep older than 6 weeks, but may be more prevalent in older animals. The clinical signs of meningoencephalitis in adult ruminants may begin with signs of depression and confusion (Low and Renton 1985). Animal holds its head to one side with drooped ears and sometimes protrusion of the tongue and salivation. Unilateral facial hyperplasia and facial paralysis are usually present, characterized by a drooping ear, dilated nostril and lowered eyelid (ptosis) on the same side. Fever (40-42°C) is

Table 10.4: Some outbreaks of human listeriosis

Region	Year/Period	Major Features	Reference
Austria	1986	Foodborne listeriosis outbreak associated with consumption of raw milk/vegetables, 28 persons affected, 5 deaths, *L. monocytogenes* serovar 1/2a isolated	Allenberger and Guggenbichler (1989)
Denmark	1989-1990	Foodborne listeriosis outbreak associated with consumption of blue-mold cheese/hard cheese, 26 persons affected, 6 died, *L. monocytogenes* serovar 4b isolated	Jensen *et al.* (1994)
France	1992	Listeriosis outbreak, a meat product factory incriminated as the source, 279 cases, 63 people died, 20 women aborted	Dorozynski (2000)
Italy	1993	Outbreak of gastroenteritis after a private supper, 39 persons affected, 78 per cent showed symptoms of gastroenteritis, 4 hospitalized with acute febrile gastroenteritis, blood cultures positive for *L. monocytogenes* in 2 cases	Salamina *et al.* (1996)
Finland	1998-1999	Foodborne listeriosis outbreak associated with consumption of butter, 25 persons affected, 6 deaths, *L. monocytogenes* serovar 3a isolated	Lyytikainen *et al.* (2000)
Ontario, Canada	2000	A small outbreak of listeriosis involving two previously healthy adults potentially linked to the consumption of imitation crab meat	Farber *et al.* (2000)
Japan	2001	Outbreak of foodborne listeriosis due to cheese in 86 persons, 38 showed symptoms of gastroenteritis or common cold	Makino *et al.* (2005)
Switzerland	2005	Listeriosis associated with tomme cheese 10 persons affected, 8 were older immunocompromised patients who became ill with bacteraemia (three deaths), two were pregnant women who had septic abortion, all cases due to serotype 1/2a	Bille *et al.* (2006)
USA	2006	A nationwide outbreak of listeriosis linked to contaminated frankfurters and deli meats made at a single facility, over 100 reported cases and 14 fatalities	Mead *et al.* (2006)
Denmark	2009	Outbreak of listeriosis caused by infected beef meat from a meals-on-wheels delivery, out of 8 patients 2 died	Smith *et al.* (2011)
Norway	2010	Nosocomial outbreak of *L. monocytogenes* infection affecting 17 patients, 3 deaths, source of infection traced to Camembert cheese made from pasteurised milk containing upto 360 million CFU per portion	Johnsen *et al.* (2010)

present during the early stage of disease, however, the body temperature remains normal when the clinical signs are apparent. Disease syndrome may be accompanied by keratitis. In the terminal stages, the animal may fall and be unable to rise. Death is due to respiratory failure. The case fatality rate is very high in sheep which is attributed to the acute nature of disease. In cattle, infections are sporadic, less acute and most survive for 4-14 days. Spontaneous recovery may occur, but permanent CNS injury is frequent in these animals (Radostits *et al.* 2007).

Septicaemic or visceral listeriosis is commonly observed among monogastric animals, including pigs, dogs, cats, rabbits and chinchillas, as well as neonatal ruminants. Septicaemia is characterized by an elevated temperature, loss of appetite, and diarrhoea. Wiedmann *et al.* (1994) observed the clinical signs of depression, fever, anorexia, respiratory distress and death in affected sheep and goats. *Listeria* has been associated with generalized septicaemia and focal necrosis of the spleen and liver in various ruminants (Gray and Killinger 1966). Death occurs due to extensive liver damage and focal pneumonia. The principle lesion is focal hepatic necrosis (Pridham *et al.* 1966). Clinical signs such as fever, depression, diarrhoea and respiratory distress were reported in neonatal foal suffering with *L. monocytogenes* septicaemia (Wilkins *et al.* 2000).

Placental tissues in mammals are the site of predilection for *L. monocytogenes*, however, it may vary in different species of animals. Abortions are more common in sheep and goats than in cattle, but in case of pigs this is a rare event (Gray and Killinger 1966). Abortions and stillbirths along with retention of placenta occur in cows in their last trimester of gestation. During this period, clinical illness can be observed with fever upto 40.5°C. In case of sheep and goats, abortions occur after 12th week of pregnancy with retention of placenta and there is bloodstained vaginal discharge for many days. In some cases, maternal metritis is complicated to fatal septicaemia resulting in death of ewes (Radostits *et al.* 2007). The aborted foetus is often autolyzed (Hovingh 2009). The organism can be demonstrated microscopically or by bacteriological culture in the aborted foetus or placenta of the affected sheep (Low and Renton 1985).

Mastitis is a rare manifestation of listeriosis. It affects only a single quarter and is unresponsive to antibiotics. Although uncommon, it does occur ranging in severity from subclinical to severe suppurative infection. Faecal matter and the farm environment may act as source of the organism and udder infection (Girdhar and Garg 2002). There is a high somatic cell count in milk from the affected quarter, but the milk appears normal (Fedio *et al.* 1990). *L. monocytogenes* can be found within neutrophils in milk from mastitic cows (Doyle and Schoeni 1987).

Disease Manifestations in Humans

Human clinical listeriosis occurs most commonly in pregnant women, often resulting in abortion, and in immunodeficient adults causing septicaemia and meningitis. In addition, outbreaks of febrile gastroenteritis are also common. Unlike invasive listeriosis, such outbreaks have frequently involved non-pregnant adults without obvious underlying immunosuppression, and the implicated food showed high level of *Listeria* organisms (Kathariou 2002).

In pregnant women, infection with L. *monocytogenes* causes foetal loss, stillbirth, premature delivery or neonatal infection. Listeriosis occurs most frequently during the third trimester of pregnancy (Bortolussi 1990). Symptoms in affected women begin with mild illness with fever lasting upto 6 days, headache, myalgia and gastrointestinal symptoms consistent with bacteremia (Schuchat *et al.* 1991). Maternal bacteremia results into infection of the foetus through transplacental transmission. Ascending spread from vaginal colonization may also cause infection of foetus. Intrauterine infection may result in preterm labor, amnionitis, spontaneous abortion, stillbirth, or early-onset neonatal infection. Neonatal infection caused by L. *monocytogenes* is a serious and often fatal disease. It is divided into two clinical forms: early onset and late onset. Infants infected *in utero* show early-onset neonatal listeriosis with symptoms of illness at birth or shortly thereafter usually within the first week of life. Between 45 and 70 per cent of neonatal listeriosis is of early-onset type characterized by sepsis (McLauchlin 1967). Signs and symptoms include respiratory distress, fever, and neurologic abnormalities. L. *monocytogenes* may be isolated from blood, CSF, oropharyngeal secretions, placenta, amniotic fluid, urine, or external sites including conjuctiva, ear, nose, or throat (Mylonakis *et al.* 2002). The late-onset type neonatal listeriosis may occur from one to several weeks after birth usually in the form of listerial meningitis. Infants are usually born healthy and full term to mothers who have had uncomplicated pregnancies. The route of infection in late-onset neonatal disease is not well understood, however, it is believed that acquisition of infection during passage through the birth canal or contamination from food is very likely. Further, nosocomial transmission has been reported in newborn nurseries (Nelson *et al.* 1985, Simmons *et al.* 1986).

Apart from pregnant women and newborn babies, *Listeria* is also responsible for meningitis in non-pregnant persons. The disease appears in the form of sepsis, meningitis and meningoencephalitis. Symptoms include fever, malaise, ataxia, seizures and altered mental status (Mylonakis *et al.* 1998). Focal infections are rare and usually result from seeding during a preceding bacteraemic phase. Endocarditis from L. *monocytogenes* occurs primarily in patients with an underlying cardiac lesion, including prosthetic or porcine valves (Bassan 1986, Makaryus *et al.* 2004). *Listeria* has also been incriminated as causative agent of endophthalmitis (Ballen *et al.* 1979), septic arthritis (Chougle and Narayanaswamy 2004), osteomyelitis (Houang *et al.* 1976), peritonitis (Nguyen and Yu 1994) and cutaneous infection (Owen *et al.* 1960).

L. *monocytogenes* is established as a cause of acute, self limiting, febrile gastroenteritis in healthy persons. The incubation period from the time of food ingestion to the onset of symptoms is usually 24 hours or less and ranges between 6 hours to 10 days. The symptoms most frequently reported are fever, diarrhoea, arthromyalgia and headache. Diarrhoea is usually nonbloody and watery (Aureli *et al.* 2000). The prevalence of asymptomatic stool carriage of L. *monocytogenes* in healthy adults is around 1 to 5 per cent (Hof 2001). A large stool survey conducted in Germany found that the recovery of L. *monocytogenes* from the faecal specimens of persons with diarrhoea (0.6 per cent) was similar to that from healthy food handlers (0.8 per cent) (Muller 1990). The carrier rate increases amongst population in close contact with the source of contamination. In Denmark, faecal specimens from 26 per cent of 34 household contacts of persons with listeriosis yielded *L. monocytogenes* (Bojsen-Moller 1972).

Diagnostic Procedures

Culture Method

L. monocytogenes is a nonfastidious organism that can be subcultured on most common bacteriological media such as tryptose agar, nutrient agar and blood agar. However, isolation of the organism from food, environmental samples and animal specimens may require the use of selective agents and enrichment procedures that restrict the growth of other contaminating microorganisms but allow multiplication of *L. monocytogenes* to levels which can be detected by plating on culture media. Cold enrichment and selective enrichment are two common methods of sample enrichment. The enriched samples are plated onto appropriate selective culture media and the colonies appearing on the media are subjected to identification.

Collection of Samples

In the case of animal listeriosis, the samples are chosen according to the clinical presentation of the disease. In septicaemic form of the disease, material from lesions in the liver, kidneys and/or spleen may be collected, while in the cases of encephalitic listeriosis, spinal fluid, pons and medulla are appropriate for examination. Placenta (cotyledons), foetal abomasal contents and/or uterine discharges may be collected from the cases of abortion. Refrigeration temperatures (4°C) must be used for handling, storing and shipping the specimens.

Human listeriosis can be diagnosed by detecting bacteria in blood and CSF. Isolation of *Listeria* from placenta and amniotic fluid in association with clinical symptoms may suggest perinatal listeriosis. Culture of *L. monocytogenes* from stool is not helpful in diagnosis, since gastrointestinal carriage of the organism may occur without clinical disease (Schuchat *et al.* 1991). Food samples are necessary if foodborne outbreak of listeriosis is under investigation.

Cold Enrichment

Cold enrichment procedure is suitable for isolation of *L. monocytogenes* from highly contaminated samples. The method is based on the ability of the organism to multiply at refrigeration temperatures in a broth medium, where other contaminating bacteria would not normally multiply. However, this procedure generally requires very long incubation times.

Selective Enrichment

Selective enrichment procedure is based on selective inhibition of other microflora through addition of certain inhibitory agents that allow the growth of *Listeria*. The selective agents include lithium chloride, nalidixic acid, acriflavin, polymyxin B, moxalactam, ceftazidime, etc.

Plating on Selective Culture Media

After enrichment of the sample in a broth medium, *Listeria* can be isolated easily on various selective/differential isolation agar media, such as McBride Listeria Agar, LPM agar, PALCAM agar, Oxford agar, modified Oxford agar, etc.

Identification of *Listeria*

Typical *Listeria* colonies appearing on the agar medium plates are isolated for identification using a battery of tests. The primary tests include Gram staining reaction, catalase, motility, haemolysis and carbohydrate use. The Christie-Atkins-Munch-Peterson (CAMP) test is a very useful tool to identify the species of *Listeria* isolate.

Rapid Methods

Rapid methods of *Listeria* detection in foods are quite important due to continuing pressure of stringent regulatory requirement and international trade. An ideal method should be specific for the target analyte (*Listeria* or *L. monocytogenes*) and sensitive enough to detect 1 CFU in a 25 g sample. The test should deliver results more rapidly than the conventional cultural methods and it should be easy to use and reproduce the results (Brehm-Stecher and Johnson 2007).

Rapid methods of analysis include colorimetric DNA probe, latex bead-based lateral flow immunoassay, enzyme-linked immunosorbent assay, enzyme-linked immunofluorescence assay, immunomagnetic separation, fluorescence *in situ* hybridization and polymerase chain reaction. Some other recent technologies include flow cytometry, biosensors, chip-based microanalytical systems and spectroscopic methods.

Prevention and Control Strategies in Animals

Strict quarantine practices should be followed to prevent entry of the pathogen in a *Listeria*-free farm whenever new animals are purchased. Any animal showing signs of listeriosis should be immediately isolated and treated. The placenta and foetus from aborted animals should be safely disposed. If female animals have listeriosis, their offspring should be fed on pasteurized colostrum, milk or a milk substitute.

Risk of transmission of *Listeria* through silage should be reduced by ensuring development of low pH in silage and absence of spoilage and mould development. It is better to discard few inches of silage that is exposed to air. Any leftover silage should be removed after feeding.

Rodents are responsible for spreading the organism, so rodent control is necessary. Similarly, wild birds may also serve as vectors; hence their entry in the farm should be checked.

In case of *Listeria* affected animals, prompt diagnosis followed by correct and early treatment is important. Intravenous injection of chlortetracycline is useful for treatment of encephalitis in cattle. Oral administration of penicillin and tetracycline and intramuscular injection of penicillin can also be used (Radostits *et al.* 2007). Supportive therapy may also be indicated and could include bicarbonate and potassium to replace losses due to excessive salivation when there is poor cranial nerve function (Schweizer *et al.* 2006).

Prevention and Control Strategies in Humans

People, particularly the high risk categories, such as pregnant women and immunosuppressed adults, should avoid consuming unpasteurized milk products, raw or partially cooked meats and raw vegetables etc. Cross-contamination between raw and cooked foods should be avoided in the kitchen and vegetables should be washed carefully. Even distribution of heat and adequate internal temperature throughout the food should be ensured while cooking, particularly in microwave oven in which heat may be unevenly distributed.

Veterinarians, animal handlers and farm workers should wear personal protective clothing and should use face protection for eyes and mouth, particularly when dealing with calving/lambing, etc. After use, all protective clothing should be washed thoroughly and disinfected.

Prompt diagnosis and treatment is important due to serious nature of the disease. In the listeriosis endemic regions, human neonatal listeriosis can be diagnosed by examining a smear of the meconium of a newborn for *Listeria* organisms.

The disease may require treatment for 2 weeks in the immunocompetent patients, but longer courses may be required in the immunocompromised persons. The course of treatment for meningitis is usually 3 weeks and for endocarditis it is 4-6 weeks. Ampicillin is the antibiotic of choice which can be used with synergistic action of gentamicin (Lorber 1997, Shakespeare 2009). Other antibiotics such as penicillin G, erythromycin, trimethoprim sulfamethoxazole, chloramphenicol, rifampin, tetracyclines and aminoglycosides can also be used.

Future Directions

There have been major global advances in the understanding of the epidemiology, taxonomy, detection methods of listeria organisms in foods and animals, and also of the molecular determinants of virulence of *Listeria*. However, diagnostic procedures of *L. monocytogenes* infections require strengthening, particularly in the developing countries, owing to the complex immunopathological aspects and varied clinical forms of the disease. It is desirable to evolve simple tests based on immune diagnostics and rapid cultural techniques.

The role of *L. monocytogenes* and *L. ivanovii* in abortion and other reproductive disorders needs to be elucidated. The status of listeriosis in organized farms, where intensive husbandry practices are in operation, needs evaluation and setting up a monitoring system to quickly record the cases of listeriosis. The carrier state of *L. monocytogenes* in healthy and diarrhoeic humans and animals should be explored in an elaborate way. The resistance of *Listeria* to antimicrobials and detergents also requires attention and continuous monitoring.

References

1. Abadias M, Usall J, Anguera M, Solsona C and Viñas I. 2008. Microbiological quality of fresh, minimally-processed fruit and vegetables and sprouts from retail establishments. *Int J Food Microbiol* 123: 121-129.

2. Abelson P, Forbes MP and Hall G. 2006. The annual cost of foodborne illness in Australia. Australian Government Department of Health and Ageing. http://www.ozfoodnet.gov.au. Accessed 27 September, 2012.

3. Al-Dughaym AM, Elmula AF, Mohamed GE, Hegazy AA, Radwan YA, Housawi FM and Gameel AA. 2001. First report of an outbreak of ovine septicaemic listeriosis in Saudi Arabia. *Rev Sci Tech* 20: 777-783.

4. Allerberger F and Guggenbichler JP. 1989. Listeriosis in Austria - Report of an outbreak in 1986. *Acta Microbiol Hung* 36: 149-152.

5. Anonymous. 1994. *Veterinary investigation diagnosis and analysis-III*. Ministry of Agriculture, Fisheries and Food, UK.

6. Anonymous. 2001. Outbreak of listeriosis associated with homemade Mexican style cheese-North Carolina-October 2000. *Morb Mort Wkly Rep* 50: 560-562.

7. Anonymous. 2012. Listeriosis Blue Book, Department of Health, Australia. http://ideas.health.vic.gov.au/bluebook/listeriosis.asp. Accessed 20 July 2012.

8. Antoniollo PC, Bandeira Fda S, Jantzen MM, Duval EH and da Silva WP. 2003. Prevalence of *Listeria* spp in feces and carcasses at a lamb packing plant in Brazil. *J Food Prot* 66: 328-330.

9. Arumugaswamy RK, Rakamat Ali GR and Hamid SNB. 1994. Prevalence of *Listeria monocytogenes* in foods in Malaysia. *Int J Food Microbiol* 23: 117-121.

10. Atanassova V, Reich F and Klein G. 2008. Microbiological quality of sushi from sushi bars and retailers. *J Food Prot* 71: 860-864.

11. Aureli P, Fiorucci GC, Caroli D, Marchiaro G, Novara O, Leone L and Salmaso S. 2000. An outbreak of febrile gastroenteritis associated with corn contaminated by *Listeria monocytogenes*. *N Engl J Med* 342: 1236-1241.

12. Awaisheh SS. 2010. Incidence and contamination level of *Listeria monocytogenes* and other *Listeria* spp in ready-to-eat meat products in Jordan. *J Food Prot* 73: 535-540.

13. Baek SY, Lim SY, Lee DH, Min KH and Kim CM. 2000. Incidence and characterization of *Listeria monocytogenes* from domestic and imported foods in Korea. *J Food Prot* 63: 186-189.

14. Ballen PH, Loffredo FR and Painter B. 1979. *Listeria* endophthalmitis. *Arch Ophthalmol* 97: 101-102.

15. Barbuddhe SB. 1996. Studies on the kinetics and efficacy of listeriolysin-O for the sero diagnosis *Listeria monocytogenes* infection. *Ph.D. thesis*, Indian Veterinary Research Institute, Izatnagar, India.

16. Barbuddhe SB, Malik SVS and Kumar P. 1999. High seropositivity against antilisteriolysin O in humans. *Ann Trop Med Parasitol* 93: 537-539.

17. Barbuddhe SB, Malik SVS, Bhilegaonkar KN, Prahlad K and Gupta LK. 2000a. Isolation of *Listeria monocytogenes* and anti-listeriolysin O detection in sheep and goats. *Small Rumin Res* 38: 151-155.

18. Barbuddhe SB, Prahlad Kumar, Malik SVS, Singh DK and Gupta LK. 2000b. Seropositivity for intracellular bacterial infections among abattoir associated personnels. *J Commun Dis* 32: 295-299.

19. Barbuddhe SB, Chaudhari SP and Malik SVS. 2002. The occurrence of pathogenic *Listeria monocytogenes* and antibodies against listeriolysin O in buffaloes. *J Vet Med B* 49: 181-184.

20. Bassan R. 1986. Bacterial endocarditis produced by *Listeria monocytogenes*: Case presentation and review of the literature. *Am J Clin Pathol* 63: 522-527.

21. Beckers HJ, Soentoro PSS and Delfgou-van-Asch EHM. 1987. The occurrence of *Listeria monocytogenes* in soft cheeses and raw milk and its resistance to heat. *Int J Food Microbiol* 4: 249-256.

22. Berzins A, Terentjeva M and Korkeala H. 2009. Prevalence and genetic diversity of *Listeria monocytogenes* in vacuum-packaged ready-to-eat meat products at retail markets in Latvia. *J Food Prot* 72: 1283-1287.

23. Berzins A, Hellström S, Siliòð I and Korkeala H. 2010. Contamination patterns of *Listeria monocytogenes* in cold-smoked pork processing. *J Food Prot* 73: 2103-2109.

24. Bhanu Rekha V, Malik SVS, Chaudhari SP and Barbuddhe SB. 2006. Listeriolysin O-based diagnosis of *Listeria monocytogenes* infection in experimentally and naturally infected goats. *Small Rumin Res* 66: 70-75.

25. Bhat SA and Willayat MM. 2011. Pathogenicity of *Listeria monocytogenes* isolated from human clinical cases and foods of animal origin. *J Pure Appl Microbiol* 5: 895-897.

26. Bhilegaonker KN, Kulshrestha SB, Kapoor KN, Ashok Kumar, Agarwal RK and Singh BR. 1997. Isolation of *Listeria monocytogenes* form milk. *J Food Sci Technol* 34: 248-250.

27. Bhilegaonker KN, Bachhil VN, Kumar A, Kapoor KN and Agarwal RK. 1999. Prevalence of *Listeria monocytogenes* in foods of animal origin. *XIX Annual conference of Indian Association of Veterinary Microbiologists, Immunologists and Specialists in Infectious Diseases*, October 22-24, 1999, Abstract. p. 16.

28. Bhujwala RA and Hingorani V. 1975. Perinatal listeriosis: a bacteriological and serological study. *Indian J Med Res* 63: 1503-1508.

29. Bhujwala RA, Hingorani V and Chandra RK. 1973. Genital listeriosis in Delhi: a pilot study. *Indian J Med Res* 61: 1284-1287.

30. Bille J, Blanc DS, Schmid H, Boubaker K, Baumgartner A, Siegrist HH, Tritten ML, Lienhard R, Berner D, Anderau R, Treboux M, Ducommun JM, Malinverni R, Genné D, Erard Ph and Waespi U. 2006. Outbreak of human listeriosis associated with tomme cheese in northwest Switzerland, 2005. *Euro Surveill* 11: 633.

31. Bojsen-Moller J. 1972. Human listeriosis: diagnostic, epidemiological, and clinical studies. *Acta Pathol Microbiol Scand* 229 (Sect B. suppl): 72-92.

32. Bortolussi R. 1990. Neonatal listeriosis. *Semin Perinatol* 14(Suppl): 44-48.

33. Breer C and Schopfer K. 1989. Listerien in Nahrungsmitteln. *Schweiz Med Wochenschr* 119: 306-311.

34. Brehm-Stecher BF and Johnson EA. 2007. Rapid methods for detection of *Listeria*. In: Ryser ET and Marth EH (eds) *Listeria, listeriosis, and food safety*, 3rd edn. CRC Press, Boca Raton, USA.

35. Bundrant BN, Hutchins T, den Bakker HC, Fortes E and Wiedmann M. 2011. Listeriosis outbreak in dairy cattle caused by an unusual *Listeria monocytogenes* serotype 4b strain. *J Vet Diagn Investig* 23: 155-158.

36. Carrique-Mas JJ, Hökeberg I, Andersson Y, Arneborn M, Tham W, Danielsson-Tham ML, Osterman B, Leffler M, Steen M, Eriksson E, Hedin G and Giesecke J. 2003. Febrile gastroenteritis after eating on- farm manufactured fresh cheese- an outbreak of listeriosis? *Epidemiol Infect* 130: 79-86.

37. CDC. 2011. National (CDC) *Listeria* Surveillance Annual Summary, 2009. Centers for Disease Control and Prevention, Atlanta, Georgia, US Department of Health and Human Services, USA.

38. Chand P and Sadana JR. 1998. Outbreak of *Listeria ivanovii* abortion in sheep in India. *Vet Rec* 145: 83-84.

39. Chaudhari SP, Malik SVS, Chatlod LR and Barbuddhe SB. 2004. Isolation of pathogenic *Listeria monocytogenes* and detection of antibodies against phosphatidylinositol-specific phospholipase C in buffaloes. *Comp Immunol Microbiol Infect Dis* 27: 141-148.

40. Chougle A and Narayanaswamy V. 2004. Delayed presentation of prosthetic joint infection due to *Listeria monocytogenes*. *Int J Clin Pract* 58: 420-421.

41. Cooper GL. 1989. A encephalitic form of listeriosis in broiler chickens. *Avian Dis* 33: 182-185.

42. Cossart P, Vicente MF, Mengaud J, Baquero F, Perez-Diaz J C and Berche P. 1989. Listeriolysin O is essential for virulence of *Listeria monocytogenes*: direct evidence obtained by gene complementation. *Infect Immun* 57: 3629-3636.

43. Cruz CD and Fletcher GC. 2011. Prevalence and biofilm-forming ability of *Listeria monocytogenes* in New Zealand mussel (*Perna canaliculus*) processing plants. *Food Microbiol* 28: 1387-1393.

44. Cumming M, Kludt P, Matyas B, DeMaria A, Stiles T, Han L, Gilchrist M, Neves P, Fitzgibbons E and Condon S. 2008. Outbreak of *Listeria monocytogenes* infections associated with pasteurized milk from a local dairy- Massachusetts, 2007. *Morb Mort Wkly Rep* 57: 1097-1100.

45. Dash PK, Malik SVS, Sharma AK and Paul S. 1998. Management of circling syndrome in pigs. *Indian J Comp Microbiol Immunol Infect Dis* 19: 102-103.

46. Davidson RJ, Sprung DW, Park CE and Rayman MK. 1989. Occurrence of *Listeria monocytogenes*, *Campylobacter* spp, and *Yersinia enterocolitica* in Manitoba raw milk. *Can Inst Food Sci Technol J* 22: 70-74.

47. Dhanda MR, Lall JM, Seth RN and Chandrasekariah P. 1959. A case of listeric abortion in an ewe with a small scale survey of the incidence of agglutinins to *Listeria* in the sera of sheep. *Indian Vet J* 36: 113-124.

48. Di Pinto A, Novello L, Montemurro F, Bonerba E and Tantillo G. 2010. Occurrence of *Listeria monocytogenes* in ready-to-eat foods from supermarkets in Southern Italy. *New Microbiol* 33: 249-252.

49. Díaz-López A, Cantú-Ramírez RC, Garza-González E, Ruiz-Tolentino L, Tellez-Luis SJ, Rivera G and Bocanegra-García V. 2011. Prevalence of foodborne pathogens in grilled chicken from street vendors and retail outlets in Reynosa, Tamaulipas, Mexico. *J Food Prot* 74: 1320-1323.

50. Dominguez RL, Garayzabel JF, Vazquez JA, Ferri RE and Fernandez SG. 1985. Isolation of de micro-oganims du genre *Listeria* a' parfir de lait cru destine a' la consommation humaine. *Can J Microbiol* 31: 939-941.

51. Donnelly CW and Baigent GJ. 1986. Method for flow cytometric, detection of *Listeria monocytogenes* in milk. *Appl Environ Microbiol* 52: 689-695.

52. Dorozynski A. 2000. Seven die in French *Listeria* outbreak. *Br Med J* 320: 601.

53. Doyle ME. 2001. Virulence characteristics of *Listeria monocytogenes*. FRI Briefings. Food Research Institute, University of Wisconsin-Medison. http://fri.wisc.edu/docs/pdf/virulencelmono.pdf. Accessed 27 July 2012

54. Doyle MP and Schoeni JL. 1987. Comparison of procedures for isolating *Listeria monocytogenes* in soft surface ripened cheese. *J Food Prot* 50: 4-6.

55. Dutta PK and Malik BS. 1978. Some epidemiological studies on listeriosis in man and animals in India. *Indian J Public Health* 22: 321-322.

56. Dutta PK and Malik BS. 1981. Isolation and characterization of *Listeria monocytogenes* from animals and human beings. *Indian J Anim Sci* 51: 1045-1052.

57. Elezebeth G, Malik SVS, Chaudhari SP and Barbuddhe SB. 2007. The occurrence of *Listeria* species and antibodies against listeriolysin-O in naturally infected goats. *Small Rumin Res* 67: 173-178.

58. El-Shenawy M, El-Shenawy M, Mañes J and Soriano JM. 2011. *Listeria* spp. in street-vended ready-to-eat foods. *Interdiscip Perspect Infect Dis* doi:10.1155/2011/968031.

59. Embarek, PKB 1994 Presence, detection and growth of *Listeria monocytogenes* in sea foods: a review *Int J Food Microbiol* 23: 17-34.

60. Esteban JI, Oporto B, Aduriz G, Juste RA and Hurtado A. 2009. Faecal shedding and strain diversity of *Listeria monocytogenes* in healthy ruminants and swine in Northern Spain. *BMC Vet Res* 5: 2.

61. Farber JM, Sanders GW and Spiers JI. 1990. Growth of *Listeria monocytogenes* in naturally contaminanted raw milk. *Lebensm Wiss Technol* 23: 252-254.

62. Farber JM, Daley EM, Mackie MT and Limerick B. 2000. A small outbreak of listeriosis potentially linked to the consumption of imitation crab meat. *Lett Appl Microbiol* 31: 100-104.

63. Fedio WM, Schoonderwoerd M, Shute RH and Jackson H. 1990. A case of bovine mastitis caused by *Listeria monocytogenes*. *Can Vet J* 31: 773-775.

64. Fenlon DR and Wilson J. 1989. The incidence of *Listeria monocytogenes* in raw milk from farm bulk tanks in North-East Scotland. *J Appl Bacteriol* 81: 641-650.

65. Flemming DW, Cochi SL, Mackonald KL, Brondum J, Hayes PS, Plikaytis BD, Holmes MB, Audurier A, Broome CV and Reingold AL. 1985. Pasteurised milk as a vehicle of infection in an outbreak of listeriosis. *N Engl J Med* 312: 404-407.

66. Francis GA and O'Beirne D. 2006. Isolation and pulsed-field gel electrophoresis typing of *Listeria monocytogenes* from modified atmosphere packaged fresh-cut vegetables collected in Ireland. *J Food Prot* 69: 2524-2528.

67. Frye DM, Zweig R, Sturgeon J, Tormey M, LeCavalier M, Lee I, Lawani L and Mascola L. 2002. An outbreak of febrile gastroenteritis associated with delicatessen meat contaminated with *Listeria monocytogenes*. *Clin Infect Dis* 35: 943-949.

68. Gebretsadik S, Kassa T, Alemayehu H, Huruy K and Kebede N. 2011. Isolation and characterization of *Listeria monocytogenes* and other *Listeria* species in foods of animal origin in Addis Ababa, Ethiopia. *J Infect Public Health* 4: 22-29.

69. Genigeorgis CA, Dutulescu D and Garayzabel JF. 1989. Prevalence of *Listeria* species in poultry meat at the supermarket and slaughter house level. *J Food Prot* 52: 618-624.

70. Girdhar OP and Garg SR. 2002. Prevalence of *Listeria* in animal farm. *Indian J Anim Sci* 72: 847-849.

71. Gitter M. 1976. *Listeria monocytogenes* in oven ready poultry. *Vet Rec* 99: 336.

72. Gitter M, Bradley R and Blampied PH. 1980. *Listeria monocytogenes* infection in bovine mastitis. *Vet Rec* 107: 390-393.

73. Gogate AA and Deodhar LP. 1981. Meningitis due to *Listeria monocytogenes*: (a case report). *J Postgrad Med* 27: 240-242.

74. Gouin E, Mengaud J and Cossart P. 1994. The Virulence Gene Cluster of *Listeria monocytogenes* is also present in *Listeria ivanovii*, an animal pathogen, and *Listeria seeligeri*, a nonpathogenic species. *Infect Immun* 62: 3550-3553.

75. Goulet V, Rocourt J, Rebiere I, Jacquet C, Moyse C, Dehaumont P, Salvat G and Veit P. 1998. Listeroisis outbreak associated with the consumption of rillettes in France in 1993. *J Infect Dis* 177: 155-160.

76. Goulet V, Jacquet C, Vaillant V, Rebière I, Mouret E, Lorente C, Maillot E, Staïner F and Rocourt J. 1995. Listeriosis from consumption of raw-milk cheese. *Lancet* 345: 1581-1582.

77. Graves LM, Helsel LO, Steigerwalt AG, Morey RE, Daneshvar MI, Roof SE, Orsi RH, Fortes ED, Milillo SR, Den Bakker HC, Wiedmann M, Swaminathan B and Sauders BD. 2010. *Listeria marthii* sp. nov, isolated from the natural environment, Finger Lakes National Forest. *Int J Syst Evol Microbiol* 60: 1280-1288.

78. Gray ML and Killinger AH. 1966. *Listeria monocytogenes* and listeric infections. *Bacteriol Rev* 30: 309-382.

79. Greenwood MH, Roberts D and Burden P. 1991. The occurrence of *Listeria* species in milk and dairy products: a national survey in England and Wales. *Int J Food Microbiol* 12: 197-206.

80. Griffith M and Deibel KE. 1990. Survival of *Listeria monocytogenes* in yogurt with varying levels of fat and solids. *J Food Saf* 10: 219-230.

81. Guerini MN, Brichta-Harhay DM, Shackelford TS, Arthur TM, Bosilevac JM, Kalchayanand N, Wheeler TL and Koohmaraie M. 2007. *Listeria* prevalence and *Listeria monocytogenes* serovar diversity at cull cow and bull processing plants in the United States. *J Food Prot* 70: 2578-2582.

82. Gunasinghe CPGL, Henderson C and Rutter MA. 1994. Comparative study of two plating media (PALCAM and Oxford) for detection of *Listeria* species in a range of meat products following a variety of enrichment procedures. *Lett Appl Microbiol* 18: 156-158.

83. Heinitz ML and Johnson JM. 1998. The incidence of *Listeria* species, *Salmonella* species and *Clostridium botulinum* in smoked fish and shellfish. *J Food Prot* 61: 318-323.

84. Ho JL, Shands KN, Friedland G, Eckind P, Fraser DW. 1986. An outbreak of type 4b *Listeria monocytogenes* infection involving patients from eight Boston hospitals. *Arch Intern Med* 146: 520-524.

85. Hof H. 2001. *Listeria monocytogenes*: a causative agent of gastroenteritis? Eur J Clin Microbiol *Infect Dis* 20: 369-373.

86. Houang ET, Williams CJ and Wrigley PFM. 1976. Acute *Listeria monocytogenes* osteomyelitis. *Infect* 4: 113-114.

87. Hovingh E. 2009. Abortions in dairy cattle: I. Common causes of abortions. Virginia Cooperative Extension publication 404-288, College of Agriculture and Life Sciences, Virginia Polytechnic Institute and State University, USA.

88. Hyslop NSG. 1975. Epidemiologic and immunologic factors in listeriosis. In: Woodbine M (ed) *Problems of listeriosis*. Leicester University Press, Leicester, UK.

89. Jacquet CB, Brosch CR, Buchrieser C, Dehaumont P, Goulet V, Lepoutre A, Veit P and Rocourt J. 1995. Investigations related to the epidemic strain involved in the French listeriosis outbreak in 1992. *Appl Environ Microbiol* 61: 2242-2246.

90. Jensen A, Frederiksen W and Gerner-Smidt P. 1994. Risk factors for listeriosis in Denmark, 1989-1990. *Scand J Infect Dis* 26: 171-178.

91. Jeyasekaran G, Karunasagar I and Karunasagar I. 1996. Incidence of *Listeria* species in tropical fish. *Int J Food Microbiol* 31: 333-340.

92. Jia J, Bi ZW, Chen YZ, Hou PB, Zhang M, Shao K and Bi ZQ. 2011. Antibiotic resistance and molecular typing of *Listeria monocytogenes* from foods in Shandong province from 2009 to 2010. *Zhonghua Yu Fang Yi Xue Za Zhi* 45: 1065-1067.

93. Johnson JL, Doyle MP and Cassens RG. 1990. Incidence of *Listeria* species in retail meat roasts. *J Food Sci* 55: 572-574.

94. Johnsen BO, Lingaasb E, Torfossc D, Strømd EH and Nordøya I. 2010. A large outbreak of *Listeria monocytogenes* infection with short incubation period in a tertiary care hospital. *J Infect* 61: 465-470.

95. Kalekar S, Rodrigues J, D'Costa D, Doijad S, Ashok Kumar J, Malik SV, Kalorey DR, Rawool DB, Hain T, Chakraborty T and Barbuddhe SB. 2011. Genotypic characterization of *Listeria monocytogenes* isolated from humans in India. *Ann Trop Med Parasitol* 105: 351-358.

96. Kalorey DR, Kurkure NV, Warke SR, Rawool DB, Malik SV and Barbuddhe SB. 2006. Isolation of pathogenic *Listeria monocytogenes* in faeces of wild animals in captivity. *Comp Immunol Microbiol Infect Dis* 29: 295-300.

97. Kalorey DR, Warke SR, Kurkure NV, Rawool DB and Barbuddhe SB. 2008. Listeria species in bovine raw milk: A large survey of Central India. *Food Control* 19: 109-112.

98. Kalyani M, Rajesh PK, Srikanth P and Mallika M. 2006. *Listeria monocytogenes* - a case report. *Sri Ramchandra J Med* 1: 45-46.

99. Kampelmacher EH. 1962. Animal products as a source of *Listeria* infection in man. In: ML Gray (ed) *Second symposium on Listeric infection*, Montana State College, Bozeman, USA.

100. Kathariou S. 2002. *Listeria monocytogenes* virulence and pathogenicity, a food safety perspective. *J Food Prot* 65: 1811-1829.

101. Katic V and Stojanovic L. 1992. The occurrence of *Listeria monocytogenes* in goat and sheep milk. *Acta Veterinaria* (Beogard) 42: 215-220.

102. Katre DD, Zade NN, Chaudhari SP, Jaulkar AD and Shinde SV. 2009. Prevalence of listeria species in foods of animal origin. *R Vet J India* 5: 5-60.

103. Kaur S, Malik SVS, Vaidya VM and Barbuddhe SB. 2007. *Listeria monocytogenes* in spontaneous abortions in humans and its detection by multiplex PCR. *J Appl Microbiol* 103: 1889-1896.

104. Kaur S, Malik SVS, Bhilegaonkar KN, Vaidya VM and Barbuddhe SB. 2010. Use of a phospholipase-C assay, *in vivo* pathogenicity assays and PCR in assessing the virulence of *Listeria* spp. *Vet J* 184: 366-370.

105. Korde NM. 1992. Studies on seroprevalence of *Listeria monocytogenes* infection in man and animals. *M.V.Sc. Thesis*, Indian Veterinary Research Institute, Izatnagar, India.

106. Krishna U, Desai MW and Daftary VG. 1966. Listeriosis: a clinico-bacteriological study. *J Obstet Gynecol India* 16: 304-310.

107. Kumar H, Singh BB, Bal MS, Kaur K, Singh R, Sidhu PK and Sandhu KS. 2007. Pathological and epidemiological investigations into listerial encephalitis in sheep. *Small Rumin Res* 71: 293-297.

108. Kwantes W and Isaac M. 1971. Listeriosis. *Br Med J* 4: 296-297.

109. Lahuerta A, Westrell T, Takkinen J, Boelaert F, Rizzi V, Helwigh B, Borck B, Korsgaard H, Ammon A and Makela P. 2011. Zoonoses in the European Union: origin, distribution and dynamics - the EFSA-ECDC summary report 2009. *Euro Surveill* 16. http://www.eurosurveillance.org/ViewArticle.aspx?ArticleId= 19832. Accessed 2.8.2011:

110. Larsson S, Cederberg A, Ivarsson S, Svanberg L and Cronberg S. 1978. *Listeria monocytogenes* causing hospital-acquired enterocolitis and meningitis in newborn infants. *Br Med J* 2: 473-474.

111. Leclercq A, Clermont D, Bizet C, Grimont PA, Le Flèche-Matéos A, Roche SM, Buchrieser C, Cadet-Daniel V, Le Monnier A, Lecuit M and Allerberger F. 2010. *Listeria rocourtiae* sp. nov. *Int J Syst Evol Microbiol* 60: 2210-2214.

112. Lee WH and McClain D. 1987. Unpublished observations.

113. Linnan MJ, Nascola L, Lou XD, Goulet V, May S, Salminen C, Hird DW, Yonekura ML, Hayes P and Weaver R. 1988. Epidemic listeriosis associated with Mexican style cheese. *N Engl J Med* 319: 823-824.

114. Little CL and Mitchell RT. 2004. Microbiological quality of pre-cut fruit, sprouted seeds, and unpasteurised fruit and vegetable juices from retail and production premises in the UK and the application of HAACP. *Commun Dis Public Health* 7: 184-190.

115. Lorber B. 1997. Listeriosis. *Clin Infect Dis* 24: 1-11.

116. Lovett J, Francis DW and Hunt JM. 1987. *Listeria monocytogenes* in raw milk: detection, incidence and pathogenicity. *J Food Prot* 50: 188-192.

117. Low JC and Renton CP. 1985. Septicaemia, encephalitis and abortions in a housed flock of sheep caused by *Listeria monocytogenes* type 1/2. *Vet Rec* 116: 147-150.

118. Lyytikäinen O, Autio T, Maijala R, Meittinen M, Hatakka M, Mikkola J, Anttila VJ, Johansson T, Rantala L, Aalto T, Korkeala H and Siitonen A. 2000. An outbreak of *Listeria monocytogenes* serotype 3a infections from butter in Finland. *J Infect Dis* 181: 1838-1841.

119. Mahajan MR. 1936. Some observations on "circling" disease of sheep in Hyderabad state. *Indian J Vet Sci Anim Husb* 5: 350-354.

120. Mahajan AK and Katoch RC. 1997. Aerobic microflora associated with endometritis in sheep and goats. *Indian J Anim Sci* 67: 290-291.

121. Makaryus AN, Yang R, Cohen R, Rosman D, Mangion J and Kort S. 2004. A rare case of *Listeria monocytogenes* presenting as prosthetic valve bacterial endocarditis and aortic root abscess. *Echocardiogr* 21: 423-427.

122. Makino SI, Kawamoto K, Takeshi K, Okada Y, Yamasaki M, Yamamoto S and Igimi S. 2005. An outbreak of food-borne listeriosis due to cheese in Japan, during 2001. *Int J Food Microbiol* 104: 189-196.

123. Malinverni, R, Bille J, Perret C, Regli F, Tanner F and Glauser MP. 1985. Listériose épidémizue-Observation de 25 cas en 15 mois au centre hospitalier universitaire vaudois. *Schweiz Med Wochenschr* 115: 2-10.

124. Maria CDS, Hofer E and Tibana A. 1998. Incidence of *Listeria monocytogenes* in cheese produced in Rio de Janeiro, Brazil. *J Food Prot* 61: 354-356.

125. Martin B, Garriga M and Aymerich T. 2011. Prevalence of *Salmonella* spp and *Listeria monocytogenes* at small-scale spanish factories producing traditional fermented sausages. *J Food Prot* 74: 812-815.

126. McGowan AP, Bowker K, McLauchlin J, Bennett PM and Reeves DS. 1994. The occurrence and seasonal changes in the isolation of *Listeria* spp. in shop bought food stuffs, human faeces, sewage and soil from urban sources. *Int J Food Microbiol* 21: 325-334.

127. McLauchlin JS. 1967. Human listeriosis in Britain. *Epidemiol Infect* 104: 181-189.

128. Mead PS, Dunne EF, Graves L, Wiedmann M, Patrick M, Hunter S, Salehi E, Mostashari F, Craig A, Mshar P, Bannerman T, Sauders BD, Hayes P, Dewitt W, Sparling P, Griffin P, Morse D, Slutsker L and Swaminathan B. 2006. Nationwide outbreak of listeriosis due to contaminated meat. *Epidemiol Infect* 134: 744-751.

129. Meredith CD and Schneider DJ. 1984. An outbreak of ovine listeriosis associated with poor flock management practices. *J S Afr Vet Assoc* 55: 55-56.

130. Miya S, Takahashi H, Ishikawa T, Fujii T and Kimura B. 2010. Risk of *Listeria monocytogenes* contamination of raw ready-to-eat seafood products available at retail outlets in Japan. *Appl Environ Microbiol* 76: 3383-3386.

131. Mokta KK, Kanga AK and Kaushal RK. 2010. Neonatal listeriosis: a case report from sub-Himalayas. *Indian J Med Microbiol* 28: 385-387.

132. Monfort P, Minet J, Rocourt J, Piclet G and Cormier M. 1998. Incidence of *Listeria* species in Briton live shellfish. *Lett Appl Microbiol* 26: 205-208.

133. Morobe IC, Obi CL, Nyila MA, Gashe BA and Matsheka MI. 2009. Prevalence, antimicrobial resistance profiles of *Listeria monocytognes* from various foods in Gaborone, Botswana. *Afr J Biotechnol* 8: 6383-6387.

134. Moura SM, Destro MT and Franco BDGM. 1993. Incidence of *Listeria* species in raw and pasteurized milk produced in São Paulo, Brazil. *Int J Food Microbiol* 19: 229-237.

135. Muller HE. 1990. *Listeria* isolations from feces of patients with diarrhoea and from healthy food handlers. *Infect* 18: 97-100.

136. Mylonakis E, Hohmann EL and Calderwood SB. 1998. Central nervous system infection with *Listeria monocytogenes*. 33 years' experience at a general hospital and review of 776 episodes from the literature. *Medicine (Baltimore)* 77: 313-336.

137. Mylonakis E, Paliou M, Hohmann EL, Calderwood SB and Wing EJ. 2002. Listeriosis during pregnancy: A case series and review of 222 cases. *Medicine (Baltimore)* 81: 260-269.

138. Nash ML, Hungerford LL, Nash TG and Zinn GM. 1995. Epidemiology and economics of clinical listeriosis in a sheep flock. *Prev Vet Med* 24: 147-156.

139. Nelson KE, Warren D, Tomasi AM, Raju TN and Vidyasagar D. 1985. Transmission of neonatal listeriosis in a delivery room. *Am J Dis Child* 139: 903-905.

140. Nguyen MH and VL Yu. 1994. *Listeria monocytogenes* peritonitis in cirrhotic patients. *Dig Dis Sci* 39: 215-218.

141. Nigam P, Katoch RC, Sharma M and Verma S. 1999. Investigations on listeriosis associated with reproductive disorders of domestic animals in Himachal Pradesh. *Indian J Anim Sci* 69: 171-173.

142. Nightingale KK, Schukken YH, Nightingale CR, Fortes ED, Ho AJ, Her Z, Grohn YT, McDonough PL and Wiedmann M. 2004. Ecology and transmission of *Listeria monocytogenes* infecting ruminants and in the farm environment. *Appl Environ Microbiol* 70: 4458-4467.

143. Ojeniyi B, Wegener HC, Jensen NE and Bisgaard M. 1996. *Listeria monocytogenes* in poultry and poultry products: epidemiological investigations in seven Danish abattoirs. *J Appl Bacteriol* 80: 395-401.

144. Owen CR, Meis A, Jackson JW and Stoenner HG. 1960. A case of primary cutaneous listeriosis. *New Engl J Med* 262: 1026-1028.

145. Pagotto F, Corneau N and Farber J. 2006. *Listeria monocytogenes* infections In: Riemann HP and Cliver DO (Eds) *Foodborne infections and intoxications*, 3rd edn, Elsevier, New York

146. Pal M. 2007. *Zoonoses*. Satyam Publishers and Distributors, Jaipur, India.

147. Pednekar DM, Kamat AS and Adhikari HR. 1997. Incidence of *Listeria* species in milk and milk products. *Indian J Dairy Sci* 50: 142-151.

148. Peer MA, Nasir RA, Kakru DK, Fomda BA, Wani MA and Hakeem QN. 2010. *Listeria monocytogenes* meningoencephalitis in an immunocompetent, previously healthy 20-month old female child. *Indian J Med Microbiol* 28: 169-171.

149. Phadke SP, Bhagwat SV, Kapshikar RN and Ghevari SD. 1979. Listeriosis in sheep and goats in Maharashtra. *Indian Vet J* 56: 634-637.

150. Pingulkar K, Kamat A and Bongirwar D. 2001. Microbiological quality of fresh leafy vegetables, salad components and ready-to-eat salads: an evidence of inhibition of *Listeria monocytogenes* in tomatoes. *Int J Food Sci Nutr* 52: 15-23.

151. Pini PN and Gillbert RJ. 1988. The occurrence in UK of *Listeria* spp. in raw chickens and soft cheese. *Int J Food Microbiol* 6: 317-326.

152. Pochop J, Kaèániová M, Hleba L, Lopasovský L, Bobková A, Zeleòáková L, Strièík M. 2012. Detection of *Listeria monocytogenes* in ready-to-eat food by step one real-time polymerase chain reaction. *J Environ Sci Health B* 47: 212-216.

153. Pramod NP and Ravindran PC. 1997. *Listeria monocytogenes* agglutinins in apparently healthy-individuals in south-India. *Med Sci Res* 25: 175-177.

154. Pridham TJ, Budd J and Karstad LHA. 1966. Common diseases of fur bearing animals: II. Diseases of chinchillas, nutria, and rabbits. *Can Vet J* 7: 84-87.

155. Radostits OM, Gay CC, Blood DC and Hinchcliff KW. 2007. *Veterinary medicine: A textbook of the diseases of cattle, horses, sheep, pigs and goats*. 10th edn. WB Saunders, Philadelphia, USA.

156. Rahman T, Sarma DK, Goswami BK, Upadhyaya TN and Choudhary B. 1985. Occurrence of listerial meningoencephalitis in pigs. *Indian Vet J* 62: 7-9.

157. Rathod PH, Shah NM, Dadawala AI, Chauhan HC, Patel SS, Singh K, Ranaware P and Chandel BS. 2010. Comparison of Standard tube agglutination test and indirect haemagglutination test in the detection of Listerial antibodies in animals. *Vet World* 3: 506-508.

158. Rawool DB, Malik SVS, Shakuntala I, Sahare AM and Barbuddhe SB. 2007. Detection of multiple virulence associated genes in pathogenic *Listeria monocytogenes* from bovines with mastitis. *Int J Food Microbiol* 113: 201-207.

159. Rosenow EM and Marth EH. 1987. Growth of *L. monocytogenes* in skim, whole and chocolate milk and in whipping cream during incubation at 4, 8, 13, 21 and 35°C. *J Food Prot* 50: 452-459.

160. Ross T, Rasmussen S, Fazil A, Paoli G and Sumner J. 2009. Quantitative risk assessment of *Listeria monocytogenes* in ready-to-eat meats in Australia. *Int J Food Microbiol* 131: 128-137.

161. Ryser ET and Marth EH. 1987. Fate of *Listeria monocytogenes* during the manufacture and ripening of carriembert cheese. *J Food Prot* 50: 372-378.

162. Ryser ET and Marth EH. 1991. *Listeria, listeriosis and food safety*. Marcel Dekker Inc, New York.

163. Sagoo SK, Little CL and Mitchell RT. 2001. The microbiological examination of ready-to-eat organic vegetables from retail establishments in the United Kingdom. *Lett Appl Microbiol* 33: 434-439.

164. Salamina G, Donne ED, Niccolini A, Poda G, Cesaroni D, Bucci M, Fini R, Maldini M, Schuchat A, Swaminathan B, Bibb W, Rocourt J, Binkin N and Salmaso S. 1996. A foodborne outbreak of gastroenteritis involving *Listeria monocytogenes*. *Epidemiol Infect* 117: 429-436.

165. Samanta A. 2009. Listeriosis in 3-month old calf from West Bengal. *Intas Polyvet* 10: 284-285.

166. Saroj SD, Shashidhar R, Dhokane V, Hajare S, Sharma A and Bandekar JR. 2006. Microbiological evaluation of sprouts marketed in Mumbai, India and its suburbs. *J Food Prot* 10: 2515-2518.

167. Scallan E, Hoekstra RM, Angulo FJ, Tauxe RV, Widdowson MA, Roy SL, Jones JL and Griffin PM. 2011. Foodborne illness acquired in the United States-major pathogens. *Emerg Infect Dis* 17: 7-15.

168. Scharff RL. 2012. Economic burden from health losses due to foodborne illness in the United States. *J Food Prot* 75: 123-131.

169. Schuchat A, Lizano C, Broome CV, Swaminathan B, Kim C and Winn K. 1991. Outbreak of neonatal listeriosis associated with mineral oil. *Pediatr Infect Dis* 10: 183-189.

170. Schwaiger K, Helmke K, Holzel CS and Bauer J. 2011. Comparative analysis of the bacterial flora of vegetables collected directly from farms and from supermarkets in Germany. *Int J Environ Health Res* 21: 161-172.

171. Schweizer G, Ehrensperger F, Torgerson PR and Braun U. 2006. Clinical findings and treatment of 94 cattle presumptively diagnosed with listeriosis. *Vet Rec* 158: 588-592.

172. Shah NM and Dholakia PM. 1983. Microflora of the cervicovaginal mucus of Surti buffaloes and their drug resistance pattern. *Indian J Anim Sci* 53: 147-150.

173. Shakespeare M. 2009. *Zoonoses*. 2nd edn. Pharmaceutical Press, London.

174. Shakuntala I, Malik SVS, Barbuddhe SB and Rawool DB. 2006. Isolation of Isolation of *Listeria monocytogenes* from buffaloes with reproductive disorders and its confirmation by polymerase chain reaction. *Vet Microbiol* 117: 229-234.

175. Sharif J, Willayat M, Sheikh GN, Roy SS and Bhat SA. 2011. Prevalence and antibiogram of *Listeria monocytogenes* in cases of abortion and stillbirths in sheep of Kashmir. *J Vet Public Hlth* 9: 43-46.

176. Sharma M, Batta MK, Katoch RC, Asrani RK, Nagal KB, Joshi VB and Sharma M. 1996. *Listeria monocytogenes* abortions among migratory sheep and goats in Himachal Pradesh. *Indian J Anim Sci* 66: 1117-1119.

177. Shirke MB, Sherikar AA, Majee SB, Das AM and Barbuddhe SB. 2000. Occurrence of rabies and *Listeria* in brains of cattle and buffaloes. *J Bombay Vet Coll* 8: 5-7.

178. Sim J, Hood D, Finnie L, Wilson M, Graham C, Brett M and Hudson JA. 2002. Series of incidents of *Listeria monocytogenes* non-invasive febrile gastroenteritis involving ready-to-eat meats. *Lett Appl Microbiol* 35: 409-413.

179. Simmons MD, Cockroft PM and Okubadejo OA. 1986. Neonatal listeriosis due to crossinfection in an obstetric theatre. *J Infect* 13: 235-239.

180. Singh J, Batish VK and Grover S. 2012. Simultaneous detection of *Listeria monocytogenes* and *Salmonella* spp. in dairy products using real time PCR-melt curve analysis. *J Food Sci Technol (Mysore)* 49: 234-239.

181. Skovgaard N and Morgen CA. 1988. Detection of *Listeria* species in faeces from animals, in feeds and in raw foods of animal origin. *Int J Food Microbiol* 6: 229-242.

182. Slade PJ, Collins-Thompson DL and Fletcher F. 1988. Incidence of *Listeria* species in Ontario raw milk. *J Can Inst Food Sci Technol* 21: 425-429.

183. Smith B, Larsson JT, Lisby M, Müller L, Madsen SB, Engberg J, Bangsborg J, Ethelberg S and Kemp M. 2011. Outbreak of listeriosis caused by infected beef meat from a meals-on-wheels delivery in Denmark. *Clin Microbiol Infect* 17: 50-52.

184. Srivastava NC and Khera SS. 1980. The prevalence of *Listeria* antibodies in farm stock. *Indian Vet J* 57: 270-272.

185. Stone J. 1997. Listeriosis. In: Wu CC (ed) Winter 1997 Newsletter. Indiana Animal Disease Diganositic Laboratory, Purdue University, USA.

186. Syne SM, Ramsubhag A and Adesiyun AA. 2011. Occurrence and genetic relatedness of *Listeria* spp in two brands of locally processed ready-to-eat meats in Trinidad. *Epidemiol Infect* 139: 718-727.

187. Thigeel H, Kendall P and Bunning M. 2011. Listeriosis fact sheet No. 9383. Colorado State University. http://wwwextcolostateedu/pubs/foodnut/09383. Accessed 27 March 2012.

188. Toaff R, Krochik N and Rabinovitz M. 1962. Genital listeriosis in the male. *Lancet* 2482-2483.

189. Todd ECD. 2007. *Listeria:* risk assessment, regulatory control, and economic impact. In: Ryser ET and Marth EH (eds) *Listeria, listeriosis, and food safety*, 3rd edn. CRC Press, Boca Raton, USA.

190. Uppal PK, Shrivastava NC, Kumar AA and Rao SH. 1981. Isolation of *Listeria mococytogenes* from neurologic disturbances in buffalo. *Indian J Comp Microbiol Immunol Infect Dis* 2: 16-18.

191. Uyttendaele MR, Neyts KD, Lips RM and Debevere JM. 1997. Incidence of *Listeria monocytogenes* in poultry and poultry products obtained from Belgian and French Abattoirs. *Food Microbiol* 14: 339-345.

192. Vaissaire J. 2000. Epidemiology of animal *Listeria* infections in France. *Bull Acad Natl Med* 184: 275-285.

193. Van Kessel JA, Karns JS, Lombard JE and Kopral CA. 2011. Prevalence of *Salmonella enterica*, *Listeria monocytogenes* and *Escherichia coli* virulence factors in bulk tank milk and in-line filters from US dairies. *J Food Prot* 74: 759-768.

194. Vasava KA, Kher HN, Chauhan HC, Chandel BS and Shah NM. 2005. Seroprevalence of *Listeria* infection in animals of North Gujarat. *Ind Vet J* 82: 254-256.

195. Vijayakrishna S, Venkata Reddy T, Varalakshmi K and Subramanyam KV. 2000. Listeriosis in broiler chicken. *Indian Vet J* 77: 285-286.

196. Vishwanathan KR and Uppal PK. 1981. Isolation of *Listeria* from sheep. *J Remount Vet Corps* 20: 127-130.

197. Viswanathan GR and Ayyer VV. 1950. Circling disease of sheep in the Madras State- etiology established. *Indian Vet J* 26: 395.

198. Wagner M, Melzner D, Bagò Z, Winter P, Egerbacher M, Schilcher F, Zangana A and Schoder D. 2005. Outbreak of clinical listeriosis in sheep: evaluation from possible contamination routes from feed to raw produce and humans. *J Vet Med* B 52: 278-283.

199. Wang F, Jiang L, Yang Q, Han F, Chen S, Pu S, Vance A and Ge B. 2011. Prevalence and antimicrobial susceptibility of major foodborne pathogens in imported seafood. *J Food Prot* 74: 1451-1461.

200. Warke R, Kamat A, Kamat M and Thomas P. 2000. Incidence of pathogenic psychrotrophs in ice creams sold in some retail outlets in Mumbai, India. *Food Control* 11: 77-83.

201. Welchman D, Hooton JK and Low JC. 1997. Ocular disease associated with silage feeding and *Listeria monocytogenes* in fallow deer. *Vet Rec* 140: 684-685.

202. Wesley IV. 2007. Listeriosis in animals. In: Ryser ET and Marth EH (eds) *Listeria, listeriosis, and food safety*, 3rd edn. CRC Press, Boca Raton, USA.

203. WHO. 1988. Food borne listeriosis (Update). *Bull WHO* 66: 421-428.

204. Wiedmann M, Czajka J, Bsat N, Bodis M, Smith MC, Divers TJ and Batt CA. 1994. Diagnosis and epidemiological association of *Listeria monocytogenes* strains in two outbreaks of listerial encephalitis in small ruminants. *J Clin Microbiol* 32: 991-996.

205. Wiedmann M, Bruce JL, Knorr R, Bodis M, Cole EM, McDowell CI, McDonough PL and Batt CA. 1996. Ribotype diversity of *Listeria monocytogenes* strains associated with outbreaks of listeriosis in ruminants. *J Clin Microbiol* 34: 1086-1090.

206. Wiedmann M, Arvik T, Bruce JL, Neubauer J, del Piero F, Smith MC, Hurley J, Mohammed HO and Batt CA. 1997. Investigation of a listeriosis epizootic in sheep in New York State. *Am J Vet Res* 58: 733-777.

207. Wilesmith JW and Gitter M. 1986. Epidemiology of ovine listeriosis in Great Britain. *Vet Rec* 119: 467-470.

208. Wilkins PA, Marsh PS, Acland H and Piero FD. 2000. *Listeria monocytogenes* septicemia in a Thoroughbred foal. *J Vet Diagn Invest* 12: 173-176.

209. Winter CH, Brockmann SO, Sonnentag SR, Schaupp T, Prager R, Hof H, Becker B, Stegmanns T, Roloff HU, Vollrath G, Kuhm AE, Mezger BB, Schmolz GK, Klittich GB, Pfaff G and Piechotowski I. 2009. Prolonged hospital and community-based listeriosis outbreak caused by ready-to-eat scalded sausages. *J Hospital Infect* 73: 121-128.

210. Wong HC, Chao WL and Lee SJ. 1990. Incidence and characterization of *Listeria monocytogenes* in foods available in Taiwan. *Appl Environ Microbiol* 56: 3101-3104.

211. Wung GH, Yan KT, Feng XM, Chen SM, Lui AP and Kokubo Y. 1992. Isolation and identification of *Listeria monocytogenes* from retail meat in Beijing. *J Food Prot* 55: 56-58.

212. Yadav MM and Roy A. 2009. Prevalence of *Listeria* spp. including *Listeria monocytogenes* from apparently healthy sheep of Gujarat State, India. *Zoonoses and Public Health* 56: 515-524.

213. Yadav MM, Roy A, Bhanderi B and Jani RG. 2011. Prevalence of *Listeria* species including *L. monocytogenes* from apparently healthy animals at Baroda Zoo, Gujarat State, India. *J Threat Taxa* 3: 1929-1935.

214. Yousif YA, Joshi BP and Ali HA. 1984. Ovine and caprine listeric encephalitis in Iraq. *Trop Anim Health Prod* 16: 27-28.

215. Yücel N and Balci S. 2010. Prevalence of *Listeria, Aeromonas* and *Vibrio* species in fish used for human consumption in Turkey. *J Food Prot* 73: 380-384.

216. Zarei M, Maktabi S and Ghorbanpour M. 2012. Prevalence of *Listeria monocytogenes, Vibrio parahaemolyticus, Staphylococcus aureus* and *Salmonella* spp. in seafood products using multiplex polymerase chain reaction. *Foodborne Pathog Dis* 9: 108-112.

2014, Zoonoses: Bacterial Diseases
Editor: **Sudhi Ranjan Garg**
Published by: **DAYA PUBLISHING HOUSE, NEW DELHI**

Pages **245–263**

11

Chlamydophila Infections

Rajesh Chahota and Mandeep Sharma
Department of Veterinary Microbiology,
Dr. G.C. Negi College of Veterinary and Animal Sciences,
C.S.K. Himachal Pradesh Agricultural University, Palampur – 176 062

Chlamydiae include a group of bacteria that are widely prevalent throughout the animal kingdom and are responsible for various disease syndromes. These are non-motile, Gram-negative, obligate intracellular pathogens of birds, animals and human beings. Their unique biphasic developmental cycle differentiates them from all other microorganisms. Chlamydiae replicate within the cytoplasm of host cells, forming characteristic intracellular inclusion bodies (Figure 11.1). They have restricted metabolic capacities and are also called as 'energy parasites' because they utilise ATP produced by the host cell (Storz 1971, Moulder *et al.* 1984, Stephens 1999).

The first disease condition of chlamydial origin, referred as pneumotyphous, was recognized as early as in 1879 by Ritter in Switzerland (Ritter 1880, Harris and Williams 1985). It was a household epidemic of an unusual pneumonia associated with exposure to tropical pet birds. After an outbreak associated with parrots occurred in Paris in 1892, the disease was named psittacosis after the Greek word for parrot, *psittakos* (Morange 1895). Chlamydial organisms were first described by Halberstaedter and von Prowazek (1907a, b), who identified intracytoplasmic inclusions containing large numbers of microorganisms within cells derived from the conjunctival scrapings of human patients with trachoma. Assuming these organisms to be protozoa, they named them chlamydozoa after the Greek word *chlamys* for mantle, because the reddish organisms appeared to be embedded in a blue matrix or mantle. The pandemic of 1929 to 1930 however drew worldwide

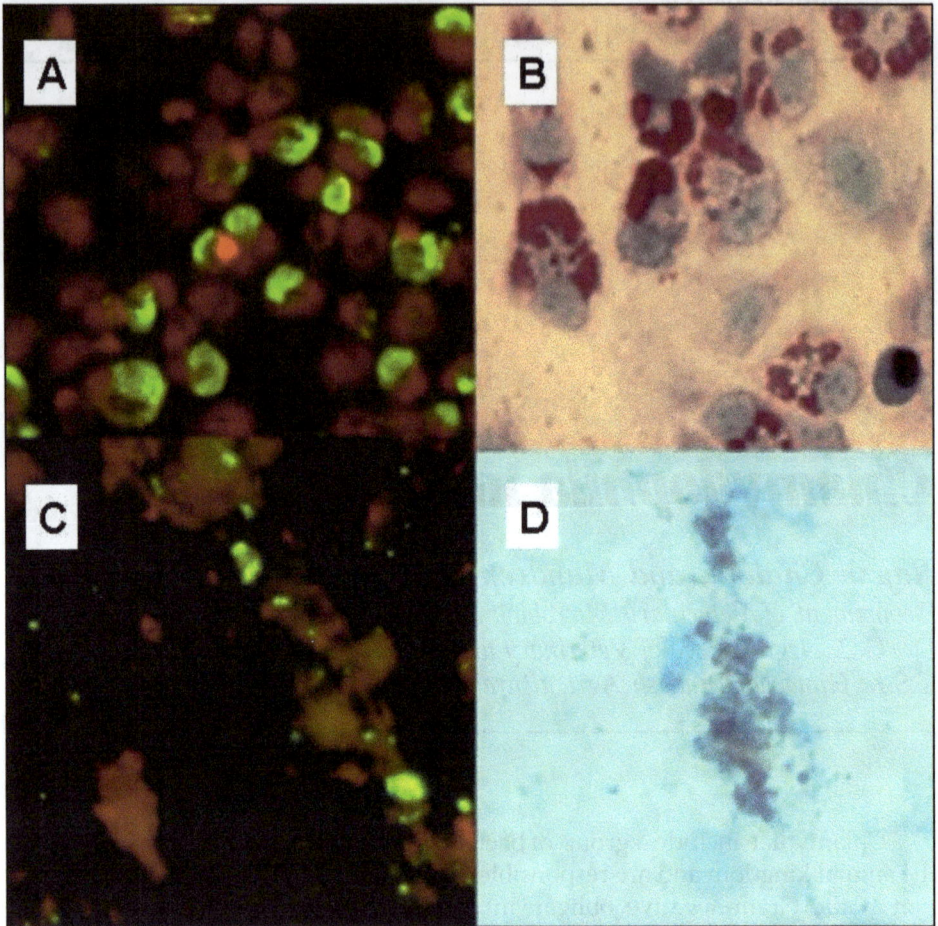

Figure 11.1: Elementary bodies (EBs) and intracellular inclusions of *C. psittaci*:
(A) FAT stained 32 h old inclusions in McCoy cell line (x 200); (B) Gimenez stained McCoy cell line with 46 h old large inclusions (x 400); (C) FAT stained EBs seen as pin point dots (x 200); (D) Gimenez stained yolk sac membrane impression smear showing EBs (x 1000).

attention to the disease. Trachoma is one of the oldest recognized human diseases, and has been described in Egyptian papyri and in ancient Chinese writings.

Chlamydiae are genetically diverse group of organisms consisting mostly of pathogens infecting birds and mammals, but some environmental species have also been detected from invertebrates like amoeba and insects in recent years. Generally chlamydial species are either strict human pathogens or pathogens of birds, domesticated animals and wildlife, however, some of the avian and animal chlamydial species are recognized as potential zoonotic pathogens. Chlamydiae are responsible for direct anthropozoonosis (animals-to-humans) without the involvement of any intermediate vertebrate or invertebrate hosts. Different wild and domestic avian and animal species are source of infection for susceptible humans and result in mild to severe, or sometimes fatal, clinical conditions depending upon the portal of entry of the infectious agent, individual's susceptibility, timely diagnosis, and therapeutic interventions.

Geographical Distribution

Human and animal chlamydiosis is prevalent worldwide. Chlamydiae are enzootic in many birds and animal species. As psittacine birds are the main source of human psittacosis, frequent outbreaks have been reported in tropical countries, which constitute their natural habitats, or in the countries where various kinds of psittacine birds are imported as pet birds.

Etiological Agents and Vectors

Earlier, all chlamydiae were placed in the Order Chlamydiales, Family Chlamydiaceae and in one genus *Chlamydia* that included four species: *Chlamydia trachomatis*, *C. psittaci*, *C. pneumoniae* and *C. pecorum* (Moulder 1984, Grayston 1989, Fukushi and Hirai 1992). Following the reclassification of the Order Chlamydiales in 1999 (Everett *et al.* 1999a) based on 16S rRNA, the Family Chlamydiacaeae is now divided into two genera, *Chlamydia* and *Chlamydophila*, containing nine species (Figure 11.2). Different chlamydial species with their typical animal hosts and important disease conditions are shown in Table 11.1.

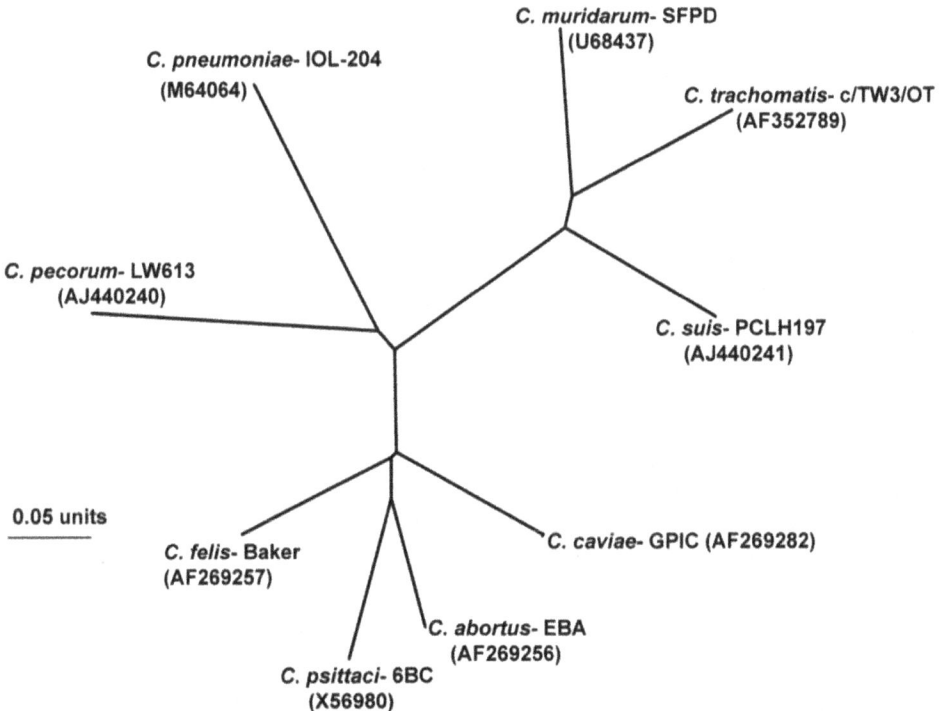

Figure 11.2: Neighbour-joining tree showing different human and animal chlamydial species and genetic distance (shown in 0.05 unit distance bar) among *Chlamydophila* and *Chlamydia* species based on *ompA* gene of representative strains of each species.

Table 11.1: Chlamydial species, their typical hosts, associated disease conditions and zoonotic potential

Species	Typical Hosts	Disease Conditions	Zoonoses
Genus Chlamydophila			
C. psittaci	Avian species	Avian chlamydiosis	Human psittacosis
C. abortus	Caprine, ovine, mammals	Enzootic abortions in ewes, reproductive infections, epididymitis in males	Abortions in pregnant women mainly
C. pecorum	Caprine, ovine, bovine, koalas, swine	Enteritis, reproductive diseases, polyarthritis, encephalomyelitis, pneumonia, infections of bladder, eye, lymphoid tissues and prostate	Not known
C. felis	Cats	Conjunctivitis, respiratory infections	Conjunctivitis and respiratory infection
C. caviae	Guinea pigs	Inclusion body conjunctivitis	Cervicitis alone or mixed with C. trachomatis
C. pneumoniae	Human, koala, horses	Acute and chronic respiratory diseases, cardiovascular diseases, asthma, infection of brain and joints	Not known
Genus Chlamydia			
C. muridarum	Hamster, mice	Genital, intestine, liver, lungs, kidney and spleen infections	Not known
C. suis	Swine	Conjunctivitis, enteritis, pneumonia, polyarthritis, reproductive disorders	Not known
C. trachomatis	Humans	Trachoma, urogenital infection and lymphogranuloma-venereum	Not reported from non-human host

All the animal chlamydial species have potential of inflicting human infections, but some species like *Chlamydophila psittaci* (formerly *Chlamydia psittaci*), *Chlamydophila abortus* (formally *Chlamydia psittaci* ewe abortion type), *Chlamydophila caviae* (formerly *Chlamydia trachomatis* GPIC) and *Chlamydophila felis* are confirmed zoonotic pathogens (Longbottom and Coulter 2003). Avian chlamydiosis caused by *C. psittaci* represents a major risk of zoonotic transmission to humans and causes human psittacosis. *C. abortus* can also be transmitted from ruminants to humans. Similarly, human infections with *C. felis* from cats have been reported over the years (Yan *et al.* 2000, Hartley *et al.* 2001, Browning 2004). The zoonotic significance of *C. caviae* has been highlighted in the recent years (Chahota 2007). The incidences of zoonotic infection by *C. pecorum*, *C. suis* and *C. muridarum* are still unknown.

Chlamydial Developmental Cycle

Chlamydiae undergo a typical biphasic developmental cycle (Figure 11.3) characterized by two morphologically distinct forms, called the elementary body (EB) and the reticulate body (RB). EBs are small (0.2-0.3 μm diameter) infectious forms that are either endocytosed or phagocytosed by host cells. EB within an intracytoplasmic inclusion transforms into larger (0.5-1.6 μm diameter) intracellular non-infectious, metabolically active RB. The RB multiplies by binary fission and fills the inclusion, which expands in size. RBs re-condense back into EBs via intermediate bodies towards the end of the cycle (24 to 72 h, depending on chlamydial species and strains) and are then released from the host cell by lysis of cell or exocytosis to initiate another cycle of infection (Storz 1971, Moulder *et al.* 1984, Stephens 1999). The development cycle shown in Figure 11.3 typically takes 2-3 days, although some species have longer cycles

Host Range

Chlamydophila psittaci has been reported from 470 avian species belonging to 30 avian Orders. It can also infect many mammalian species. *Psittaciformes* are predominant reservoirs of these organisms. The commonly infected avian species include parrots, parakeets, macaw, cockatiels and budgerigars and also non-psittacine birds like pigeons, sparrows, chickens, ducks, doves and gulls as well as game birds and birds of prey. The incidence of infection in canaries and finches is believed to be lower than that in psittacine birds. There are 8 known serovars (variants) of *C. psittaci* and all should be considered to be transmissible to humans. Each serovar is associated with a different Order or group of birds *e.g.* serovars A and F in psittacines, B in pigeons and doves, and C and D in poultry, ducks and geese. Serovar E isolates have been obtained from a variety of human or avian hosts worldwide. Two other serovars, M56 and WC, have been isolated during outbreaks in mammals (Andersen and Vanrompay 2000, Kaleta and Taday 2003, Chahota *et al.* 2006).

All mammals are potentially susceptible to *C. abortus* but it is commonly detected in small and large ruminants and occasionally in horse, swine, carnivores, mouse, guinea pig and rabbit (Everett 2000). Domestic and stray cats act as hosts for *C. felis*, whereas guinea pig is a specific host for *C. caviae* (Gordon *et al.* 1966, Yan *et al.* 2000, Hartley *et al.* 2001). The increasing number of reports of isolation of the strains of

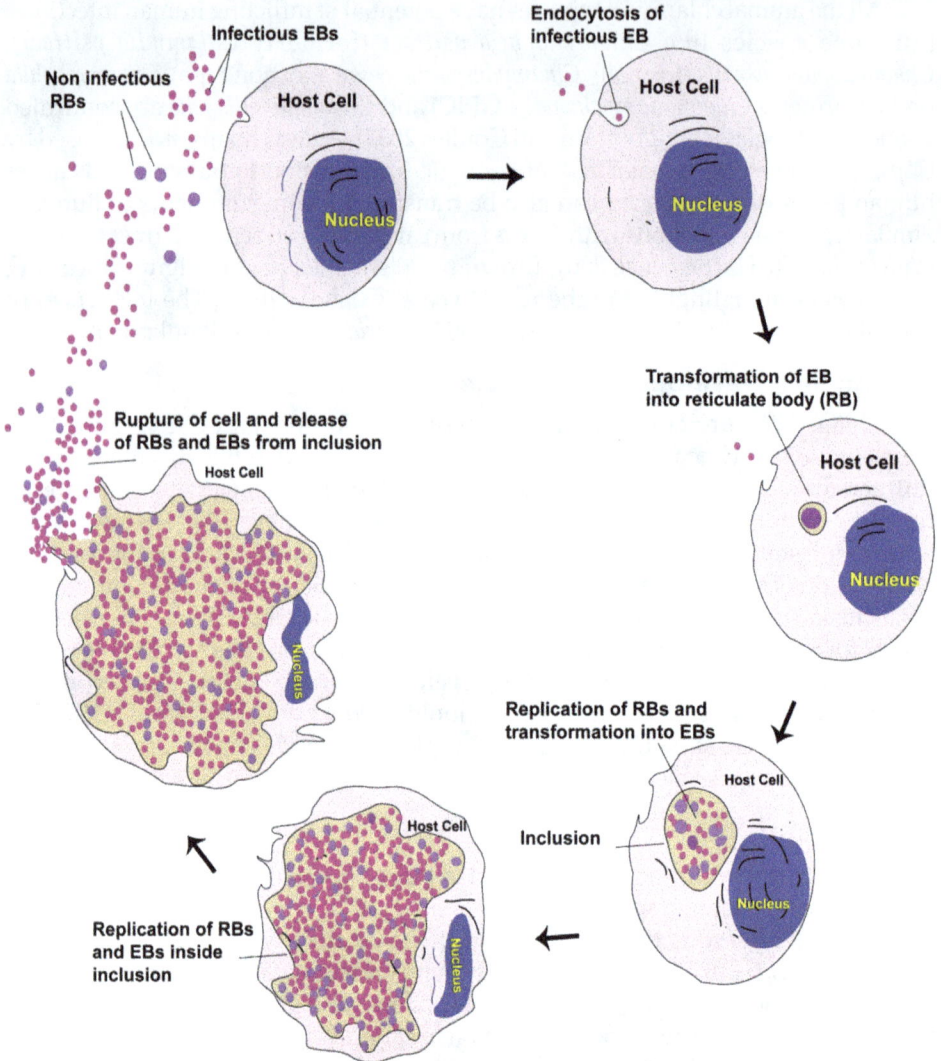

Figure 11.3: Development cycle of chlamydiae showing infection of host cell and replication

human pathogen *C. pneumoniae* from frogs (Berger *et al.* 1999), koalas (Glassick *et al.* 1996) and horses (Wills *et al.* 1990) are of particular interest because of their potential hazard to human health. Environmental chlamydial species need to be assessed for their zoonotic potential.

Disease Transmission

Human psittacosis is mainly acquired by exposure to infected or clinically normal synanthropic avian species harbouring *C. psittaci*. Subclinical or inapparent infections are predominant in avian species and such birds shed organisms intermittently or at low levels. Shedding can be activated by stress factors such as transport, relocation, crowding, chilling and breeding. Most reported cases are the result of exposure to pet caged birds (especially parrots, parakeets). The organism is shed in the faeces, urine, nasal, oral or ocular secretions. Such dried or aerosolized material from body secretions/excretion and infectious viscera or carcasses contaminate the environment. Humans or susceptible birds get infection by inhaling airborne contaminated dust particles/aerosols. Infectious EBs of *C. psittaci* can survive for months in the environment. Besides inhalation, rarely bird bites and mouth-to-beak contact can also spread infection to humans. In birds, faecal-oral transmission is possible in crowded conditions and in nest boxes. Generally, inhaled chlamydiae cause severe disease, while ingested chlamydiae tend to develop carrier birds. Vertical transmission of chlamydiae has been found in chickens, ducks, parakeets, seagulls and snow geese. Bloodsucking ectoparasites, which include arachnids, lice and simulid flies, have been shown to transmit chlamydiae in turkeys, but probably act as mechanical vectors rather than biological vectors (Shewen 1980, Andersen and Vanrompay 2000, Smith *et al.* 2010).

C. abortus is endemic among ruminants. Infected animals excrete EBs at its highest at the time of kidding/abortion or at the time of ovulation leading to oral, nasal or sexual transmission to other mammals. Bacteria are also carried in the intestine and in some lymphoid tissue, creating carrier animals. About 20 per cent of sheep, which have had chlamydial disease, remain infective. *C. abortus* is transmitted to human beings, especially pregnant women, leading to abortions, particularly in those directly exposed to infected sheep or goats. Transmission most probably occurring by oro-nasal route following the handling of an infected ewe or lamb or by inhalation of aerosols generated by water bags or uterine fluids at the time of lambing/kidding or by aborted foetus, placenta, vaginal discharges and infected faeces. Rarely infection is by contaminated clothing or food or smoking with unwashed hands, or mouth-to-mouth resuscitation of weak lambs (Winter and Charnley 1999, Aitkin 2000, OIE 2011a). Cases of *C. felis* infection in humans are well established now by way of molecular techniques and serological studies (Yan *et al.* 2000, Hartley *et al.* 2001). The infection may spread through direct inoculation of eyes with chlamydia contaminated hands or through infectious aerosols/dust particles from infected pet cats. Route of transmission of *C. caviae* to human genital tract resulting in cervicitis in women is still unclear (Chahota 2007).

Environmental and Other Predisposing Factors

Physiological conditions like breeding or any stress that increase the shedding of infectious EBs by host species leads to increased chances of human infection. Human psittacosis is a significant occupational hazard to the workers in poultry industry, slaughterhouses, aviaries, and zoos, pet shop owners, farmers, laboratory workers and veterinarians due to close association with the source of infection. Similarly, passion of feeding birds from close proximity, nesting of birds, especially pigeons in human dwellings may expose the human beings to psittacosis. Sometime human infection can result from brief, passing exposure to high doses of EBs from infected birds or their dried contaminated droppings; therefore people with no identified occupational or recreational risk can become infected. *C. abortus* infection may be acquired by sheep and goat farmers during routine management activities or by shepherds who stay along with their flocks during the seasonal migrations. The chances of zoonotic infections may rise at the time of lambing or kidding season. Owning pet birds, cats and guinea pigs may expose the owners to *C. psittaci, C. felis* and *C. caviae* infections. Some pet avian species and young birds are highly susceptible to avian chlamydiosis leading to high mortality, particularly under stressful conditions, leading to higher chances of transmitting infection to the humans.

Prevalence of Disease in Animals

Chlamydiosis in animals and human is usually under diagnosed and underestimated due to vague clinical symptoms that may be confused with other clinical conditions, and also because of easy curability with commonly used antibiotics.

In India, chlamydial diseases are known to be prevalent since long and have been found responsible for abortions, infertility, conjunctivitis, pneumonitis, enteritis and encephalomyilitis in sheep, goats, cattle, buffaloes, equines and yaks (Jain *et al.* 1975, Krishna 1988, Dhingra *et al.* 1994, Chahota *et al.* 2001, Katoch *et al.* 2002). *Chlamydophila psittaci* has been isolated from 10.93 per cent chicken, 18.18 per cent crows, 16.36 per cent pigeons and 26.31 per cent parrots (Chahota *et al.* 1997) and also from wild rats and zoo animals (Chahota *et al.* 1997, Rattan *et al.* 2005). Chlamydiae have been reported to cause reproductive problems *i.e.* abortions and/or endometritis in goats (33.47 per cent), sheep (17.41 per cent) and cattle (23.68 per cent), followed by pneumonia, enteritis, mastitis in goats, keratoconjunctivitis in goats and cattle (Katoch *et al.* 2002). In poultry, concurrent avian chlamydiosis with Newcastle disease as well as aflatoxicosis have been reported (Chahota *et al.* 2000, Mahajan *et al.* 2002). Recent molecular epidemiological studies based on specific *ompA* gene, based PCR test and nucleotide sequence analysis, revealed high prevalence of *C. abortus* and *C. psittaci* in cases of abortion, respiratory infections and enteritis among migratory sheep and goats in Himachal Pradesh (Bhardwaj 2011). All these studies recognized chlamydial diseases as one of the major health problems in livestock, avian species and wild fauna in India. Outbreaks of abortions in sheep and goats and avian chlamydiosis have also been frequently reported from other developing countries of Asia and Africa (OIE 2011 a,b).

Chlamydiphila psittaci infections are not restricted to Asia or Africa alone; these occur worldwide in both in wild and domesticated birds (Andersen and Vanrompay 2003). Enzootic abortions in ewes (EAE) or ovine enzootic abortion (OEA), caused by *C. abortus*, is recognized as a major cause of lamb loss in sheep throughout Europe and is encountered in most sheep-rearing areas of the world, including Africa and North America. The disease is not considered to be a problem in Australia and New Zealand, although sporadic cases have been reported in Australia (Seaman 1985). In Northern European countries, including UK, OEA is the most common infectious cause of abortions in lowland flocks that are intensively managed at lambing time and it accounts for around 45 per cent of all diagnosed abortions and estimated to affect 8.6 per cent of flocks, equating to approximately 1.5 million sheep annually

(Aitken 2000, Longbottom and Coulter 2003). Enzootic abortion of goats is similar in severity in most parts of the world, but its spread and economic impact is less defined because of inadequate epidemiological data. Cattle, pigs and horses are also infected with *C. abortus*, but such infections are thought to occur less frequently. *C. felis* is also endemic among house cats worldwide.

Prevalence of Disease in Human Beings

There are many reports of human chlamydiosis caused by *C. trachomatis* and *C. pneumoniae* from India, based on serological or isolation studies and PCR (Mittal *et al.* 1995, Vats *et al.* 2004, Pandey *et al.* 2005), but the prevalence of chlamydial zoonoses in the country is not known. India has a vast variety of avian fauna like psittacine birds, poultry, pigeons, ducks, doves, etc. The cases/outbreaks of avian chlamydiosis have been detected in birds; so it is most likely that human infections also exist in the country. However, these have not been diagnosed or reported yet, probably due to lack of diagnostic facilities, technical expertise or public awareness.

Human psittacosis is notifiable disease in many countries. In the 1950s, the importance of *C. psittaci* infections (ornithosis) in poultry was recognized, and human psittacosis was described as an important occupational hazard to workers in poultry processing plants. Since the first detection of psittacosis in 1879, many sporadic cases or large outbreaks have been reported in different countries (Yung and Grayson 1988, Andersen and Vanrompay 2003, Telfer *et al.* 2005). The World Organisation for Animal Health (OIE) database on avian chlamydiosis reveals that from 1996 to 2010, up to 17 OIE member countries reported chlamydial zoonoses annually (OIE 2011c). In 2010, total 12 countries reported 309 cases of human psittacosis. It includes 6 European nations (101 cases), Japan (11 cases), Argentina (144 cases), Russia (48 cases) and USA (4 cases). Hong Kong and New Zealand reported a single case each. In UK, over the last decade, there have been approximately 150 cases of human psittacosis annually, 70 per cent of which could be traced to exotic birds and 20 per cent to turkey farms and poultry dressing plants. But many more cases might have gone undiagnosed owing to confusing symptoms and difficult diagnosis.

In contrast to *C. psittaci* infection, *C. abortus* infection is relatively rare in human beings. In a review of 1157 cases of human chlamydiosis (thought to be psittacosis) in Scotland between 1967 and 1987, only 11 cases were linked with sheep and cattle. Human infections with *C. abortus* have been reported from UK, France, Netherlands and USA. In Japan and in some European countries, evidence of human *C. felis* infection has been reported. Seroepidemiological studies suggest that human infection might be more frequent as antibodies have been found present in 1.7 per cent of the general population and 8.8 per cent of small animal clinic veterinarians in Japan (Yan *et al.* 2000).

Disease Manifestations in Animals

Chlamydiae are responsible for many disease conditions in birds and mammals. All types of *C. psittaci* infections in birds are now referred as avian chlamydiosis (AC) that was earlier known as psittacosis or parrot disease, parrot fever or ornithosis. In old literature, the term psittacosis was used when the disease was carried by psittacine birds, whereas it was called ornithosis when other birds carried the disease. AC is characterized by diarrhoea, anorexia, respiratory distress, sinusitis, rhinitis and conjunctivitis. The clinical symptoms can be variable, depending on the host species infected, the virulence of the agent, the degree of exposure, portal of entry and concurrent stress. Typically, diseased bird has ruffled plumage and labored breathing, it is depressed, shows nasal and ocular discharge, and neither eats nor vocalizing. The appearance of lime-green or yellow droppings, especially when the urine component is discolored, is highly suggestive although not diagnostic of the malady. Infrequently, nervous form of AC with manifestation of central nervous system signs like tremors, shaking, head twisting, paralysis of the limbs and convulsions may occur. This clinical peculiarity has been recognized in Amazons parrots, African greys and cockatoos. Additionally, cockatiels and neophemas with low-grade infections may seem to have an eye disease resembling a sty. The financial losses resulting from the disease, particularly those incurred in the poultry and pet industries, combined with the fact that this is the most common animal chlamydiosis transmissible to humans, highlights the economic importance and public health significance of avian chlamydiosis (Andersen and Vanrompay 2003, OIE 2011b).

C. abortus organisms colonize the placenta and are primarily responsible for abortions and weak neonates in sheep and goats are and, also responsible for reproductive failures in bovine, equine and porcine. Chlamydial abortion in sheep and goats typically occurs in the last 2 to 3 weeks of pregnancy with the appearance of stillborn lambs, congested or necrotic cotyledons and inflamed placentas. However, infection can also result in the delivery of full-term stillborn lambs or weak lambs that do not survive longer than 48 hours. Infection is generally established in a 'clean' flock through the introduction of infected replacements and results in a small number of abortions in the first year, which is followed by an 'abortion storm' in the second year that can affect up to around 30 per cent of ewes (Aitken 2000, Everett 2000, OIE 2011a). *C. felis* is responsible for feline chlamydiosis, which is characterized by rhinitis and conjunctivitis in cats and respiratory problems particularly in kittens. It can be recovered from the stomach and reproductive tract. In Japan, *C. felis* has been found more widespread in stray cats than in pet cats (Yan *et al.* 2000, Hartley *et al.* 2001). *C. caviae* causes either self-limiting or asymptomatic inclusion conjunctivitis among young (4-8 weeks old) guinea pigs, usually observed as ocular inflammation and eye discharge (Gordon *et al.* 1966, Everett 2000).

Important clinical conditions caused by other chlamydial species are shown in Table 11.1.

Disease Manifestations in Humans

C. psittaci infection in humans is called human psittacosis. In humans, lungs are primarily involved as infectious EBs of *C. psittaci* gain entry via the respiratory tract, however, the infection may spread to other site through blood stream infecting a wide variety of cells and damaging most anatomic sites, probably due to the cytocidal effect of the infection. The incubation period is usually between 7 and 14 days. Infections are often subclinical and may be mild, resembling common cold or a mild influenza attack, but severe pneumonitis may occur. Atypical pneumonia is a common presentation. Overt clinical disease is almost always accompanied by fever, low pulse rate, chills and severe headache. Cough, when present, is usually nonproductive. Lung infections are often complicated by secondary bacterial infections that obscure the characteristic pathologic picture. Many complications with involvement of multiple organs are recognized, including meningoencephalitis, endocarditis, myocarditis and hepatitis, and may result in deaths (Longbottom and Coulter 2003, Smith *et al.* 2010). Radiographs may show more extensive lung involvement than is expected on the basis of respiratory difficulty. Alternatively, the disease may present with a general toxic, febrile state without respiratory findings. Hepatosplenomegaly is common. Person-to-person transmission is uncommon (Yung and Grayson 1988, Telfer *et al.* 2005). In the last few years, association of *C. psittaci* with ocular adnexal lymphoma has been reported from many countries and the incidence is on rise (Moslehi *et al.* 2006).

C. abortus can cause severe life-threatening disease in pregnant women. Infection can result in spontaneous abortion or stillbirths, which are typically preceded by several days of acute influenza-like illness, as well as renal failure, hepatic dysfunction, disseminated intravascular coagulation, and possibly death. A number of pregnant women exposed to *Chlamydia*-infected sheep have had spontaneous abortions. The organisms have been recovered from placental tissues and from other organs in fatal cases. Cases of *C. felis* infection associated with conjunctivitis and atypical pneumonia in humans (Browning 2004) or seropositive cases without clinical signs have been reported (Yan *et al.* 2000). Involvement of *C. caviae* in human cervicitis has indicated wider implications of chlamydial zoonoses (Chahota 2007).

Diagnostic Procedures

Presumptive diagnosis of chlamydial infection, particularly in farm animals and birds, can often be made on the basis of history, clinical symptoms and presenting pathology. Clinical signs are not pathognomonic in case of human psittacosis but the history about exposure to birds or aborted animals plays crucial role in diagnosis. Accurate diagnosis requires confirmation through laboratory investigations. There are essentially two main approaches to diagnosing chlamydial infections in mammals and birds. The first one involves direct detection of the agent in tissue or swab samples, while the second approach consists of serological screening of blood samples for the presence of anti-chlamydial antibodies. Different techniques are used to accomplish the diagnosis in humans, animals and birds.

1. Conventional Methods

a) Direct Impression Smears

The presence of chlamydiae (EBs or inclusions) in tissue and swab samples can be demonstrated by cytochemical staining of smears from appropriate clinical specimens (excretions, faecal/cloacal swabs, vaginal swabs, placental or foetal tissues etc.) for rapid diagnosis of infection (Andersen and Vanrompay 2003, Longbottom and Coulter 2003). Prepared smears can be stained with Machiavello or modified Gimenez, Giemsa, or modified Ziehl-Neelsen (MZN) methods for detecting EBs, which take red/pink colour against blue or green cellular background (Sachse *et al.* 2009).

b) Histopathological Detection

Chlamydial antigen detection in clinical tissue samples can be accomplished with formalin-fixed or formalin-fixed and paraffin-embedded tissue specimens using histochemical staining techniques, but these techniques are nonspecific. Therefore, care must be taken while interpreting the results. In human psittacosis, inclusions of *C. psittaci* can be demonstrated in alveolar macrophages and epithelial cells.

c) Isolations

Being obligate intracellular bacteria, chlamydiae require tissue culture or embryonated chicken eggs for isolation and propagation. Suitable samples must be collected, carefully in sucrose phosphate glutamate (SPG) containing antibiotics like streptomycin (200 µg/ml), gentamicin (50 µg/ml), vancomycin (75 µg/ml) and nystatin (25 units/ml), but penicillin, tetracycline and chloramphenicol should not be used.

(i) Embryonated Chicken Eggs

Conventionally, 6 to 8 days old embryonated chicken eggs are inoculated by intra yolk sac route and screened by preparing smears from yolk-sac membranes at the time of death, or from surviving eggs on completion of the experiment. Then yolk sac (membranes) impression smears are stained using an appropriate procedure, such as Gimenez, MZN, modified Machiavello or Giemsa to visualize EBs (Figure 11.1D).

(ii) Cell Culture

Several cell lines have been used to grow chlamydiae. The success of propagation is dependent on the cell line and the chlamydial species being tested. Staining of cell monolayers with Giemsa or indirect immunouorescence or with immunoperoxidase method reveals characteristic inclusions and morphology of the organism. Isolation of *C. abortus* can be done in McCoy, HeLa, buffalo green monkey kidney (BGMK), baby hamster kidney (BHK) and the mouse fibroblast L cells. Most cell types are susceptible to *C. psittaci* infection; BGMK, African green monkey kidney (Vero), McCoy and L cells are commonly used, while *C. felis* appears to grow well in McCoy cells. Infection of chlamydiae in cell culture can be enhanced by centrifugation and/or by chemical treatment of cultured cells with cycloheximide (1 µg/ml) or emetine (1 µg/ml). For confirmation of human psittacosis, the patient's blood or sputum is used for recovery of *C. psittaci*.

d) Agglutination, Precipitation and Complement Fixation Tests

Agar gel precipitation test (AGPT) and complement fixation test (CFT) are most commonly used in veterinary laboratories for serodiagnosis of animal chlamydiosis. In EAE, CFT serum titres rise at the time of abortion and remain high for at least 6 weeks. A CFT titre greater than 1/32 is considered as an indication for the presence of EAE, whereas lower values can be due to subclinical enteric infections by *C. abortus* or cross-reactivity with *C. pecorum* and other bacterial LPS. Modified CFT is used for detection of avian chlamydiosis. A simple latex agglutination test that detects specific IgM is also quite handy (Moore *et al.* 1991, Andersen and Vanrompay 2003).

2. Modern Techniques

a) Fluorescent Antibody Tests (FAT)

FAT using Chlamydiaceae-specific anti-LPS (lipopolysaccharide) antibodies or species-specific monoclonal antibodies (mAbs) to major outer membrane protein (MOMP), which are either directly conjugated with fluorescein or combined with a fluorescein-conjugated anti-mouse antiserum, improve the sensitivity of detection of chlamydial EBs in smears. Microimmunofluorescence (MIF) is currently used in routine serodiagnosis. In case of human psittacosis, serodiagnosis is generally considered to be the method of choice and a four-fold rise in chlamydial antibody titers in acute and convalescent serum as demonstrated by CF or MIF tests is diagnostic.

b) Immunohistochemical Staining

Immunohistochemical staining procedures like immunoperoxidase method that utilise mAbs directed against chlamydial surface antigens, such as LPS or MOMP, are more sensitive and produce more striking results in comparison to histochemical staining in formalin-fixed tissues.

c) PCR-Based Detection

For detecting and identifying specific chlamydial species/strain, polymerase chain reaction (PCR) tests, targeting 16S rRNA gene or 23S rRNA gene or spacer region between them or *ompA* gene coding MOMP, are now routinely used (Kaltenboeck, *et al.* 1991, Everett *et al.* 1999b, Chahota *et al.* 2005). For quantifying chlamydial antigen present in samples, real-time PCR targeting the 23S rRNA or *ompA* genes is used (Everett *et al.* 1999b, Sachse *et al.* 2009).

d) DNA Microarray Technology

In many instances, mere identification of the involved microorganism's species is no longer sufficient; the information on subspecies, serotype or genotype, toxins and other virulence factors is also expected. For this, DNA microarray technology is beneficial for laboratory diagnosis of chlamydial diseases. Sachse *et al.* (2005) developed a microarray assay for the detection and differentiation of *Chlamydia* sp. and *Chlamydophila* sp.

d) Enzyme-Linked Immunosorbent Assay (ELISA)

Different types of ELISA targetting proteins like MOMP, LPS and polymorphic outer membrane proteins (POMP) have been developed. ELISA, based on recombinant MOMP with 100 per cent sensitivity and 100 per cent specificity, has been developed, evaluated and successfully applied in cases of human psittacosis (Verminnen *et al.* 2006). Competitive ELISA (cELISA), based on the binding of specific mAbs against the MOMP variable domain 1 (VD1) or VD2 and POMPs specific mAbs, has been developed for specific diagnosis of OEA. Many chlamydial LPS-based immunodiagnostics developed primarily for detecting *C. trachomatis* infections in human clinical specimens should also be suitable for detecting chlamydial infections in animals because they are based on the family-specific LPS antigen. One of the main advantages of using immunoassays over cell culture for diagnosing infection, other than the shorter time it takes to complete the test, is that they are not dependent on viability of EBs.

Prevention and Control Strategies

Chlamydiae can infect humans, animals and birds; so its prevention and control require participation of both public health and animal health agencies. It is essential to control animal/avian chlamydiosis in order to reduce infections in humans. Prevention and control measures in animals include clean management practices that include housing, breeding, sanitation of premises or enclosures, quarantine, surveillance and chemoprophylaxis. Secondly, appropriate measures must be taken to prevent the spread of infection among animals/birds by isolation of newly acquired animals or birds. The implementation of rigorous disinfection measures and disposal of infected material is also essential.

Transmission of *C. psittaci* from birds to humans and precipitation of disease may be dose-dependent; therefore, cages should be regularly cleaned, ensuring that excretions do not accumulate for long enough to dry out and become aerosolised (airborne). It is good to minimize the circulation of feathers and dust by wet-mopping the floor frequently with disinfectant or spraying the floor with disinfectant or water before sweeping, and preventing air currents in the area. Vacuum cleaner should not be used as it aerosolizes the infectious particles. Protective clothing and masks etc. are quite useful. For disinfection 1:1000 dilution of quaternary ammonium compounds is effective, as are 70 per cent isopropyl alcohol, 1 per cent lysol, 1:100 dilution of household bleach, and chlorophenols. Many disinfectants are respiratory irritants and should only be used in well-ventilated areas. In wild birds, psittacosis is controlled naturally by the inability of sick birds to keep up with the flock. Chemoprophylaxis for exotic birds has been developed. Chlortetracycline is usually delivered as impregnated seed. Small pet birds are usually treated with millet containing the antibiotic at the rate of 0.5 mg/g for 30-45 days. Blood levels of tetracycline can be enhanced with citric acid in the birds' drinking water.

Tetracycline and its derivatives, mainly Vibramycin, are used to treat sick patients as well as carriers by intravenous or intramuscular injections, or by administering orally or by mixing in proper ratio with palatable food. Intramuscular or subcutaneous administration of tetracyclines is only recommended for up to one week due to the risk of local tissue damage. Chlamydiosis patients need intense, supportive care as well as therapy for concurrent problems.

As far as prognosis of human psittacosis is concerned, complete recovery is the rule (even some time without antibiotic use), Prior to antibiotic therapy the case fatality rate was quite high (>20 per cent), with most fatalities seen in those above the age of 50 years. Treatment of psittacosis with tetracycline via intravenous infusion is almost always successful and fatalities are rare, although the clinical response may not be rapid and recovery may be prolonged. Tetracycline (40 mg/kg/d) or doxycycline (100 mg b.i.d.) in children >8 years of age or erythromycin (40 mg/kg/d) in children <8 years of age are used. Azithromycin is also used to treat psittacosis. Fever and other symptoms begins to subside 48 to 72 hours after initiation of the antibiotic treatment, but treatment must continue for 2 weeks after the temperature returns to normal. Untreated patients may have severe pulmonary symptoms for 1 to 3 weeks.

Infections in pet birds continue to present a significant hazard to human health; however, eradication or elimination of chlamydial infections in birds is impossible as these infections are carried by a wide range of avian hosts throughout the world. Improved methods of diagnosis and treatment, quarantining of imported birds, and improvements in hygiene in the poultry industry have resulted in reduction in the number of outbreaks in poultry since the 1960s. Many countries have prohibited importation of psittacine birds. Screening to select psittacosis-free birds for breeding has been used to establish uninfected flocks of small pet birds that can be bred in captivity.

Vaccination for human psittacosis and avian chlamydiosis is not available. However, a vaccine for sheep against EAE using mutant strain of *C. abortus* has been tried with limited success. Also an attenuated live vaccine for cats has been marketed, but its use should be restricted to cats at risk of exposure (Andersen and Vanrompay 2003, Smith *et al.* 2010, OIE 2011a, b).

Future Directions

The Order Chlamydiales consists of a group of organisms that are diverse in epidemiology and genetic makeup and are of considerable significance to both human and animal health. They are also of major economic importance worldwide. The zoonotic potential of these organisms requires further elucidation through epidemiological investigations of animal chlamydioses, especially the host range including domestic, wild and synanthropic animals and birds and the association of the organism with upcoming new clinical conditions in human beings like recently reported ocular adnexal lymphoma due to *C. psittaci*. There is still much to learn about these organisms in terms of their biology, immunology and pathogenesis if improvements in diagnosis and vaccines are to be made to effectively control and prevent infections in humans and animals.

As with the increasing economic prosperity, people are more inclined to pet keeping or outdoor recreation, predisposing themselves to chlamydiosis. The availability of new molecular biological techniques have made it possible for veterinarians and laboratory technicians to carry out diagnostic work without the risk of exposure to zoonotic infections, however, human and animal chlamydiosis is generally overlooked due to lack of awareness and indistinct clinical picture in both developed and developing countries. Therefore, the individuals at the high-end risk due to their professional or pastime activities must be sensitized to this malady and must be tested frequently for timely therapeutic intervention. Further in India and other developing countries, scientific research must be given the impetus in this specialized area to ascertain the epidemiology and genetic diversity of natively prevalent chlamydial species/strains. This will help to develop indigenous diagnostics, control and preventive measures according to local needs.

References

1. Aitken ID. 2000. Chlamydial abortion. In: Martin WB and Aitken ID (ed) *Diseases of sheep* 3rd edn, Blackwell Scientific Ltd., Oxford, UK. p. 81-86.

2. Andersen AA and Vanrompay D. 2000. Avian chlamydiosis. *Rev Sci Tech* 19: 396-404.

3. Andersen AA and Vanrompay D. 2003. Avian chlamydiosis (psittacosis, ornithosis). In: Saif YM, Barnes HJ, Glisson JR, Fadly AM, McDougald LR and Swayne DE (ed.). *Diseases of poultry*, 11th ed. Iowa State Press, A Blackwell Publishing Company, Ames Iowa.

4. Berger L, Volp K, Mathews S, Speare R and Timms P. 1999. *Chlamydia pneumoniae* in a free-ranging giant barred frog (*Mixophyes iteratus*) from Australia. *J Clin Microbiol* 37: 2378-2380.

5. Bhardwaj B. 2011. Molecular characterization of different chlamydiae associated with the infections of ruminants. *The MVSc thesis*, Faculty of Veterinary and Animal Sciences, C.S.K. Himachal Pradesh Krishi Vishvavidyalaya, Palampur.

6. Browning GF. 2004. Is *Chlamydophila felis* a significant zoonotic pathogen? *Australian Vet J* 82: 695-696.

7. Chahota R. 2007. Molecular characterization of the *Chlamydophila caviae* strain OK135, isolated from human genital tract infection, by analysis of some structural and functional genes. In: Pathogenesis of chlamydial infections. *PhD Thesis*, United Graduate School of Veterinary Science, Gifu University, Japan.

8. Chahota R and Katoch RC. 1997. Screening of rats, wall lizards and frogs for the prevalence of *Chlamydia psittaci*. *Indian J Anim Sci* 67: 489-490.

9. Chahota R, Katoch RC and Batta MK. 1997. Isolation of *Chlamydia psittaci* from domestic poultry (*Gallus gallus*) in Himachal Pradesh, India. *Indian J Poult Sci* 32(2): 115-118.

10. Chahota R, Katoch RC, Singh SP, Verma S and Mahajan A. 2000. Concurrent outbreak of chlamydiosis and aflatoxicosis among chickens in Himachal Pradesh, India. *Veterinarski Arhiv* 70(4): 207-213.

11. Chahota R, Katoch RC, Verma S, Sharma M, Deswal RS, Asrani RK, Katoch R and Mahajan A. 2001. Aetio-pathology of respiratory distress syndrome (RDS) among equines in Himachal Pradesh. *Indian J Vet Path* 25: 38-40.

12. Chahota R, Ogawa H, Mitsuhashi Y, Ohya K, Yamaguchi T, and Fukushi H. 2006. Genetic diversity and epizootiology of *Chlamydophila psittaci* prevalent among the captive and feral avian species based on VD2 region of *ompA* gene. *Microbiol Immunol* 50: 663-678.

13. Dhingra PN, Agarwal LP, Mahajan VM and Adalakha SC. 1994. Isolation of chlamydia from pneumonic lungs of buffaloes, cattle and sheep. *Zentralblatt fur Veterinarmedizin* 27B(8): 680-682.

14. Everett KD. 2000. *Chlamydia* and *Chlamydiales*: more than meets the eye. *Vet Microbiol* 75: 109-126.

15. Everett KD, Bush RM and Andersen AA. 1999a. Emended description of the order *Chlamydiales*, proposal of *Parachlamydiaceae* fam. nov. and *Simkaniaceae* fam. nov., each containing one monotypic genus, revised taxonomy of the family *Chlamydiaceae*, including a new genus and five new species, and standards for the identification of organisms. *Int J Syst Bacteriol* 49 Pt 2: 415-440.

16. Everett KD, Hornung LJ and Andersen AA. 1999b. Rapid detection of the *Chlamydiaceae* and other families in the order *Chlamydiales*: three PCR tests. *J Clin Microbiol* 37: 575-580.

17. Fukushi H, and Hirai K. 1992. Proposal of *Chlamydia pecorum* sp. nov. for *Chlamydia* strains derived from ruminants. *Int J Syst Bacteriol* 42: 306-308.

18. Glassick T, Giffard P and Timms P. 1996. Outer membrane protein 2 gene sequences indicate that *Chlamydia pecorum* and *Chlamydia pneumoniae* cause infections in koalas. *Syst Appl Microbiol* 19: 457-464.

19. Gordon FB, Weiss E, Quan AL and Dressler HR. 1966. Observations on guinea pig inclusion conjunctivitis agent. *J Infect Dis* 116: 203-207.

20. Grayston JT, Kuo CC, Campbell LA and Wang SP. 1989. *Chlamydia pneumoniae* sp. nov. for *Chlamydia* sp. strain TWAR. *Int J Syst Bacteriol* 39: 88-90.

21. Halberstaedter L and von Prowazek S. 1907a. UÈ ber ZelleinschluÈ sse parasitaÈ rer Natur beim Trachom. *Arbeiten aus dem Kaiserlichen Gesundheitsamte* 26: 44-47.

22. Halberstaedter L and von Prowazek S. 1907b. Zur Aetiologie des Trachoms. *Deutsche Medizinische Wochens-chrift* 33: 1285-1287.

23. Harris RL and Williams TW Jr. 1985. Contribution to the question of Pneumotyphus: a discussion of the original article by J. Ritter in 1880. *Rev Infect Dis* 7: 119-122.

24. Hartley JC, Stevenson S, Robinson AJ, Littlewood JD, Carder C, Cartledge J, Clark C and Ridgway GL. 2001. Conjunctivitis due to *Chlamydophila felis* (*Chlamydia psittaci* feline pneumonitis agent) acquired from a cat: case report with molecular characterization of isolates from the patient and cat. *J Infect* 43: 7-11.

25. Jain SK, Rajya BS, Mohanty GC, Paliwal OP, Mehrotra ML and Saha RL. 1975. Pathology of chlamydial abortions in ovine and caprine. *Curr Sci* 44: 209-210.

26. Kaleta EF and Taday EM. 2003. Avian host range of *Chlamydophila* spp. based on isolation, antigen detection and serology. *Avian Pathol* 32: 435-462.

27. Kaltenboeck B, Kousoulas KG and Storz J. 1991. Detection and strain differentiation of *Chlamydia psittaci* mediated by a two-step polymerase chain reaction. *J Clin Microbiol* 29: 1969-1975.

28. Katoch RC, Chahota R, Sharma M and Nagal KB. 2002. Chlamydiosis in livestock of H.P. In: Kapoor AC (ed). *Proceedings of International Symposium on Sustainable Agriculture in Hill Areas*, Palampur, 29-30 October 1998. p. 225-232.

29. Krishna L. 1988. Chlamydial conjunctivitis in Jersey calves. *Indian J Anim Sci* 58: 780.

30. Longbottom D and Coulter LJ. 2003. Animal chlamydiosis and zoonotic implications. _J Comp Pathol_ 128: 217-244.

31. Mahajan A, Katoch RC, Chahota R, Verma S and Manuja S. 2002. Concurrent outbreak of infectious disease (IBD), aflatoxicosis and secondary microbial infection in broiler chicks. _Veterinarski Arhiv_ 72(2): 81-90.

32. Mittal A, Kapur S and Gupta S. 1995. Infertility due to _Chlamydia trachomatis_ infection: What is the appropriate site for obtaining samples? _Genitourin Med_ 71: 267-269.

33. Moore FM, McMillan MC and Petrak ML. 1991. Comparison of culture, peroxidase-antiperoxidase reaction, and serum latex agglutination methods for diagnosis of chlamydiosis in pet birds. _J Am Vet Med Assoc_ 199: 71-73.

34. Morange A. 1895. De la psittacose, ou infection spéciale déterminée par des perruches. _PhD Thesis_, Academie de Paris, Paris, France.

35. Moslehi R, Devesa S, Schairer C and Fraumeni FJ. 2006. Rapidly increasing incidence of ocular non-Hodgkin lymphoma. _J Natl Cancer Inst_ 98: 936-939.

36. Moulder JW, Hatch TP, Kao CC, Schachter J and Storz J. 1984. Genus _Chlamydia_. In: N. R. Krieg (ed.) _Bergey's Manual of Systematic Bacteriology_, vol. 1. Baltimore: Williams and Wilkins. p. 729-739.

37. OIE. 2011a. Chapter 2.3.1 Enzootic Abortion of ewes (ovine chlamydiosis). In: Manual of diagnostic tests and vaccines for terrestrial animals, 2011. World Organisation for Animal Health, Paris. http://www.oie.int/fileadmin/Home/eng/Health_standards/tahm/2.07.07_ENZ_ABOR.pdf.

38. OIE. 2011b. Chapter 2.3.1 Avian Chlamydiosis. In: Manual of diagnostic tests and vaccines for terrestrial animals, 2011. World Organisation for Animal Health, Paris. http://www.oie.int/fileadmin/Home/eng/Health_standards/tahm/2.03.01_AVIAN_CHLAMYD.pdf.

39. OIE. 2011c. World Animal Health Information Database, World Organisation for Animal Health, Paris. http://web.oie.int/wahis/public.php?page=country_zoonoses. Accessed 5 December 2011.

40. Pandey A, Chaudhry R, Kapoor L and Kabra SK. 2005. Acute lower respiratory tract infection due to _Chlamydia_ species in children under five years of age. _Indian J Chest Dis Allied Sci_ 47: 97-101.

41. Rattan S, Katoch RC, Sharma M, Dhar P, Chahota R and Singh M. 2005. Isolation of _Chlamydia psittaci_ from _Cervidae_ in north-western Himalayan region. _Indian J Comp Microbiol Infect Dis_ 26: 103-104.

42. Ritter J. 1880. Beitrag zur Frage des Pneumotyphus (Eine Hausepidemie in Uster [Schweiz] betreffend). Deutsches Archiv für Klinische Medizin 25: 53-96.

43. Sachse K, Hotzel H, Slickers P, Ellinger T, Ehricht R. 2005. DNA microarray-based detection and identication of _Chlamydia_ and _Chlamydophila_ spp. _Mol Cell Probes_ 19: 41-50.

44. Sachse K, Vretou E, Livingstone M, Borel N, Pospischil A and Longbottom D. 2009. Recent developments in the laboratory diagnosis of chlamydial infections. *Vet Microbiol* 135: 2-21.

45. Seaman JT. 1985. Chlamydia isolated from abortion in sheep. *Australian Vet J* 62: 436.

46. Shewen PE. 1980. Chlamydial infections in animals: a review. *Can Vet J* 21: 2-11.

47. Smith KA, Campbell CT, Murphy J, Stobierski MG and Tengelsen LA. 2010. Compendium of measures to control *Chlamydophila psittaci* infection among humans (psittacosis) and pet birds (Avian Chlamydiosis). http://www.nasphv.org/Documents/Psittacosis.pdf.

48. Stephens RS. 1999. *Chlamydia: intracellular biology, pathology, and immunity*. American Society for Microbiology, Washington, DC.

49. Storz J. 1971. *Chlamydia and chlamydia-induced diseases*. Charles C Thomas Publisher Limited, Springfield.

50. Telfer BL, Moberley SA, Hort KP, Branley JM, Dwyer DE, Muscatello DJ, Correll PK, England J and McAnulty JM. 2005. Probable psittacosis outbreak linked to wild birds. *Emerg Infect Dis* 11: 391-397.

51. Vats V, Rastogi S, Kumar A, Ahmed M, Singh V, Mittal A, Jain RK and Singh J. 2004. Detection of *Chlamydia trachomatis* by polymerase chain reaction in male patients with non-gonococcal urethritis attending an STD clinic. *Sex Transm Infect* 80: 327-328.

52. Verminnen K, Duquenne B, Keukeleire D, Duim B, Pannekoek Y, Braeckman L, Vanrompay D. 2008. Evaluation of a *Chlamydophila psittaci* diagnostic platform for zoonotic risk assessment. *J Clin Microbiol* 46: 281-285.

53. Wills JM, Watson G, Lusher M, Mair TS, Wood D and Richmond SJ. 1990. Characterisation of *Chlamydia psittaci* isolated from a horse. *Vet Microbiol* 24: 11-19.

54. Winter AC and Charnley JG. 1999. Zoonoses. In: *The sheep keeper's veterinary handbook*, The Crowood Press Ltd, Ramsbury, Marlborough, UK. p. 199-202.

55. Yan C, Fukushi H, Matsudate H, Ishihara K, Yasuda K, Kitagawa H, Yamaguchi T and Hirai K. 2000. Seroepidemiological investigation of feline chlamydiosis in cats and humans in Japan. *Microbiol Immunol* 44: 155-160.

56. Yung AP and Grayson ML. 1988. Psittacosis- a review of 135 cases. *Med J Aust* 148: 228-233.

2014, Zoonoses: Bacterial Diseases   *Pages* **264–280**
Editor: **Sudhi Ranjan Garg**
Published by: **DAYA PUBLISHING HOUSE, NEW DELHI**

12

Erysipeloid

Kuldeep Dhama[1], Mohd. Yaqoob Wani[2], Ruchi Tiwari[3]
and Senthilkumar Natesan[4]
[1]*Division of Pathology,* [2]*Division of Veterinary Biotechnology,*
Indian Veterinary Research Institute, Izatnagar – 243 122
[3]*Department of Microbiology and Immunology,*
College of Veterinary Sciences, Pandit Deen Dayal Upadhyaya
Veterinary University, Mathura – 281 001
[4]*Institute of Human Virology and Department of Medicine,*
University of Maryland School of Medicine, Baltimore, MD 21201, USA

Erysipeloid, caused by a bacterium *Erysipelothrix rhusiopathiae*, is a destructive contagious disease of a variety of animals. The microorganism was first established as a human pathogen late in the nineteenth century. It is of ubiquitous nature with almost worldwide distribution and is able to persist for a long period of time in the environment, including marine locations. It remains in a wide variety of wild and domestic animals, birds and fish as a pathogen or as a commensal, particularly associated with respiratory, digestive and lymphatic systems (Wood and Henderson 2006). The disease was reported in India in 1945 and it has been encountered in many central, southern and eastern states.

The disease in pigs is commonly called swine erysipelas. The disease in acute form in pigs causes sudden death with high mortality rates and is generally characterized by diamond-shaped skin lesions and sometimes with arthritis and endocarditis (Wood and Henderson 2006, Veraldi *et al.* 2009). The disease is of zoonotic importance and in humans three different forms of the disease have been recognized.

These include a localized cutaneous lesion form called erysipeloid (so called to distinguish it from the human streptococcal disease erysipelas), a generalized cutaneous form, and a septicaemic form which is often associated with endocarditis (Wang *et al.* 2010). Erysipeloid is the most common form in humans and in the past was named Rosenbach's disease, Baker-Rosenbach disease or pseudoerysipelas.

The chronic form of the disease in swine causes arthritis which resembles rheumatoid arthritis in many respects, making erysipelas polyarthritis an interesting model for studying the immunopathological mechanisms of chronic inflammation of joints. The chronic form also causes endocarditis in swine. Although bacteremia and endocarditis are relatively uncommon, these types of diseases appear to be increasing in incidence in human and animals (Brooke and Riley 1999, Edwards *et al.* 2009, Harada *et al.* 2011). Diseases in other animals include erysipelas of farmed turkeys, chickens, ducks and emus, and polyarthritis in sheep and lambs (Griffiths *et al.* 1991, Wood and Henderson 2006). Morbidity and mortality of the disease vary greatly from one region to another, perhaps due to differences in the virulence of the etiologic agent. In humans, the infections by the organism possibly remain under diagnosed due to the resemblance it bears to other infections, and the problems encountered in isolation and identification (Wang *et al.* 2010).

Geographic Distribution

E. rhusiopathiae is distributed in all continents among many species of domestic and wild mammals and birds. It is said that over 30 species of wild birds and at least 50 species of wild mammals harbour *E. rhusiopathiae*. In humans, it does not cause a notifiable disease and little is known of its incidence, mostly it occurs as an occupational disease (Veraldi *et al.* 2009, Wang *et al.* 2010). In the former Soviet Union, nearly 3,000 cases were reported between 1956 and 1958 in 13 slaughterhouses in Ukraine, and 154 cases were reported in Tula region in 1959. From 1961 to 1970, the Centers for Disease Control and Prevention confirmed the diagnosis of 15 cases in the USA. Few isolated cases have occurred in Latin America. Some epidemic outbreaks have occurred in the former Soviet Union, in the United States, and on the southern Baltic coast.

Among animals, swine and turkey are the most important affected species. The disease in swine is important in Asia, Canada, Europe, Mexico and the United States and has also been confirmed at low incidence levels in Brazil, Chile, Guatemala, Guyana, Jamaica, Peru and Suriname (Brooke and Riley 1999, Wood and Henderson 2006, Kurian *et al.* 2011, Coutinho *et al.* 2011). In India, the disease has wide range occurrence in Uttar Pradesh and Punjab in north, Andhra Pradesh in south and West Bengal in the east. In the USA, 19 of the 22 serotypes have been found and the most frequent are serotypes 1, 2, 5, 6, and 21 (Wood and Shuman 1981). In a survey in Japan, the organism was found in the tonsils of 10 per cent healthy slaughter pigs: 54 per cent were serotype 7, 32 per cent serotype 2, 9.5 per cent serotype 6 and 1.6 per cent each of serotypes 11, 12, and 16. Serotypes 1 (subtypes 1a and Ib), 2, 5, 6, and 4 have been reported in Puerto Rico (Hassanen *et al.* 2003). Serotypes 1a or 2 have been found most commonly in pigs in Australia, less commonly in sheep and infrequently in other animals (Eamens *et al.* 1988).

All serotype 2 isolates have been reported highly virulent to pigs, while the other serotypes as only weakly virulent. Members of the other nonvirulent or weakly virulent groups, mainly serotype 7 strains, are considered to be resident in porcine tonsils. The Australian field isolates of *E. rhusiopathiae* and *E. tonsillarum* indicate widespread genetic diversity. Those recovered from sheep or birds were more diverse than those isolated from pigs, and isolates of serovar 1 were more diverse than those of serovar 2.

Etiological Agent

Erysipelothrix rhusiopathiae is a facultative, non-spore forming, non-acid-fast, non-motile, small, Gram-positive bacillus (Brooke and Riley 1999, Veraldi *et al.* 2009). Its genome (1,787,941 bp) is one of the smallest genomes in the phylum *Firmicutes*. It contains 1,704 CDSs, of which 1,332 CDSs were assigned with biological functions with seven recognizable pseudogenes (Ogawa *et al.* 2011). Recent analysis based on genomic phylogenetic tree using the 31 universal genes that are conserved among all *E. rhusiopathiae* and other *Erysipelotrichia* strains, with the exception of *Turicibacter sanguinis*, showed that they are distinct from other *Firmicutes* species and placed at the position closest to *Mollicutes* (Ogawa *et al.* 2011). The microorganism is negative for catalase, oxidase, methyl red, indole, esculin, nitrate reduction, Voges-Proskauer and liquefaction of gelatin (Ewald 1981, Veraldi *et al.* 2009). It produces acid from many commonly used sugars.

E. rhusiopathiae has demonstrated a great serological, biochemical and antigenic variation between strains. The isolates have been serotyped into serotypes 1 to 23 and type N (Takahashi *et al.* 1987, 1992). Using molecular tools like DNA base composition and DNA-DNA homology, avirulent strains predominantly from pig tonsils have been found to be genetically distinct from *E. rhusiopathiae*. These strains formed the basis of a new species, *E. tonsillarum*. The genus *Erysipelothrix*, therefore, now contains two accepted species, *E. rhusiopathiae* which include serotypes 1a, 1b, 2, 4, 5, 6, 8, 9, 11, 12, 15, 16, 17, 19, 21 and N; and *E. tonsillarum* containing serotypes 3, 7, 10, 14, 20, 22 and 23 (Takahashi *et al.* 1987, 1992, Wang *et al.* 2010). Two unclassified *Erysipelothrix* groups representing serotypes 13 and 18 are also recognized (Takahashi *et al.* 1992). Previously *E. tonsillarum* was considered morphologically and biochemically identical to *E. rhusiopathiae*, but recently it was shown that *E. tonsillarum* could ferment sucrose while *E. rhusiopathiae* could not.

Virulence Factors

E. rhusiopathiae strains are known to vary considerably in virulence and very little is known about its mechanisms of pathogenicity as no toxin has been associated with this organism. Also, no specific virulence factors have been found but the application of modern genetic methods has provided new insights to its pathogenesis and it has been revealed that it has a capsule which plays an important role with regard to its virulence by protecting it from phagocytosis (Shimoji 2000). The other virulence factors which play accessory roles are enzymes like neuraminidase and hyaluronidase, adhesion and other surface proteins such as SpaA.

It has been suggested that neuraminidase of *E. rhusiopathiae* plays an important role in bacterial attachment and subsequent invasion into host cells (Wang *et al.* 2005, Abrashev and Orozova 2006). The role of hyaluronidase in the pathogenesis of *E. rhusiopathiae* is still controversial. Shimoji *et al.* (2002) observed that most of the experimentally constructed hyaluronidase-negative mutants were avirulent for mice with concurrent lost capsule, whereas a virulent hyaluronidase-negative mutant still possessed the capsule. They suggested that the lack of virulence of most of the hyaluronidase-negative mutants could be attributed to loss of the capsule and not due to hyaluronidase. Lachmann and Deicher (1986) demonstrated that *E. rhusiopathiae* has heat-labile capsule which is now considered as a crucial factor in the pathogenesis of infection. The virulent parent strain resisted phagocytosis by murine polymorphonuclear leukocytes (PMNs), whereas all the mutants were susceptible to phagocytosis.

The other virulence factors which play conjunctional role in the pathogenesis of *Erysipelothrix* infections include adhesions proteins and some surface proteins. The two adhesive surface proteins that have been identified and cloned from *E. rhusiopathiae* are RspA and RspB (Shimoji *et al.* 2003). Participation of both proteins in early adherence to an inert surface and binding to extracellular matrix suggests that they may play an important role in the virulence of disease. Several other surface proteins which are located on the cell surface of the organism like P64 (64 and 66 kDa) and P43 (43 kDa) proteins, obtained by alkaline treatment of whole cells, are regarded as associated virulence factors. Another surface protein is the SpaA antigen (Makino *et al.* 1998).

Host Range

E. rhusiopathiae is ubiquitous and encountered wherever nitrogenous substances decompose (Klauder 1938, Wang *et al.* 2010). The microorganism and the infections caused by it are worldwide in distribution and affect a wide variety of vertebrate and invertebrate species, including swine, sheep, cattle, horses, dogs, cats, rodents, seals, bears, kangaroos, reindeer, mice, sea lions, cetaceans, mink, chipmunks, crustaceans, fresh and salt water fish, crocodiles, caymen, stable flies, houseflies, ticks, mites, mouse lice, turkeys, chickens, ducks, geese, guinea fowl, pigeons, sparrows, starlings, eagles, parrots, pheasants, peacocks, quail, parakeets, mud hens, canaries, finches, siskins, thrushes, blackbirds, bottlenose dolphin, turtledoves and white storks (Brooke and Riley 1999, Eriksson *et al.* 2009, Wang *et al.* 2010, Lee *et al.* 2011). Human disease can originate from an animal or environmental source (Harada *et al.* 2011, Werner *et al.* 2011).

Transmission

Domestic pig is the most important reservoir of *E. rhusiopathiae*. It is estimated that 30-50 per cent of healthy swine harbour the organism in their tonsils and other lymphoid tissues (Stephenson and Berman 1978, Wood and Henderson 2006). Soil, food, water and bedding contamination occurs from the faeces, urine, saliva, nasal and other secretions of the affected and carrier pigs, leading to indirect transmission of the organism (Wood and Handerson 2006). Young pigs coming in contact with carrier sows rapidly acquire the status of carriers and act as shedders. The organism can survive in faeces for several months, however, its persistence in soil is variable and may be governed by many factors including temperature, pH and the presence of other bacteria. It can survive for 60 months in frozen or refrigerated media, 4 months in flesh and 90 days in highly alkaline soil. The organism can be isolated from the effluent of commercial piggeries and from the soil and pasture of effluent disposal sites for up to 2 weeks after application of the effluent containing the organism. It has been found to survive for long periods in marine environments by growing on the exterior mucoid slime of fish without causing disease. *E. rhusiopathiae* is highly susceptible to moist heat and is killed at 55°C for 15 minutes. However, it is resistant to many food preservation methods, such as salting, pickling and smoking (Wang *et al.* 2010). Other sources of infection include infected animals of other species, and birds. It has been proposed that poultry red mite *D. gallinae* may act as a reservoir and possibly as a vector for *E. rhusiopathiae* (Eriksson *et al.* 2009, Eriksson 2010). It has been isolated from a horse affected with vegetative endocarditis and is postulated that cattle may harbour strains that are pathogenic for swine.

Erysipeloid in humans is an occupational hazard. It occurs mostly in those people whose jobs are closely related with contaminated animals, their products or wastes, or soil. The disease has been found mainly in butchers, abattoir workers, animal breeders, veterinarians, farmers, furriers, fishermen, fish handlers, housewives, cooks and grocers (Veraldi *et al.* 1992, Wang *et al.* 2010, Harada *et al.* 2011). Veterinarians in particular are exposed to the infection while vaccinating with virulent culture. It usually produces a swollen finger and is initiated either by an injury to the skin with infective material or when a previous injury is contaminated. Most cases in humans and other animals may occur via scratches or puncture wounds of the skin (Veraldi *et al.* 1992, 2009, Joo *et al.* 2011).

Risk Factors

There is considerable variation in the severity and the ease with which the disease can be reproduced. Many factors affect the disease outcome including age, health and intercurrent diseases, heredity and immune status of the animal. Pigs of all ages are susceptible, particularly those which have recently farrowed (Wood and Henderson 2006, Eriksson *et al.* 2009). Piglets from an immune sow may get sufficient antibodies in the colostrums to give them immunity for some weeks. It is likely that the animals are immune to the strains that are normally found in their particular environment as it is thought that 30-50 per cent of pigs may carry the organism in the tonsil (Wood and Henderson 2006, Wang *et al.* 2010). Sudden diet changes such as accidental access to tankage or to a field of corn, feeding new corn, or placing pigs on new pasture and the environmental factors like exposure to extremes of heat and cold also predisposes to the disease by activating the latent infections (Wood and Henderson 2006).

Although the factors that lead to stress predispose to the condition, virulence of the strain is probably the most important factor for final disease outcome. Serotypes 1 and 2 are the most common types isolated from swine affected with clinical erysipelas and are generally believed to be the only serotypes that cause the acute disease (Shimoji 2000, Eriksson *et al.* 2009). The other serotypes are relatively uncommon and none of them has yet been a cause of acute epidemics, but some have been isolated from lesions of chronic erysipelas. Serotypes 1a, 3, 5, 6, 8, 11, 21, and type N have been isolated from pigs with chronic erysipelas, mainly arthritis and lymphadenitis (Takahash *et al.* 1987, 1992). Smooth strains can be used successfully to produce the disease experimentally but rough strains are non-pathogenic. This variation in virulence between strains of the organism has been utilized in the production of living, avirulent vaccines.

Pathogenesis

E. rhusiopathiae is generally regarded as an opportunistic animal pathogen and the routes of infection are believed to be digestive and cutaneous through abrasions and wounds (Wang *et al.* 2005). The long survival of the agent in the environment ensures endemism in affected areas. The invasion of *E. rhusiopathiae* in pig can occur under particular circumstances like hot and humid weather conditions. The microorganism is best known as a facultative intracellular pathogen and can survive inside polymorphonuclear leukocytes and macrophages after its phagocytosis (Shimoji *et al.* 1996). The presence of capsule by pathogenic serotypes has the ability to resists phagocytosis and is regarded as main pathogenic factor (Yamazaki *et al.* 1999). Genome analysis revealed that *E. rhusiopathiae* possesses nine antioxidant factors and nine phospholipases which facilitate intracellular survival in phagocytes. The antioxidant factors include a superoxide dismutase, two thioredoxins, two thioredoxin reductases, a thiol peroxidase and a glutaredoxin which potentially confer resistance from reactive oxidative metabolites (Ogawa *et al.* 2011). However, the precise mechanisms for intracellular survival of the organism are still unknown (Shimoji 2000, 2004).

The virulence factors like neuraminidase cleave the mucopolysaccharides and cause vascular damage leading to haemorrhage and thrombosis which subsequently ensures the spread of the bacterium. Invasion of the bloodstream occurs in all infected animals in the first instance. Septicemia generally results within 1-7 days. The subsequent development of either an acute septicemia or a bacteremia with localization in organs and joints as chronic form is dependent on undetermined factors. Concurrent viral infection, especially hog cholera in pig, may increase susceptibility of the host. Localization in the chronic form is commonly in the skin and joints, and on heart valves, with probable subsequent bacteremic episodes. The heart lesions may begin with early inflammatory changes associated with emboli. Selective adherence of some strains to heart valves may be a factor in the pathogenesis of endocarditis (Bratberg 1981, Harada *et al.* 2011). This attribute may be due to adhesion proteins. In joints, the initial lesion is an increase in synovial fluid and hyperemia of the synovial membrane that occurs after several weeks by the proliferation of synovial villi, followed by thickening of the joint capsule and enlargement of the local lymph nodes. Diskospondylitis also occurs in association with chronic polyarthritis due to erysipelas. Amyloidosis may occur in pigs with chronic erysipelas polyarthritis (Pedersen *et al.* 1984, Wood and Henderson 2006). The microscopic lesions include vasculitis in capillaries and venules in many sites, including glomeruli, pulmonary capillaries, and the skin. Sometimes, it is possible to see emboli of bacteria without specific stains to demonstrate bacteria.

Clinical Manifestations in Animals

The disease manifests almost in a similar way in different host animal species. Erysipelas, the skin lesions and polyarthritis are typical forms of infection in animals. Swine are the most affected animals where it is seen in three forms: acute, subacute and chronic (Wood and Henderson 2006). Sometimes hyperacute stages are also found. Hyperacute form occurs in pigs approaching market weight and animal is found dead without showing other symptoms. Morbidity and mortality vary from one region to another, perhaps due to differences in the virulence of the etiologic agent and the immune status of the animal. The incubation period lasts from one to seven days.

The acute form begins suddenly with a high fever (up to 42°C). Some animals suffer from prostration, anorexia and vomiting, while others continue to feed despite the high fever (Wood and Henderson 2006). Skin lesions are almost pathognomonic but may not always be apparent (Brooke and Riley 1999, Shimoji 2000). These may take the classical diamond shape and may be red, urticarial plaques about 2.5-5 cm square or a more diffuse edematous eruption with the same appearance. The lesions are most common on the belly, inside the thighs and on the throat, neck, and ears, and usually appear about 24 hours after the initial signs of illness. Sometimes they can be felt rather than seen. The mortality rate may reach 75 per cent but wide variation occurs. Pregnant animals may sometime abort and it is thought that this is due to the fever but it may be that there is a direct fetal action, as congenital infections and isolations of the organism from the fetus have occurred. Conjunctivitis with ocular discharge may be present. There is splenomegaly and swelling of the lymph nodes. In the final phase of septicemic erysipelas, dyspnea and diarrhoea are the most common and obvious symptoms. The disease has a rapid course and mortality is usually very high (Timoney *et al.* 1988, Brooke and Riley 1999, Harada *et al.* 2011).

The subacute form is characterized by urticaria, which initially appears as reddish or purple rhomboid-shaped spots on the skin but are less severe than the acute form. The plaques later become necrotic, dry up, and fall off. It has been found that the intensity of skin lesions has a direct relation to the prognosis. Light pink to light purplish red lesions will disappear within several days whereas deep-purplish red lesions can precede either death or necrosis of the skin (Shimoji 2000, 2004).

The chronic form of infection may follow acute or subacute disease and is characterized most commonly by signs of local arthritis or proliferative pathological changes in the heart (endocarditis) (Boo *et al.* 2003, Harada *et al.* 2011). Chronic arthritis results in joints showing various degrees of stiffness and enlargement. This is the most important clinical manifestation of swine erysipelas from an economic standpoint considerable because the animals' development and weight gain are affected and they may be confiscated from the abattoirs. There may be alopecia, sloughing of the tail and tips of the ears, and dermatitis in the form of hyperkeratosis of the skin of the back, shoulders and legs. The affected animals develop a stiff gait and the lesions are commonest in the elbow, hip, hock, stifle, and knee joints

(Shimoji 2000, 2004). The involved joint is usually swollen and the joint capsule is thickened. Animals seldom die from the infection. Paraplegia may occur when intervertebral joints are involved or when there is gross distortion of limb joints. The cardiac impulse is usually markedly increased, the heart rate is faster, and a loud murmur is audible on auscultation if the valves are badly damaged.

E. rhusiopathiae causes polyarthritis in sheep and lambs. It is seen in lambs beginning from 2 to 3 months of age usually after tail docking or sometimes as a result of an umbilical infection. Lesions are only found in the joints and one or several may be affected. The main symptoms are difficulty in movement and stunted growth. Serotype 1b is the most common of the isolates found and less frequent are 1a and 2 (Eamens *et al.* 1988). Other forms of erysipelas in sheep are valvular endocarditis, septicemia, and pneumonia (Griffiths *et al.* 1991). Fthenakis *et al.* (2006) reported *E. rhusiopathiae* associated abortion in ewes.

On rare occasions, infection has been seen in cattle, causing arthritis in calves. The agent has been isolated from the tonsils of healthy adult cows and is said to be the source of infection for swine and human (Sawada *et al.* 2001, Hassanein *et al.* 2003). Wild birds, including turkeys, chickens, geese, ducks, parrots, mud hens, pigeons and quail, have also been reported to be affected (Eriksson *et al.* 2009, Galindo-Cardiel *et al.* 2011). Turkeys are the birds most often and most seriously affected. They exhibit a cyanotic skin that is most obvious as blue comb. Other symptoms include general weakness, diarrhoea, cyanosis, and a reddish-purple swollen comb. The disease tends to attack males in particular. Mortality can vary between 2.5 per cent and 25 per cent. The lesion consists of massive haemorrhages and petechiae in the breast, leg muscles serous membranes, intestine, and gizzard. The birds become droopy, develop diarrhoea and die. There are reports of large outbreaks in ducks and starlings. Disease in chicken flocks has occurred but is rarely reported.

Clinical Manifestations in Humans

Clinical manifestations of the disease in humans closely resemble those in swine (Brooke and Riley 1999). The clinical categories include the localized cutaneous form called true erysipeloid and the diffuse/generalized variety (Boo *et al.* 2003, Veraldi *et al.* 2009). The cutaneous form is known by the name erysipeloid as to distinguish it from erysipelas caused by a hemolytic streptococcus and is the most common form of human infection. Erysipeloid localizes primarily in the hands and fingers and consists of an erythematous, edematous skin lesion with violet coloration around a wound (the inoculation point) that may be a simple abrasion. The palms, forearms, arms, face and legs are rarely involved (Veraldi *et al.* 1992). Arthritis in the finger joints occurs with some frequency. Swelling of the fingers may be severe, leading to the common terms 'whale finger' and 'seal finger'. Localized erysipeloid is characterized clinically by an inflammatory plaque, with well-defined and raised borders. The patient experiences a burning sensation, a pulsating pain, and at times an intense pruritus. The course of the disease is usually benign and the patient recovers in two to four weeks. The absence of suppuration, lack of pitting oedema, and disproportionate pain help distinguish erysipeloid from staphylococcal or streptococcal infection.

The disease is self-limiting and usually resolves in 3-4 weeks without therapy. Chronic and recurrent cases of erysipeloid have rarely been reported. The generalized form of the disease involves lesions that progress from the initial site to other locations on the body or appear at remote areas, however, is much rarer than the cutaneous form. It is characterized clinically by widespread erythematous-oedematous lesions, often accompanied by fever, lymphadenitis, arthralgia and myalgia (Veraldi *et al.* 2009). The principal symptoms include fever, splenomegaly and hematuria. It is possible that these forms occur because of immunosuppressed status of the patient (Meric and Ozcan 2012). More important are the complications of encephalitis, meningitis, endocarditis, renal failure, peritonitis, and sepsis (Joo *et al.* 2011).

Diagnosis

Diagnosis of the disease can be made on clinical signs, based on the patient's occupation and later confirmed by isolation and identification of the etiologic agent. The conventional techniques that are used for diagnosis include isolation, morphological and biochemical tests, protection assay, serologic tests such as agglutination, growth inhibition, passive hemagglutination, complement fixation tests, etc. and the modern techniques include identification by fluorescent antibody test and molecular techniques like PCR and real time PCR (To *et al.* 2009, Pal *et al.* 2010, Wang *et al.* 2010).

Isolation

Liver, spleen, kidney, heart and synovial tissue can be taken from necropsy examination. In septicemic cases, the etiologic agent can be isolated even from the blood by directly culturing on blood and from internal organs. Routine isolation involves the use of sheep or ox blood agar with a MacConkeys plate to aid in detection of any gram negative pathogen or contaminant. In cases of arthritis or skin infections, cultures are made from localized lesions but are many times difficult. Isolations from contaminated materials are accomplished through inoculation of mice, which are very susceptible. Commercially available blood culture media are also satisfactory for primary isolation from blood. A number of selective media for the isolation include Erysipelothrix selective broth (ESB) which is a nutrient broth containing serum, tryptose, kanamycin, neomycin, and vancomycin. Modified blood azide medium (MBA) is a selective agar containing sodium azide and horse blood or serum (Harrington and Hulse 1971); Packer's medium containing sodium azide and crystal violet (Packer 1943); Bohm's medium containing sodium azide, kanamycin, phenol and water blue (Ewald 1981) and Shimoji's selective enrichment broth containing tryptic soy broth, tween 80, tris-aminomethane, crystal violet and sodium azide (Shimoji 2004, Wang *et al.* 2010).

Morphological and Cultural Characters

E. rhusiopathiae is a Gram-positive bacterium. The colonies grown appear as clear, circular and very small, with a diameter of 0.1-0.5 mm after 24 hour incubation at 37°C, or 0.5-1.5 mm after 48 hour. Two distinct morphological forms, which grow on solid agar media, include smooth (S) colonies as bluish, transparent, and convex, and rough (R) colonies as larger having a flat rough surface with irregular edges.

Long filaments up to 60 mm or more, in chains, can be seen under the microscope. Morphology changes appear with changing growth conditions especially with regard to alterations in pH and incubation temperature. A pH of 7.6-8.2 favours the S-form while the R-form predominates at pH less than 7. The S-form grows better at 33°C and the R-form favours 37°C (Jones 1986, Wang *et al.* 2010). Most strains exhibit a narrow zone of alpha haemolysis on blood agar and beta-haemolysis has not been yet observed (Jones 1986).

Biochemical Reactions

Biochemically the members of the genus *Erysipelothrix* are relatively inactive. These are negative for catalase, oxidase, methyl red, indole, esculin, nitrate reduction, Voges-Proskauer and liquefaction of gelatin (Ewald 1981, Brooke and Riley 1999, Wang *et al.* 2010). However, it produces acid from glucose, fructose, galactose and lactose but not from maltose, xylose, and mannitol. Sucrose is fermented by most strains of *E. tonsillarum* but not by *E. rhusiopathiae* and is used as one of the differentiating tests between the species. Hydrogen sulfide (H_2S) is produced by 95 per cent of strains on triple sugar iron (TSI) agar. *E. rhusiopathiae* can be differentiated from other Gram-positive bacilli, in particular, from *Arcanobacterium* (*Corynebacterium*) *pyogenes* and *Arcanobacterium* (*Corynebacterium*) *haemolyticum*, which are beta haemolytic on blood agar and do not produce hydrogen sulfide in TSI agar slants, and from *Listeria monocytogenes*, which is catalase positive, motile and is sensitive to neomycin.

Animal Inoculation and Protection Assays

It is not used routinely but both mice and pigeons will die within 4 days after intraperitoneal inoculation with 0.1-0.4 ml of a broth cultured from a virulent *E. rhusiopathiae* strain. In animal protection assay, the suspected *E. rhusiopathiae* in broth culture is administered subcutaneously to mice along with a dose of commercially available antiserum or the equine hyperimmune *E. rhusiopathiae* antiserum. Control mice that do not receive antiserum die within 5-6 days, but those receiving antiserum are protected (Wang *et al.* 2010).

Serological Methods

Various serological methods for the diagnosis of chronic swine erysipelas, which are based on detection of maternal antibody and acquired antibody before and after vaccination, were reported. The three most important include growth agglutination tests, the latex agglutination test and enzyme-linked immunosorbent assay (ELISA) (Dahms *et al.* 1989, Chin *et al.* 1992, Sato *et al.* 1998). The growth agglutination test requires culture of live pathogenic bacteria, which can be hazardous to laboratory workers and for this reason, now-a-days latex agglutination kits are increasingly being used. On the other hand, ELISA is the test of choice among existing serological procedures because it is simple, permits the testing of large numbers of samples in a short time, and gives precise, objective results. The major protective antigen of *E. rhusiopathiae* used in ELISA is the 64- to 66-kDa protein antigen. Genes encoding the surface protective antigens (Spa) have been successfully cloned, and nucleotide sequences have been also determined in order to produce bulk pure recombinant proteins. Based on Spa proteins, E. rhusiopathiae can be classified into three molecular species: SpaA, SpaB, and SpaC (Shen et al. 2010). The SpaA protein was identified in E. rhusiopathiae serotypes 1a, 1b, 2, 5, 8, 9, 12, 15, 16, 17, and N, the SpaB protein was identified in E. rhusiopathiae serotypes 4, 6, 11, 19, and 21, and the SpaC protein was identified only in serotype 18 (Shen et al. 2010, Wang et al. 2010). It was found that both fully mature protein and the N-terminal 416 amino acids (SpaA416) were having sufficient antigenicities and results indicate that the SpaA416-based ELISA is an effective method not only for evaluating pigs for the presence of protective antibody levels resulting from vaccination or maternal antibody, but also for detecting antibody produced by natural infection and has important potential for the effective control of swine erysipelas (Imada et al. 2003).

Molecular Methods

The polymerase chain reaction (PCR) is a high-power molecular biology technique which permits specific and exponential amplification of a given DNA sequence. Amplification of a single DNA molecule can reach 10^9 copies and thus can be detected. PCR methods have been developed for detecting Erysipelothrix species. The Makino PCR uses primers based on a region of the 16S rRNA gene and is genus-specific (Makino et al. 1994), while the Shimoji PCR is species specific and is designed from sequences associated with virulence of E. rhusiopathiae (Shimoji et al. 1998). The sensitivity of the Makino PCR is less than 20 bacteria, while the Shimoji PCR detects a minimum of 1000 bacteria per reaction mixture. An improved PCR system, based on DNA sequence coding for the rRNA gene cluster including 16S, 23S, and 5S rRNAs and the noncoding region, has been established to distinguish four species of the genus Erysipelothrix (Takeshi et al. 1999). This species specific PCR method may provide a new approach for the rapid diagnosis of Erysipelothrix infection in animals and humans. Recently, quantitative real-time PCR and multiplex PCR methods were also developed for detection of and discrimination between E. rhusiopathiae and other Erysipelothrix species (Yamazaki 2006, To et al. 2009, Pal et al. 2010).

Treatment

Penicillin and anti-erysipelas serum comprise the standard treatment and these are often administered together by dissolving the penicillin in the serum. Most strains are highly susceptible to penicillins, cephalosporins, clindamycin, imipenem, and ciprofloxacin (Takahashi et al. 1984, Venditti et al. 1990, Fidalgo et al. 2002), and resistant at vancomycin. Penicillin alone is usually adequate when the strain has only mild virulence. Standard dose rates give a good response in the field but experimental studies suggest that 50,000 iu/kg body weight of procaine penicillin intramuscularly for 3 days are required for complete chemotherapeutic effect. Most animals are significantly improved within 2 days. However, due to indiscriminate use of antimicrobials, resistant strains are now emerging. Chuma et al. (2010) reported number of strains resistant to oxytetracycline, erythromycin, lincomycin, ofloxacin and enrofloxacin to be 56 (37.6 per cent), 4 (2.7 per cent), 18 (12.1 per cent), 21 (14.1 per cent) and 19 (12.8 per cent), respectively, but all strains were found susceptible to ampicillin.

In human infections, although the lesions of erysipeloid may resolve spontaneously, the time to resolution may be prolonged and relapses may occur and antimicrobial therapy is therefore recommended. Additionally, this may prevent progression to bacteraemia and/or endocarditis (Haddad *et al.* 2009). Given the relative infrequency of *E. rhusiopathiae* infection, it is not surprising that there have been no clinical trials to determine optimal antimicrobial therapy. Chronic cases do not respond well to either treatment because of the structural damage that occurs to the joints and the inaccessibility of the organism in the endocardial lesions.

Prevention and Control

Successful prevention and control of the infection depends on rapid diagnosis, quarantine, good hygiene, biosecurity, reduction of stress and vaccination policy. Some measures which may be followed are listed below:

☆ Persons exposed as a result of their occupations should follow good hygiene, namely frequent hand washing with disinfectant and proper treatment of wounds.

☆ Control of swine erysipelas depends mostly on vaccination which involves the use of two different types of vaccines: a bacterin adsorbed on aluminum hydroxide and a live avirulent vaccine (EVA: erysipelas vaccine avirulent).

☆ As vaccination confers immunity for five to eight months, regular and proper vaccination should be followed. The bacterin is first administered before weaning, followed by another dose two to four weeks later. The avirulent vaccine can be administered orally via drinking water.

☆ In the case of an outbreak of septicemic erysipelas, it is important to destroy the whole carcasses immediately, disinfect the premises, and to treat sick animals with penicillin and the rest of the herd with anti-erysipelas serum.

☆ Rotation of animals to different pastures and environmental hygiene helps in controlling the disease by reducing the burden of etiological agent.

☆ Control of rodent populations is necessary and should be carried out in establishments where foods of animal origin are processed.

☆ As eradication is virtually impossible because of the ubiquitous nature of the organism and its resistance to adverse environmental conditions, it is important to dispose clinically affected animals quickly and all introductions should be isolated and examined for signs of arthritis and endocarditis. However, this procedure will not prevent the introduction of clinically normal carrier animals.

☆ Suckling pigs in herds where the disease is endemic should receive 10 mL of anti-erysipelas serum during the first week of life and at monthly intervals until they are actively vaccinated, which should be done as early as 6 weeks, provided the sows have not been vaccinated.

☆ A routine vaccination prior to farrowing should be followed to provide persisting maternal passive immunity to piglet so as to boost their immune system when they are young with less developed immune axis.

☆ In herds where sows are routinely vaccinated prior to farrowing, vaccination can be delayed until 10-12 weeks of age for effective active immunity.

☆ The use of live-culture vaccines should be limited as there is risk of variation in virulence of the strains used and the possibility of spreading infection.

☆ For turkey-raising establishments, where the infection is endemic, vaccination is necessary like by the use of bacterins. A live vaccine administered orally via drinking water has yielded good results in tests.

Future Directions

Erysipelas of animals is a bacterial zoonotic disease. The causative organism has been found to be associated with a variety of diseases in many species of amphibians, fish, reptiles, birds and mammals. Among the domestic animals, swine and turkey erysipelas are of the major economic importance. While it has been suggested that the incidence of human infection could be declining due to technological advances in animal industries, infection still occurs in specific environments. Additionally, infection by the organism is possibly under-diagnosed due to the resemblance it bears to other infections, and problems encountered in isolation and identification.

The future areas of research are concerned mostly with its pathogenicity and epidemiology. The pathogenicity of organism appears to be related mostly with the intracellular survival properties of the bacterium and probably utilizes several different host cell receptors to gain access to the host intracellular niche. The receptor-ligand interactions involved in entry into macrophages constitute a complex but provocative area of research. A clear understanding of the molecular basis of cell surface molecule(s) of the bacterium involved in entry of the organism into the cells, and to clarify the receptor-ligand interactions that decide the fate of the organism are the unrevealed areas. The serotype classification of the causative organism is based on soluble peptidoglycan antigens of the cell surface. This 64-66 kDa cell surface protein, surface protective antigen A (spaA) has been found to responsible for eliciting highly protective antibodies and is considered to be the major immunogenic antigen of the specie. However, the results from PCR assays and western blotting with a spaA-specific monoclonal antibody demonstrated that spa-type is not confined to specific serotype groups. Also, the serotyping results are often inconsistent among laboratories and without international guidelines. Molecular-based systems can be more efficient and reliable method of organizing *E. rhusiopathiae* strains and can be anticipated in future.

References

1. Abrashev I and Orozova P. 2006. *Erysipelothrix rhusiopathiae* neuraminidase and its role in pathogenicity. *Z Naturforsch C* 61: 434-438.

2. Boo TW, Hone R and Hurley J. 2003. *Erysipelothrix rhusiopathiae* endocarditis: a preventable zoonosis? *Ir J Med Sci* 172: 81-82.

3. Bratberg AM. 1981. Selective adherence of *Erysipelothrix rhusiopathiae* to heart valves of swine investigated in an *in vitro* test. *Acta Vet Scand* 22: 39-45.

4. Brooke CJ and Riley TV. 1999. *Erysipelothrix rhusiopathiae*: bacteriology, epidemiology and clinical manifestations of an occupational pathogen. *J Med Microbiol* 48:789-799.

5. Chin JC, Turner B and Eamens GJ. 1992. Serological assay for swine erysipelas using nitrocellulose particles impregnated with an immunodominant 65-kDa antigen from *Erysipelothrix rhusiopathiae*. *Vet Microbiol* 31: 169-180.

6. Chuma T, Kawamoto T, Shahada F, Fujimoto H and Okamoto K. 2010. Antimicrobial susceptibility of *Erysipelothrix rhusiopathiae* isolated from pigs in Southern Japan with a modified agar dilution method. *J Vet Med Sci* 72: 643-645.

7. Coutinho TA, Moreno AM, Imada Y, Lopez RP and Neto JS. 2011. Characterization of *Erysipelothrix rhusiopathiae* isolated from Brazilian Tayassupecari. *Trop Anim Health Prod* Sep 14. [Epub ahead of print]

8. Dahms H, Schilow WF, and Hagemann G. 1989. Detection of *Erysipelothrix rhusiopathiae* specific antibodies in the serum of experimentally infected swine by ELISA and immunoblotting. *Arch Exp Veterinarmed* 43: 907-916.

9. Eamens GJ, Turner MJ and Catt RE. 1988. Serotypes of *Erysipelothrix rhusiopathiae* in Australian pigs, small ruminants, poultry, and captive wild birds and animals. *Aust Vet J* 65: 249-252.

10. Edwards GT, Schock A and Smith L. 2009. Endocarditis in a British heifer due to *Erysipelothrix rhusiopathiae* infection. *Vet Rec* 165: 28-29.

11. Eriksson H, Jansson SS, Johansson K, verud B, Chirico J and Aspan A. 2009. Characterization of *Erysipelothrix rhusiopathiae* isolates from poultry, pigs, emus, the poultry red mite and other animals. *Vet Microbiol* 137: 98-104.

12. Eriksson H. 2010. Characterization of *Erysipelothrix rhusiopathiae* isolates from laying hens and poultry red mites (*Dermanyssus gallinae*) from an outbreak of erysipelas. *Avian Path* 39: 505-509.

13. Ewald FW. 1981. The genus *Erysipelothrix*. In: Starr MP, Stolp H, Truper HG, Balows A, Schlegel HG. (Eds.) *The Prokaryotes: A handbook on habitats, isolation, and identification of bacteria*. Spring-Verlag, New York. p. 1688-1700.

14. Fidalgo S, Longbottom C and Riley T. 2002. Susceptibility of *Erysipelothrix rhusiopathiae* to antimicrobial agents and home disinfectants. *Pathol* 34: 462-465.

15. Fthenakis GC, Christodoulopoulos G, Leontides L, Tzora A. 2006. Abortion in ewes associated with *Erysipelothrix rhusiopathiae*. *Small Ruminant Res* 63: 183-188

16. Galindo-Cardiel I, Opriessnig T, Molina L and Juan-Salles C. 2011. Outbreak of mortality in Psittacine birds in a mixed species aviary associated with *Erysipelothrix rhusiopathiae* infection. *Vet Pathol* Aug 30. [Epub ahead of print]

17. Griffiths IB, Done SH and Readman S. 1991. *Erysipelothrix* pneumonia in sheep. *Vet Rec* 128: 382-383.

18. Haddad V, Lupi O, Lonza J and Tyring S. 2009. Tropical dermatology: marine and aquatic dermatology. *J Am Acad Dermatol* 61: 733-750.

19. Harada K, Amano K, Akimoto S, Yamamoto K, Yamamoto Y, Yanagihara K, Kohno S, Kishida N and Takahashi T. 2011. Serological and pathogenic characterization of *Erysipelothrix rhusiopathiae* isolates from two human cases of endocarditis in Japan. *New Microbiol* 34: 409-412.

20. Harrington Jr R and Hulse DC. 1971. Comparison of two plating media for the isolation of *Erysipelothrix rhusiopathiae* from enrichment broth culture. *Appl Microbiol* 22: 141-142.

21. Hassanein R, Sawada T, Kataoka Y, Gadallah A, Suzuki Y, Takagi M and Yamamoto K. 2003. Pathogenicity for mice and swine of *Erysipelothrix* isolates from the tonsils of healthy cattle. *Vet Microbiol* 91: 231-238.

22. Imada Y, Mori Y, Daizoh M, Kudoh K and Sakano T. 2003. Enzyme-linked immunosorbent assay employing a recombinant antigen for detection of protective antibody against swine erysipelas. *J Clin Microbiol* 41: 5015-5021.

23. Jones D. 1986. Genus *Erysipelothrix rhusiopathiae*. In: Sneath PH, Mair NS and Sharpe ME. (Eds.) *Bergey's Manual of Systematic Bacteriology*, vol. 2. Williams and Wilkins, Baltimore. p. 1245-1249.

24. Joo EJ, Kang CI, Kim WS, Lee NY, Chung DR, Peck KR and Song JH. 2011. Acute meningitis as an initial manifestation of *Erysipelothrix rhusiopathiae* endocarditis. *J Infect Chemother* 17: 703-705.

25. Klauder JV. 1938. Erysipeloid as an occupational disease. *J Am Med Assoc* 111:1345-1348.

26. Kurian A, Neumann E, Hall W and Marks D. 2011. Serological survey of exposure to *Erysipelothrix rhusiopathiae* in poultry in New Zealand. *N Z Vet J* Nov 16. [Epub ahead of print]

27. Lachmann PG and Deicher H. 1986. Solubilization and characterization of surface antigenic components of *Erysipelothrix rhusiopathiae* T28. *Infect Immun* 52: 818-822.

28. Lee JJ, Kim DH, Lim JJ, Kim DG, Chang HH, Lee HJ, Kim SH, Rhee MH, Endale M, Imada Y, Kim OJ and Kim S. 2011. Characterization and identification of *Erysipelothrix rhusiopathiae* isolated from an unnatural host, a cat, with a clinical manifestation of depression. *J Vet Med Sci* 73: 149-154.

29. Makino SI, Okada Y, Maruyama T, Ishikawa K, Takahashi T, Nakamura M, Ezaki T and Morita H. 1994. Direct and rapid detection of *Erysipelothrix rhusiopathiae* DNA in animals by PCR. *J Clin Microbiol* 32: 1526-1531.

30. Makino SI, Yamamoto K, Murakami S, Shirahata T, Uemura K, Sawada T, Wakamoto H, Morita H and Morita Y. 1998. Properties of repeat domain found in a novel protective antigen, SpaA, of *Erysipelothrix rhusiopathiae*. *Microb Pathog* 25: 101-109.

31. Meric M and Ozcan KS. 2012. *Erysipelothrix rhusiopathiae* pneumonia in an immunocompetent patient. *J Med Microbiol* 61: 450-451.

32. Ogawa Y, Ooka T, Shi F, Ogura Y, Nakayama K, Hayashi T and Shimoji Y. 2011. The genome of *Erysipelothrix rhusiopathiae*, the causative agent of swine erysipelas, reveals new insights into the evolution of firmicutes and the organism's intracellular adaptations. *J Bacteriol* 193: 2959-2971.

33. Packer RA. 1943. The use of sodium azide and crystal violet in a selective medium for streptococci and *Erysipelothrix rhusiopathiae*. *J Bacteriol* 46: 343-349.

34. Pal N, Bender JS and Opriessnig T. 2010. Rapid detection and differentiation of *Erysipelothrix* spp. by a novel multiplex real-time PCR assay. *J Appl Microbiol* 108: 1083-1093.

35. Pedersen KB, Henrichsen J and Perch B. 1984. The bacteriology of endocarditis in slaughter pigs. *Acta Pathol Microbiol Immunol Scand* 92: 237.

36. Sato H, Yamazaki Y, Tsuchiya K, Aoyama T, Akaba N, Suzuki T, Yokoyama A, Saito H and Maehara N. 1998. Use of the protective antigen of *Erysipelothrix rhusiopathiae* in the enzyme-linked immunosorbent assay and latex agglutination. *Zentbl Veterinarmed* B 45: 407-420.

37. Sawada T, Hassanein R, Yamamoto T and Yoshida T. 2001. Distribution of antibody against *Erysipelothrix rhusiopathiae* in cattle. *Clin Diagn Lab Immunol* 8: 624-627.

38. Shen HG, Bender JS and Opriessnig T. 2010. Identification of surface protective antigen (spa) types in *Erysipelothrix* reference strains and diagnostic samples by spa multiplex real-time and conventional PCR assays. *J Appl Microbiol* 109: 1227-1233.

39. Shimoji Y. 2000. Pathogenicity of *Erysipelothrix rhusiopathiae*: virulence factors and protective immunity. *Microbes Infect* 2: 965-972.

40. Shimoji Y. 2004. *Erysipelothrix rhusiopathiae*, In: Gyles GL, Prescott JF, Songer JG and Thoen CO. (ed.) *Pathogenesis of bacterial infections in animals*, 3rd ed. Blackwell Publishing, Ames, IA.

41. Shimoji Y, Mori Y, Hyakutake K, Sekizaki T and Yokomizo Y. 1998. Use of an enrichment broth cultivation-PCR combination assay for rapid diagnosis of swine erysipelas. *J Clin Microbiol* 36: 86-89.

42. Shimoji Y, Asato H, Sekizaki T, Mori Y and Yokomizo Y. 2002. Hyaluronidase is not essential for the lethality of *Erysipelothrix rhusiopathiae* infection in mice. *J Vet Med Sci* 64: 173-176.

43. Shimoji Y, Ogawa Y, Osaki M, Kabeya H, Maruyama S, Mikami T and Sekizaki T. 2003. Adhesive surface proteins of *Erysipelothrix rhusiopathiae* bind to polystyrene, fibronectin, and type I and IV collagens. *J Bacteriol* 185: 2739-2748.

44. Stephenson EH and Berman DT. 1978. Isolation of *Erysipelothrix rhusiopathiae* from tonsils of apparently normal swine by two methods. *Am J Vet Res* 39: 187-188.

45. Takahashi T, Sawada T and Ohmae K. 1984, Antibiotic resistance of *Erysipelothrix rhusiopathiae* isolated from pigs with chronic swine erysipelas. *Antimicrob Agents Chemother* 25: 385-386.

46. Takahashi T, Sawada T, Muramatsu M, Tamura Y, Fujisawa T, Benno Y and Mitsuoka T. 1987. Serotype, antimicrobial susceptibility, and pathogenicity of *Erysipelothrix rhusiopathiae* isolates from tonsils of apparently healthy slaughter pigs. *J Clin Microbiol* 25: 536-539.

47. Takahashi T, Fujisawa T, Tamura Y, Suzuki S, Muramatsu M, Sawada T, Benno Y and Mitsuoka T. 1992. DNA relatedness among *Erysipelothrix rhusiopathiae* strains representing all twenty-three serovars and *Erysipelothrix tonsillarum*. *Int J Syst Bacteriol* 42: 469-473.

48. Timoney JF and Yarkoni U. 1976. Immunoglobulins IgG and IgM in synovial fluids of swine with *erysipelothrix* polyarthritis. *Vet Microbiol* 1: 467-474.

49. To H, Koyama T, Nagai S, Tuchiya K and Nunoya T. 2009. Development of quantitative real-time polymerase chain reaction for detection of and discrimination between *Erysipelothrix rhusiopathiae* and other *Erysipelothrix* species. *J Vet Diagn Invest* 21:701-706.

50. Venditti M, Gelfusa V and Tarasi A. 1990. Antimicrobial susceptibilities of *Erysipelothrix rhusiopathia*. *Antimicrob Agents Chemother* 34: 2038-2040.

51. Veraldi S, Rizzitelli G and Schianchi-Veraldi R. 1992. Occupational cutaneous infections. *Clin Dermatol* 10: 225-230.

52. Veraldi, S, Girgenti V, Dassoni F and Gianotti R. 2009. Erysipeloid: a review. *Clin Exp Dermatol* 34: 859-862.

53. Wang Q, Chang BJ, Mee BJ and Riley TV. 2005. Neuraminidase production by *Erysipelothrix rhusiopathiae*. *Vet Microbiol* 107: 265-72.

54. Wang Q, Chang BJ, Mee BJ and Riley TV. 2010. *Erysipelothrix rhusiopathiae*. *Vet Microbiol* 140: 405-417

55. Werner K, Gartrell B and Norton SA. 2011. Erysipeloid (*Erysipelothrix rhusiopathiae* infection) acquired from a dead kakapo. *Arch Dermatol* 147: 1456-1458.

56. Wood RL and Henderson LM. 2006. Erysipelas. In: Straw BE, Zimmerman JJ, D'Allaire S and Taylor DJ (eds.) *Diseases of swine*, 9th ed. Blackwell Publishing Professional, Ames, IA.

57. Yamazaki Y, Sato H, Sakakura H, Shigeto K, Nakano K, Saito H and Maehara N. 1999. Protective activity of the purified protein antigen of *Erysipelothrix rhusiopathiae* in pigs. *Zen Fur Vet Reihe* B 46: 47-55.

58. Yamazaki Y. 2006. A multiplex polymerase chain reaction for discriminating *Erysipelothrix rhusiopathiae* from *Erysipelothrix tonsillarum*. *J Vet Diagn Invest* 18: 384-387.

2014, Zoonoses: Bacterial Diseases
Editor: Sudhi Ranjan Garg
Published by: DAYA PUBLISHING HOUSE, NEW DELHI

Pages **281–300**

13

Pasteurellosis

Sathish B. Shivachandra and K.N. Viswas
Indian Veterinary Research Institute, Mukteswar – 263 138

Pasteurellosis is a term used to denote several infectious disease manifestations among animals, birds as well as humans caused mainly by *Pasteurella multocida* belonging to family Pasteurellaceae. World Health Organization (WHO) considered pasteurellosis as an important zoonoses as early as in 1959 (WHO 1959).

The history of *Pasteurella* organism dates back to 1782, when it was first studied by Chabert in France. In 1836, Mailet first used the term 'fowl cholera' for an outbreak of disease in fowls, however, the causative bacterium was first described in 1878-79 by Revolta and Revolee. About the same time, Bollinger from Germany described a disease in cattle and buffaloes. In 1885, Kitts isolated the causative organism. A septicaemic disease in rabbits was described by Gaffky in 1881 and a similar disease in swine by Loeffler in 1886. Later, in 1887, Oreste and Armani from Italy proposed the name *Bacillus septicaemiae*. In 1887, Trevisan proposed the name *Pasteurella* in recognition of pioneering work done in early 1880s by Louis Pasteur on the etiology of fowl cholera. Initially, the bacterium was named according to the host species in which it caused the disease. The isolates from cattle were named *Boviseptica*, those from pigs as *Suiseptica*; and from poultry as *Aviseptica*. Subsequently, there were several changes in the nomenclature until in 1939. Rosenbach and Merchant proposed the name *Pasteurella multocida* (L. adj. *multus* - many; L. adj. suf. *cidus* - kill; *multocida* - many killing, referring to pathogenicity for many species of animals). The species is further divided into three subspecies: *P. multocida* subsp. *multocida*, *P. multocida* subsp. *septica* (*septic* - poisoning, infecting) and *P. multocida* subsp. *gallicida* (*gallina* - hen; *cida* - kill; *gallicida* - hen killing, referring to pathogenicity for poultry). At present, the organism *P. multocida* belongs to the family Pasteurellaceae, which includes a large group of Gram-negative bacteria (Christensen and Bisgaard 2000).

Based on phenotypic and genotypic analysis, about 20 named and unnamed species of *Pasteurella* have been identified till date. Of these, *P. multocida* and *P. haemolytica* are the most common pasteurellae causing severe diseases and economic losses in livestock sector. Being genetically heterogeneous, *P. multocida*, the type species of genus *Pasteurella*, is known to colonize the mucous membrane of upper respiratory tract and intestinal tract of a wide variety of mammals and avian species. The infectious complications are known to precipitate during environmental stress or concurrent infections or immune-compromised status of animals/humans. Several subspecies of *P. multocida* have been identified from different animal and avian hosts with different predilection sites of infection and associated with infections of differing severity. Of several diseases, haemorrhagic septicaemia (HS) in cattle and buffaloes, fowl cholera in poultry, atrophic rhinits in pigs, pneumonic and septicaemic pasteurellosis in sheep and goats, and snuffles in rabbits have been well documented.

Since *P. multocida* is a colonizer of mucous membrane of upper respiratory tract of a wide variety of wild and domesticated animals, especially dogs and cats; human exposure to this species is frequent (Shivachandra *et al.* 2003b). The prevalence of *P. multocida* is 50-60 per cent in dogs and 50-90 per cent in cats. However, despite the fact that there exists a very frequent contact between domestic animals and human beings, the incidence of reported *P. multocida* infections in humans is generally low suggesting that the risk of its transmission from domestic animals to humans is probably very low. Moreover, it is noteworthy that this microorganism is not as frequent colonizer of the human upper respiratory tract as in animals. Many members of family Pasteurellaceae make excellent natural models for the study of bacterial pathogenesis and host-pathogen interactions, thus giving valuable insights into related human diseases. Research especially with regard to human infection is at an infant stage. Based on the direction of transmission, pasteurellosis is considered a zooanthroponosis. Its transmission from infected humans to companion animals has not yet been established.

Geographical Distribution

P. multocida is a ubiquitous pathogen and has a global distribution but the occurrence of particular subgroups and subspecies may depend upon the distribution of host species and other environmental factors. Pathogenic *Pasteurellae* have been isolated from diverse host species, both in developing as well as developed countries. Comparative analysis of worldwide distribution indicates the occurrence of the pathogen and disease in highly varied agro-climatic geographical regions ranging from temperate to tropical environment with different livestock densities involving different breeds and age groups. The region-based husbandry practices are known to influence the disease outbreaks.

Etiological Agent

Although most human infections are associated with *P. multocida*, which is acquired through some form of animal contacts, other species involved occasionally are: *P. haemolytica, P. dogmatis, P. pneumotropica* and *P. ureae.* Among such human infections, *P. multocida* subsp. *multocida* represents the majority, followed by a smaller proportion of *P. canis* and *P. multocida* subsp. *septica*. Among serotypes, capsular serotype A (A:5, A:6) has been found as the common type of *P. multocida* isolated from dogs and cats (Arashima *et al.* 1992).

P. multocida are non-motile, small, coccoid to rod-shaped, pleomorphic bacteria. They occur either singly or in pairs or short chains depending on the growth stage. These are Gram negative, non-acid fast and non-spore forming. Surface colonies on agar plates are smooth round, grayish or yellowish, glistening, translucent, and grow nearly 1-2 mm in diameter after 24 hours of incubation at 37°C. In nutrient broth, there is a slight clouding after 18 hours of incubation at 37°C. However, a good growth is achieved in brain heart infusion (BHI) broth. The organism doesn't grow on MacConkey's agar plates. Fresh cultures, blood smears, exudates and infected tissues show bipolar staining of the organism with Leishman, methylene blue or Giemsa stain.

Pasteurella are mesophilic, chemoorganotrophic with both oxidative and fermentative types of metabolism. Oxidase, alkaline phosphatase and catalase positive, most species are V-factor (β-nicotinamide adenine dinucleotide) and X-factor (protohemin or proto porphyrin IX) independent, although V-factor requiring species and strains do occur. A positive fermentative reaction is generally observed with D-glucose, D-galactose, D-fructose, D-mannose and sucrose; and acid production from L-sorbose, L-rhamnose, m-inositol and adonitol. The absence of hydrolysis is noticed with starch, salicin/aesculin and negative reaction with arginine dihydrolase, lysine decarboxylase, gelatin liquefaction and urease test. The genome of *P. multocida* contains ~37.7 to 45.9 mol per cent G+C, with molecular weight ranging from 1.4-1.9x10^9 (Mutters *et al.* 1985). The strains of *P. multocida* originated from bovine, avian, caprine, leporine, ovine, and felines/canines have been found to vary phenotypically (Ekundayo *et al.* 2008).

Despite several attempts made in the past to classify *P. multocida* and to correlate these with a specific disease manifestation, currently, the most acceptable and widely used serotype designation system is a combination of Carter capsular typing and Heddleston somatic typing (Carter and Chengappa 1981), which recognizes five capsular serogroups (A, B, D, E and F) based on indirect haemagglutination assay (IHA) and sixteen (1-16) somatic serotypes based on agar gel precipitation test (Carter 1967, Heddleston *et al.* 1972). Various virulent factors, including capsule, fimbriae (Ruffalo *et al.* 1997), outer membrane proteins, LPS, extracellular enzymes and toxins, have been attributed towards virulence of a pathogen (Harper *et al.* 2006, Hatfaludi *et al.* 2010).

Host Range

Pasteurellae are known to be present in almost all domestic animals, birds as well as wild animals (Hunt *et al.* 2000). Some species are involved in causing highly pathogenic infections whereas the majority of other species remain as commensals, especially on the mucous membrane of upper respiratory tract. In view of these, it could be presumed that almost all species of animals act as reservoir hosts, although animals that harbour potentially transmissible pasteurellae need to be thoroughly investigated. Current literature indicates transmission of *P. multocida* from pet animals. All age groups of humans can be affected but young children, elderly as well as immunocompromised persons, are most likely to encounter pasteurellosis. In recent times, since wild animals are also being reared as pet animals, both in developed and developing countries, there is a greater need to study the *Pasteurella* species from diverse animals for potential human pathogenicity.

Disease Transmission

P. multocida organisms are generally prevalent among domestic animals as well as wild animals; and are known to result in highly infectious disease manifestations with varied clinical signs and mortality depending on the species affected. Persons at risk for infection related to animal exposure include veterinarians, farmers, livestock handlers, pet owners and food handlers.

Canines and felines have been potentially considered as major pet animals transmitting *P. multocida* in recent reports. It has been estimated that as many as 66 per cent of dogs and 90 per cent of cats are colonized with this organism, typically in the respiratory and gastrointestinal tracts. The organisms are transmitted to humans, mainly through scratches and bite wounds and through airway infection. Although the mode of infection in most reported cases is not clear, most cases are thought to result from inadvertent, direct inoculation of organisms or from upper respiratory tract colonization with subsequent dissemination via lymphatics or hematogenously to the target organs, causing skin and soft tissue infections, bone and joint infections, pneumonia, meningitis, endocarditis and septicemia (Holst *et al.* 1992). The mode of acquisition of *P. multocida* into the lower respiratory tract may involve inhalation of contaminated aerosols or ingestion of cat or dog secretions.

Apart from animal bites/scratches, a significant proportion of *P. multocida* infections in humans is associated with animal exposure without any injury. Moreover, some systemic infections can even not be associated with any animal exposure at all. In a non-bite, non-scratch exposure, the occurrences of infections are probably due to contact with oral secretions of cats/dogs that contained the pathogen. Inhalation or licking and oral transmission have been suggested as the possible infection routes for those patients without injury. The danger of causal contact between humans and pets has been perceived due to certain unusual circumstances, such as dogs visiting the hospitalized people becoming commonplace in Canada, USA and other developed countries (Lefebvre *et al.* 2006). However, little is known about the potential health risks of introducing dogs to healthcare settings.

Pigs kept as pets in Western countries have been found to contribute to the transmission of zoonotic *P. multocida* (Patton *et al.* 1980). Ruminants (cattle, buffalo, sheep, and goats) are presumed as potential transmitters of the pathogen, especially in one report which indicated that ruminants had been found to act as reservoirs for *P. multocida* as they are known to persist in the crypts of tonsils and transmitted to the farmer who was handling cattle (Umemori *et al.* 2005). This was a very rare case suggesting transmission from cattle. Despite the prevalence of *P. multocida* in swine herds, the relationship between porcine colonisation and human disease is poorly established. However, there is a report of unidentified mode of zoonotic transmission in respiratory pasteurellosis among domestic cooking of pig trotters (Henderson *et al.* 2010). Although, there are no reports of direct transmission of pasteurella from wild animals, it is presumed that carnivores, especially those belonging to feline family, would be able to transmit the pathogen to humans through bite wounds, if they harbour some *P. multocida* strains. Some possible modes of zoonotic transmission of pasteurellosis have been illustrated in Figure 13.1.

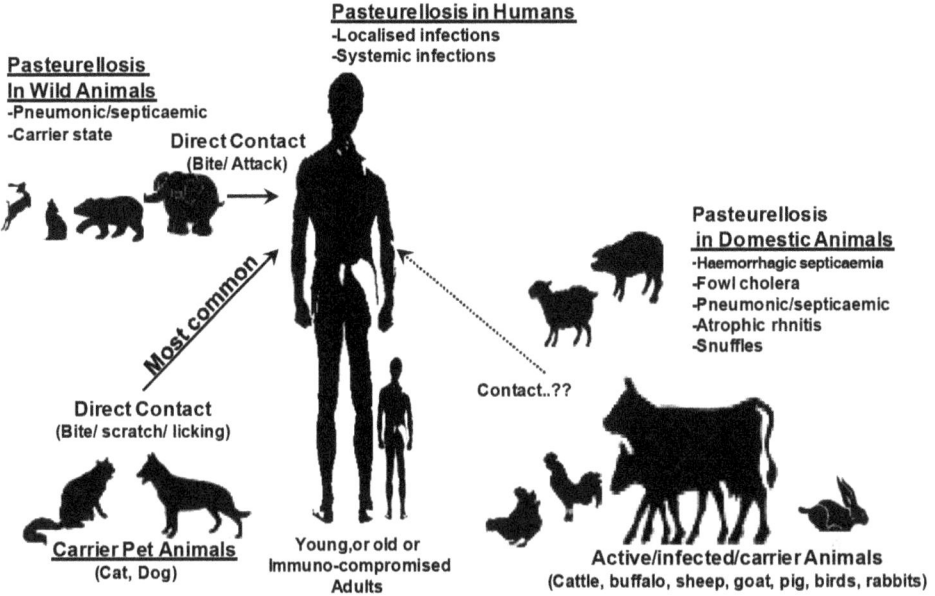

Figure 13.1: Possible modes of transmission of zoonotic pasteurellosis

Vertical transmission of pasteurellosis has been identified in neonates. Horizontal human-to-human spread has not been documented but recently, a case of neonatal sepsis and meningitis resulting from horizontal transmission of *P. multocida* has been described (Siahanidou *et al.* 2012).

Environmental and Other Predisposing Factors

Studies have shown that the type and number of bacteria associated with dogs and cats is influenced by their living environment. The number and species isolated from the hair and front paw samples from dogs kept outdoors and from cats were greater and more varied than those from the samples from dogs kept indoors.

Since *P. multocida* often acts as an opportunistic pathogen with a predilection for causing bacteraemia, very young or elderly persons who are less immunocompetent to fight the microbial invasion may be affected more. It has been well documented by several clinicians that immunosuppressive diseases such as AIDS, autoimmune diseases, etc. are likely to predispose a person to pasteurellosis (Elsey *et al.* 1991, Drabick *et al.* 1993). Diabetes mellitus and liver dysfunction in humans are also considered potential predisposing conditions associated with zoonotic pasteurellosis. The occurrence of *P. multocida* and *P. dagmatis* septicaemia in diabetic patient after contact with two pet dogs has been reported (Fajfar-Whetstone *et al*. 1995). Apart from these, contaminated surgical instruments either by contact with pet animals or by accidental exposure could also potentially transmit pasteurella to susceptible or immunocompromised patients (Elsey *et al*. 1991). The increased susceptibility to *P. multocida* septicaemia during pregnancy and *in utero* transmission in women has also been noticed (Rasaiah *et al*. 1986).

Exposure of *P. multocida* in a slaughterman who developed meningitis illustrates the role of occupational factors in the transmission of infection (Fell 1984).

Prevalence of Disease in Animals

Family Pasteurellaceae includes diverse range of microorganisms: opportunistic pathogens, human and animal pathogens, and avirulent commensal strains. Varied disease conditions are observed among different animal and avian species.

HS, a major disease caused by *P. multocida*, is an endemic bacterial disease of cattle and buffaloes in India. Fowl cholera affects a wide range of avian species, namely, chickens, ducks, turkeys, quails and geese in the country (Shivachandra *et al*. 2006c). Pasteurellosis has also been reported in sheep, pigs, rabbits, goats and poultry. Despite relatively little information about pasteurellosis in Indian wild animals, the presence of *P. multocida* organisms in different wild species *viz*. lion, tiger, spotted deer, hippopotamus and leopard has been reported (Prabhakar *et al*. 1998).

Pasteurellosis in animals has been recorded in most countries of the world but the developed nations seem to be less affected with the disease. HS has been reported in the Middle East and Southern Europe. In the United States, the disease was reported among bison in various national parks in 1912, 1922 and 1965-67. Outbreaks were also reported among dairy cattle and beef calves in the US and Canada. However, apart from sporadic outbreaks, the disease is not endemic in North America. Oceanic countries such as Australia, New Zealand and others like Japan and Western Europe have not yet reported the disease (De Alwis 1999). Fowl cholera is widely reported from almost all developed countries of the world. Sporadic cases of pasteurellosis in sheep and goats as well as in pigs have been recorded in all intensive farms of developed countries. Isolated cases of HS have also been reported from the Central and South American states, and from Southern and Eastern European countries. It is presumed that HS may occur in developed countries with high water buffalo population and at the places where geo-climatic conditions are similar to that of tropical Asia.

Compared to *P. multocida* infections in Asia, *P. haemolytica* infections are known to be prevalent more among the livestock in the developed countries. The disease conditions like shipping fever, transit fever, enzootic pneumonia and bovine respiratory disease (BRD) complex have been reported (Dabo *et al*. 2007). In the US, it is believed that the losses to beef and dairy industries from this disease complex are greater than the losses due to all other diseases put together. The disease complex is associated with multiple factors with involvement of more than one species and serotype of pasteurellae. The organisms, which reside normally in the healthy hosts, are known to play a secondary role to other infections due to either bacteria or viruses or stress. Following the stress and other incriminating factors, pasteurellae are known to undergo rapid proliferation resulting in a disease complex with high severity and fatality.

Pasteurellosis in sheep and goats in temperate countries is mainly due to *P. haemolytica* causing pneumonia in all ages of sheep and septicaemia in lambs. Pasteurellosis in pigs is known to involve *Bordetella bronchiseptica,* a normal inhabitant of upper respiratory tract.

Prevalence of Disease in Human Beings

P. multocida infections occur worldwide. In humans, the initial case of pasteurellosis was puerperal fever reported in 1913. The incidence is mainly reported from urban areas, where intimate relationship/contact between humans and pet animals has increased over the past few decades, especially among children and elderly persons. It has been found that *P. multocida* infection caused respiratory infectious disease in more than 50 per cent of human pasteurellosis cases, and local infection through bite wounds and scratches by dogs and cats occurred in approximately 27 per cent of clinical cases, followed by other diseases.

The occurrence of zoonotic *P. multocida* associated infections in humans has not yet been reported in India, however, it doesn't rule out the possibilities of their existence (Shivachandra *et al.* 2003b). Lack of thorough investigation of clinical cases might have led to limited literature in the country. Growing number of pet owners and rising association of pet owners with companion animals in recent times makes it an imminent need to re-investigate the chronic/recurring human clinical cases that are in association with animals. Other developing countries also have a similar scenario.

Overall, increasing incidences of zoonotic pasteurella infections in humans have been reported from most of the developed nations. For instance, in Japan, the mean increase rate of 25 per cent in the incidence of pasteurellosis in humans has been observed (Arashima *et al.* 1993).

Disease Manifestations in Animals

P. multocida is known to cause a variety of diseases in a wide range of hosts. Some major diseases occurring in livestock are described below:

Cattle and Buffalo

HS in cattle and buffalo is a highly fatal, acute septicaemic disease caused by two specific serotypes of *P. multocida* designated as B:2 (Asian Serotype) and E:2 (African Serotype). Incidence and distribution of HS varies greatly from a few cases to large number of reports depending on the geographical area and agroclimatic conditions prevailing in different regions.

It occurs in acute, sub-acute and chronic forms. A typical clinical case of HS is characterized by an often unnoticed initial phase of temperature elevation (41-42°C), followed by a phase of respiratory involvement, and a terminal phase of septicaemia and recumbence leading to death. The incubation period ranges from 1 to 3 days, and the course of the disease may range from sudden death without clinical signs to a protracted course extending up to 5 days. The acute form is manifested by sudden onset and death within 24 hours. However, some animals show thin pulse, rapid and shallow respiration with cyanosis of visible mucous membranes. Restlessness, mild colic pain, muscular tremors, lacrymation, nasal discharge and diarrhoea are also observed. In sub-acute form (subcutaneous edematous form), animal survives for 2-3 days with oedematous swelling of throat and brisket region and bronchopneumonia. In chronic form, course of infection is longer with signs of short, rapid painful respiration, coupled with mucopurulent or blood stained nasal discharge (OIE 2008, Kumar and Shivachandra 2011).

Birds

Fowl cholera in birds may result from infection with several strains of *P. multocida*. However, sero group A strains are regarded as the primary cause of the most severe form of the disease. The strains, belonging to serogroup D and F, have also been isolated from diseased birds (Shivachandra *et al.* 2006c), but with low incidence compared with serogroup A. In birds, clinical signs are often classified into two forms, acute and chronic, but disease outbreaks are often known to present clinical signs and lesions that are intermediate between these two forms. In the classic acute form, death may be the only indication of the disease; however, a few hours before death, the birds can reveal signs such as fever, anorexia, ruffled feathers, mouth discharge, diarrhoea and increased respiratory rate. In the chronic form, signs of localized infection appear *viz.* swelling of sinuses, wattles and leg and/or wing joints (Christensen and Bisgaard 2000).

Pigs

Atrophic rhinitis and pneumonia are two well defined disease manifestations caused by *P. multocida* in pigs. It is known to cause septicaemia and death in young pigs. However, it is more commonly associated with pneumonia in pigs of two to six months age, usually secondary to other infections (Ross 2006). *P. multocida* sero group A and D are mostly associated with the pneumonic condition. It has been well documented that some strains of *P. multocida* sero group D and sero group A are also toxigenic and release exotoxins (PMT, also known as dermonecrotic toxin), following rapid proliferation causing atrophic rhinitis, which is characterized by the atrophy of nasal turbinates, resulting in shortening or twisting of snout accompanied by sneezing and epistaxis. It is rarely known to cause death but significantly reduces the growth rate of the infected pig (Hungerford 1989).

Rabbits

P. multocida is the most common respiratory pathogen in the domestic rabbit. It is a very contagious respiratory disease that is commonly known as 'snuffles'. Different strains of *P. multocida* are involved in causing the upper respiratory signs in rabbits. Some strains are commonly found in the nasal tract of rabbits, but may not cause infection unless the animal is stressed or has suppressed immune system. Serogroups

A and D are frequently isolated but serogroup A:1, 2 is predominant. Susceptibility of individual rabbits varies greatly though all age groups are susceptible. The disease is known to cause acute pneumonia, septicaemia, otitis media, encephalitis, abscesses and pyometra. Once rabbits are infected, they become carriers without exhibiting clinical signs. The bacteria can be transmitted via aerosol or contact with an infected rabbit, either directly or through fomites.

Small Ruminants and Wild Animals

Septicaemic and pneumonic menifestations are observed among sheep, goats and wild animals. Climatic changes and stressful management practices, such as transport, dipping and shearing may precipitate the disease outbreaks in sheep and goats. Occasional or sporadic outbreaks of pasteurellosis have been recorded among different wild animals like deer, yaks, horses, minks, elephants, bison, snow leopards, lions, tigers, camels and monkeys. The types of infections recorded are highly variable ranging from septicaemia to respiratory infections. Clinical signs of pneumonic and septicaemic forms are observed and the severity of infection varies depending on the serotype involved and species affected. At necropsy, serofibrinous peritonitis, fibrinous pleurisy with areas of pneumonia, congestion, oedema, and widespread haemorrhages are seen. Moreover, latent infection is known to occur in wild as well as many domestic animals.

Dogs and Cats

P. multocida is known to be common component of the normal oral flora in the majority of cats and dogs, but particularly in cats (Shivachandra *et al.* 2003b). Different species of *Pasteurella* organisms normally inhabit the nasal, gingival, and tonsillar regions of many dogs and cats. The infection in dogs and cats can result in rhinitis, conjunctivitis, pneumonia, abscesses, genital tract infections and septicemia. *P. multocida* initially colonizes the pharynx and then moves to the nasal cavity and surrounding tissue, with possible spread to the rest of the body. Colonization of the nasal cavity may take two weeks to occur, after which clinical signs may or may not appear. Affected/carrier animals apparently seem to be healthy without showing any signs of infection which could potentially deceive the persons at risk.

Disease Manifestations in Humans

In humans, the clinical signs widely vary depending upon the site of infection. Most human infections ensue subsequent to traumatic introduction of *P. multocida* through the skin, secondary to dog or cat bites or scratch. As most of the time, *P. multocida* infections are in association with other major diseases, there is every chance of mixed clinical signs or masked signs as reported in earlier cases. Sometimes, the manifestation goes unnoticed in the shadow of major bacterial/viral diseases of humans. However, some of the afflictions and their signs have been well documented in the past. *P. multocida* infections appear to occur most commonly in patients with chronic illness who have some type of underlying structural and/or functional alterations of various systems. An increase in the incidences of *P. gallinarum* induced sepsis developing food poisoning-like symptoms (Arashima *et al.* 1999) and *P. hemolytica* induced sepsis, which was considered to be splenic abscess, have been reported (Takeda *et al.* 2003). Human cases of *P. multocida* induced lung abscess have been rarely reported worldwide. Most of the cases occurred in middle-aged to elderly patients with underlying diseases, and in many cases, patients were in close contact with animals. Apart from elders, *P. multocida* infection has been noticed in neonates also (Nakwan *et al.* 2010). Of 38 reported cases of *P. multocida* meningitis during infancy, 18 were less than or equal to 1 month of age and had a documented exposure to household dogs and/or cats, most only incidentally (Kobayaa *et al.* 2009). Several Pasteurella species are associated with dog and cat bites, including *P. multocida* subspecies *multocida, P. multocida* subsp. *septica, P. stomatis, P. dogmatis* and *P. canis.* A wide range of conditions from which *P. multocida* has been isolated and its involvement has been confirmed could be broadly classified as below:

A) Local Infections

P. multocida infection usually presents as an infection that complicates an animal bite or injury. Following animal bite, severe swelling and erythema may surround the traumatized tissue. If the penetrating animal bite has punctured the periosteum, osteomyelitis may result. *P. multocida* most commonly causes soft tissue and wound infections after receipt of an animal bite or scratch, but the organism has also been reported to cause deep-seated infections as well as systemic infections. Complications of animal bites include various manifestations and lead to metastatic foci in various organs and joints, which include rapidly progressive cellulitis, abscesses, tenosynovitis, osteomyelitis and septic arthritis. The latter two are particularly common, following cat bites because of their small, sharp and penetrative teeth. Clinically, erythema, warmth, pain and tenderness, purulent discharge, lymphangitis, joint swelling, decreased range of motion are noticed in humans. Although localized *P. multocida* infections in humans have shown to be a good prognosis, the disseminated *P. multocida* infections have been found to carry a 25-30 per cent overall mortality risk.

B) Systemic Infections

Serious systemic infections in humans with *Pasteurella* species are unusual but may occur in individuals with predisposing underlying illnesses involving other bacterial and/or viral infections causing varied disease manifestations. Some of these, which have been noticed in the past, are discussed below:

1. **Respiratory system:** Clinically, respiratory infectious disease caused by *P. multocida* has no characteristic feature, and various symptoms develop from pharyngolaryngitis to dyspnea. Bronchitis, acute pneumonia, lung abscess and pyothorax have been reported, and acute pneumonia has been most frequently reported (Weber *et al.* 1984). In rare cases, lower respiratory tract infections include pneumonia, tracheo-bronchitis, lung abscess, and empyema, especially in individuals with underlying pulmonary disease. Clinical signs following various conditions such as sinusitis, otitis media, mastoiditis, epiglottitis, pharyngitis, and Ludwig angina include sinus tenderness, hoarseness, pharyngeal erythema, rales and rhonchi upon chest auscultation, dullness to percussion, and changes in vocal fremitus. Chronic obstructive pulmonary disease is a risk factor for *P. multocida* respiratory tract infection, which carries a mortality rate of about 30 per cent.

2. **Cardiovascular system:** The conditions include native valve and prosthetic valve endocarditis, pericarditis, mycotic aneurysms, vascular graft infections, central venous catheter infections, bacteremia, sepsis, and septic shock. Clinically, signs of hypotension, tachycardia, cardiac murmur and embolic phenomenon are noticed.

3. **Central nervous system:** *P. multocida* is an uncommon cause of meningitis, subdural empyema and brain abscess. However, *P. multocida* meningitis has been associated with cat licks and bites occurring on the face in aged persons. The signs of focal neurologic deficits and meningeal irritation are noticed.

4. **Gastrointestinal system:** Although *P. multocida* rarely causes gastrointestinal problems, it has been associated with appendicitis, hepatosplenic abscesses, and spontaneous bacterial peritonitis. Noticeable clinical signs include abdominal tenderness, guarding and rebound, hepato-splenomegaly and costo-vertebral angle tenderness. *P. multocida* has also been isolated in patients with polymicrobial peritoneal dialysis catheter–associated peritonitis.

5. **Ocular system:** The afflictions include periocular abscess, conjunctivitis, corneal ulcers, and endophthalmitis (McNamara *et al.* 1988). Clinically, signs of corneal ulcer, conjunctivitis and visual acuity are noticed (Ho and Rapuano 1993).

6. **Genito-urinary tract:** Pyelonephritis, renal abscess, epididymitis and cervicitis have been reported in rare cases.

7.. **Musculoskeletal system:** Following an animal bite, a rapidly progressive cellulitis may develop; deeper structures, including tendons, joints and bones, can become affected as a result of pathogen dissemination. Degenerative joint disease, rheumatoid arthritis, and prosthetic joints have been associated with the development of *P. multocida* septic arthritis.

Diagnostic Procedures

Pasteurellosis in animals or humans could be diagnosed either by employing conventional or molecular tools. However, for confirmative diagnosis, isolation of organism from specimens, followed by characterization using conventional as well as molecular tools, is considered reliable.

Conventional Approaches

Traditionally, diagnosis of *Pasteurella* is based on clinical findings, culture and / or serological testing. Although culture identification methods are definitive, they are time consuming and costly. False-negative culture results are frequently observed due to the fact that *P. multocida* dies easily during transport to the laboratory or is overgrown by other bacteria (nasal flora or contaminants) in the culture. Serology has been used when infection is suspected in organs for which cultures are not obtainable or when the culturing yields no results. However, a serological test (ELISA) positive for *P. multocida* can indicate either current infection or previous exposure to the organism. Because many animals have been exposed to this organism, a current diagnosis of pasteurellosis cannot be based on the results of conventional methods alone. Since, several molecular methods offer a more sensitive, specific and timely method for detecting *Pasteurella* and confirming infection, they could be effectively employed in parallel to conventional approaches (Dziva *et al.* 2008).

For diagnosis, specimens collected from infected live animals are blood samples and nasal secretions, whereas from dead animals, heart blood, liver, spleen, lungs and bone marrow of long bones may be collected. Similarly, clinical samples should also be collected from various lesions which appear in the affected humans for diagnosis. The diagnosis is usually done by the tests given below:

1. **Clinical diagnosis:** It is based on the characteristic signs, gross pathological lesions, herd history, morbidity and mortality pattern, species susceptibility and age group affected.

2. **Laboratory confirmation:** Laboratory tests include blood smear examination by staining with Gram's, Giemsa, Leishman's or methylene blue stain, isolation by culture or mice inoculation, identification by colony morphology, pathogenicity test, biotyping and antibiogram studies. Gram stain of purulent material or other fluid specimens including blood, sputum, and cerebrospinal fluid may show small, Gram-negative, nonmotile and non-spore forming pleomorphic coccobacilli, indicating the presence of *Pasteurella* species. However, other bacterial species such as *Haemophilus*, *Neisseria*, *Moraxella* and *Acinetobacter* have a morphology that is similar to that of *P. multocida* infection and can, therefore, be easily confused with *Pasteurella* species. Further, Wright's and Giemsa stains enhance bipolar staining of *P. multocida*.

   Despite the fact that most of the wound infections associated with animal bites in humans usually have a multi-microbial etiology comprising of both aerobic and anaerobic Gram-negative bacteria, *Pasteurella* species are commonly isolated in most animal bites, especially in dog-related and cat-related injuries.

3. **Immunological methods of identification:** These include rapid slide agglutination test, indirect haemagglutination test (IHA); agar gel immunodiffusion test (AGID); counter-immunoelectrophoresis (CIE).

Serological tests are for detecting antibodies but are not normally used for diagnosis. However, IHA test with high titres from 1/160 upto 1/1280 or higher among the in-contact animals surviving in the affected herds are indicative of recent exposure to HS. Other serological tests like passive haemagglutinations, AGID and ELISA have been used with varying degrees of success in attempts to predict the level of immunity. The ease of obtaining definite diagnosis through isolation and identification of organisms precludes the need for serodiagnosis. Non-serological methods such as acriflavine flocculation test and hyaluronidase test could be employed. However, all these conventional methods of diagnosis with potential problems limit their application in clinical diagnosis of pasteurellosis (Dziva *et al.* 2008, Kumar and Shivachandra 2011).

Molecular Approaches

Due to lack of phenotypic identification tests applicable in a routine diagnostic laboratory, many times clinician might misidentify the bacterial isolates. Polymerase chain reaction (PCR) assays for the specific identification of *P. multocida* isolates have been developed, which can be applied directly to the bacterial colony, bacterial culture lysate and mixed bacterial culture for rapid and specific identification within 3-4 hours (Choi and Chae 2001, Shivachandra *et al.* 2003a, 2004b, 2006b). PCR assays have also been employed directly on clinical samples to augment rapid disease diagnosis. Although, molecular tools have been in use for more than a decade for rapid detection and epidemiological investigation of pasteurellosis in animals, their use in humans is still limited which could be due to lack of awareness about the occurrence of pasteurella as a zoonotic agent.

P. multocida-specific PCR (PM-PCR) assay, developed by Townsend *et al.* (1998), could be effectively employed for initial species identification originated either from animals or humans. Further, the classification of capsular serogroup of *P. multocida* could be accomplished by the application of Multiplex Capsular PCR Typing system as described by Townsend *et al.* (2001). Unfortunately, there is no availability of species specific PCR assays for different species of *Pasteurella* which are involved in zoonosis.

Nucleic acid-based typing tools like pulsed field gel electrophoresis (PFGE), restriction endonuclease analysis (REA), ribotyping, enterobacterial repetitive intergenic consensus-PCR (ERIC-PCR) (Shivachandra *et al.* 2008), random amplification of polymorphic DNA-PCR (RAPD-PCR) (Shivachandra *et al.* 2007), amplified fragment length polymorphism (AFLP) (Shivachandra *et al.* 2006a), and sequencing have been successfully employed for epidemiological tracing of the occurrence of *P. multocida* infections in animals as well as humans (Shivachandra *et al.* 2003a, Dziva *et al.* 2008). Among several molecular typing tools, REA (Drabick *et al.* 1993) and RE-/ERIC-PCR (Loubinoux *et al.* 1999) assays have been found to be the most robust and reliable in early investigation of pathogen and its source. Several molecular assays have been successfully employed in detection and differentiation of disease outbreaks (Biswas *et al.* 2004, Shivachandra *et al.* 2005). Although PFGE and AFLP are considered highly sensitive and discriminatory, they are more laborious than other tools. In a recent study, PFGE showed that the clinical isolate recovered from humans was identical to the *P. multocida* isolate recovered from the pharyngeal swab specimen culture obtained from the patient's dog (Satomura *et al.* 2010). PFGE data strengthened the evidence of zoonotic transmission of *P. multocida* through casual contact in a patient with a respiratory infection. PFGE should prove useful in both epidemiological tracing and preventing pasteurellosis and clarifying its pathology.

A novel loop-mediated isothermal amplification (LAMP) assay has been developed recently targeting the kmt1 gene from the species-specific region of the *P. multocida* genome (Sun *et al.* 2010). Being a simple, rapid, sensitive and highly specific assay, *P. multocida* specific-LAMP (Pm-LAMP) could be potentially employed as a rapid diagnostic tool at field level in most underdeveloped and developing countries for initial bacterial species identification.

Apart from confirmation of causative agent, the differential diagnosis of pasteurellosis in humans should also be conducted to rule out other conditions which include abdominal abscess, amoebic hepatic abscesses, animal bites, brain abscess, cat scratch disease, cellulitis, *Haemophilus influenzae* infections, intra-abdominal sepsis, liver abscess, lung abscess, meningitis, *Meningococcal* infections, perinephric abscess, pneumonia caused by other bacterial infections, community-acquired pneumonia, and acute pyelonephritis. In humans, other sophisticated medical diagnostic devices such as CT scanning, MRI, echocardiography, etc. could also be effectively employed in differential diagnosis of various disease manifestations.

Prevention and Control Strategies

The rise in the number of reports in recent times indicating zoonotic potential of *P. multocida* from animals warrants imminent strategy to tame the spread of the pathogen by implementing control measures. The control of zoonotic pasteurellosis is a long term pragmatic plan which requires a proper blueprint with timely scheduled milestones to be accomplished. Some of the multi-pronged strategies are given below:

- ☆ Application of newer methods and typing tools for rapid identification and differentiation of field isolates from human, animal/avian hosts for epidemiological studies to track their origin.

- ☆ Selection of sensitive or a relatively newer broad spectrum antibiotics based on the antibiogram for treatment of pasteurellosis.

- ☆ Informing susceptible persons about the risk of contracting zoonotic agents

- ☆ Implication of basic hygiene rules while keeping pet animals.

- ☆ Avoiding the misguided concept of promoting bonding between the baby and the family's pets by the parents especially in urban areas.

- ☆ Guiding the elderly people to avoid intimacy with pets harbouring deadly pathogen.

- ☆ Guiding the immunocompromised patients to stay away from contact with pet animals and about the basic precautions.

- ☆ Considering a history of pet exposure, whether occupational or recreational, even without bites or scratches, while treating the affected individuals.

- ☆ Safety monitoring of biological products that derive from susceptible animals.

- ☆ Minimizing personnel exposure to *Pasteurella* infected animals.

- ☆ Ensuring *Pasteurella*-free animal facilities for domestic and pets animals.

- ☆ Establishment of national repository of zoonotic *P. multocida* isolates and maintenance of regional data base of prevalent serotypes, especially in endemic urban areas.

- ☆ Strengthening the disease investigation process by streamlining the process of sample collection, fast transport of specimens to laboratory, and rapid laboratory processing.

☆ Effective network of specialists, physicians, veterinarians and pet owners, to have continued monitoring and surveillance of pathogenic pasteurellae circulating among different animal species.

Future Directions

The panacea for pasteurellosis in animals as well as humans lies in the concerted efforts and multidisciplinary approach involving public, farmers, veterinarians, medical practitioners, scientists, policymakers and administrative personnel aiming to control pasteurellosis. Some suggested key areas for strategic control of zoonotic pasteurellosis are:

☆ Mass awareness extension programmes to educate people at risk, and strict adherence to the basic hygienic measures at home.

☆ Mass vaccination programmes of animals to strengthen herd immunity.

☆ Strict enforcement of legislation on pet/domestic animal movement/ quarantine/disposal of contaminants.

☆ Effective implementation of improved or advanced animal husbandry practices that lessen stress and incriminating factors which promote pasteurellosis.

☆ Funding the research activities for better understanding of the intricacies of epidemiology, pathogenesis, mode of transmission, pathogen and host interaction, cross species transmission, virulence factors, emergence of antibiotic resistance pattern and human exposure in the disease process.

☆ Exploring the possibilities of development of broadly active vaccine for human use.

☆ Identification of potential or candidate strains of *P. multocida* for inclusion in either monovalent, bivalent or multivalent vaccines for pasteurellosis in animals.

☆ Development of species specific PCR assays for rapid identification of different species of pasteurellae prevalent among pet/domestic animals.

☆ Development of improved diagnostics, mass applicable cost-effective vaccines and drugs to treat the infected animals or human population at risk.

Since the number of pets is increasing in India, patients with pasteurellosis may increase in the days ahead. Insufficient data on incidence, occurrence and distribution of pasteurellosis, coupled with inefficient prophylactic and control measures, make an insidious impact on the livestock sector as well as public health. The need of the hour is to focus on the key thrust areas such as strengthening the monitoring and surveillance system for zoonotic pasteurellosis, networking of well-equipped regional laboratories, supporting the region/species-based basic and applied research, effective implementation of rapid disease diagnosis/treatment/control measures as per the international standards and mass awareness programmes by multi-pronged extension activities at urban as well as rural levels. Moreover, the farmers and pet owners should adhere to the basic rules of hygienic measures, coupled with advanced management practices and continuous monitoring of the general health status of animals as well as persons at risk of exposure on routine basis. It would ultimately help in control of zoonotic pasteurellosis in India.

References

1. Arashima Y, Kumasaka K, Okuyama K, Tsuchiya T, Kawano K, Koizumi F, Sano K and Kaya H. 1992. The first report of human chronic sinusitis by *Pasteurella multocida* subsp. *multocida* in Japan. *Kansenshogaku Zasshi* 66: 232-235.

2. Arashima Y, Kumasaka K, Tsuchiya T, Kawano K and Yamazaki E. 1993. Current status of *Pasteurella multocida* infection in Japan. *Kansenshogaku Zasshi* 67: 791-794.

3. Arashima Y, Kato K, Kakuta R, Fukui T, Kumasaka K, Tsuchiya T and Kawano K. 1999. First case of *Pasteurella gallinarum* isolation from blood of a patient with symptoms of acute gastroenteritis in Japan. *Clin Infect Dis* 29: 698-699.

4. Biswas A, Shivachandra SB, Saxena MK, Kumar AA, Singh VP and Srivastava SK. 2004. Molecular variability among strains of *Pasteurella multocida* isolated from an outbreak of haemorrhagic septicaemia in India. *Vet Res Commun* 28: 287-298.

5. Boerlin p, Siegrist H H, Burness A P, Kuhnert p, Mendez P Pretat G, Lienhard r and Nicolet J. 2000. *J Clin Microbiol* 38: 1235-1237.

6. Buma R, Maeda T, Kamei M and Kourai H. 2006. Pathogenic bacteria carried by companion animals and their susceptibility to antibacterial agents. *Biocontrol Sci* 11: 1-9.

7. Byrd RP Jr and Roy TM. 2003. *Pasteurella multocida* respiratory infection: an important cat-associated zoonosis. *Arch Intern Med*. 26; 163(10): 1239.

8. Carter GR and Chengappa MM. 1981. Recommendation for a standard system of designating serotype of *Pasteurella multocida*. *Am Assoc Vet Lab Diag* 24: 37-42.

9. Carter GR. 1955. Studies on *Pasteurella multocida*: A haemagglutination test for the identification of serological types. *Am J Vet Res* 16: 481-484.

10. Choi C and Chae C. 2001. Enhanced detection of toxigenic *Pasteurella multocida* directly from nasal swabs using nested polymerase chain reaction. *Vet J* 162: 255-258.

11. Christensen JP and Bisgaard M. 2000. Fowl cholera. *Rev Sci Tech* 19: 626-637.

12. Drabick JJ, Gasser RA Jr, Saunders NB, Hadfield TL, Rogers LC, Berg BW and Drabick CJ. 1993. *Pasteurella multocida* pneumonia in a man with AIDS and nontraumatic feline exposure. *Chest* 103: 7-11.

13. Dziva F, Muhairwa AP, Bisgaard M and Christensen H. 2008. Diagnostic and typing options for investigating diseases associated with *Pasteurella multocida*. *Vet Microbiol* 128: 1-22.

14. Elsey RM, Carson RW and DuBose TD Jr. 1991. *Pasteurella multocida* peritonitis in an HIV-positive patient on continuous cycling peritoneal dialysis. *Am J Nephrol* 11: 61-63.

15. Ewers C, Lübke-Becker A, Bethe A, Kiebling S, Filter M and Wieler LH. 2006. Virulence genotype of *Pasteurella multocida* strains isolated from different hosts with various disease status. *Vet Microbiol* 114: 304-317.

16. Fajfar-Whetstone CJ, Coleman L, Biggs DR and Fox BC. 1995. *Pasteurella multocida* septicemia and subsequent *Pasteurella dagmatis* septicemia in a diabetic patient. *J Clin Microbiol* 33: 202-204.

17. FAO. 2005. *Production Yearbook* 2005. Statistics Division, FAO.

18. FAO-WHO-OIE 1994. *Animal Health Yearbook*, 1994. FAO.

19. Fell HW. 1984. *Pasteurella multocida* infections in West Suffolk. *J Infect* 9: 83-86.

20. Griego RD, Rosen T, Orengo IF and Wolf JE. 1995. Dog, cat and human bites: a review. *Am Acad Dermatol* 33: 1019-1029.

21. Harper M, Boyce JD and Adler B. 2006. *Pasteurella multocida* pathogenesis: 125 years after Pasteur. *FEMS Microbiol Lett* 265: 1-10.

22. Hatfaludi T, Al-Hasani K, Boyce JD and Adler B. 2010. Outer membrane proteins of *Pasteurella multocida. Vet Microbiol* 144: 1-17.

23. Heddleston KL, Gallagher JE and Rebers PA. 1972. Fowl cholera: Gel diffusion precipitation test for serotyping *Pasteurella multocida* from avian species. *Avian Dis* 16: 925-936.

24. Henderson SR, Shah A, Banford KB and Howard LS. 2010. Pig trotters lung-novel domestic transmission of *Pasteurella multocida. Clin Med* 10: 517-518.

25. Ho AC and Rapuano CJ. 1993. *Pasteurella multocida* keratitis and corneal laceration from a cat scratch. *Ophthalmic Surg* 24: 346-348.

26. Hungerford TG. 1989. *Diseases of livestock*, 9th ed. McGraw-Hill Book Company, New Delhi.

27. Hunt ML, Adler B and Townsend KM. 2000. The molecular biology of *Pasteurella multocida. Vet Microbiol* 72: 3-25.

28. Kawashima S, Matsukawa N, Ueki Y, Hattori M and Ojika K. 2010. *Pasteurella multocida* meningitis caused by kissing animals: a case report and review of the literature. *J Neurol* 257: 653-654.

29. Kobayaa H, Souki RR, Trust S and Domachowske JB. 2009. *Pasteurella multocida* meningitis in newborns after incidental animal exposure. *Pediatr Infect Dis J* 28: 928-929.

30. Kumar AA and Shivachandra SB. 2011. Action plan for control of haemorrhagic septicaemia and pasteurellosis. In: Garg SR (ed) *Veterinary and livestock sector: A blueprint for capacity building*. Satish Serial Publishing House, Delhi. p. 381-426.

31. Kumar AA, Shivachandra SB, Biswas A, Singh VP, Singh VP and Srivastava SK. 2004. Prevalent serotypes of *Pasteurella multocida* isolated from different animal and avian species in India. *Vet Res Commun* 28: 657-667.

32. Lefebvre SL, Waltner-Toews D, Peregrine AS, Reid-Smith R, Hodge L, Arroyo LG and Weese JS. 2006. Prevalence of zoonotic agents in dogs visiting hospitalized people in Ontario: implications for infection control. *J Hosp Infect* 62: 458-466.

33. Liu W, Chemaly RF, Tuohy, MJ, LaSalvia MM and Procop GW. 2003. *Clin Inf Dis* 36: e58-e60.

34. Loubinoux J, Lozniewski A, Lion C, Garin D, Weber M and Le Faou AE. 1999. Value of enterobacterial repetitive intergenic consensus PCR for study of *Pasteurella multocida* strains isolated from mouths of dogs. *J Clin Microbiol* 37: 2488-2492.

35. May BJ, Zhang Q, Li LL, Paustian ML, Whittam TS and Kapur V. 2001. Complete genomic sequence of *Pasteurella multocida*, Pm70. *PNAS* 98: 3460-3465.

36. McNamara MP, Richie M and Kirmani N. 1988. Ocular infections secondary to *Pasteurella multocida*. *Am J Ophthalmol* 106: 361-362.

37. Mutters R, Ihm P, Pohl S, Frederiksen W and Mannheim W. 1985. Reclassification of the genus *Pasteurella* Trevisan 1887 on the basis of deoxyribonucleic acid homology, with proposals for the new species *Pasteurella dagmatis*, *Pasteurella canis*, *Pasteurella stomatis*, *Pasteurella anatis* and *Pasteurella langaa*. *Int J Syst Bacteriol* 35: 309-322.

38. Nakwan N, Nakwan N and Chokephaibulkit K. 2010. *Pasteurella multocida* infection in the neonates. *Pediatr Infect Dis* J 29: 192.

39. Pace D and Attard-Montalto S. 2008. Quest for the diagnosis. Case 1: a neonatal zoonosis. Neonatal *Pasteurella multocida* septicaemia. *Acta Paediatr* 97: 250-252.

40. Patton F, Dumas M and Cannon NJ. 1980. *Pasteurella multocida* septicemia and peritonitis in a cirrhotic cock trainer with a pet pig. *N Engl J Med* 303: 1126-1127.

41. Rasaiah B, Otero JG, Russell IJ, Butler-Jones DA, Prescott JF, West MM, Maxwell BE and Beaver J. 1986. *Pasteurella multocida* septicaemia during pregnancy. *CMAJ* 135: 1369-1372.

42. Rimler RB. 2000. Restriction endonuclease analysis using *Hha*I and *Hpa*II to discriminate among serogroup B *Pasteurella multocida* associated with haemorrhagic septicaemia. *J Med Microbiol* 49: 81-87.

43. Ruffalo GC, Carmel G, Tennet JM, Michalski WP and Adler B. 1997. Identification, purification and characterization of type 4 fimbriae of *Pasteurella multocida*. *Infect Immun* 65: 339-343.

44. Satomura A, Yanai M, Fujita T, Arashima Y, Kumasaka K, Nakane C, Ito K, Fuke Y, Maruyama T, Maruyama N, Okada K, Nakayama T and Matsumoto K. 2010. Peritonitis associated with *Pasteurella multocida*: molecular evidence of zoonotic etiology. *Ther Apher Dial* 14: 373-376.

45. Shivachandra SB, Kumar AA, Srivastava SK and Choudhury P. 2003a. Nucleic acid based typing techniques: emerging tools for epidemiological studies. *Livestock International* 7(2): 20-23.

46. Shivachandra SB, Biswas A and Kumar AA. 2003b. *Pasteurella multocida*: a pathogen of zoonotic importance. *The Veterinarian* 27(8): 9-10.

47. Shivachandra SB, Kumar AA, Biswas A, Ramakrishnan MA, Singh VP and Srivastava SK. 2004a. Antibiotic sensitivity patterns among Indian strains of avian *Pasteurella multocida*. *Trop Anim Health Prod* 36: 743-750.

48. Shivachandra SB, Kumar AA, Gautam R, Singh VP, Chaudhuri P and Srivastava SK. 2004b. PCR assay for rapid detection of *Pasteurella multocida* serogroup-A in morbid tissue materials from experimentally induced Fowl cholera in chicks. *Vet J* 168: 349-352.

49. Shivachandra SB, Kumar AA, Gautam R, Saxena MK, Chaudhuri P and Srivastava SK. 2005. Detection of multiple strains of *Pasteurella multocida* in fowl cholera outbreaks by PCR- based typing. *Avian Path* 34: 456-462.

50. Shivachandra SB, Kumar AA, Gautam R, Joseph S, Saxena MK, Chaudhuri P and Srivastava SK. 2006a. Characterization of avian strains of *Pasteurella multocida* by restriction endonuclease and amplified fragment length polymorphism. *Res Vet Sci* 81: 8-18.

51. Shivachandra SB, Kumar AA, Gautam R, Joseph S, Chaudhuri P, Saxena MK, Srivastava SK and Singh N. 2006b. Detection of *Pasteurella multocida* in experimentally infected embryonated chicken eggs by PCR assay. *Ind J Exp Biol* 44: 321-324.

52. Shivachandra SB, Kumar AA, Gautam R, Singh VP, Saxena MK and Srivastava SK. 2006c. Identification of avian strains of *Pasteurella multocida* in India by conventional and PCR assays. *Vet J* 172: 561-564.

53. Shivachandra SB, Kumar AA, Saxena MK and Srivastava SK. 2006d. Biochemical characterization of avian strains of *Pasteurella multocida* in India. *Ind J Anim Sci* 76: 429-432.

54. Shivachandra SB, Kumar AA and Chaudhuri P. 2007. Differentiation of avian *Pasteurella multocida* strains by single primer PCR. *Vet Res Commun* 31: 941-949.

55. Shivachandra SB, Kumar AA and Chaudhuri P. 2008. Molecular characterization of avian strains of *Pasteurella multocida* serogroup-A based on amplification of repetitive regions by PCR. *Comp Immun Microbiol Infect Dis* 31: 47-62.

56. Siahanidou T, Gika G, Skiathitou AV, Oikonomopoulos T, Alexandrou-Athanassoulis H, Koutouzis EI and Syriopoulou VP. 2012. *Pasteurella multocida* infection in a neonate: Evidence for a human-to-human horizontal transmission. *Pediatr Infect Dis J* 31: 536-537.

57. Strand CL and Helfman L. 1971. *Pasteurella multocida* chorioamnionitis associated with premature delivery and neonatal sepsis and death. *Am J Clin Pathol* 55: 713-716.

58. Sun D, Wang J, Wu R, Wang C, He X, Zheng J and Yang H. 2010. Development of a novel LAMP diagnostic method for visible detection of swine *Pasteurella multocida*. *Vet Res Commun* 34: 649-657.

59. Takeda S, Arashima Y, Kato K, Ogawa M, Kono K, Watanabe K and Saito T. 2003. A case of *Pasteurella haemolytica* sepsis in a patient with mitral valve disease who developed a splenic abscess. *Scand J Infect Dis* 35: 764-765.

60. Tjen C, Wyllie SA and Pinto A. 2007. *Pasteurella* meningo-encephalitis - a risk of household pets. *J Infect* 55: 479-480.

61. Townsend KM, Frost AJ, Lee CW, Papadimitriou JM and Dawkins HJS. 1998. Development of PCR assays for species and type specific identification of *Pasteurella multocida* isolates. *J Clin Microbiol* 36: 1096-1100.

62. Townsend KM, Boyce JD, Chung JY, Frost AJ and Adler B. 2001. Genetic organisation of *Pasteurella multocida* cap loci and development of a multiplex capsular PCR typing system. *J Clin Microbiol* 39: 924-929.

63. Umemori Y, Hiraki A, Murakami T, Aoe K, Matsuda E, Makihara S and Takeyama H. 2005. Chronic lung abscess with *Pasteurella multocida* infection. *Intern Med* 44: 754-756.

64. Weber DJ, Wolfson JS, Swartz MN and Hooper DC. 1984. *Pasteurella multocida* infections. Report of 34 cases and review of the literature. *Medicine (Baltimore)* 63(3): 133-154.

65. WHO. 1959. Second report of joint FAO/WHO expert group on zoonoses, Technical Report Series No. 169. World Health Organization, Geneva.

2014, Zoonoses: Bacterial Diseases *Pages* **301–313**
Editor: **Sudhi Ranjan Garg**
Published by: **DAYA PUBLISHING HOUSE, NEW DELHI**

14

Glanders

P.K. Kapoor[1] and S.K. Khurana[2]
[1]*Department of Veterinary Public Health and Epidemiology,
Lala Lajpat Rai University of Veterinary and Animal Sciences,
Hisar – 125 004*
[2]*National Research Centre on Equines, Hisar – 125 001*

Glanders is a highly contagious disease primarily of equids, particularly horses, donkeys and mules. Caused by Gram-negative bacteria, *Burkholderia mallei* (previously named *Pseudomonas mallei*), it is a disease of high zoonotic and public health significance. The infection in humans usually results in fatal disease. It is primarily an animal disease and humans get infected accidentally. Based on the direction of transmission of disease, it is a direct anthropozoonosis. During recent times, glanders has been reported in 2011 from many parts of world, including India, Iran, Afghanistan, Mongolia, Lebanon and Brazil. In India, several cases were reported during the years 2006-2010.

Considering that the disease in India is restricted to certain pockets with sporadic cases, if any case of glanders is diagnosed in the field on clinical symptoms, it needs a timely confirmation for applying early mitigation and control strategy. The disease is a notifiable disease in India under the old Glanders and Farcy Act 1899, now substituted by the Infectious Animal Disease Act 2009.

Glanders organisms are believed to have been used as biological warfare agent during the World War I to infect horses and mules that carried supplies and also during the World War II on horses and humans. Thus, the organism has a clear potential as bioterrorism agent in the present-day scenario. The Center for Disease

Control and Prevention (CDC) classifies glanders as a category B disease of concern for bioterrorism because of its easy spread.

Geographic Distribution

Although glanders was once widespread throughout the world, it has been eradicated from many countries by diligent test and slaughter programmes. It is currently limited to parts of Asia and Africa, possibly the Balkan states, former Soviet republics, Mexico and South America. The affected Asian countries include Turkey, Syria, Iraq, Iran, Pakistan, India, Myanmar (Burma), Indonesia, Philippines, China and Mongolia. In India, several cases have been reported during the past few years (Malik *et al.* 2009, 2010, 2012).

The geographic distribution of *B. mallei* can be difficult to determine precisely, as cross-reactions with *B. pseudomallei* interfere with serological surveys. In countries that have eradicated glanders, cases may occur in researchers who work with this agent. In 2000, the disease was reported in a researcher in the United States (CFSPH 2007).

Etiological Agent and Vectors

The causative agent of glanders is a Gram-negative rod-shaped bacterium known as *Burkholderia mallei*. The organism is non-motile. Since its discovery, this microorganism has undergone large taxonomic changes. It was earlier placed in other genera including *Bacillus, Corynebacterium, Mycobacterium, Loefflerella, Pfeifferella, Malleomyces, Actinobacillus* and *Pseudomonas*. The organism was finally assigned to the current genus *Burkholderia* in 1992. Former names of this pathogen include *Loefflerella mallei, Pfeifferella mallei, Malleomyces mallei, Actinobacillus mallei, Corynebacterium mallei, Mycobacterium mallei,* and *Bacillus mallei.* The organism is closely related to *B. pseudomallei,* the etiological agent of melioidosis. Genetic homology between these is about 70 per cent. Because of this many consider them to be biotypes or isotypes.

The organism is destroyed by direct sunlight and is sensitive to desiccation. It may survive for up to 6 weeks in infected stables. It is susceptible to many common disinfectants, including benzalkonium chloride, 1 per cent sodium hypochlorite, 70 per cent ethanol, 2 per cent glutaraldehyde, iodine, mercuric chloride in alcohol and potassium permanganate. The organism is less susceptible to phenolic disinfectants. It is also destroyed on heating to 55°C (131°F) for 10 minutes, or with ultraviolet irradiation.

Host Range

Glanders primarily affects the equids. Donkeys generally experience acute form of the disease, horses suffer a more chronic form, while mules face an intermediate form because of the differences in their susceptibility to the disease. Carnivores are susceptible to disease if they consume meat affected by glanders; felids appear to be more susceptible than canids. Outbreaks of glanders in captive wild felids have been reported. Humans are also susceptible to infection making glanders an important occupational disease of veterinarians, farriers, animal farm workers and laboratory personnel. Swine and cattle are resistant to infection with *B. mallei*, but goats can be infected. Several laboratory animals including hamsters and guinea pigs are susceptible. Experimental inoculation of the organism in guinea pigs results in severe peritonitis and orchitis (Strauss reaction).

Disease Transmission

The disease gets introduced into the horse population from the diseased or latently infected animals. Major route of infection is the ingestion of the pathogen present in secretions of infected animals. Inhalation of the organism is less likely to cause typical disease. Invasion by way of skin lesion is considered of minor importance in natural spread of disease. Close proximity alone may not usually be sufficient for transmission of glanders. Transmission is facilitated if the animals share feeding troughs and watering facilities or if they nuzzle each other.

Transmission through ocular mucous membranes and abrasions in the skin is also possible. Vertical transmission from mare to foal has occurred naturally in horses. *In utero* transmission from sow guinea pig to pup has been reported in laboratory animals. Sexual transmission from stallion or jack to mare or jenny has also been recorded. The use of asymptomatic stallions for breeding purpose resulted in the spread of glanders near the turn of the 20th century. Horses, donkeys and mules can transmit disease to other animals, including human beings. Nasal discharge and exudates from various lesions contain large number of bacteria and are potent source of infection.

Humans are infected by contact with sick animals, contaminated fomites, tissues or bacterial cultures. Transmission is often through small wounds and abrasions in the skin. Ingestion or inhalation can also cause infection. Transmission through unbroken skin has been reported, but not proven. Most laboratory-acquired infections have occurred during routine handling of cultures or samples, rather than after injuries or accidents. Rare cases of person-to-person transmission have been reported in family members who nursed the sick individuals. Two cases, thought to be sexually transmitted, are also on record. Aerosols may be the major route of infection in a bioterrorist attack.

Environmental and Other Predisposing Factors

Although *B. mallei* organisms are inactivated by heat and sunlight, their survival is prolonged in wet or humid environments. The organisms remain viable in water at room temperature for up to a month. Some sources suggest that it might be able to survive for more than a year in the environment under some circumstances. Others state that it may survive for up to few months in favourable environments, but it is likely to be inactivated within two weeks in unfavourable conditions.

Occurrence of Disease in Animals

Once prevalent throughout the world, glanders has now been eradicated by rigorous control programmes in many advanced countries like USA, England, Australia and Canada. However, the disease is still endemic in many countries of Middle East, Central and South America, Asia and Africa (Al-Ani *et al.* 1998, Antonov *et al.* 2008). During the last two decades, the occurrence of glanders in equines has shown steady rise and thus it is currently considered as a re-emerging disease (Wittig *et al.* 2006).

As regards India, several cases of the disease among horses occurred in Bareilly (Uttar Pradesh) during the period 1881-1884 (Verma 1981). During 1988 to 1991 and in 1998, sporadic cases of glanders were recorded from Haryana, Punjab and Uttar Pradesh (Anon. 1988, 1990, 1991, Kumar *et al.* 1999). A major outbreak of glanders occurred in India in 2006 in Maharashtra (26 cases). The cases of the disease were later reported in six other states too, namely, Uttar Pradesh (70), Uttarakhand (21), Punjab (3), Andhra Pradesh (16), Himachal Pradesh (6) and Haryana (1) till 2007. The disease in equines was confirmed on the basis of clinical signs and symptoms, and microbiological and serological investigations. Fourteen isolates indistinguishable from *B. mallei* were obtained from clinical samples while serum samples of 140 equines were found positive by complement fixation test (Malik *et al.* 2009).

In India, several cases have been reported during 2006-07 in Maharashtra (Malik *et al.* 2010), Punjab, Uttarakhand and Uttar Pradesh during 2007-08 (Malik *et al.* 2009), from Uttarakhand, Andhra Pradesh, Himachal Pradesh and Haryana during 2009-10 (Malik *et al.* 2012), from Chhattisgarh, Uttar Pradesh and Himachal Pradesh.

Occurrence of Disease in Human Beings

Though not a common disease of human beings, outbreaks of glanders have been reported in Austria and Turkey, no human epidemic has been recorded (Neubauer *et al.* 1997). Infection occurs rarely and the probability is more in certain risk groups as this is primarily an occupational disease in humans. In India, human glanders was first reported as early as in the first half of 20th century (Steele 1979).

Zoonotic transmission of *B. mallei* from equid to humans is uncommon even on close and frequent contact with infected animals, which might be due to low concentrations of organisms in the infection sources and the species-specific differences in susceptibility to virulent strains. During the World War II, human glanders was rare despite 30 per cent prevalence in horses in China. Between 5 per cent and 25 per cent of the animals tested in Mongolia were reactive, yet no human cases were reported. However, humans are susceptible and the probability of acquiring infection is more in certain risk groups. Humans exposed to infected equids have contracted glanders in occupational, hobby, and lifestyle set-tings. Veterinarians and veterinary students, farriers, flayers (hide workers), transport workers, soldiers, slaughterhouse personnel, farmers, horse fanciers and caretakers, and stable hands have been naturally infected. Subclinical or inapparent infections in horses and mules pose a hidden risk to humans. Human-to-human transmission is rare. Infection by ingesting contaminated food and water has occurred; however, it does not appear to be a significant route of entry for human infections. Laboratory workers have also been rarely and sporadically infected. In contrast to zoonotic transmission, culture aerosols are highly infectious to laboratory workers. The six infected work-ers in the Howe and Miller case series represented 46 per cent of the personnel actually working in the laboratories during the year of occurrence (Darling 2004).

Laboratory acquired cases of glanders with 7 deaths due to accidental ingestion through mouth pipetting and centrifuge accident have been reported (Collins 1983). Wilkinson (1981) has mentioned that many early investigators became infected during the course of their studies and died of glanders but were not reported. A series of cases of occupational glanders in horse owners and veterinarians was reported with five deaths including the veterinarian who conducted the postmortem of infected horse and the physician who performed autopsy of the corpse of veterinarian in Czech Republic during 1923 (Pospisil 2000). A microbiologist researching *B. mallei* at the US Army Medical Research Institute of Infectious Disease contracted glanders in March 2000 (Srinivasan *et al.* 2001).

A veterinary pathologist at the Punjab Veterinary College, Lahore (in undivided India) became infected during 1910s while autopsying a diseased horse and suffered chronic illness and 45 surgeries including amputation of an arm. In 1970s, human glanders was reported in veterinary microbiologists in Mathura and Mukteswar (Steele 1979).

Disease Manifestations in Animals

Glanders is a contagious and fatal disease of horses, donkeys and mules. The disease causes nodules and ulcerations in the upper respiratory tract and lungs. Typical nodules, ulcers, scars and a debilitated condition can be sufficient to diagnose glanders. Unfortunately, many cases of glanders are latent and clinically inapparent. The disease is commonly classified into three types:

1. *Pulmonary glanders*: This form is characterized by formation of round, grayish or white, firm, encapsulated nodules throughout the lung tissue. The nodules contain yellowish and cheesy pus.

2. *Nasal glanders*: There is typical formation of nodules in mucous membrane of nasal cavity, particularly of septum. Many of these nodules rupture and liberate mucopurulent exudates which become mixed with the serous or mucopurulent exudates already discharging from the nostrils. The yellowish green exudate is highly infectious. The ruptured nodules form ulcers with irregular, raised and hyperemic borders. The ulcers heal slowly to form a stellate scar. Both nodules and ulcers may be seen simultaneously. Regional lymph nodes are commonly swollen.

3. *Cutaneous glanders or Farcy*: Multiple nodules formation occurs along the lymph vessels between the affected lymph nodes. These nodules often rupture through skin to discharge yellowish exudate and form deep ulcers which heal slowly. Typical cutaneous lymphatic vessels may be referred to as Farcy pipes.

In most outbreaks, these forms may not be clearly distinct and may occur simultaneously in an animal. Chronic forms are more common. The acute form typically progresses to death within about a week. The acute form is more common in donkeys and mules than in horses.

Disease Manifestations in Human Beings

Symptoms of glanders in humans include fever, muscle aches, chest pain, muscle tightness, weakness and headache. Additional symptoms of glanders largely depend on the way the infection was acquired:

1. *Skin infections*: If the bacteria enters through a cut or scratch in the skin, a localized infection with ulceration will develop at the infected site within 1 to 5 days. Swollen lymph nodes are also common.

2. *Pulmonary infections*: Infection of the lungs can lead to pneumonia, pulmonary abscesses and collection of fluid inside the chest cavity around the lung. Symptoms include chest pain, cough and shortness of breath.

3. *Bloodstream infections*: The infection can progress to bloodstream infection which is usually fatal within 7 to 10 days. Symptoms include very high fever, rapid heart rate and often a rash.

4. *Chronic infections*: Symptoms of chronic infection become apparent more slowly than other types of infection. The usual symptoms are abscesses in the skin and muscles of the arms and legs, or rarely, in the spleen or liver.

Diagnostic Procedures

Tentative diagnosis of the disease is obtained through observation of clinical signs and symptoms in an animal. The following field and laboratory tests help in confirmatory diagnosis of the disease.

Mallein Test

Mallein test is the main test for field diagnosis of glanders and is a prescribed test for international trade. Mallein PPD is a lysate (solution of water soluble protein fractions of heat-treated *B. mallei*) containing both endotoxins and exotoxins elaborated by the organism. Infected animals are allergic/hypersensitive to mallein and exhibit local and systemic hypersensitivity after mallein inoculation similar to that exhibited in tuberculin testing. Inoculation with mallein may trigger a humoral serologic reaction to the complement fixation test. This seroconversion is thought to be transient but may be permanent if the animal undergoes repeated mallein testing. This is extremely important to consider if samples from the animals are destined for complement fixation test. Further, advanced clinical cases in horses and acute cases in donkeys and mules may give inconclusive results requiring additional methods of diagnosis to be employed.

Intradermo-palpebral Test

This is the most sensitive, reliable and specific test. The preferred method of application of mallein is intradermo-palpebral. The concentrated mallein PPD (0.1 ml) is injected intradermally into the lower eyelid. The test is usually read at 24 and 48 hours after injection. A positive reaction is visualized by marked oedematous swelling of the eyelid, and there may be a purulent discharge from the inner canthus or conjunctiva, accompanied by photophobia, pain and depression within 12 to 72 hours. There is usually a rise in temperature. In a negative response, there is usually no reaction or only a little swelling of the lower lid.

Ophthalmic Test

In this test, few drops of mallein are instilled into the eye at the canthus. In an infected animal, the eyelids, and sometimes the side of the face, become swollen. Development of severe conjunctivitis within 6-12 hrs may also be evident. The reaction may also occur to a lesser extent in the opposite eye. This is less reliable than the intradermo-palpebral test.

The Subcutaneous Test

In this test, a 10 cm square skin patch in the middle of the neck is clipped and disinfected. Dilute mullein (2.5 ml) is injected subcutaneously into the centre of the patch. In a positive test, the horse develops pyrexia of 104°F (40°C) or over during the first 15 hours, and a firm painful swelling with raised edges develops within 24 hours at the injection site. In nonglandered horses, there is no, or minimal, transient local swelling. Doubtful reactors may be retested after 14 days using a double dose of mallein. The horse's temperature has to be under 102°F (38.8°C) on the day before the test, at the time of the injection, and at 9, 12 and 15 hours after the injection. This test interferes with subsequent serological diagnosis.

Microscopic and Cultural Examination

Laboratory examination is done of a whole or section of a lesion, exudate and serum collected aseptically. The samples are required to be shipped to laboratory under refrigeration as soon as possible. Air-dried smears of exudate on glass slides may also be submitted for microscopic examination. The causative organism of the disease may be cultured from fresh lesions or lymph nodes. It may also be demonstrated microscopically in films made from this material.

Morphological Features of *Burkholderia mallei*

The organisms are fairly numerous in smears from fresh lesions, but in older lesions they are scanty. The organisms are mainly extracellular, fairly straight Gram-negative rods with rounded ends, 2-5 µm long and 0.3-0.8 µm wide with granular inclusions of various sizes. The organisms often stain irregularly and do not have capsules or form spores. These have no flagellae and are therefore non-motile. The organisms are difficult to demonstrate in tissue sections, where they may have a beaded appearance. In culture media, they vary in appearance depending on the age of the culture and type of medium. In older cultures, there is much pleomorphism. Branching filaments form on the surface of broth cultures.

Cultural Characteristics

It is preferable to attempt isolation from unopened uncontaminated lesions. The organism is aerobic and facultatively anaerobic only in the presence of nitrate, growing optimally at 37°C. It grows well, but slowly, on ordinary culture media and 48 hour incubation of cultures is recommended. Glycerol enrichment is particularly useful. After a few days on glycerol agar, there is a confluent, slightly cream coloured growth that is smooth, moist and viscid. With continued incubation, the growth thickens and becomes dark brown and tough. The organism also grows well on glycerol potato agar and in glycerol broth, wherein a slimy pellicle forms. The growth is much less luxuriant on nutrient agar while it is poor on gelatin. Glycerin dextrose agar (GDA) containing 5 per cent glycerin, 1 per cent dextrose in nutrient agar is described as suitable for isolation (Verma 1989). In contaminated samples, supplementation of media with substances that inhibit the growth of Gram-positive organisms (*e.g.* crystal violet, proflavine) has proved useful, as has pretreatment with penicillin (1000 units/ml for 3 hours at 37°C).

Alterations to characteristics may occur *in vitro*, so fresh isolates should be used for identification reactions. The organism reduces nitrates. Although some workers have claimed that glucose is the only carbohydrate that is fermented (slowly and inconstantly), other workers have shown that if appropriate medium and indicator are used, glucose and other carbohydrates, such as arabinose, fructose, galactose and mannose, are consistently fermented by *B. mallei*. Indole is not produced, horse blood is not haemolysed and no diffusible pigments are produced in cultures. Lack of motility is of special relevance.

Laboratory Animal Inoculation

Guinea pigs, hamsters and cats have been used for diagnosis when necessary. The Strauss reaction is observed when infectious material from glanders patients is injected intraperitoneally into male guinea pigs. In positive cases, the guinea pig develops localized peritonitis involving the scrotal sac. Glanderous orchitis follows with painful enlargement of the testis. The testis becomes enlarged and painful and ultimately necrotic and is discharged through the scrotal skin. Individual variations have been reported. The number of organisms and their virulence determines the severity of the lesions. The Strauss reaction is not specific for glanders, and other organisms can elicit it. Bacteriological examination of infected testes should confirm the specificity of the response obtained.

Complement Fixation Test

The complement fixation test (CFT) is widely used and a prescribed test for international trade. Although not as sensitive as the mallein test, CFT is an accurate serological test that has been used for glanders diagnosis for many years. It is reported to be about 95 per cent accurate, serum being positive within 1 week of infection and remaining positive in the case of exacerbation of the chronic process.

Enzyme Linked Immunosorbent Assays (ELISA)

Recently a dot ELISA has been developed and found to be superior to all previously described tests in its sensitivity. This test is inexpensive, rapid and easy to perform, and is not influenced by anticomplement activity. Both plate ELISA and membrane (blot) ELISA have been reported for the serodiagnosis of glanders but none of these procedures has been shown to differentiate serologically between *B. mallei* and *B. pseudomallei*. Blotting approaches have involved both dipstick dot-blot and electrophoretically separated and transferred western blot methods. A competitive ELISA that uses an anti-lipopolysaccharide monoclonal antibody has also been developed and found to be similar to the CFT in performance. Continuing development of monoclonal antibody reagents specific for *B. mallei* antigenic components offers the potential for more specific ELISAs in the foreseeable future that will help resolve questionable test results of quarantined imported horses. At this time, none of these tests has been validated.

The avidin-biotin dot ELISA has been described but has not yet been widely used or validated. The antigen is heat-inactivated bacterial culture that has been concentrated and purified. A dot of this antigen is placed on a nitrocellulose dipstick that is then used to test for antibody against *B. mallei* in equine serum. Using antigen-dotted, preblocked dipsticks, the test can be completed in about an hour. Serum or whole blood can be used for the test, and partial haemolysis does not impart any background colour to the antigen-coated area on the nitrocellulose.

Other Serological Tests

A counter-immunoelectrophoresis test has been described. Rose Bengal Plate Agglutination Test (RBT) has been evolved for the diagnosis of glanders in horses and other susceptible animals; the test has been validated in Russia only. The antigen is a heat-inactivated bacterial suspension coloured with Rose Bengal, which is used in a plate agglutination test.

The accuracy of other agglutination tests and precipitin is unsatisfactory for use in control programmes. Horses with chronic glanders and those in a debilitated condition give negative or inconclusive results. Cross-reaction with *B. pseudomallei* is encountered in all serological tests for glanders resulting in false-positive reactions in animals from areas where melioidosis is endemic.

Polymerase Chain Reaction

In the past few years, several PCR and real-time PCR assays for the identification of *Burkholderia mallei* have been developed, but only a PCR and a real-time PCR assay were evaluated using samples from a recent outbreak of glanders in horses (OIE 2008).

A polymerase chain reaction (PCR) for specific detection of *B. mallei* DNA has been developed that allows differentiation between *B. mallei* and *B. pseudomallei*. Two *B. mallei*-specific real-time PCR assays targeting the *B. mallei* $bimA_{ma}$ gene (*Burkholderia* intracellular motility A; BMAA0749), which encodes a protein involved in actin polymerization, were developed (Ulrich *et al.* 2006). The PCR primer and probe sets were tested for specificity against a collection of *B. mallei* and *B. pseudomallei* isolates obtained from numerous clinical and environmental (*B. pseudomallei* only) sources. The assays were also tested for cross-reactivity using template DNA from 14 closely related *Burkholderia* species. The relative limit of detection for the assays was found to be 1 pg or 424 genome equivalents. The applicability of assays to detect *B. mallei* within infected BALB/c mouse tissues was also analysed. Beginning 1 hour post-aerosol exposure, *B. mallei* was successfully identified within the lungs, and starting at 24 hour post-exposure in the spleen and liver. Surprisingly, the organism was not detected in the blood of acutely infected animals (Ulrich *et al.* 2006).

The technique has not yet been fully validated or gained wide acceptance. However, PCR has the potential to be a safe, fast method to confirm the infection. Methods to differentiate between *B. mallei* and *B. pseudomallei* bear great importance.

Differential Diagnosis

Cases for specific glanders investigation should be differentiated on clinical grounds from other chronic infections of the nasal mucosae or sinuses and also from strangles (*Streptococcus equi* infection), ulcerative lymphangitis (*Corynebacterium pseudotuberculosis*), pseudotuberculosis (*Yersinia pseudotuberculosis*), sporotrichosis (*Sporotrichium* spp.) and the other forms of pneumonia. Glanders should be excluded positively from suspected cases of epizootic lymphangitis (caused by *Histoplasma farciminosum*), with which it has many clinical similarities. In humans in particular, glanders should be distinguished from melioidosis (*B. pseudomallei* infection), which is caused by an organism with close similarities to *B. mallei*

Purulent sinusitis, guttural pouch empyema, and other causes of nasal catarrh should also be considered. Skin lesions may be similar to those of dermatophilosis or dermatomycoses such as sporotrichosis. Knowledge of the progressive debilitating nature of glanders and the application of serological or mallein tests will serve to distinguish glanders from other similar diseases.

Prevention and Control Strategies

Animals that test positive for glanders are euthanized except in endemic areas. In an outbreak, the premises should be quarantined, thoroughly cleaned and disinfected. All contaminated bedding and food should be burnt or buried, and equipment and other fomites should be disinfected. Carcasses should also be burnt or buried. Whenever possible, susceptible animals should be kept away from the contaminated premises for several months. In endemic areas, susceptible animals should be kept away from communal feeding and watering areas since glanders is more common where animals congregate. Routine testing and euthanasia of positive animals can eradicate the disease (CFSPH 2007, OIE 2008).

At presenthourthere is no preventive vaccination available for glanders. Definite recommendations for post-exposure prophylaxis also do not exist. In endemic countries, identification and elimination of infection in animals is done to prevent disease in human population.

Strict measures and precautions are essential in handling the infected animals and contaminated fomites. Protective clothing and face masks are required while handling infected animals. Protection from aerosols is also required. Biosafety level 3 facilities and practices are required while handling with cultures infected tissues or other biological material.

Although person-to-person transmission is rare, human glanders patients should be isolated. Infection control precautions should be taken, and disposable surgical masks, face shields, and gowns should be used appropriately during nursing (CFSPH 2007, OIE 2008).

Treatment

Some antibiotics may be effective against glanders but treatment is given only in endemic areas. Treatment is risky even in these regions, as infections can spread to humans and other animals, and treated animals can become asymptomatic carriers (CFSPH 2007). Few studies have been published on the antibiotic susceptibility of *B. mallei*, but some treatment recommendations are available. This organism is usually resistant to some classes of antibiotics. Long-term treatment or multiple drugs may be necessary. Abscesses may need to be drained (OIE 2008).

Future Directions

One Health approach, combining the efforts of public health specialists and veterinary experts, is required to mitigate the problem. Strengthening the following should be considered for better future outlook:

☆ Surveillance for glanders in susceptible animal populations with reporting of all suspected cases supported by appropriate legislation,

☆ General registration system for equids,

☆ Powerful veterinary services,

☆ Identification and humane euthanasia of infected animals,

☆ Compensation of animal owners,

☆ Quarantine measures,

☆ Cleaning and disinfection of infected farms,

☆ Destruction by incineration of euthanized animals and any contaminated material, and

☆ Cooperation of horse owners with veterinarians in disease detection and control.

References

1. Al-Ani FK, Al-Rawashdeh OF, Ali AH and Hassan FK. 1998. Glanders in horses: clinical, biochemical and serological studies in Iraq. *Vet Arhiv* 68: 155-162.

2. Anonymous. 1988. Glanders. *Annual Report*. National Research Centre on Equines, Hisar. p. 48-51.

3. Anonymous. 1990. Glanders. *Annual Report*, 1989-90. National Research Centre on Equines, Hisar. p. 58-59.

4. Anonymous. 1991. Glanders. *Annual Report*, 1990-91. National Research Centre on Equines, Hisar. p. 49.

5. Anonymous. 2006. Glanders. In: *Manual of diagnostic tests and vaccines for terrestrial animals*. http://www.oie.int/eng/normes/MMANUAL/A_00086.htm.

6. CFSPH. 2007. Glanders. The Center for Food Security and Public Health www.cfsph.iastate.edu/Factsheets/pdfs/glanders.pdfShare.

7. Collins CH. 1983. *Laboratory acquired infections: History, incidence, cases and prevention*. Butterworth Ltd., London. p. 6-35

8. Darling P and Woods J (eds.) 2004. *Medical management of biological casualties handbook*. 5th ed.: US Army Medical Research Institute of Infectious Diseases, Fort Detrick, Md.

9. Jana AM, Gupta AK, Pandya G, Verma RD and Rao KM. 1982. Rapid diagnosis of glanders in equines by counter-immunoelectrophoresis. *Indian Vet J* 59: 5-9.

10. Kumar S, Malik P, Jindal N and Garg DN. 1999. Cutaneous glanders in a mule: A case study. *J Remount Vet Corps* 38: 131-133.

11. Malik P, Khurana SK, Singh BK and Dwivedi SK. 2009. Recent outbreak of glanders in India. *Indian J Anim Sci* 79: 1015-1017.

12. Malik P, Khurana SK and Dwivedi SK. 2010. Re-emergence of glanders in India - Report of Maharashtra state. *Indian J Microbiol* 50: 1345-1348.

13. Malik P, Singha H, Khurana SK, Kumar R, Kumar S, Raut, AA, Riyesh T, Vaid RK, Virmani N, Singh BK, Pathak SV, Parkale DD, Singh B, Pandey SB, Sharma TR, Chauhan BC, Awasthi V, Jain S, and Singh RK. 2012. Emergence and re-emergence of glanders in India: a description of outbreaks from 2006 to 2011. *Veterinaria Italiana* 48:167-178.

14. Neubauer H, Meyer H and Finke EJ. 1997. Human glanders. *International Review of the Armed Forces Medical Services* 70: 258-265.

15. OIE. 2008. Glanders. Chapter 2.5.11. *OIE terrestrial manual*. World Organization for Animal Health (Office international des Epizooties), Paris. p. 919-928.

16. Pospisil L. 2000. Contribution to the history of glanders. WAHVM Congress. 31st international congress on the history of Veterinary medicine, Veterinary University, Bruno, Czech Republic. http://www.wahm.vet.uu.nl/specific/activities/congresses/brno.html#23.

17. Srinivasan A, Kraus CN, DeShazer D, Becker PM, Dick JD, Spacek L, Bartlett JG, Byrne WR and Thomas DL. 2001. Glanders in a military research microbiologist. *New Engl J Med* 345: 256-258.

18. Steele JH. 1979. Glanders. In: Steele JH (ed) *CRC Handbook Series in Zoonoses*. Vol. I. CRC Press, Florida. p. 339-362.

19. Ulrich MP, Norwood DA, Christensen DR and Ulrich RL. 2006. Using real-time PCR to specifically detect *Burkholderia mallei*. *J Med Microbiol* 55: 551-559.

20. Verma RD. 1981. Glanders in India with special reference to incidence and epidemiology. *Indian Vet J* 58: 177-183.

21. Verma RD. 1989. Research on epizootic, diagnosis and control of glanders with a view to eradicate the disease from India. *Project Report*, ICAR Research Scheme. 54 pp.

22. Wilkinson L. 1981. Glanders: medicine and veterinary medicine in common pursuit of a contagious disease. *Med Hist* 25: 363-384.

23. Wittig MB, Wohlsein P, Hagen RM, Al Dahouk S, Tomaso H, Scholz HC, Nikolaou K, Wernery R, Wernery U, Kinne J, Elschner M and Neubauer H. 2006. Glanders - a comprehensive review. *Deutsche Tierarztliche Wochenschrifte* 113: 323-330.

2014, Zoonoses: Bacterial Diseases   *Pages* **314–324**
Editor: **Sudhi Ranjan Garg**
Published by: **DAYA PUBLISHING HOUSE, NEW DELHI**

15

Melioidosis

Subhash Verma and Mandeep Sharma
Dr. G.C. Negi College of Veterinary and Animal Sciences,
C.S.K. Himachal Pradesh Agricultural University, Palampur – 176 062

Captain Alfred Whitmore and C. S. Krishnaswami, an Indian bacteriologist, first described melioidosis as a 'glanders-like' disease among morphine addicts, malnourished and ne-glected inhabitants in Rangoon (Myanmar) in 1911. The organism could be distinguished from the glanders causing bacterium by its relatively rapid growth, motility, and lack of the "Strauss reaction" when injected into guinea pigs. The term melioidosis for the disease was derived from the Greek words 'melis' (distemper of asses) and 'eidos' (resemblance) by Stanton and Fletcher in 1921. The causative organism is a bacterium called *Burkholderia pseudomallei* that was used to be known as *Pseudomonas pseudomallei* until mid 1990s. The organism has also been known by other older names like *Malleomyces pseudomallei, Bacillus whitmori, Pfeiferella pseudomallei* and *Whitmorella pseudomallei.* It is an environmental organism found in soils and water, and the disease is frequently reported from tropical and subtropical zone between 20°S and 20°N, particularly Thailand and Northern Australia.

Melioidosis has been known by different names such as Whitmore's disease or morphine injector's septicemia (for the contribution of Whitmore in understanding the disease in morphine addicts); Vietnamese time bomb (due to possibility of re-activation of latent infections in soldiers who returned from war in Vietnam); pseudomallens; and pseudoglanders. The disease is important from the human health, animal health and zoonotic standpoint. In humans, it can exist in acute and chronic forms and can lead to significant mortality. The infection in animals may lead to notable morbiditiy, followed by deaths and substantial economic losses.

Geographical Distribution

Melioidosis is endemic in several Southeast Asian countries like Vietnam, Cambodia, Laos, Thailand, Malaysia and Singapore, but also occurs in other regions including Africa, Australia, the Middle East, India and China. The disease is most commonly reported in northeast Thailand and northern Australia. Isolation of the organism has also been made from the temperate regions of southwest Australia and France. Isolated cases have also occurred in South America, western hemisphere in Aruba, Brazil, Mexico, Panama, Ecuador, Haiti, Peru and Guyana, and in the states of Hawaii and Georgia in the United States. *B. pseudomallei* are generally found in water or moist soil.

Etiological Agent

Burkholderia pseudomallei are Gram-negative, bipolar, aerobic, flagellated, motile bacteria that are rod-shaped, non-sporulating and measure 2-5 µm in length and 0.4-0.8 µm in diameter. The bacteria can grow in a number of artificial nutrient media such as blood agar, MacConkey Lactose agar, eosine methylene blue agar, etc. Their optimal proliferation temperature is around 40°C in neutral or slightly acidic environments (pH 6.8-7.0). These constitute environmental organisms that as such have no requirement such as vectors to pass through an animal host in order to replicate.

Host Range

Burkholderia pseudomallei organisms affect both humans and animals. These have a broad host range and can infect nematodes, amoebae, dolphins, birds, swine, gorillas and domestic animals. Several livestock species, including goats, sheep, camels and alpacas, are considered particularly sensitive to the infection. Pigs and deer are moderately sensitive, while cats, dogs, crocodiles, birds and cattle are only partially sensitive to it. Laboratory animals such as hamsters, guinea pigs, rabbits, mice and rats are also affected by melioidosis. The introduction of naive livestock to endemic regions may predispose them to disease as seen in sheep, goats, pigs and camelids.

Disease Transmission

Humans and animals are believed to acquire the infection by inhalation of contaminated dust or water droplets, ingestion of contaminated water, and contact with contaminated soil and surface water, especially through skin abrasions. During warfare, persons can become infected through contamination of war wounds. The bacteria live below the soil surface during the dry season and may rise to groundwater during the wet season increasing the chances of infection in animals and humans. Though melioidosis is not highly contagious, care must be taken when handling potentially infected animals. The bacteria can be shed from an animal in urine, milk and nasal secretions. Unpasteurised milk from an infected animal thus is a possible source of infection and it should be avoided. Direct human-to-human and animal-to-human transmission is rare but can occur after contact with blood or body fluids. While a few cases have been documented through direct contact, contaminated soil and surface muddy water remain the primary way in which people become infected.

Rodents also constitute a possible source of infection. Infected animals can contaminate and maintain organisms in the soil for a very long period of time. *B. pseudomallei* have also been introduced to new environments with the export of animals and shipments of contaminated soil and water.

Environmental and Other Predisposing Factors

Burkholderia pseudomallei is a saprophytic organism which could be isolated from top layer of soil and muddy water in endemic areas but mostly found below 25 cm in deeper layers up to 120 cm in dry months (Thomas *et al.* 1979, Brook *et al.* 1997). Watering of fields can result in surfacing of the bacteria thus contaminating the surrounding environment as happens after agricultural interventions, heavy rains or floods after dry seasons. Under such situations, the risk of infection through inhalation, ingestion of contaminated water or contact (through open wounds) becomes a frequent possibility. Higher incidence of infection is associated with reconstruction work of water bodies, soil erosion and use of drinking water or untreated water for spray cooling of animals during summer season in tropical areas when there is heavy contamination with soil particles (Thomas *et al.* 1981, Ketterer *et al.* 1986, Inglis *et al.* 1999).

Further, it has been shown that the bacteria could be recovered from water for up to 8 weeks at room temperature, for up to 7 months from muddy water and for as long as up to 30 months from the soil kept under laboratory conditions (Thomas and Forbes-Faulkner 1981). Bacterial multiplication is very well supported in soils having wide range of temperature (4-42°C) and pH (4.0-8.0) together with an average minimal humidity of 10-15 per cent. It is believed that dispersion of *B. pseudomallei* is often influenced by natural environmental factors such as temperature, rainfall and humidity, and also by the soil micro environment (Frederick 1997). In hyper-endemic areas of northern Australia and north-eastern Thailand, the disease incidence in humans peaks in the rainy season. The primary mode of infection shifts from the percutaneous inoculation to inhalation after heavy rainfalls, which leads to more severe illness. A history of contact with soil or surface water is, therefore, almost invariable in human subjects with melioidosis.

The bacteria are shed in the infected animal's urine, milk and nasal secretions. These organisms can withstand sunlight, are resistant to dry heat and can survive for a month in dry soil, water or rodent excrement. One of the most interesting aspects of melioidosis epidemiology has been its association with extreme weather events. In northern Australia, the acute form of infection is strongly associated with the onset of summer rainy season (Currie *et al.* 2003). Further, it is believed that the Tsunami caused additional cases of melioidosis around the rim of the Indian Ocean, when many people were forcefully swept through mud and surface water by the tidal wave (Allworth *et al.* 2005).

The presence of certain other disease conditions in people also influences the occurrence of melioidosis. The persons suffering from diabetes mellitus, alcoholism, cirrhosis, thalassemia, or other immunosuppressive syndromes are at an increased risk of developing symptomatic melioidosis infection. Chronic lung diseases, excess kava consumption, and cystic fibrosis are other risk factors. Diabetes appears to be the most important of all the known risk factors because up to 50 per cent of human patients with melioidosis have this condition.

Prevalence of Disease in Animals

Cases of melioidosis in animals have been reported from many countries around the globe, including Australia, China, Thailand, Taiwan, Singapore, Iran, Saudi Arabia, United Arab Emirates, Chad, South Africa, Brazil, France and Spain. In Western Europe, the melioidosis outbreak in a zoo in Paris in 1975 was quite unusual as it occurred in a non-endemic region and subsequently spread to other zoos in Paris and Mulhouse and equestrian clubs throughout France. Some of the classical episodes of melioidosis, that were recorded from animals, have been presented in Table 15.1. Cattle and water buffaloes seem to be relatively resistant to infection despite the fact that these are exposed to mud.

In India, though there are no recent published reports regarding the prevalence of melioidosis in animals, but given the fact that the disease has been reported in humans from various parts of India including the coastal areas, there is no reason to believe that animals in these areas do not remain susceptible to autochthonous melioidosis. As early as 1930, a reference to animal melioidosis in the Indian subcontinent has been provided by Nicholls (1930).

Prevalence of Disease in Human Beings

The global epidemiology of melioidosis in human beings has been very comprehensively reviewed by Dance (1991 and 2000). Melioidosis is endemic in the countries of Southeast Asia and northern Australia. The human cases of melioidosis, which have been reported outside the Southeast Asia, are mainly among the travellers visiting these endemic areas. In addition to this, sporadic autochthonous cases have been reported throughout the world, including Africa, the Caribbean, Central and South America, and the Middle East.

Many cases of sporadic melioidosis have been reported from distinct regions of India, and the disease is now increasingly being recognized as an emerging infectious disease in the country. Cases have been reported but not limited to the states largely comprising of south Indian and coastal territories like Karnataka, Tamil Nadu, Andhra Pradesh, Odisha, Kerala, Maharashtra, West Bengal, Tripura, etc. According to a published report, there has been a six-fold increase in the number of cases in 2006 and 2007 as compared to 2001 in humans in the state of Karnataka (Saravu *et al.* 2012). The disease has been noted predominantly in males, probably due to their greater involvement in outdoor activities. The other factors such as rains and >45 years age have been also associated with higher risk of the disease. Farmers have been more commonly affected probably because of their exposure to soil. A study on the persons involved in farming in rural rice growing areas around Vellore has shown as high as 7 per cent seroprevalence of the infection (Kang *et al.* 1996). In Malaysia also, high exposure to *B. pseudomallei* has been indicated by serological results. Around 17-22 per cent rice farmers and 26 per cent in blood donors have antibodies against the microorganism (Vadivellu *et al.* 1995). In North Australia, 0.6

Table 15.1: Episodes of melioidosis in animals

Region	Year/Period	Major Features of the Event	References
Caribbean islands/Aruba	1957	Epizootic cluster infection of sheep, goats, pigs	Sutmoller et al. (1957)
South-Western Australia	1967	Melioidosis in lambs	Ketterer and Bamford (1967)
Hong Kong	1975	Death of 24 dolphins in oceanarium	Huang (1976)
South-Eastern Australia	1975	Melioidosis in cattle	Ketterer et al. (1975)
Paris zoo and other parts of France	1970s	"l'affaire du jardin des plantes," due to either importation of horses from Iran or an infected panda donated by Mao Tse-Tung	Dance (1991), White (2003)
South- Eastern Queensland, Australia	1981-1983	Melioidosis in 159 pigs in 8 intensive piggeries	Ketterer et al. (1986)
Britain	1992	Outbreak of melioidosis in primates imported from Philippines and Indonesia	Dance et al. (1992)
UAE	1997	First case of melioidosis confirmed in Dromedary	Wernery et al. (1997)

to 16 per cent of children have shown evidence of *B. pseudomallei* infection (Currie *et al.* 2000).

Disease Manifestations in Animals

Not only several epidemiological features but also the clinical features of melioidosis are similar in animals and humans. Both animals and humans acquire infection from the organisms present in soil and surface water, and they can be infected by the same environmental clone of *B. pseudomallei*. The disease is manifested more in pigs, goats and sheep as compared to cattle, horses, dogs, rodents, birds, dolphins, tropical fish, primates and various wild animals. Laboratory animals such as hamsters, guinea pigs and rabbits can be infected experimentally. The incubation period can be variable, being as short as few days or may extend up to months or years.

Depending on the site of infection and location of lesions, the symptoms vary in the animals and include fever followed by depression and weight loss. Occasionally, swelling of the joints leading to lameness is also noted. Symptomatic melioidosis may mimic many other diseases. *B. pseudomallei* infection results in suppurating or caseous lesions in lymph nodes or other organs. Sometimes the infection may be asymptomatic and abscesses may be found in clinically normal goats, sheep and pigs. The clinical signs may include fever, loss of appetite and lymphadenopathy, often involving the submandibular nodes in pigs. Some species exhibit lameness or posterior paresis, encephalitis, nasal discharge, gastrointestinal symptoms and/or respiratory signs. Irrespective of the species involved, extensive abscessation of vital organs proves fatal.

In sheep and goats, lung abscesses and pneumonia are common. Affected sheep are pyretic, exhibit coughing along with ocular and nasal discharge. Sheep may also suffer from neurologic disease and could be lame due to swollen joints. All this eventually leads to gradual emaciation that precedes death. Some animals may display only weakness and fever. Mastitis is sometimes seen in goats and the superficial lymph nodes and udder may contain palpable abscesses. Mastitis and secretion of organism in milk necessitates pasteurization of commercial goat milk in the tropics. Pulmonary lesions in goats are usually less severe than in sheep and coughing is not prominent. Infections in pigs are usually chronic and asymptomatic; but if acute infection happens it may result in septicaemia with fever, anorexia, coughing, nasal and ocular discharges. Abortions, stillbirths in sows and orchitis in boars may also occur. Cattle are rarely affected, but may develop pneumonia or neurologic signs. In horses, neurologic, respiratory and gastrointestinal disease (colic and diarrhoea) has been reported.

Disease Manifestations in Humans

Melioidosis has a broad spectrum of clinical manifestation showing acute, subacute and chronic forms earning it a reputation of 'the remarkable imitator'. Septicemia with metastatic lesions is the hallmark of actue disease; whereas cellulitis, lymphangitis and tuberculosis-like pneumonia is reported in subacute forms. Chronic cellulitis involving localised areas is seen in chornic form of disease. However, as suggested by serology, most infections in humans are asymptomatic, and a part of these could be latent infection with a potential for reactivation.

The clinical disease occurs mostly in the individuals who have some associated risk factors such as diabetes mellitus, alcoholism, cirrhosis, thalassemia or other immunosuppressive states. Among the important disease manifestations, pneumonia comes at top. Other clinical presentations include septic arthritis, osteomyelitis, and abscesses in internal organs, particularly spleen, kidneys, prostate and liver.

Intracellular survival of *B. pseudomallei* in hosts (animals and humans alike) equips the organism with the property of latency. It has long been recognized that analogous to tuberculosis, *B. pseudomallei* has the potential for re-activation. Hence there was a concern of the 'Vietnamese time bomb' in returned US soldiers who could have possibly been exposed to infected environment during the war in Vietnam.

Diagnosis

The samples for laboratory diagnosis include throat swabs, sputum, blood, urine, abscess material and wound swabs. Bacteriological culture represents the gold standard for diagnosis. Culture media such as blood agar, MacConkey Lactose agar and Ashdown agar (selective medium) are used for isolation of *B. pseudomallei*. Ashdown selective medium contains certain dyes and gentamicin to suppress the growth of fast growing contaminating microflora present in sputum.

The exudate/morbid material swabs can directly be streaked onto the culture media and incubated for a minimum of 4 days at 37°C. The organism reveals small, smooth creamy colonies in the first 1-2 days, which gradually change after a few days to dry, wrinkled colonies on blood agar (Figure 15.1). The size of the colony can vary depending upon the strain and culture medium. *B. pseudomallei* often produces a distinctive musty or earthy odour which is very pronounced on opening a petri dish harbouring the microorganism; however "sniffing" of plates containing *B. pseudomallei* should never be practised.

On Gram staining, the organisms appear pale pink thin rods exhibiting non-specific bipolar staining, occurring singly but occasionally in clusters. Filamentous forms may be seen particularly in patients receiving antibiotic therapy. Suspected isolates should be subjected to biochemical phenotyping. The organism are oxidase positive, resistant to gentamicin and colistin, and motile.

Molecular methods such as polymerase chain reaction, ribotyping, pulse field gel electrophoresis, multilocus sequence typing, variable number of tandem repeats are useful in strain differentiations in epidemiological studies to identify the source of infection. Ribotyping has proved its value in the disease outbreak investigations in animals and humans by identifying the source of contamination and showing that melioidosis is saprozoonosis and animal-to-human transmission occurs only occasionally.

The diagnostic value of serological tests is questionable in endemic areas, as healthy individuals may show persistent IgG levels. In non-endemic areas, the tests might be useful for detection of chronic infections. Some of the serological assays that

Figure 15.1: *Burkholderia pseudomallei* colonies on sheep blood agar after 72 hour incubation (Photo courtesy: CDC, Larry Stauffer, Oregon State Public Health Laboratory).

could be used are indirect haemagglutination assay, fluorescent antibody technique and enzyme-linked immunosorbant assay.

Prevention and Control

Given the nature of disease, prevention and control of melioidosis is difficult. In endemic areas, avoiding any kind of stress which puts the animals to risk of contracting the infection is helpful. Limiting the access of animals to high risk areas may reduce the disease incidence. As the infected animals can shed the organism in nasal secretions, milk, faeces, urine, and in wound exudates, it is advisable to prevent transmission from these sources. Pasteurization of milk from infected dairy farms is an important preventive strategy. Proper condemnation and disposal of the infected animal carcasses at abattoir is also important to prevent the contamination of the premises.

Rain water entering into yards with earthen floors could put animals to risk of acquiring infection. Construction of yards with concrete floors to ensure dry solid ground and practice of feeding in troughs rather than from the floor in regions where melioidosis may occur should be followed. Excreta of infected animals should be removed several times a day, their movement should be restricted; and disinfection may be carried out with potassium hypochlorite solution. Usage of less water on the premises may also be helpful in reducing the incidence of infection.

In areas, where the disease is not endemic, it is advisable to cull the infected animals to prevent any possibility of their acting as subclinical carriers and potential source of infection for humans and other animals since the organism evidences latency and antibiotic resistance. Where environmental conditions are not conducive for the survival of the organism, animals may provide a suitable niche. Some other preventable strategies include making provision of clean water at farms, proper sewage disposal, removing potentially infected animals, hygienic processing of meat, etc.

It is believed that most places from where the organisms could be isolated are cleared, irrigated, cultivated agricultural sites. Therefore, persons engaged in agricultural and outdoor activities should take precautionary measures by wearing protective shoes and clothing and not exposing open wounds to soil and water. Safe work practices must be ensured in laboratories which handle this organism so as to avoid aerosol formation for their own safety and that of others against contracting the infection. Under circumstances of suspected exposure to this organism in the laboratory through aerosol or through experimental animal bite wounds, antibiotic therapy should be undertaken. Currently, no commercial vaccines are available either for humans or for animal use.

Future Directions

Melioidosis is an endemic problem in many countries and is now emerging in other parts of the world. However, its true incidence is unknown in many developing countries because of inadequate diagnostic facilities and lack of awareness of the disease. These issues need to be addressed. Further, as *B. pseudomallei* could usually be mistaken for *Pseudomonas* species, updating the microbiologists and veterinary public health experts on this aspect could help in proper diagnosis. The creation of better culture and diagnostic facilities in rural tropics may provide insights into the actual prevalence of this organism.

Surveillance programmes are also required to understand the actual geographic and demographic distribution of the infection and the disease. It is also important to find out as to how does melioidosis become established by different routes of infection. It will be though challenging but very useful to find out the reasons for the innate resistance of certain species of animals to melioidosis.

Finally, due to the inherent resistance of the causative organism to many antibacterial agents such as penicillin, ampicillin, first and second generation cephalosporins, gentamicin, tobramycin and streptomycin (Mandell *et al.* 2005) and the possibility of reactivation of latent infections, there is urgent need of devising vaccines that are able to induce good protective immunity.

References

1. Allworth AM. 2005. Tsunami lung: a necrotising pneumonia in survivors of the Asian tsunami. *Med J Aust* 182: 364.

2. Brook MD, Currie B and Desmarchelier PM. 1997. Isolation and identification of *Burkholderia pseudomallei* from soil using selective culture techniques and the polymerase chain reaction. *J Appl Microbiol* 82: 589-596.

3. Currie BJ and Jacups SP. 2003. Intensity of rainfall and severity of Melioidosis, Australia. *Emerg Infect Dis* 9: 1538-1542.

4. Currie BJ, Fisher DA, Howard DM, Burrow JN, Selvanayagam S and Snelling PL. 2000. The epidemiology of melioidosis in Australia and Papua New Guinea. *Acta Trop* 74: 121-127.

5. Dance DAB. 1991. Melioidosis: the tip of the iceberg? *Clin Microbiol Rev* 4: 52-60.

6. Dance DA. 2000. Melioidosis as an emerging global problem. *Acta Trop* 74: 115-119

7. Dance DA, King C, Aucken H, Knott CD, West PG and Pitt TL. 1992. An outbreak of Melioidosis in imported primates in Britain. *Vet Rec* 130: 525-529.

8. Frederick AM. 1997. Melioidosis. In: Daniel HC, Francis WC, Jerbert JM, Kevin J, David AS and Ernest EL (eds.) *Pathology of infectious diseases*. Library of Congress: Simon and Schuster Company.

9. Huang CR. 1976. What is *Pseudomonas pseduomallei. Elixir*: 70-72.

10. Inglis TJ, Garrow SC, Adams C, Henderson M, Mayo M and Currie BJ. 1999. Acute melioidosis outbreak in Western Australia. *Epidemiol Infect* 123: 437-443.

11. Kang G, Rajan DP, Ramakrishna BS, Aucken HM and Dance DA. 1996. Melioidosis in India. *Lancet* 347: 1565-1566.

12. Ketterer PJ and Bamford VW. 1967. A case of Melioidosis in lambs in south western Austrlaia. *Aust Vet J* 411: 79-80.

13. Ketterer PJ, Donald B and Rogers RJ. 1975. Bovine melioidosis in south eastern Queensland. *Aust Vet J* 51: 395-398.

14. Ketterer PJ, Webster WR, Shield J, Arthur RJ, Blackall PJ and Thomas AD, 1986: Melioidosis in intensive piggeries in south eastern Queensland. *Aust Vet J* 63: 146-149.

15. Mandell GL, Bennett JE and Dolin RD. 2005. Mandell, Douglas and Bennett's *principles and practice of infectious diseases*. 6th ed. Elsevier Churchill Livingstone, Philadelphia (PA).

16. Nicholls L. 1930. Melioidosis with special reference to the dissociation of *Bacillus whitmori*. *Br J Exp Pathol* 11: 393-399.

17. Saravu K, Mukhopadhay C, Vishwanath S, Valsaln R, Docherla M, Vandana KE, Shastry BA, Bairy I and Rao SP. Melioidosis in southern India: epidemiological and clinical profile. *Southeast Asian J Trop Med Public Health* 41:401-409

18. Stanton AT and Fletcher W. 1921. Melioidosis, a new disease of the tropics. *Trans Fourth Congr Far East Assoc Trop Med* 2: 196-198.

19. Sutmoller P, Kraneveld FC and van der Schaaf A. 1957. Melioidosis in sheep, goats and pigs on Aruba. *J Am Vet Med Assoc* 130: 415-417.

20. Thomas AD and Forbes-Faulkner JC. 1981. Persistence of *Pseudomonas pseudomallei* in soil. *Aust Vet J* 57: 535.

21. Thomas AD, Forbes-Faulkner JC and Parker M. 1979. Isolation of *Pseudomonas pseudomallei* from clay layers at defined depths. *Am J Epidemiol* 110: 515-521.

22. Thomas AD, Norton JH, Forbes-Faulkner JC and Woodland G. 1981: Melioidosis in an intensive piggery. *Aust Vet J* 57: 144-145.

23. Vadivellu J, Puthucheary SD, Gendeh GS and Parasakthi N. 1995. Serodiagnosis of melioidosis in Malaysia. *Singpore Med J* 36: 299-302.

24. Wernery R, Kinne J, Haydn-Evans J and Ul-Haq A. 1997. Melioidosis in a seven year old camel, a new disease in the United Arab Emirates (UAE). *J Camel Res* 4: 141-143.

25. White NJ. 2003. Melioidosis. *Lancet* 361: 1715-1722.

26. Whitmore A and Krishnaswami CS. 1912. An account of the hitherto undescribed infective disease occurring among the population of Rangoon. *Indian Med Gazette* 47: 262-267.

2014, Zoonoses: Bacterial Diseases
Editor: **Sudhi Ranjan Garg**
Published by: **DAYA PUBLISHING HOUSE, NEW DELHI**

Pages **325–334**

16

Tetanus

Subhash Verma and Geetanjali Singh
Dr. G.C. Negi College of Veterinary and Animal Sciences,
C.S.K. Himachal Pradesh Agricultural University, Palampur – 176 062

The word tetanus is derived from Greek word *tetanos* which means to stretch. This disease is characterized by tautness of skeletal muscles. It is commonly called as 'lockjaw' as the stiffness of muscles around the jaw causes difficulty in the movement of jaw and is the first symptom of the disease. It is a disease caused by a bacterium, *Clostridium tetani*, which readily forms spores. Soil is the natural habitat of these bacteria. The occurrence of tetanus in humans and animals has been recorded since 5th century BC. In the year 1884, Arthur Nicolia demonstrated production of the disease in rabbits by injecting them with soil. In the same year, Antonio Carle and Giorgio Rattone did the same by injecting pus from a human patient suffering from tetanus. In 1899, *C. tetani* was isolated for the first time from a human patient by Kitasato Shibasaburo.

High incidence of deaths in neonates due to tetanus is a result of poor hygiene practised in midwifery and non-immunization of expectant mothers with the tetanus toxoid. In developing countries, the adult population suffers high incidence of tetanus while working in the agricultural fields that are heavily manured with animal dung. The infection may generally occur through deep wounds or injuries caused by sharp objects such as nails, rods, glass splinters and barbed wire.

Geographical Distribution

Tetanus is prevalent throughout the world, but its incidences are higher in developing countries, both in humans and animals. Many parts of Africa, Asia Central and South America are affected with deadly tetanus. Although now rare in Europe, United States and Australia, tetanus used to be a cause of about 50,000 deaths each year worldwide, mainly of newborn babies contracting the disease during birth time, before 1960 (Bytchenko 1966). Tetanus remains a leading cause of maternal and neonatal morbidity and mortality in developing countries (Roper *et al.* 2007).

From the available data, it is comprehended that the morbidity and mortality rates from tetanus are the highest in tropical countries. Within a country that comprises of various geographical features and climatic zones, this disease tends to be more in areas with a warm, damp climate and fertile soil as compared to mountains or deserts. In the developed countries, such as European countries, Canada, USA, Japan and Australia, there are reduced incidences of tetanus due to tremendous improvement in living standards, extensive industrialization, urbanization, mechanization of agriculture, and widespread use of chemical fertilizers instead of animal manure. This reduction occurred even before the start of mass immunization against tetanus (Bytchenko 1966). Further decline in the incidence of tetanus occurred consequent to the introduction of tetanus toxoid as a means to provide active immunization to the population against tetanus. There is more than 95 per cent decline in the reported cases of human tetanus and 99 per cent decline in the deaths due to tetanus since 1947.

In the developing countries, particularly in African continent, tetanus remains as the most serious disease amongst newborns. It is more prevalent in rural areas where illiteracy and poor environmental conditions prevail, along with the non-availability of antenatal facilities. According to a report of the Ministry of Health and Family Welfare in the year 2010, there is a decreasing trend in the occurrence of neonatal tetanus in India. Similarly in China, there is a large decrease in neonatal tetanus incidence due to the tetanus prophylactic initiatives. However, in rural China, the incidence is comparatively higher than that in the urban area. The prevalence of tetanus depends upon the tetanus immunization coverage. This coverage is higher in developed countries; hence low tetanus occurrence, whereas in developing countries, this coverage is variable and the incidence of human tetanus is high. According to the WHO/UNICEF database of prevalence of human maternal and neonatal tetanus (MNT) as of 2011, 21 countries eliminated MNT between 2000 and 2011, which include 15 out of 33 states in India, all of Ethiopia except Somali region, 29 out of 33 provinces in Indonesia, leaving 38 countries yet to eliminate it.

Etiological Agent

Tetanus is a disease caused by the toxin of the bacteria *Clostridium tetani*. The organism is rod-shaped, Gram positive, obligate anaerobe and motile. The sporulated form has spores usually in the terminal end of the rod-giving it the appearance of a spoon or drumstick when viewed under microscope (Figure 16.1). The organism is commonly found in the soil and in the faeces of most animals including humans from where it can contaminate deep wounds. The presence of the tetanus organisms has been shown in the environment, air and dust at a variety of places such as hospitals, warehouses, factories, playgrounds, fields, etc. Its presence has also been

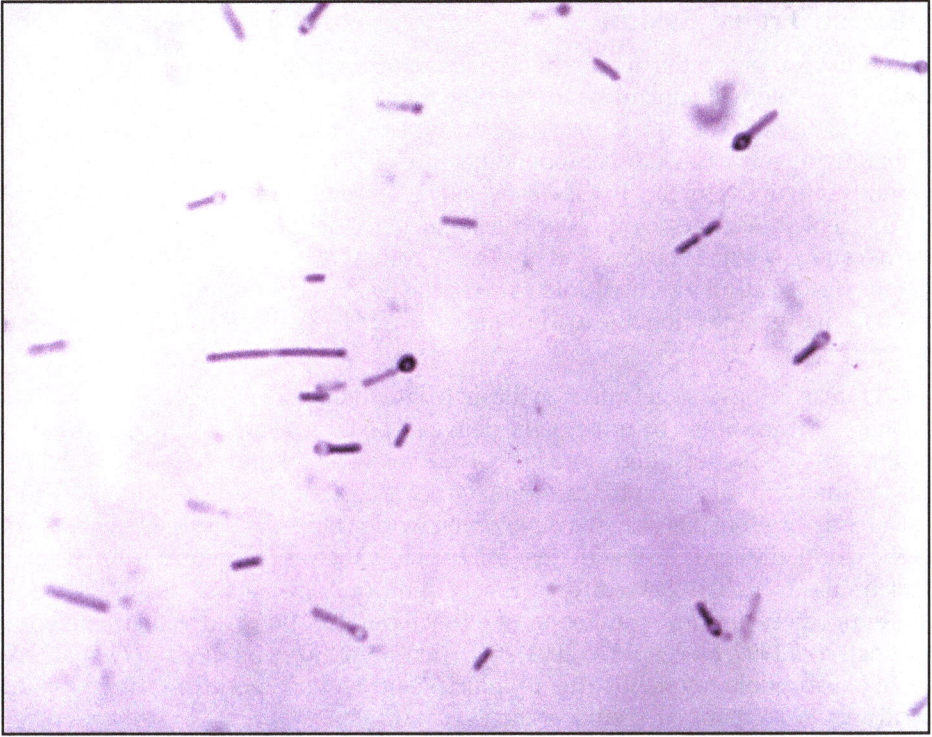

Figure 16.1: Gram positive *Clostridium tetani* bacteria
(Photo courtesy: CDC, Dr. Holdeman)

reported in the surgical dressing, raw catgut, talcum powder, surgical and dental instruments, castrators and dehorners for animals.

The microorganism is widely distributed in soil. It is also present in the gastrointestinal tract of domestic animals and human beings. The spores of the organism can pass through the gastric tract of various animals without causing any disease. The spores also remain unharmed by the gastric juices and enzymes and are also not destroyed by heat, light, strong chemicals, antiseptics and moisture. They do not germinate inside normal living tissue, but germinate in environments such as deep wounds, where a low oxygen tension is there due to necrosis, tissue damage or chemical changes.

Host Range

Tetanus occurs not only in humans but also in other animal species such as horses, pigs, donkeys, mules, sheep, goats, camels, dogs and cats. Almost all mammals are susceptible to tetanus. Among farm animals, horses are most susceptible, while dogs and cats are resistant to the disease. All birds including poultry are also very resistant. Rare cases have also been reported in captive Rhesus monkeys. Wild animals such as elephants, deer and big apes such as gorilla, orangutan, chimpanzee and bonbon are also susceptible to tetanus.

Disease Transmission

Adult animals often acquire the infection through parturition wound, dental caries, wire cuts, nail punctures wounds, shearing cuts, dehorning, castration, scissor wound or any type of deep wound. Certain types of wounds that favour the growth of tetanus organisms include compound fractures, injuries, deep penetrating wounds, wounds containing foreign bodies (especially wood splinters), wounds complicated by pyogenic infections, wounds with extensive tissue damage (*e.g.* contusions or burns) and any superficial wound obviously contaminated with soil, dust or horse manure (especially if topical disinfection is delayed more than 4 hours). Caged exotic animals may get infection due to the injuries caused by faulty cages and protruding wires.

Human beings also get infection through wounds and injuries. Tetanus in human neonates is transmitted by unhygienic delivery, poor cord care, harmful traditional health practices, superstitions, and lack of anti-tetanus immunization. The umbilical cord cutting and care of umbilical stump varies according to accepted practices and cultures at different places in the developing world (Elhassani 1984). In many parts of the world, the cord is cut with unsterilized tools such as razors and scissors, after which substances like charcoal, grease, cow dung, or dried banana are applied on it. These practices are important sources of bacterial infection and neonatal tetanus (Bennett *et al.* 1999, Meegan *et al.* 2001). Both non-immunized mother and neonate are highly susceptible to tetanus due to injuries and wounds occurring during child birth and the contaminated environmental conditions.

Scratches or bite injuries to humans by domestic or wild animals are also a potential portal of entry for the bacterium. Similarly, domestic animals are also at a risk of contracting tetanus due to injuries caused by other wild or domestic animals. Tetanus can also occur in humans and animals following natural disasters such as floods, tsunamis, earthquakes, tropical cyclones, hurricanes and tornadoes. An outbreak of 106 cases of tetanus was reported in humans in one of the most catastrophic tsunami in Aceh in Indonesia in 2006 (Jeremijenco *et al.* 2007). Recently, concerns have also been raised over chronic wounds in the form of varicose ulcers, dermatosis and necrosed tumours in old and debilitating patients, which could be the point of entry for *C. tetani* spores (Farnworth *et al.* 2012).

As *C. tetani* is ubiquitous and obligate anaerobe, it finds suitable conditions for growth in the deep wounds, where it can multiply favourably due to the low partial pressure of oxygen and the spores germinate to vegetative bacilli. The proliferating bacteria release two types of toxins, one a haemolysin and the other a neurotoxin. Haemolysin called as tetanolysin is of minor importance in the disease, while the neurotoxin called as tetanospasmin is a more important toxin. The tetanus toxin apparently passes by diffusion from the wound to the surrounding area and then spreads to the central nervous system. This toxin inhibits the release of neurotransmitters from inhibitory motor neurons resulting in spastic paralysis.

The incubation period of tetanus is about 7-10 days, but may vary from 2-30 days. The period may be shorter in neonates as compared to the adults. In some cases, there may not be any recognizable wound or portal of infection but tetanus may appear. These cases are probably due to small un-noticed wounds that heal before the appearance of tetanus. The organism has also been reported to show dormancy for weeks or months. There are even reports of development of tetanus up to 14 years after the presumed entry of *C. tetani* spores into the tissues (Topley *et al.* 1990). The multiplication of bacteria depends upon many factors inside the wound such as trauma, tissue necrosis, haemorrhage, foreign bodies, chemical environment and infection by the other microbes. The presence of favourable environment enables the spores to germinate and multiply in the tissue. It has been noted that the contamination of wound with the soil is an important factor in the development of tetanus. It has been observed that ionizable calcium salts in soil might be important in promoting tetanus in the soil contaminated wound. Calcium ions have been found to be crucial in causing damage to the wound (Topley *et al.* 1990).

Prevalence of Tetanus in Animals

The occurrence of tetanus in domestic animals is sporadic and depends on the species of animal and nature of injury to the animal. A wide variety of animals, including domesticated as well as wild animals, are susceptible to the disease. Its prevalence is the highest in horses. Its prevalence is also higher in young animals as compared to older animals that acquire deep wounds during castration, dehorning or tail docking. The prevalence of tetanus is common in all the countries. However, since this is a non-communicable disease, its occurrence is sporadic. In developed countries, it is mostly due to accidental injury. Published reports of individual or mass occurrence of tetanus in domestic animals are not readily available, however, the incidents of tetanus in domestic animals do occur but remain unreported.

Prevalence of Tetanus in Humans

In contrast to animals, the incidents of tetanus are frequently documented, however with reporting artifacts. According to WHO (2006) data, there is decline in the reported tetanus incidence worldwide. The incidence in African continent decreased from 17,241 cases in 1980 to 2,754 in 2005. During the same period, in North American continent, the number of cases decreased from 7,055 to 944, in Europe from 1,715 to 192, and in South East Asian region from 62,172 to 6,791. The overall incidence (number of reported cases) worldwide was 114,248 in 1980 as compared to 15,561 in 2005. The reduction in the reported incidents of human tetanus is due to coverage of human population with tetanus toxoid.

Disease Manifestations in Animals

Tetanus is very important in horses. Its typical symptoms may appear in a horse within a few days of an injury but can also take several months. The first sign is usually muscle stiffness of the jaw and rear legs. After these initial signs, the symptoms progress further within 24 hours and the body becomes oversensitive with sporadic spasms noticeable in muscles. The animal is not able to move its head up and down, thus preventing it from eating feed and fodder. The eyes, ears and tail all appear rigid and nostrils typically flare up. The gait is abnormal too. The appearance of third eyelid in the inner corner of eye is the tell tale sign (Figure 16.2). As the disease

Figure 16.2: Prolapsed third eyelid in horse
(Photo courtesy: Dr. Ankur Sharma, COVAS, Palampur)

progresses, the temperature increases and there is sweating. The prognosis of a horse recovering from tetanus is good if it was previously immunized with tetanus toxoid. Nevertheless, an unimmunized horse can also be saved with proper therapeutic intervention and care, provided that it responds to the therapy and does not become recumbent.

In bovine tetanus, cattle become very sensitive to noise and light. They are also irritable and excitable. A high temperature may also be present and there is stiffness in the limbs, neck, jaw and tail. Additionally, the rumen motility is absent leading to bloat. Protrusion of third eyelid occurs when head is raised and it leads to diagnosis of tetanus. The prognosis of bovine tetanus is almost always poor when the infection in unvaccinated animal occurs and this usually leads to death. Similar symptoms occur in sheep as in cattle which are stiff gait, lock jaw and protrusion of third eye lid. The animal usually falls down with all four legs held out straight and stiff and head drawn back. Convulsions may also occur. Prognosis of the disease is poor in sheep and treatment with antiserum and antibiotic is not very successful. Pigs are also susceptible to tetanus that usually occurs through a wound upon castration in young pigs. The incubation period of *C. tetani* in pigs is about 1-10 weeks. The affected pigs are hypersensitive, show stiffness of the leg muscles, an erect tail and muscular spasms of ear and face.

Dogs and cats are relatively resistant to tetanus and as a result, rare cases of tetanus are encountered in these species. Therefore, no immunoprophylaxis is recommended for them. Cats suffer from localized form of tetanus, while dogs evidence a generalized form which results in severe muscular contraction. The mortality rate in dogs is 50 per cent, however, with an early detection and intensive therapeutic intervention, prognosis is good and up to 90 per cent cases may survive (Adamantos and Cherubini 2010). Clinical signs develop between 3-14 days after infection in dogs. In cats, clinical signs may be delayed up to three weeks and range from muscle stiffness, inability to move the neck and jaw, and increased sensitivity.

Disease Manifestations in Humans

In humans, most tetanus cases occur in adults who were never vaccinated against this infection or who have an incomplete vaccination. The incubation period in adults is 4-21 days, averaging 10 days. Appearance of stiffness in jaw muscles is the first symptom of tetanus in human beings. It further progresses from jaw to neck and may lead to inability to swallow and breath. Other symptoms of tetanus in humans are high temperature above 100°F, sweating, tachycardia and hypertension. If not treated timely, it can be fatal.

Worldwide the most reported cases of human tetanus are the neonate types. The disease occurs in neonates who have no passive immunity because of non-immunization of the expectant mothers against tetanus. Unsterile birth conditions and unsanitary cord care increases the risk of neonate tetanus. The signs and symptoms of neonate tetanus include difficulty in sucking milk, muscle spasms, rigidity (Figure 16.3), and convulsions. The fatality rate of neonate suffering from the disease is 70-100 per cent.

Figure 16.3: Neonate tetanus displaying body rigidity (Photo courtesy: CDC)

Diagnosis of Tetanus

Conventionally, diagnosis of tetanus is based on detection of clinical symptoms and it can be diagnosed easily. The knowledge about the clinical features of tetanus is, therefore, very crucial. Sometimes in animals, it can be confused with strychnine or lead poisoning, grass tetany, eclampsia or rabies. However, these diseases can be differentiated from tetanus on the basis of classical signs and symptoms. There are no major gross lesions in tetanus internally or externally. The knowledge about occurrence of an injury and the presence of a contaminated wound along with a history of recent castration and dehorning or shearing in animals helps in arriving at a diagnosis.

Similarly, the presence of a clinical signs along with a history of injury helps in diagnosis of tetanus in humans. In human neonates, history of non-immunized status of mother and characteristic symptoms of tetanus helps in its diagnosis. Laboratory detection of the organism is seldom done because of difficulty in recovering the bacteria from the infected site. However, culturing of pus or wound scraping or tissue from a necrotic wound on blood agar may demonstrate the presence of *C. tetani.* Some advanced laboratories detect the presence of protective antitoxin in the blood. Detection of high level of antitoxin in patients suggests an unlikely presence of tetanus.

Prevention and Control Strategies

Behring and Kitasato identified the toxin in 1980 and subsequently showed that repeated inoculation of animals with small quantities of toxin led to the development of antibodies in the surviving animals. In 1924, Descombey prepared toxoid, which is a chemically altered toxin that is able to induce neutralizing antibodies without causing illness. This ultimately paved the way to active immunization and tetanus prophylaxis on a larger scale. Tetanus is thus a highly preventable non-communicable disease which can be easily curtailed by following an active immunization schedule. The active immunization is done with tetanus toxoid *i.e.* formaldehyde-inactivated toxin of *C. tetani*, which has lost toxicity but retains the ability to stimulate the production of antitoxin.

Neonatal and maternal cases are prevented by immunization of expectant mothers with tetanus toxoid, hygienic delivery, sterile cutting of cord and its after care. The neonates should be subsequently immunized against tetanus. Worldwide, a combination vaccine against diphtheria, tetanus and pertussis is used in human pediatric population. An adsorbed toxoid is used now-a-days that gives a better lasting immunity than the plain one. The basic course of immunization adopted in many countries consists of three spaced injections with the intervals of 6-8 weeks and 4-8 months. In some countries a fourth dose is also given one year after the third dose. The protective antibodies last life long, however, as their concentration falls and becomes stationary, a dose after 10 years is also usually recommended to keep the body protected. Patients without a clear history of at least three tetanus vaccinations and with any wound that is deep and contaminated need tetanus antitoxin and not just tetanus toxoid. Antitoxin, which contains antibodies against tetanus toxin, provides immediate protection to any unimmunized individual who is wounded and likely to contract tetanus. It is worthwhile to note that both in animals and humans, antigenic stimulation by an attack of tetanus will not produce sufficient antibodies that could immunize the individual to further attack of tetanus.

Immunization of Farm Animals

Active immunization in farm animals is done with tetanus toxoid. Horses are very susceptible to tetanus; hence these may be protected by administration of tetanus toxoid and yearly booster doses. Mares should be vaccinated during the last 6 weeks of pregnancy and foals vaccinated at 5-8 weeks of age. In high risk areas, foals may be vaccinated with antitoxin at birth and every 2-3 weeks until they are three months old when tetanus toxoid is given. In horses, the treatment includes 300,000 IU of tetanus antitoxin repeated two times in 24 hours in conjunction with broad spectrum antibiotics, sedatives or tranquilizers. Similar course of therapy is followed in case of dogs and cats. A multivalent *Clostridium* vaccine containing tetanus toxoid could be used in pregnant sows for preventing this disease. Tetanus in sheep can be prevented by vaccinating pregnant ewes with tetanus toxoid and by administering tetanus antitoxin to lambs at the time of docking or castrating.

Many exotic animal species that are kept in zoos and captivity are susceptible to tetanus. Therefore, primates, exotic equines, elephants, big apes, deer, camels and wild sheep and goats should be immunized against tetanus. Exotic equines and elephants are vaccinated following the same schedule as in domestic horses. Big apes are often vaccinated using the diphtheria, tetanus toxoid, and phase 1 pertussis vaccines intended for use in human children or monovalent human tetanus toxoid. Primary immunization consists of 0.5 ml vaccine (Kahn and Line 2000). Wild sheep, goats and cervids are immunized beginning at 10-12 week of age with multivalent clostridial bacterin-toxoids containing *C. tetani* antigens in areas of high exposure risk. The initial dose of 5 ml is followed in 6 week by a 2 ml dose which is administered subcutaneosly. This is followed by annual 2 ml booster dose (Kahn and Line 2010).

Future Directions

The occurrence of tetanus in human population can be largely curtailed by implementing effective programme of active immunisation. The disease can be effectively avoided by spreading the knowledge about the prophylactic immunization to the masses. The improvement in socio-economic conditions of the population has been shown to cause lowering of the incidence of tetanus. Some of these measures include the provision of proper footwear, safer farm and industry practices, proper disposal of sharp wastes and adequate knowledge and facility for tetanus vaccination. Tetanus immunization should supplement good practices of midwifery and proper attention to wound injuries.

In farm animals also, there should be prompt attention to wound injuries, followed by prophylactic administration of antitoxin to prevent tetanus. In tetanus prone areas, expensive animals especially horses should be vaccinated yearly. Passive immunistion by protective antitoxin should be followed after tail docking, castration and dehorning in animals.

References

1. Adamantos S and Cherubini GB. 2009. Tetanus in dogs. *Companion Animal* 14(8): 56-60.

2. Bennett J, Ma C, Traverso H, Agha SB and Boring J. 1999. Neonatal tetanus associated with topical umbilical ghee: cover role of cow dung. *Internat J Epidemio* 28: 1172-1175.

3. Bytchenko B. 1966. Geographical distribution of tetanus in the world, 1951-60: A review of the problem. *Bull World Health Organ* 34: 71-104.

4. Elhassni SB. 1984. The umbilical cord: care anomalies, and diseases. *South Med J* 77: 730-736.

5. Farnworth E, Roberts A, Rangaraj A, Minhas U, Holloway S, Harding K. 2012. Tetanus in patients with chronic wounds - are we aware? *Int Wound J* 1: 93-99.

6. Jeremijenko A, McLaws ML and Kosasih H. 2007. A tsunami related tetanus epidemic in Aceh, Indonesia. *Asia Pac J Public Health* 19: 40-44.

7. Kahn CM and Line S. 2010. *Merck Veterinary Manual*, 10th Edition. Merck Publishing Group.

8. Meegan ME, Conroy RM, Ole Lengeny S, Renhault K and Nyangole J 2001. Effect on neonatal tetanus mortality after a culturally-based health promotion programme. *Lancet* 358: 640-641.

9. Roper MH, Vandelaer JH and Gasse FL. 2007. Maternal and neonatal tetanus. *Lancet* 370: 1947-1959.

10. Topley WWC, Wilson G, Parker MT, Collier LH and Timbury MC. 1990. *Topley and Wilson's principles of bacteriology, virology and immunity*, 8th Edition, Vol. 3. Edward Arnold, London.

11. WHO. 2006. *WHO vaccine preventable disease monitoring system, 2006 global summary*. World Health Organization, Geneva.

2014, Zoonoses: Bacterial Diseases   *Pages* **335–353**
Editor: **Sudhi Ranjan Garg**
Published by: **DAYA PUBLISHING HOUSE, NEW DELHI**

17

Lyme Disease

Kuldeep Dhama[1], Pradeep Mahadev Sawant[2],
Ruchi Tiwari[3] and Senthilkumar Natesan[4]
[1]*Division of Pathology,* [2]*Division of Veterinary Biotechnology,*
Indian Veterinary Research Institute, Izatnagar – 243 122
[3]*Department of Microbiology and Immunology,*
College of Veterinary Sciences, Pandit Deen Dayal Upadhyaya
Veterinary University, Mathura – 281 001
[4]*Institute of Human Virology and Department of Medicine,*
University of Maryland School of Medicine, Baltimore, MD 21201, USA

Lyme disease is a non-contagious infectious, tick-borne zoonotic disease caused by spirochetes belonging to the *Borrelia burgdorferi* Sensu Lato complex and more often transmitted by the bite of infected ticks of the genus *Ixodes*. The disease is characterized by a diverse clinical profile, which can trigger cutaneous, articular, neurologic and cardiac manifestations (Gayle and Ringdahl 2001, Stanek *et al.* 2012). The first cases of Lyme disease were described with single or multiple erythematous plaques of centrifugal growth, which were designated as erythema chronicum migrans (ECM) (Afzelius 1910, Lipschutz 1914). Later on treatment of ECM cases with procaine penicillin suggested bacterial etiology for the disease (Hollstrom 1958). In the city of Lyme (USA), cases of ECM with arthritis were called Lyme disease, Lyme arthritis or Lyme borreliosis (Steere 1977). Burgdorfer *et al.* (1982) noticed the presence of spirochetes in the gut of ticks of the *Ixodes dammini* species, which were called *B. burgdorferi* Sensu Lato. The association of this spirochete with ECM was established after identification of organisms in skin biopsies, and detection by polymerase chain reaction (PCR) and sequencing of skin, spinal fluid and synovial fluid samples from patients with Lyme disease (Barbour and Hayes 1986). Currently, Lyme disease is considered as a multisystem disease (Stanek and Strle 2003).

Lyme disease has been detected in USA, Europe, Australia, parts of Asia, the province of Ontario (Canada), and the Amazon region of Brazil. The importance of borreliosis as a zoonotic disease is increasing globally. Although its incidence in a geographic area is similar in animals and humans, animals, especially dogs, are at significantly higher risk (Popovic *et al.* 1993).

The disease, in initial stage, is recognized as unusual rash which can be readily cured with antibiotics in human. However, untreated infections sometimes develop chronic arthritis, neurological signs and other syndromes, called as persistent Lyme disease. Clinical signs attributed to Lyme disease have also been reported in dogs, horses and cattle but no distinctive rash occurs. The illness is best characterized in the dogs, where arthritis and nephropathy appear to be the most common sequelae.

Etiology

The causative agent of Lyme disease is *B. burgdorferi* Sensu Lato complex in the family Spirochaetaceae (Stanek *et al.* 2012). *B. burgdorferi* Sensu Lato complex is a diverse group of worldwide distributed bacteria that includes 18 named spirochete species and an unnamed group - genospecies 2. Eleven species from this complex were identified and strictly associated with Eurasia (*B. afzelii, B. bavariensis, B. garinii, B. japonica, B. lusitaniae, B. sinica, B. spielmanii, B. tanukii, B. turdi, B. valaisiana,* and *B. yangtze*), while another 5 (*B. americana, B. andersonii, B. californiensis, B. carolinensis, and B. kurtenbachii*) were previously believed to be restricted to the USA only. *B. burgdorferi* Sensu Stricto, *B. bissettii* and *B. carolinensis* share the distinction of being present in both the Old and the New World (Rudenko *et al.* 2011). Of these species, four are associated with Lyme disease: a) *B. burgdorferi* Sensu Stricto, b) *B. garinii* c) *B. afzelii*, and d) *B. spielmanii*. Some genospecies in the *B. burgdorferi* Sensu Lato complex appear to be associated with disseminated forms of the illness, while others are associated only with skin lesions. *B. garinii* circulating between both mammalian and avian reservoir hosts is divided into birds related strains and mammals or humans related strains. However, this division is questionable as bird and human isolates are similar. *B. burgdorferi* Sensu Lato has also been divided into eight serotypes, with *B. burgdorferi* Sensu Stricto corresponding to serotype 1, *B. afzelii* corresponding to serotype 2, and *B. garinii* containing serotypes 3 to 8. *B burgdorferi* was the first spirochaete for which the complete genome was sequenced (Fraser *et al.* 1997). Genetic studies suggest a nearly complete absence of biosynthetic pathways, making the microorganism dependent on its environment for nutritional requirements. Nevertheless, Lyme borrelia can be grown *in vitro* in highly enriched culture media (Barbour 1984, Stanek *et al.* 2012).

Epidemiology

In North America, Lyme disease results from infection with *B. burgdorferi* Sensu Stricto. *B. burgdorferi* Sensu Lato is endemic in Ontario, Canada. In Europe, several genospecies can cause Lyme disease including *B. afzelii* in northern Europe, *B. burgdorferi* Sensu Stricto in Western Europe and *B. lusitaniae* in the Mediterranean basin region. In Asia, the disease is mainly caused by *B. garinii* and *B. afzelii*. Moreover, *B. burgdorferi* Sensu Stricto (Taiwan), *B. bissettii*, *B. valaisiana* and *B. lusitaniae* have also been found in Asian countries (Stanek and Strle 2003, Stanek *et al.* 2012). Seroprevalence of the organism has been recorded in USA (2-12 per cent), Europe (26 per cent), Japan (5-21 per cent), China (5.06-26.2 per cent) and Malaysia (4.1 per cent) (Praharaj *et al.* 2008). In India, a study shows seroprevalence of Lyme disease in Arunachal Pradesh (17.8 per cent), Meghalaya (9.09 per cent), Nagaland, Manipur (8.46 per cent), and Assam (9.6 per cent), which indicates the existence of the disease in the country. Higher prevalence rate was observed in females (15.86 per cent) than in males (10.95 per cent).

Vectors, Reservoirs and Transmission Cycle

The principal vector of Lyme disease in Europe is Ixodes ricinus, whereas Ixodes persulcatus is the main vector in Asia, Ixodes scapularis in northeastern and upper mid-western USA and Ixodes pacificus in western USA (Gray 1998). These ticks have a four stage life cycle: egg, larva, nymph and adult, feeding only once during every active stage (Figure 17.1). Unfed (flat) ticks attach to the skin of a host animal using specialized mouthparts as the animal passes through vegetation. After feeding for a

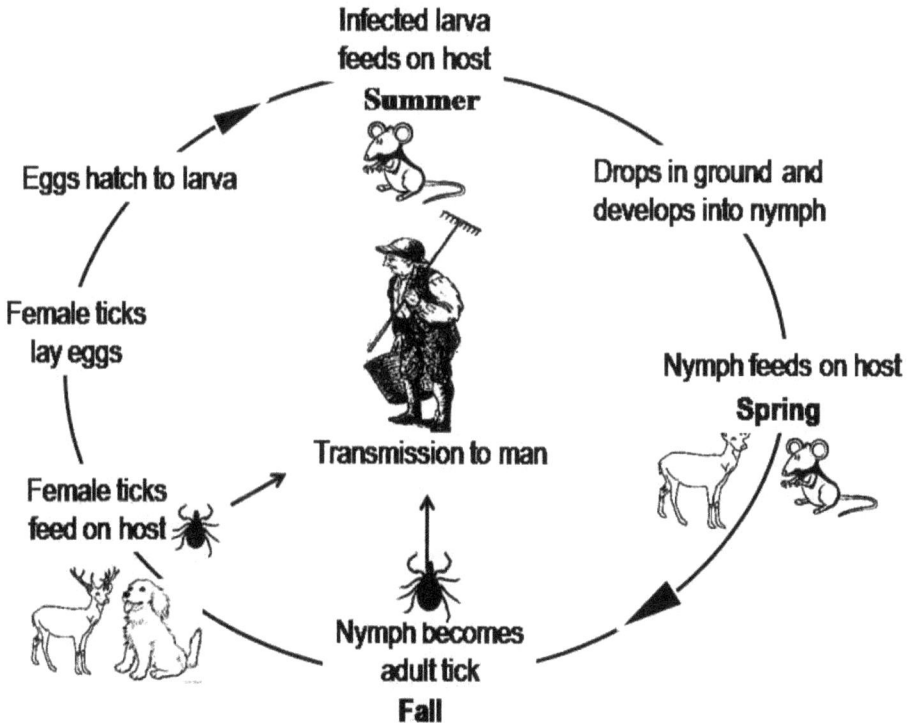

Figure 17.1: Life cycle of *Borrelia burgdorferi*

few days (about 3 days for larvae, 5 for nymphs, and 7 days for adult females, male ticks rarely feed and never engorge), the ticks drop off their host and remain on or near the soil surface, where they need a minimum relative humidity of 80 per cent for survival. The ticks take several months to develop into the next developmental stage and the adult females lay about 2000 eggs. The length of a tick's life cycle varies between 2 years and 6 years depending on climate, host availability, and the effects of development-delaying diapause mechanisms (Piesman and Gern 2004, Stanek *et al.* 2012).

B. burgdorferi mainly survives in a tick-mouse cycle. In addition, deer serve as hosts, especially during the adult tick stages (Fraser *et al.* 1997). *B. burgdorferi* microorganisms occur in temperate regions because *Ixodes* ticks can survive only in these climates. The ticks are widely distributed and feed on a variety of large and small mammals, birds and reptiles. The risk of infection in a given area depends largely on the environmental density of ticks, their feeding habits and their animal hosts (Nahimana *et al.* 2004). The predominant tick species that transmit *B. burgdorferi* to humans are the Eurasian species *I. ricinus* and *I. persulcatus*, and the North American species *I. dammini*, *I. pacificus* and *I. scapularis* (Anderson 1991, Schuijt *et al.* 2011).

The *Ixodes* tick can become infected after blood meal in all stages of life which include larvae, nymph, and adult (Murray and Shapiro 2010). In late summer, larvae feed on the white-footed mouse which serves as a continuous source for infection. Then larvae molt into nymphs that feed in spring to early summer which coincides with the increased incidence of Lyme disease in humans. In the fall, nymphs molt into adult ticks, which feed and mate on deer. The deer, while feeding, spreads ticks to the leafy areas, where new larvae are hatched in summer (Elbaum-Garfinkle 2011). Control of deer populations, which supply blood meal and serve as a mating ground, has proven successful in decreasing the prevalence of *I. scapularis* ticks and Lyme disease (Barbour and Fish 1993, Rand 2004).

The main vertebrate reservoirs for Lyme disease are small mammals, such as mice and voles, and some species of birds (Radolf *et al.* 2012; Stanek *et al.* 2012). Deer, cattle and sheep are essential for the maintenance of tick populations as they can feed sufficient number of adult ticks, but they are not true reservoirs for spirochaetes. Dogs and wild carnivores may also occasionally transmit *B. burgdorferi* Sensu Lato to ticks, but they are not true reservoir hosts. Transplacental transmission of the organism from infected dams to their foetuses also occurs through *in utero* infection and can be a cause of mortality in foals and calves (Burgess 1988). Co-feeding, a phenomenon where bacteria are transmitted between ticks feeding on a host even if the host is not infected has been demonstrated in sheep. Naturally occurring Lyme disease in dogs, horses and cattle has been reported while serologic evidence has been shown in cats. Experimental infection in dogs, horses, cats, rabbits, mice, hamsters, gerbils and guinea pigs can be established.

Pathophysiology

B. burgdorferi is a highly adaptable organism. It has complex genetic structure, a linear chromosome and 21 plasmids (Porcella *et al.* 2001). The spirochete can spread throughout the body during the course of the disease, and has been found in the skin, heart, joint, peripheral nervous system, and central nervous system (Pachner and Steiner 2007). Many of the signs and symptoms of Lyme disease are a consequence of the immune response to the spirochete in those tissues. Different mechanisms are used by *B. burgdorferi* for evasion of immune response in humans which involves immunosuppression, the genetic, phase and antigenic variation, physical seclusion, and secreted factors (Ramamoorthi *et al.* 2005, Stricker 2007, Berende *et al.* 2010). Feeding tick introduces *B. burgdorferi* and the saliva containing analgesic, anticoagulant, and immunosuppressive factors, which initially help the spirochete penetrate the skin and evade local immune response (Hannier *et al.* 2003, Berende *et al.* 2010). After invasion, the spirochete induces immunosuppression by complement inhibition, induction of inhibitory cytokines (IL-10), monocyte and lymphocyte tolerization and antibody sequestration in immune complexes (Liang *et al.* 2004). *B. burgdorferi* engages in genetic phase, and antigenic variation that shares various features with other organisms (Stricker 2007). Neutrophil calprotectin may induce dormant state in *B. burgdorferi*, allowing it to persist in tissue and providing the means to avoid antibiotics (Montgomery *et al.* 2006), and the gene mutations in the Lyme spirochete have been demonstrated to confer *in vitro* resistance to various antibiotics (Criswell *et al.* 2006).

The intracellular survival of spirochete in multiple cell types, including synovial cells, endothelial cells, fibroblasts, macrophages, Kupffer cells and neuronal cell, helps evade the immune response (Livengood and Gilmore 2006). Physical seclusion at extracellular sites, including the joints, eyes and CNS, may also promote survival of the spirochete. Moreover, *B. burgdorferi* can bind to proteoglycan, collagen, plasminogen, integrin, and fibronectin, masking by these substances makes the spirochete undetectable to the immune system. The factors secreted by *B. burgdorferi* or the immune response against the infection play an important role in pathology of Lyme disease (Behera *et al.* 2006, Fallon *et al.* 2010). Role of hemolysin in pathogenesis is unknown (Williams and Austin 1992, Stricker 2007), however, porin and adhesin proteins help bacteria to adhere to cells and pierce the cell wall to gain entry (Stricker 2007, Berende *et al.* 2010). The pheromones secreted (AI-2) indicates that the sprochaete engages in autoresuscitation like other dormant organisms, such as the tubercle bacillus. Aggrecanase enzyme induces damage and inflammation in joints (Behera *et al.* 2006). All these mechanisms help the spirochete establish infection. Cell wall-deficient (CWD) spirochetal (Borrelial) forms, termed as "cysts or round bodies", have been detected. A link between CWD *Borrelia* and neurodegenerative diseases has been suggested recently, and resistance of cystic forms to antibiotic therapy has been reported. The formation of biofilms is another mechanism accountable for chronic infection to be established. Coordinated steps in the elaboration of biofilms have been demonstrated in other bacteria, whether similar molecular processes having role in persistent infection occur in *Borrelial* strains is to be found out (Stricker and Johnson 2011).

Immunity

Shortly after the tick attaches, an increasing number of *B. burgdorferi* around the site of the tick bite may cause clinical signs of cutaneous inflammation, and bacteria disseminate to distant locations of the host's body. In various body sites, they can induce specific immune responses that may result in chronic lesions due to inflammatory reactions. Innate defenses play important role in the early stage of *B. burgdorferi* infection. *B. garinii* (OspA serotypes 5 and 6) and *B. valaisianna* are sensitive to complement-mediated killing, *B. burgdorferi* Sensu Stricto displays intermediate sensitivity, while *B. afzelii* is complement-resistant (van Dam *et al.* 1997, Berende *et al.* 2010). Humoral immunity provides effective defense against *B. burgdorferi* infection. Rapid humoral response of the host controls *Borreliae* infection, which requires additional signaling by innate immune receptors such as Toll-like receptors for optimal T cell-dependent and T cell-independent antibody responses (Dickinson and Alugupalli 2012). The development of *B. burgdorferi* specific antibodies follows the classic pattern of IgM preceding IgG and IgA, and is characterized by the initial recognition of a limited number of antigens *viz.*, OspC (22 kD), flagellar (41kD), heat shock protein (58kD), OspA (31kD) and OspB (38kD), followed by a marked expansion in the number of antigens recognized later in the disease. The early serological response is mostly directed against 22 kD OspC and p22 of the spirochete. In dog, there is high agreement between antibodies to OspF and C6 as robust markers for knowing stage of infection, treatment outcomes and vaccination status while OspC is as an early infection marker (Wagner *et al.* 2012). Specific cellular immune response arises 1-3 weeks after *B. burgdorferi* infection and may persist during the entire course of disease. The T-cell mediated immune response may have role in immune protection and in pathogenesis of Lyme disease. Both Th1-type (*e.g.* IL-1β, IL-2, IL-6, TNF-α, IL-11, IL-12, IFNα) and Th2- type (*e.g.* IL-4 and IL-10) cytokines are induced in T lymphocytes, monocytes, macrophages or mast cells by *B. burgdorferi* lipoporoteins *in vitro* and are detected in the synovial fluid, cerebrospinal fluid (CSF) and cardiac lesions of patients and infected mice (Anguita *et al.* 1999). *Borrelia* uses various mechanisms to avoid the host defense system, including inhibition of innate immune cells such as macrophages and the complement system. Apart from binding to endothelial cells in blood vessels via p66, spirochetes are also able to enter the host tissues, form cysts to hide from attacking cells or change their outer surface antigens to protect themselves (Berende *et al.* 2010).

Clinical Signs in Animals

The syndromes caused by *B. burgdorferi* Sensu Lato complex in animals are poorly characterized and infections are predominantly asymptomatic (Parker and White 1992, Popovic *et al.* 1993). In dogs, inappetence, lethargy, lyphadenopathy, stiffness or pain in joint constitute the most common clinical syndrome. Renal borreliosis characterized by uremia, hyperphosphatemia, and severe protein-losing nephropathy, often accompanied by peripheral edema is the second most common canine syndrome and is generally fatal (Popovic *et al.* 1993). Conduction abnormality with bradycardia is typical of the rare cardiac form. Horse show chronic weight loss, sporadic lameness, laminitis, low grade fever, swollen joints, muscle tenderness and anterior uvetitis. In addition to these clinical sings, neurological signs such as depression, behavioral changes, dysphagia and encephalitis can be seen in chronic cases (James *et al.* 2010). Cattle affected with acute Lyme disease often show fever, stiffness and swollen joints as well as decreased milk production. Chronic weight loss, laminitis and abortion are also a possible outcome of borreliosis in cattle (Popovic *et al.* 1993). Experimentally infected rabbits show erythema migrans skin lesions, polyarthritis and carditis.

Clinical Signs in Humans

The incubation period in humans is typically 7 to 14 days, but can vary from 1 to 36 days. The progression of untreated disease may advance in stages from mild symptoms to serious, long-term disabilities. The disease can be roughly divided into three stages, early localized, early disseminated and late persistent, although there is considerable overlapping between the stages.

Stage 1: Early Localized Infection

Some people with Lyme disease develop a typical rash, called erythema migrans "bulls eye" rash, after one to two weeks at the site of the tick bite, which is usually circular and gets larger over time, but it can be absent in up to 50 per cent of patients. This is frequently associated with flu-like symptoms, which include lack of energy, headache and stiff neck, fever and chills, muscle and joint pain and regional lymphadenopathy (Hengge *et al.* 2003, Nau *et al.* 2009, Stricker and Johnson 2011). In some cases, the person does not notice any symptoms during this stage. If left untreated, erythema migrans usually resolves spontaneously in a few days to weeks (median, 4 weeks). Some children infected with *B. afzelii* or *B. garinii* develop, at the site of the tick bite, a painless bluish-red or reddish-purple nodule or plaque called borrelia lymphocytoma, occurring concurrently with erythema migrans, and is most often seen on the ear, nipple, or scrotum, but also on the nose, arms and other areas.

Stage 2: Early Disseminated Infection

If the disease remains asymptomatic, undetected or untreated during early stage, *B. burgdorferi* might spread from the site of the bite into the bloodstream, causing clinical signs of early dissemination. This stage affects skin, joints, nervous system and heart within weeks to months after the initial infection (Singh and Girschick 2004). Symptoms at this stage comprises of tiredness; as the infection spreads, additional skin rashes develop, pain, weakness, or numbness in the arms or legs, recurring headaches or fainting, poor memory and reduced ability to concentrate, conjunctivitis (pinkeye) or sometimes damage to deep tissue in the eyes and occasional rapid heartbeats or, in rare cases serious heart problems.

Neural manifestations during this stage are meningo radiculoneuritis (inflammation of the meninges and nerve roots, Bannwarth's syndrome), meningitis, plexus neuritis, inability to control the muscles of the face (paralysis of the facial nerves) and mononeuritis multiplex. After erythema migrans, Bannwarth's syndrome characterised by cerebrospinal fluid lymphocyte pleocytosis and radiculitis (inflammation of the spinal nerve roots) is the second most common manifestation of acute Lyme disease in adults. Relatively high protein concentration in the CSF distinguishes neuroborelliosis from a viral infection of the central nervous system. In the peripheral nervous system, spirochaetes have been detected in the perineurium, the connective tissue sheath surrounding each bundle of peripheral nerve fibres.

In Lyme disease, heart block can occur at multiple levels often involving the atrioventricular node, and sometimes myopericarditis develops (Steere 1989). Carditis can occur in early stages of the disease. Arthralgia and myalgia indicate early musculoskeletal involvement. Frank arthritis and myositis can also be observed occasionally in the first few months of the disease. Regional lymphadenopathy and generalized lymphadenopathy may develop.

Stage 3: Late Persistent Infections

If Lyme disease is not promptly or effectively treated, some patients develop a subjective syndrome that includes considerable fatigue, musculoskeletal aches, and neurocognitive dysfunction after receiving standard antibiotic courses for the treatment of Lyme disease (Aucott *et al.* 2012). Chronic arthritis, which causes recurring episodes of swelling, redness, and fluid buildup in one or more joints that last up to 6 months at a time. Sometimes it is called as chronic Lyme disease believing that persistent infection with *B. burgdorferi* is responsible. According to Joseph and Burrascano (2008), the condition chronic Lyme disease should fulfill following three criteria:

1. Illness present for at least one year;

2. Persistent major neurologic involvement or active arthritic manifestations; and

3. Active infection with *B. burgdorferi*, regardless of prior antibiotic therapy (if any).

Currently, the term post-Lyme disease syndrome is preferred over chronic Lyme disease, which better reflects the post-infectious nature of this condition.

Diagnosis

The diagnosis of clinical Lyme disease is difficult and it depends on a successful recognition of clinical signs, a history of possible exposures to the infection and on serologic testing (Popovic *et al.* 1993, O'Connell 2010). Routine diagnosis is based on clinical signs and supported, where possible, by laboratory tests. A variety of laboratory techniques have been developed for direct detection of *B. burgdorferi* Sensu Lato complex, including microscope-based assays, detection of *B. burgdorferi*-specific proteins or nucleic acids, and cultural identification (Bratton *et al.* 2008, Stanek *et al.* 2012). Diagnosis using the 2-tier antibody testing method is recommended, but the test may give false positive results in patients with early disseminated Lyme disease and a small percentage of patients with early Lyme disease who are treated with antibiotics (Ogden *et al.* 2008). The gold standard for the detection of antibodies to *B. burgdorferi* is a two-step procedure of an enzyme-linked immunosorbent assay (ELISA), followed by confirmatory Western blotting (Wagner *et al.* 2012). Recently, many methods have been in practice but they are not recommended for diagnostic purposes. These include antigen tests in body fluids, PCR of urine and lymphocyte transformation tests (CDC 2005). The majority of patients with post-Lyme disease syndrome show predominantly involvement of nervous system. Therefore, evaluation of this condition should include neuropsychiatric testing, SPECT and MRI brain scans, CSF analysis when appropriate, and regular input from Lyme-aware neurologists and psychiatrists, pain clinics, and occasionally specialists in psychopharmacology (Joseph and Burrascano 2008).

Direct Detection Methods

Culture and Identification

B. burgdorferi is a Gram negative, motile spirochete, with a length of 10 to 30 µm and a width of 0.2 to 0.5 µm. The organism is fastidious and microaerophilic, and it must be cultured on enriched bacteriologic media such as Barbour-Stoenner-Kelly or modified Kelly-Pettenkofer media (Wilske and Schriefer 2003). *B. burgdorferi* can be cultured from erythema migrans lesions (20 per cent to 90 per cent with an average of 50 per cent), blood samples (less than 5 per cent), CSF samples (less than 10 per cent), and joint fluid samples (1 per cent). Direct culture is very time-consuming (generation time about 7–20 h), and shows low sensitivity in body fluids (Zore *et al.* 2002). Nevertheless, it helps detect seronegative Lyme disease, *e.g.* in atypical erythema migrans, suspected acute neuroborreliosis without detection of intrathecal antibodies or suspected Lyme disease in patients with immune deficiencies (Wilske *et al.* 2007). The spirochete can be visualized using dark-field or phase-contrast microscopy, immunofluorescent microscopy, silver staining and Giemsa or acridine orange.

PCR

PCR allows the diagnosis of the pathogenic *Borrelia* species. Various target sequences have been used like plasmid-borne genes (ospA and ospB), chromosomal genes (flagellar protein or p66), or gene segments of the 16S rRNA or 5S/23S rRNA gene intergenic spacer region (Dumler 2003, Zore *et al.* 2002, Wilske *et al.* 2007). The sensitivity of PCR is about the same as that of culture except synovial fluid where it seems to surpass culture (Nocton *et al.* 1994). Borreliae are detected with much more difficulty from body fluids than from tissue specimens by either PCR or culture (Arnez *et al.* 2001, Wilske *et al.* 2007). Recently, equine uveitis associated with *B. burgdorferi* has been identified within ocular fluids with PCR testing (Priest *et al.* 2012). A real-time PCR assay has been employed to quantitate spirochetes in field mice, dogs, field-collected or laboratory-infected tick vectors, and clinical specimens of patients with LB (Aguero-Rosenfeld *et al.* 2005). A novel molecular beacon-based multiplex real-time quantitative PCR assay can be used to identify and detect small numbers of *B. burgdorferi* in infected mouse tissues (Saidac *et al.* 2009).

Indirect Detection Methods

ELISA

The ELISA is the simplest, least expensive, easiest to perform, and most common Lyme disease test used for screening as at least second generation tests that have been improved with respect to cross-reactivity with other bacteria (*e.g.* antigen extract with previous Reiter treponema adsorption) (Wilske and Preac-Mursic 1993, Wilske *et al.* 2000) or by using purified intact flagella as antigen (Hansen *et al.* 1988). The C6 *B. burgdorferi* ELISA test can distinguish between antibodies developed as a result of exposure and those produced as a result of vaccination against Lyme disease. As with the other antibody tests, however, the C6 test will not distinguish between exposure to *Borrelia* and actual infection (Liang *et al.* 1999). The C6-peptide ELISA is frequently used on serum and the same has been tested with a high sensitivity and good specificity for the diagnosis of Lyme neuroborreliosis patients in CSF (van Burgel *et al.* 2011).

Immunoblot

Western blots show which bands are reactive, 41kD bands appear the earliest but can cross react with other spirochetes. The species-specific bands which appear later or may not appear at all include 18kD, 23-25kD (Osp C), 31kD (Osp A), 34kD (Osp B), 37kD, 39kD, 83kD and the 93kD. There should be at least the 41kD and one of the specific bands for Lyme disease diagnosis. 55kD, 60kD, 66kD, and 73kD are non-specific and non-diagnostic (Joseph and Burrascano 2008). As a confirmatory assay, the immunoblot should have high specificity (at least 95 per cent). If a whole cell lysate is used as antigen, diagnostic bands must be defined by monoclonal antibodies. For the whole cell lysate blot, strains expressing immunodominant variable antigens (OspC, DbpA= Osp17) in culture should be used (Wilske *et al.* 2000). The sensitivity and standardization of immunoblots have been enhanced by the use of recombinant antigens, instead of whole cell lysates. Improved sensitivity has resulted from the use of recombinant proteins primarily expressed *in vivo* (*e.g.* VlsE) and the combination of homologous proteins from different strains (*e.g.* DbpA) (Schulte-Spechtel *et al.* 2003).

CD-57 Test

Chronic Lyme disease infections suppress the immune system and can decrease the quantity of the CD-57 subset of the natural killer cells. The degree of decrease indicates how active the Lyme infection is and also help in prediction of relapses after end of treatment. A normal count is remains above 200 while Lyme patients should measure above 60. The count remains low until the Lyme borelliosis infection is controlled. At the end of the antibiotic course, if the CD-57 count is not within the normal range, then a relapse of Lyme disease will almost certainly occur (Joseph and Burrascano 2008).

Prevention and Control Strategies

Lyme disease is a preventable disease. There are several approaches for prevention, including personal protection, tick control, post-exposure antibiotics, and early diagnosis and treatment (Stanek and Strle 2003, Hengge *et al.* 2003, Murray and Shapiro 2010, O'Connell 2010, Santos *et al.* 2010, Stanek *et al.* 2012).

Personal Protection from Tick Bites

☆ Walk in the center of trails to avoid contact with overgrown grass, brush, and leaf litter at trail edges.

☆ Spray repellent containing a 20 per cent concentration of DEET (diethyltoluamide) on clothes and on exposed skin.

☆ Treat clothes with permethrin, which kills ticks on contact or buy clothes that are pre-treated.

☆ Permethrin can also be used on tents and some camping gear. Do not use permethrin directly on skin.

☆ Wear long pants, long-sleeved shirts and tuck trousers into socks.

☆ Wear light-coloured clothing to make it easier to see and remove any ticks that come in contact.

☆ Bath or shower as soon as possible after coming indoors to wash off and more easily find ticks that are crawling on you.

☆ After being outdoors, wash and dry clothing at a high temperature to kill any ticks that may remain on clothing.

☆ Self-examination for attached ticks after being in a potential tick habitat. If infected and feeding ticks are removed within 36 hours of attachment, transmission of *B. burgdorferi* is prevented in most cases.

☆ Remove attached ticks with tweezers. Grasp the tick's head and mouth parts as close to the skin as possible, and pull slowly until the tick is removed.

☆ After removing the tick, use soap and water to wash the site where the tick was attached.

☆ If clinical signs appear, seek treatment early. Treatment is much easier in early Lyme disease and appropriate treatment might prevent the more severe and debilitating symptoms of disseminated Lyme disease.

Tick Control

☆ Tick-safe landscaping: *Ixodes* ticks need higher humidity; they die quickly in drier environments. Increasing exposure to sun and air by removing leaf litter and clearing tall grass and brush around houses and at the edges of lawns will reduce the numbers of ticks that transmit Lyme disease. Laying down gravel where lawns and recreational areas butt up against wooded areas can reduce the number of ticks on grassy areas by creating a drying barrier.

☆ The use of pesticides to control tick populations when nymphal *Ixodes* populations are at their local peaks is another highly effective option.

☆ Removing plants that attract deer and constructing physical barriers may help discourage tick-infested deer from coming near homes.

☆ Organic approach to control ticks involves the use of domesticated guinea fowl which are voracious consumers of insects and arachnids, and have a particular fondness for ticks (Duffy *et al.* 1992).

☆ Use plant extracts, essential oils, etc. exhibiting noxious or toxic activity (Pohlit *et al.* 2011).

Vaccines

There is no vaccine available for human beings (Stanek *et al.* 2012). The recombinant outer surface protein (OspA) based two vaccines, namely LYMErix (Glaxo

Smith Kline) and ImuLyme (Pasteur Me´rieux Connaught), were licensed but later on withdrawn from market because of various reasons (Poland 2011). Developing a vaccine to prevent Lyme disease remains a challenge. The way forward in the development of an effective Lyme disease vaccine could be combining different vaccinogens, whether these are multiple *Borrelia* antigens (OMPs), tick antigens (salivary protein, Salp15) or a combination, eliciting synergistic anti-Borrelia and anti-tick immune responses (Dai *et al.* 2009). These approaches might not only be applicable for the prevention of transmission of *B. burgdorferi* from the tick to the host but could also prove to be instrumental for the prevention of transmission of arthropod-borne pathogens in general.

It has shown that the disulfide-bridging of 2 OspC molecules via their N termini forms a complex which is more suitable for the determination of IgM-OspC and is a promising candidate for a monovalent vaccine (Probst *et al.* 2012). Recently, a Vaccinia virus (VV)-based OspA vaccine is proven stable in an oral bait preparation and provides protection against infection for both the natural reservoir and the tick vector of Lyme disease (Bhattacharya *et al.* 2011). In dogs, killed, whole-cell bacterins and Osp A vaccine are available. Immunized dogs induce strong antibody titers to Osp A and Osp B and produce antibodies that kill *B. burgdorferi in vitro*. Antibodies to Osp A block transmission of *B. burgdorferi* from the midgut of infected, engorging ticks and prevent infection. In endemic areas, young dogs should be vaccinated before natural exposure to ticks to attain the highest percentages of protection while in newly emerging areas two doses of either type of vaccine are administered SC to dogs ≥ 9 wk old at 3-wk intervals, and annual revaccination is recommended.

Treatment

Lyme disease is rarely life-threatening. There is no universally effective antibiotic for treating Lyme disease, so combinations of several antibiotics must be used. In general, early Lyme disease is treated for four to six weeks, and late Lyme disease usually requires a minimum of four to six months of continuous treatment (Joseph and Burrascano 2008). Systemic antibiotic treatment of Lyme disease is effective during the early stages of the infection, while chronic manifestations of the disease may remain refractory and difficult to treat (Knauer *et al.* 2011). According to the Infectious Diseases Society of America (IDSA) guidelines, the antibiotics of choice are doxycycline (in adults), amoxicillin (in children), erythromycin (for pregnant women) and ceftriaxone, with treatment lasting 10 to 28 days (Bratton *et al.* 2008). Several days after the onset of appropriate antibiotic therapy, symptoms often flare due to lysis of the spirochetes with release of increased amount of antigenic material and possibly bacterial toxins. This is referred to as a Jarisch Herxheimer-like reaction. Because it takes 48 to 72 hours of therapy to initiate bacterial killing, the Herxheimer reaction is therefore delayed. A novel approach of controlling the disease at early stage is a topical application of antibacterial agents. This is practical because *B. burgdorferi* organisms remain localized for >2 days in the mammalian skin at the site of tick bite. The infection at this stage can be effectively terminated by the topical application of antibiotics (*e.g.* tetracycline hydrochlorides, penicillin G, amoxicillin, ceftriaxone, doxycycline and azithromycin) directly on areas of tick bites (Knauer *et al.* 2011). Repeated treatment failures occur. One reason for persisted or worsened symptoms may be due to coinfection with *Babesia, Anaplasma, Ehrlichia,* or *Bartonella*. Other three things that will predict treatment failure, regardless of regimen chosen, are noncompliance, alcohol use, and sleep deprivation (Joseph and Burrascano 2008).

A few patients may have persistent or recurrent symptoms and may require a second 4-week course of antibiotic treatment. Longer courses of antibiotics have not been shown to be beneficial in patients who have been previously treated and have chronic symptoms. Treatment of post-Lyme disease syndrome usually requires combinations of antibiotics. It is because the bacteria are found in both fluid and tissue compartments (azithromycin plus penicillin), and they survive intracellularlly (erythromycin derivative or metronidazole) and occur in L-forms (spheroplast) (tetracyclines and the advanced erythromycin derivatives) and cystic form (metronidazole and tinidazole) (Joseph and Burrascano 2008). Post-Lyme disease syndrome has inhibitory effect on the immune system. This is why, steroids and other immunosuppressive medications are absolutely contraindicated including intra-articular steroids (Joseph and Burrascano 2008). The therapy in animals is based upon the principal therapy of Lyme disease in human medicine (Popovic *et al.* 1993). Procaine penicillin at 30000-45000 iu/kg IM daily for 10 days, followed by benzathine penicillin every other day, has been recommended for horses (Parker and White 1992). Penicillin or oxytetracyclines daily for 3 weeks have also been recommended for use in cattle. In infected mice treated with antibiotics, the persistence of nonpathogenic residual spirochetes needs to be understood regarding their biological nature and the lack of pathogenicity, irreversible or not (Wormser and Schwartz 2009).

Future Directions

Lyme disease has risen from relative insignificance to become a global public health problem and a prototype of an emerging infection. Lyme disease in domesticated animals is still poorly understood. Many aspects of the pathophysiology of the disease remain unsolved, and the nature of the immune response to the pathogen is only partly understood (Berende *et al.* 2010). Also, diagnosis and treatment is stalled due to inadequate clinical disease appearance, short of tests with higher sensitivity, and treatment regimens not being highly successful (Stricker and Johnson 2010).

Currently, diagnosis and treatment is hindered by the lack of a uniform case definition that adequately reflects the clinical presentation of the disease, poor laboratory test sensitivity, and high treatment failure rates. Preventive measures aimed at minimizing tick-bite risk are promoted as one of the best ways to avoid *Borrelia* infection. Researchers in this field agree that a complete eco-epidemiological knowledge is also needed to develop accurate risk prediction models. Further work in this area should be focused on information about geographic distribution of ticks, alternative strategies and chemicals to protect against them, development of more effective vaccines and diagnostics, improved antibiotic treatment and in-depth understanding of post-Lyme disease syndrome.

Extensive evidence now shows that persistent symptoms of Lyme disease are due to chronic infection with the Lyme spirochete in conjunction with other tick-borne coinfections. Present knowledge shows that prolonged antibiotic therapy does not offer lasting or substantive benefit in treating patients with post-Lyme disease syndrome. Recent studies revealed that *B. burgdorferi* infected animals treated with antibiotic therapy demonstrated the presence of PCR positivity in the absence of culture positivity. Persistence of nonpathogenic residual spirochetes has been found in infected mice treated with antibiotic therapy, the biological nature of these residual spirochetes is unclear, and it is also unclear whether the lack of pathogenicity is irreversible. Novel therapeutic strategies for the complete eradication of these spirochetes from infected patients are required in future. There is greater need to clarify the role of inflammation as a mediator of the chronic persistent symptoms. Moreover, the potential importance of CWD forms and biofilm in persistent infection, newer antibiotics aimed at this evasive mutant and biofilm regulatory process are very much needed to eradicate persistent infection in Lyme disease. Therefore, it is time to move forward to test other approaches that may help these patients. This opens research area of the development of new drugs targeting novel infectious processes and more effective treatment of Lyme disease over the next decade.

Greater efforts are needed to provide education for prevention and early diagnosis to avoid late complications. Standardized diagnoses, worldwide reporting systems and datasets concerning all aspects of the molecular ecology and epidemiology are crucial to treating and fighting Lyme borrelliosis.

References

1. Afzelius A. 1910. Verhandlungen den dermatologischen Gesellschaft zu Stockholm. *Acta Derm Venereol* 101: 404.

2. Aguero-Rosenfeld ME, Wang G, Schwartz I and Wormser GP. 2005. Diagnosis of Lyme borreliosis. *Cl Microbiol Rev* 18 (3): 484-509.

3. Anderson JF. 1991. Epizootiology of Lyme borreliosis. *Scand J Infect Dis* 77: 23-34.

4. Anguita J, Barthold SW, Samanta S. Ryan J and Fikrig E. 1999. Selective anti-inflammatory action of interleukin-11 in murine Lyme disease : arthiritis decrease while carditis persists. *J Infect Dis* 179: 734-737.

5. Arnez M, Ruzic-Sabljic E, Ahcan J, Radsel-Medvescek A, Pleterski-Rigler D and Strle F. 2001. Isolation of *Borrelia burgdorferi* sensu lato from blood of children with solitary erythema migrans. *Pediatr Infect Dis J* 20: 251-255.

6. Aucott JN, Rebman AW, Crowder LA and Kortte KB. 2012. Post-treatment Lyme disease syndrome symptomatology and the impact on life functioning: is there something here? Qual Life Res doi: 10.1007/s11136-012-0126-6.

7. Barbour AG. 1984. Isolation and cultivation of Lyme disease spirochetes. *Yale J Biol Med* 57: 521-525

8. Barbour AG and Fish D. 1993. The biological and social phenomenon of Lyme disease. *Science* 260(5114): 1610-1616.

9. Barbour AG and Hayes S. 1986. Biology of *Borrelia* species. *Microbiol Rev* 50: 381-400.

10. Behera AK, Hildebrand E, Szafranski J, Hung H, Grodzinsky AJ, Lafyati R, Koch AE, Kalish R, Perides G, Steere AC and Hu LT. 2006. Role of aggrecanase 1 in Lyme arthritis. *Arthritis Rheum* 54: 3319-3329.

11. Berende A, Oosting M, Kullberg BJ, Netea MG and Joosten LAB. 2010. Activation of innate host defense mechanisms by *Borrelia*. *Eur Cytokine Netw* 21(1): 7-18.

12. Bhattacharya D, Bensaci M, Luker KE, Luker G, Wisdom S, Telford SR and Hu LT. 2011. Development of a baited oral vaccine for use in reservoir-targeted strategies against Lyme disease. *Vaccine* 29(44): 7818-7825.

13. Bratton RL, Whiteside JW, Hovan MJ, Engle RL and Edwards FD. 2008. Diagnosis and treatment of Lyme disease. *Mayo Clin Proc* 83 (5): 566-571.

14. Burgdorfer W, Barbour AG, Hayes SF, Benhach JL and Davis JP. 1982. Lyme disease, a thick-borne spirochetosis? *Science* 216: 1317-1319.

15. Burgess EC. 1988. *Borrelia burgdorferi* infection in Wisconsin horses and cows. *Ann NY Acad Sci* 539: 235-243.

16. CDC. 2005. Caution regarding testing for Lyme disease. Department of Health and Human Services, Centers for Disease Control and Prevention. *Morb Mortal Wkly Rep* 54: 125.

17. Criswell D, Tobiason VL, Lodmell JS and Samuels DS. 2006. Mutations conferring aminoglycoside and spectinomycin resistance in *Borrelia burgdorferi*. *Antimicrob Agents Chemother* 50: 445-452.

18. Dai J, Wang P, Adusumilli S, Booth CJ, Narasimhan S, Anguita J and Fikrig E. 2009. Antibodies against a tick protein, Salp15, protect mice from the Lyme disease agent. *Cell Host Microbe* 6: 482-492.

19. Dickinson GS and Alugupalli KR. 2012. Deciphering the role of Toll-like receptors in humoral responses to Borreliae. *Front Biosci (Schol Ed)* 1(4): 699-712.

20. Duffy DC, Downer R and Brinkley C. 1992. The effectiveness of Helmeted Guinea fowl in the control of the deer tick, the vector of Lyme disease. *The Wilson Bulletin* 104 (2): 342-345.

21. Dumler JS. 2003. Molecular methods for Ehrlichiosis and Lyme disease. *Clin Lab Med* 23: 867-884.

22. Elbaum-Garfinkle S. 2011. Close to home: A history of Yale and Lyme Disease. *Yale J Biol Med* 84: 103-108.

23. Fallon BA, Levin ES, Schweitzer PJ and Hardesty D. 2010. Inflammation and central nervous system Lyme disease. *Neurobiol Dis* 37(3): 534-541.

24. Fraser CM, Casjens S, Huang WM, Sutton GG, Clayton R, Lathigra R, White O, Ketchum KA, Dodson R, Hickey EK, Gwinn M, Dougherty B, Tomb JF, Fleischmann RD, Richardson D, Peterson J, Kerlavage AR, Quackenbush J, Salzberg S, Hanson M, van Vugt R, Palmer N, Adams MD, Gocayne J, Weidman J, Utterback T, Watthey L, McDonald L, Artiach P, Bowman C, Garland S, Fuji C, Cotton MD, Horst K, Roberts K, Hatch B, Smith HO and Venter JC. 1997. Genomic sequence of a Lyme disease spirochaete, *Borrelia burgdorferi*. *Nature* 390: 580-586.

25. Gayle A and Ringdahl E. 2001. Tick-borne Diseases. *Amer Fam Phys* 64: 461-466.

26. Gray, JS. 1998. The ecology of ticks transmitting Lyme borreliosis. *Exp Appl Acarol* 22: 249-258.

27. Hannier S, Liversidge J, Sternberg JM and Bowman AS. 2003. *Ixodes ricinus* tick salivary gland extract inhibits IL-10 secretion and CD69 expression by mitogen-stimulated murine splenocytes and induces hyporesponsiveness in B lymphocytes. *Parasite Immunol* 25: 27-37.

28. Hansen K, Hindersson P and Pedersen NS. 1988. Measurement of antibodies to the *Borrelia burgdorferi* flagellum improves serodiagnosis in Lyme disease. *J Clin Microbiol* 26: 338-346.

29. Hengge UR, Tannapfel A, Tyring SK, Erbel R, Arendt G and Ruzicka T. 2003. Lyme borreliosis. *Lancet Infect Dis* 3: 489-500.

30. Hollstrom B. 1958. Penicillin treatment of erythema chronicum migrans. *Acta Dermatol Venerol* 38: 285-289.

31. James FM, Engiles JB and Beech J. 2010. Meningitis, cranial neuritis, and radiculoneuritis associated with *Borrelia burgdorferi* infection in a horse. *J Am Vet Med Assoc* 237: 1180-1185.

32. Joseph J and Burrascano JR. 2008. Advanced topics in Lyme disease diagnostic hints and treatment guidelines for Lyme and other tick borne illnesses. Sixteenth Edition, e-Book http://www.lymenet.org/BurrGuide200810.pdf

33. Knauer J, Krupka I, Fueldner C, Lehmann J and Straubinger RK. 2011. Evaluation of the preventive capacities of a topically applied azithromycin formulation against Lyme borreliosis in a murine model. *J Antimicrob Chemother* 66(12): 2814-2822.

34. Liang FT, Steere AC, Marques AR, Johnson BJ, Miller JN and Philipp MT. 1999. Sensitive and specific serodiagnosis of Lyme disease by enzyme-linked immunosorbent assay with a peptide based on an immunodominant conserved region of *Borrelia burgdorferi* vlsE. *J Clin Microbiol* 37(12): 3990-3996.

35. Liang FT, Brown EL, Wang T, Iozzo RV and Fikrig E. 2004. Protective niche for *Borrelia burgdorferi* to evade humoral immunity. *Am J Pathol* 165: 977-985.

36. Lipschutz B. 1914. Ubereine Seltene Erythemform (Erythema cronicum migrans). *Arch Dermatol Syphilol* 118: 349-356.

37. Livengood JA and Gilmore RD. 2006. Invasion of human neuronal and glial cells by an infectious strain of *Borrelia burgdorferi*. *Microbes Infect* 8: 2832-2840.

38. Montgomery RR, Schreck K, Wang X and Malawista SE. 2006. Human neutrophil calprotectin reduces the susceptibility of *Borrelia burgdorferi* to penicillin. *Infect Immun* 74: 2468-2472.

39. Murray TS and Shapiro ED. 2010. Lyme disease. *Clin Lab Med* 30(1): 311-328.

40. Nahimana I, Gern L, Blanc DS, Praz G, Francioli P and Péter O. 2004. Risk of *Borrelia burgdorferi* infection in western Switzerland following a tick bite. *Eur J Clin Microbiol Infect Dis* 23: 603-608.

41. Nau R, Christen H and Eiffert H. 2009. *Lyme Disease-Current State of Knowledge*. Deutsches Ärzteblatt International, Dtsch Arztebl Int 106(5): 72-82.

42. Nocton JJ, Dressler F, Rutledge BJ, Rys P N, Persing DH and Steere AC. 1994. Detection of *Borrelia burgdorferi* DNA by polymerase chain reaction in synovial fluid from patients with Lyme arthritis. *N Engl J Med* 330: 229-234.

43. O'Connell S. 2010. Lyme borreliosis: current issues in diagnosis and management. *Curr Opin Infect Dis* 23(3): 231-235.

44. Ogden NH, Artsob H, Lindsay LR and Sockett PN. 2008. Lyme disease: A zoonotic disease of increasing importance to Canadians. *Canadian Family Physician* 54: 1381-1384.

45. Pachner AR and Steiner I. 2007. Lyme neuroborreliosis: infection, immunity, and inflammation. *Lancet Neurol* 6 (6): 544-552.

46. Parker JL and White KK. 1992. Lyme borreliosis in cattle and horses: a review of the literature. *Cornell Vet* 82: 253.

47. Piesman J and Gern L. 2004. Lyme borreliosis in Europe and North America. *Parasitology* 129: 191-220.

48. Pohlit AM, Rezende AR, Lopes Baldin EL, Lopes NP and Neto VF. 2011. Plant extracts, isolated phytochemicals, and plant-derived agents which are lethal to arthropod vectors of human tropical diseases: a review. *Planta Med* 77(6): 618-630.

49. Poland GA. 2011. Vaccines against Lyme disease: What happened and what lessons can we learn? *Clin Infect Dis* 52(S3): S253-S258.

50. Popovic N, Djuricic B and Valcic M. 1993. The importance of Lyme borreliosis in veterinary medicine. *Glas Srp Akad Nauka Med* 43: 277-285.

51. Porcella SF and Schwan TG. 2001. *Borrelia burgdorferi* and *Treponema pallidum*: a comparison of functional genomics, environmental adaptations, and pathogenic mechanisms. *J Clin Invest* 107: 651-656.

52. Praharaj AK, Jetley S and Kalghatgi AT. 2008. Seroprevalence of *Borrelia burgdorferi* in North Eastern India. *Medical J Armed Forces India* 64(1): 26-28.

53. Priest HL, Irby NL, Schlafer DH, Divers TJ, Wagner B, Glaser AL, Chang YF and Smith MC. 2012. Diagnosis of Borrelia-associated uveitis in two horses. *Vet Ophthalmol. doi*: 10.1111/j.1463-5224.2012.01000.x.

54. Probst C, Ott A, Scheper T, Meyer W, Stöcker W and Komorowski L. 2012. N-terminal disulfide-bridging of *Borrelia* outer surface protein C increases its diagnostic and vaccine potentials. *Ticks Tick Borne Dis* 3(1): 1-7.

55. Radolf JD, Caimano MJ, Stevenson B and Hu LT. 2012. Of ticks, mice and men: understanding the dual-host lifestyle of Lyme disease spirochaetes. *Nat Rev Microbiol* 10(2): 87-99.

56. Ramamoorthi N, Narasimhan S, Pal U, Bao F, Yang XF, Fish D, Anguita J, Norgard MV, Kantor FS, Anderson JF, Koski RA and Fikrig E. 2005. The Lyme disease agent exploits a tick protein to infect the mammalian host. *Nature* 436: 573-577.

57. Rand PW, Lubelczyk C, Holman MS, Lacombe EH and Smith RP Jr. 2004. Abundance of *Ixodes scapularis* (Acari: Ixodidae) after the complete removal of deer from an isolated offshore island, endemic for Lyme Disease. *J Med Entomol* 41(4): 779-784.

58. Rudenko N, Golovchenko M, Grubhoffer L and Oliver JH Jr. 2011. Updates on *Borrelia burgdorferi* sensu lato complex with respect to public health. *Ticks Tick Borne Dis* 2(3): 123-128.

59. Saidac DS, Marras SA and Parveen N. 2009. Detection and quantification of Lyme spirochetes using sensitive and specific molecular beacon probes. *BMC Microbiol* 9: 43.

60. Santos M, Haddad Jr. V, Ribeiro-Rodrigues R and Talhari S. 2010. Lyme borreliosis. *An Bras Dermatol* 85(6): 930-938.

61. Schuijt TJ, Hovius JW, van der Poll T, van Dam AP and Fikrig E. 2011. Lyme borreliosis vaccination: the facts, the challenge, the future. *Trends Parasitol* 27(1): 40-47.

62. Schulte-Spechtel U, Lehnert G, Liegl G, Fingerle V, Heimerl C, Johnson BJ and Wilske B. 2003. Significant improvement of the recombinant Borrelia-specific immunoglobulin G immunoblot test by addition of VlsE and a DbpA homologue derived from *Borrelia garinii* for diagnosis of early neuroborreliosis. *J Clin Microbiol* 41: 1299-1303.

63. Singh SK and Girschick HJ. 2004. Lyme borreliosis: from infection to autoimmunity. *Clin Microbiol Infect* 10: 598-614.

64. Stanek G and Strle F. 2003. Lyme borreliosis. *Lancet* 362: 1639-1647.

65. Stanek G, Wormser GP, Gray J and Strle F. 2012. Lyme borreliosis. *Lancet* 379(9814): 461-473.

66. Steere AC. 1989. Lyme disease. *N Engl J Med* 321: 586-596.

67. Steere AC, Malawista SE and Syndman DR. 1977. Lyme arthritis: an epidemic of oligoarticular arthritis in children and adults in three Connecticut communities. *Arthritis Rheum* 20: 7-17.

68. Stricker RB. 2007. Counterpoint: long-term antibiotic therapy improves persistent symptoms associated with lyme disease; Antibiotic therapy and Lyme disease. *Clin Infect Dis* 45: 149-157.

69. Stricker RB and Johnson L. 2010. Lyme disease diagnosis and treatment: lessons from the AIDS epidemic. *Minerva Med* 101(6): 419-425.

70. Stricker RB and Johnson L. 2011. Lyme disease: the next decade. *Infect Drug Resist* 4: 1-9.

71. van Burgel ND, Brandenburg A, Gerritsen HJ, Kroes AC and van Dam AP. 2011. High sensitivity and specificity of the C6-peptide ELISA on cerebrospinal fluid in Lyme neuroborreliosis patients. *Clin Microbiol Infect* 17(10): 1495-1500.

72. van Dam AP, Oei A, Jaspars R. Fijen C, Wilske B, Spanjaard L and Dankert J. 1997. Complement mediated serum sensitivity among spirochetes that cause Lyme disease. *Infect Immun* 65: 1228-1236.

73. Wagner B, Freer H, Rollins A, Garcia-Tapia D, Erb HN, Earnhart C, Marconi R and Meeus P. 2012. Antibodies to *Borrelia burgdorferi* OspA, OspC, OspF and C6 antigens as markers for early and late infection in dogs. *Clin Vaccine Immunol* doi: 10.1128/CVI.05653-11.

74. Williams LR and Austin FE. 1992. Hemolytic activity of *Borrelia burgdorferi*. *Infect Immun* 60: 3224-3230.

75. Wilske B and Preac-Mursic V. 1993. Microbiological diagnosis of Lyme borreliosis. In: Weber K and Burgdorfer W. (eds) *Aspects of Lyme borreliosis*. Springer Verlag, Berlin. p. 267-300.

76. Wilske B and Schriefer M. 2003. *Borrelia*. In: Murray PR, Baron EJ, Jorgensen JH, Pfaller MA and Yolken RH. (eds) *Manual of Clinical Microbiology*. ASM Press, Washington, DC. p. 937-954.

77. Wilske B, Zoller L, Brade V, Eiffert H, Gobel UB, Stanek G and Pfister HW. 2000. MIQ 12, Lyme-Borreliose. In: Mauch H and Lutticken R. (eds) *Qualitatsstandards in der mikrobiologisch-infektiologischen diagnostik*. Urban and Fischer Verlag, Munich. p. 1-59.

78. Wilske B, Fingerle V and Schulte-Spechtel U. 2007. Microbiological and serological diagnosis of Lyme borreliosis. *FEMS Immunol Med Microbiol* 49: 13-21.

79. Wormser GP and Schwartz I. 2009. Antibiotic Treatment of Animals Infected with *Borrelia burgdorferi*. *Clin Microbiol Rev* 22: 387-395.

80. Zore A, Ruzic-Sabljic E, Maraspin V, Cimperman J, Lotric-Furlan S, Pikelj A, Jurca T, Logar M and Strle F. 2002. Sensitivity of culture and polymerase chain reaction for the etiologic diagnosis of erythema migrans. *Wien KlinWochenschr* 114: 606-609.

2014, Zoonoses: Bacterial Diseases
Editor: Sudhi Ranjan Garg
Published by: DAYA PUBLISHING HOUSE, NEW DELHI

Pages **354–366**

18

Tularaemia

S.R. Garg

Department of Veterinary Public Health and Epidemiology,
College of Veterinary Sciences, Lala Lajpat Rai University of
Veterinary and Animal Sciences, Hisar – 125 004

Tularaemia is a bacterial septicaemia that affects a large number of species of wild and domestic mammals, birds, reptiles, fish and humans. Although many wild and domestic animals have been infected, rabbits are most often involved in disease outbreaks; hence the disease is commonly known as rabbit fever too. The other common names for the disease are deer fly fever, meat cutter's disease, Ohara disease and Francis disease. Tularaemia is caused by the bacterium *Francisella tularensis.* It is a classic zoonosis, capable of being transmitted by aerosol, direct contact, ingestion, or arthropods. The organism is also listed as a biological weapon due to its potential for airborne dissemination. The disease can range from mild to severe, flu-like illness in humans and many animals (CFSPH 2009, Aiello 2012).

Geographical Distribution

With rare exceptions, tularaemia occurs only in the Northern Hemisphere. The disease has been reported from North America, Europe, parts of Asia, the Middle East, and northern Africa. It appears that the organism is present in Australia but is yet to be isolated from any wildlife species. A single human case, caused by *F. tularensis* subsp. *novicida*, has been reported in Australia in 2003 (Whipp *et al.* 2003, CFSPH 2009).

Etiological Agent

The causative organism of tularaemia is *Francisella tularensis*, which is a Gram negative, nonsporulating coccobacillus in the family Francisellaceae. The bacterium was formerly known as *Pasteurella tularensis*. It is a facultative intracellular parasite. *Francisella* currently has four recognized subspecies. All of these cause tularaemia but they tend to be associated with different animal hosts, and the severity of the disease varies. *F. tularensis* subsp. *tularensis* (also known as type A) has been found almost exclusively in North America. It is the most virulent species, particularly in humans. *F. tularensis* subsp. *holarctica* (type B) is much more widely distributed in the Northern Hemisphere, but it is less virulent. It is most commonly isolated from aquatic animals and water-associated infections in North America and Eurasia. The other two species *F. tularensis* subsp. *mediasiatica* and *F. tularensis* subsp. *novicida* have been recognized in limited geographical regions. These are rarely found in people, and seem to cause relatively mild disease when they occur. Additional subspecies have also been proposed (CFSPH 2009, Aiello 2012).

Vectors

A variety of arthropods can transmit *F. tularensis*. Ixodid ticks are biological vectors. Species that are important in transmission include *Dermacentor andersoni* (the wood tick), *Amblyomma americanum* (the lone star tick), *D. variabilis* (the dog tick) in North America, and *Haemaphysalis flava* and *Ixodes japonesi* in Japan. *F. tularensis* is transmitted transstadially, and ticks can remain infected throughout their lifetimes; however, the occurrence of transovarial transmission is controversial (CFSPH 2009, Aiello 2012). Biting flies in the family Tabanidae, particularly the deer fly (*Chrysops discalis*), can act as mechanical vectors. Individual flies can carry the organism for two weeks. Mosquitoes can also be vectors. The arthropods that are most important in transmission vary with the geographical area. *F. tularensis* also occurs in other arthropods, although their role in transmission is not confirmed in many cases. In Europe, *F. tularensis* subsp. *holarctica* has been isolated from mites (family Gamasidae) collected from rodents, and mites have been shown to transmit tularaemia between rodents in the laboratory (CFSPH 2009). Ceratopogonids (biting midges) and Simulidae (blackflies) have also been proposed as potential vectors. *F. tularensis* subsp. *holarctica* seems to be associated with lakes, streams and aquatic animals in some areas, and it might also be maintained in aquatic protozoans (CFSPH 2009).

Host Range

More than 250 species of animals can be infected with *F. tularensis*. Common wild animal hosts include lagomorphs (cottontail rabbits, jack rabbits and hares), muskrats, beavers, and a variety of rodents, including voles, field mice, vole rats, squirrels and lemmings. In domestic animals, sheep are the primary host. Clinical infection has been reported in dogs, pigs, horses, primates, ranched mink and captive prairie dogs, but cattle appear to be resistant. Cats are at increased risk due to predatory behaviour and appear to have an increased susceptibility. Infections have been reported occasionally in birds, reptiles, amphibians, crayfish, mollusks and fish.

Each subspecies of *F. tularensis* tends to be associated with certain host species (CFSPH 2009, Aiello 2012).

Disease Transmission

F. tularensis can be transmitted by direct contact with mucous membranes and broken skin, ingestion, inhalation and through the bites of infected arthropod vectors. The most common source of infection for people and herbivores is the bite of infected vectors. Infected animals can transmit tularaemia to arthropod vectors. The organism is found in the blood and tissues of infected animals, so the common route of infection in humans includes inoculation of the skin or mucous membranes with blood or tissue while handling infected animals or meat. Ingestion of improperly cooked infected meat can cause infection in human beings. Similarly, carnivores may acquire infection from ingestion of an infected carcass. *F. tularensis* has also been reported in the faeces and urine of some animals. Drinking contaminated water and inhaling contaminated dust may also cause infection. Aquatic animals may develop tularaemia after being immersed in contaminated water (CFSPH 2009, Aiello 2012). Person-to-person transmission has not been reported. However, infectious organisms can be found in the blood, lesions and tissues (CFSPH 2009).

Predisposing and Environmental Factors

Hunters, butchers, farmers, fur/wool handlers, veterinarians, laboratory workers and others who come in contact with infected animals are at greater risk of acquiring tularaemia. Some outbreaks in humans have been linked to epizootics among animals. Hunting lagomorphs, hamsters or other species during epizootics in these animals has been reported to cause epidemics. Epidemics can be seen during or after wars because there may be rodent population explosions during such situations due to the availability of ample unharvested grain and other food sources. These animals contaminate human food (CFSPH 2009).

In a study in a tularaemia-endemic boreal forest area of Sweden, the predicted mosquito abundance was correlated with the annual number of human cases. The predicted mosquito peaks consistently preceeded the median onset time of human tularaemia. This study suggested that a high prevalence of mosquitoes in late summer was a prerequisite for outbreaks of tularaemia in that area and that environmental variables could be used as risk indicators (Rydén *et al.* 2012). Geospatial analysis in Azerbaijan identified associations between village-level tularaemia prevalence and suitable tick habitats, annual rainfall, precipitation in the driest quarter, and altitude (Clark *et al.* 2012).

Prevalence of Disease in Animals

Natural foci of infection exist in North America and Eurasia, where the organism circulates between arthropod vectors and various mammals, birds, reptiles and fish. Though little is known of the true incidence of infection and clinical disease in animals, particularly the domestic animals, tularaemia is relatively common and highly fatal in some species of wild animals; epizootics occur regularly among rabbits and rodents (CFSPH 2009). In lower Austria, Burgenland and Styria, 271 cases of tularaemia in hares were recorded from 1994 to 2005 (Deutz *et al.* 2009).

The tularaemia seroprevalence studies in wild boars in north-eastern Germany revealed anti-Francisella antibodies in 3.1 per cent of the wild boars. The high seroprevalence of tularaemia in wild boars indicated that its natural foci is present in wildlife in Germany (Al Dahouk *et al.* 2005). In another recent study, 2475 animals from Germany, both captive and wild, were tested for antibodies against *Francisella tularensis* to obtain more knowledge about the presence of this pathogen in Germany. Of the zoo animals sampled between 1992 and 2007 (n=1122), three (0·3 per cent) were seropositive. From 1353 serum samples of wild foxes (*Vulpes vulpes*), raccoon dogs (*Nyctereutes procyonoides*) and wild boars (*Sus scrofa*) collected between 2005 and 2009 in the federal state of Brandenburg (surrounding Berlin), 101 (7·5 per cent) tested positive for antibodies to *F. tularensis* lipopolysaccharide. The studies indicated a higher seroprevalence of the organism in wildlife in eastern Germany than commonly assumed (Kuehn *et al.* 2012).

In Hungary, it was shown that among hares captured for live animal export during 1984-2009 (2.8-40 thousand exported hares/year), the prevalence of tularaemia ranged between 0.31 per cent and 20.2 per cent (Gyuranecz 2011). Very recently, in Japan, seroprevalence of tularaemia among wild black bears and hares was determined. Blood samples collected from 431 Japanese black bears (*Ursus thibetanus japonicus*) and 293 Japanese hares (*Lepus brachurus*) between 1998 and 2009 were examined for antibodies against *F. tularensis*. Eight sera from Japanese black bears were definitely shown to be seropositive. All of these eight bears were residents of the northeastern part of main-island of Japan, where human tularaemia had been reported. On the other hand, no seropositive Japanese hares were found (Hotta *et al.* 2012).

Prevalence of Disease in Humans

F. tularensis has been recognized as a human pathogen for almost 100 years, however, the disease outbreaks of tularaemia during the last two decades have led to a renewed interest in the disease. A plague-like disease in California ground squirrels was described by McCoy in 1911. The causative agent was named Bacterium tularense (McCoy and Chapin 1912). The human disease was recognized and described by Edward Francis (Francis 1921) as tularaemia, and the agent was later renamed *Francisella tularensis* in his honour. About 120 cases of tularaemia are reported each year in the United States (CDC 2012). In Canada, there were 289 reported cases of tularaemia and 12 deaths between 1940 and 1981 (PHAC 2012).

In most European countries, tularaemia exists endemically. In some areas, such as Finland and Sweden, outbreaks comprising hundreds of cases are recorded at least once a decade. In other areas, outbreaks of such a magnitude occur only occasionally, except in times of war (Tärnvik *et al.* 2004). According to the Eurostat, the European Centre for Disease Prevention and Control, during the years 1997 to 2006, every year between 280 and 1737 cases of tularaemia were reported in the European Union countries (European Commission 2012). In the European Union (EU), European Economic Area (EEA) and European Free Trade Association (EFTA) countries, 568, 1230 and 850 confirmed cases of tularaemia were reported during the years 2006, 2007 and 2008, respectively (ECDC 2012). In a study in Hungary, it was shown through retrospective data collection that the number of human cases ranged between 20 and 148 per year during 1984-2009 (Gyuranecz 2011). In a tularaemia-endemic boreal forest area of Sweden, 370 humans were diagnosed with tularaemia between 1981 and 2007 (Rydén *et al.* 2012).

Tularaemia has recently emerged in some areas. In Spain, it was recognized in 1997-1998, when two outbreaks affected more than 500 people. One of these outbreaks was associated with hares and the other was linked to crayfish (CFSPH 2009). Similarly, tularaemia had not been registered in Kosovo before an outbreak in 1999 and 2000. In 2000, a human epidemic in Kosovo was associated with a population explosion and epizootic among rodents, which occurred after large a number of people had been displaced from their homes by the conflict. In 2001 and 2002, a second outbreak occurred with 327 serologically confirmed cases. From 2001 to 2010, 25 to 327 cases were registered per year (mean annual incidence 5.2/100,000 population). By 2010, the disease had spread throughout Kosovo (Grunow *et al.* 2012).

F. tularensis has never been found in Australian animals (Ladds 2009). *F. tularensis novicida* was isolated from a man in the Northern Territory in 2003 (Whipp *et al.* 2003). In February and September 2011, a diagnosis of *F. tularensis* subsp. *holarctica* biovar *japonica* was made based on PCR, supported by typical clinical presentation, in two women who were scratched and bitten by possums between Queenstown and Zeehan in Tasmania (Brown and McGregor unpublished), however, the organism was not cultured and follow-up surveys of possums from both areas failed to identify any evidence of tularaemia infection or exposure. The Australian status with respect to tularaemia remains thus unclear. It appears that the organism is present in Australia, but is yet to be isolated from any wildlife species (AWHN 2012).

A representative study in 40 selected villages in northern Azerbaijan to screen the people (n=796) for evidence of prior infection showed that *F. tularensis* seropositivity was prevalent in 15.5 per cent volunteers, but there was minimal reporting of signs and symptoms consistent with clinical tularaemia, suggesting that mild or asymptomatic infection commonly occurred (Clark *et al.* 2012).

Tularaemia has become a re-emerging disease in Turkey. Recently several cases and epidemics of tularaemia have been reported in the northwest areas of Turkey. Yazqi *et al.* (2011) carried out a study to determine *F. tularensis* antibody seropositivity in the population at risk living in both rural and urban area of Erzurum city. Blood samples from 240 volunteer subjects, whose occupations were farming and animal husbandry, were included in the study, of which 2.1 per cent were evaluated as seropositive for tularaemia.

Disease Manifestations in Animals

Tularaemia is a highly contagious disease occurring principally in wild animals but it may transmit to farm animals, causing septicaemia and high mortality. The incubation period is 1 to 10 days. The full spectrum of clinical signs is not known in animals, but syndromes corresponding to the typhoidal, respiratory, ulceroglandular and oropharyngeal forms have been reported. Septicaemia is often reported in susceptible animals, but asymptomatic infections also seem to be common, especially in resistant species such as dogs and cattle. The disease is relatively common and highly fatal in some species of wild animals. Epizootics occur regularly among rabbits and rodents. Signs of septicaemia may occur in wild mammals, domesticated rabbits and pet rodents, but many animals are found dead. Rabbits and rodents may be depressed, with anorexia and ataxia, a roughened coat and a tendency to huddle (CFSPH 2009).

Cats seem to be relatively susceptible to tularaemia. Infections often begin with the sudden onset of fever, lethargy, anorexia, and regional or generalized lymphadenopathy. The lymph nodes may suppurate and drain. The submandibular lymph nodes are often affected, and oral lesions including white patches or ulcers may be found. Pneumonia, icterus, hepatomegaly, splenomegaly, weight loss and vomiting have also been reported. Sick cats often have severe clinical signs, and the mortality rate is high if the disease is not treated early. However, milder cases are reported occasionally, and some cats with no history of disease are seropositive (Radostits *et al.* 2007, CFSPH 2009). Dogs, in contrast, seem to be relatively resistant, and may have milder clinical signs and may recover spontaneously. Clinical signs include anorexia, depression, mild fever, lymphadenopathy, draining abscesses, and mucoid ocular discharge or conjunctivitis (CFSPH 2009).

Among domesticated animals, susceptibility varies and it is most commonly recorded in sheep and pigs. Sheep and pigs of all ages are susceptible but most losses occur in lambs, and in pigs clinical illness occurs only in piglets. There is a sharp seasonal incidence, the bulk of cases occurring during the spring months. The morbidity rate in affected flocks of sheep is usually about 20 per cent but may be as high as 40 per cent, and the mortality rate may reach 50 per cent, especially in young animals (Radostits *et al.* 2007, CFSPH 2009). In sheep, the onset of the disease is slow with a gradually increasing stiffness of gait, dorsiflexion of the head and a hunching of the hindquarters. The affected animals lag behind the group. The pulse and respiratory rates are increased, the temperature is elevated up to 42°C (107°F), and a cough may develop. Outbreaks in sheep are usually characterized by late term abortions in ewes, and illness and deaths among lambs. Fever, listlessness and regional lymphadenopathy may be seen. There is diarrhoea, the faeces being dark and foetid, and urination occurs frequently with the passage of small amounts of urine. Body weight is lost rapidly, and progressive weakness and recumbency develop after several days, but there is no evidence of paralysis, the animal continuing to struggle while down. Death occurs usually within a few days but a fatal course may be as long as 2 weeks. Animals that recover commonly shed part or all of the fleece but are solidly immune for long periods (Radostits *et al.* 2007, CFSPH 2009).

The disease is latent in adult pigs but young piglets show fever up to 42°C (107°F), accompanied by depression, profuse sweating and dyspnoea. The course of the disease is about 7-10 days (Radostits *et al.* 2007). In horses, fever (up to 42°C, 107°F) and stiffness and oedema of the limbs occur. Foals are more seriously affected and may also show dyspnoea and incoordination (Radostits *et al.* 2007). Calves appear more resistant but can be infected in association with heavy tick infestation. Although serological evidence suggests that infections are fairly common in cattle, a specific syndrome has not been described, and many cases may be asymptomatic. Tularaemia has been diagnosed rarely in sick calves (CFSPH 2009).

In nonhuman primates, nonspecific signs including lethargy, anorexia, vomiting, diarrhoea, generalized lymphadenopathy, pale mucous membranes and cutaneous petechiae have been reported. Some animals have died acutely (CFSPH 2009).

Disease Manifestations in Humans

The incubation period in humans can be 3 to 15 days; most often, clinical signs appear after 3 to 5 days. The symptoms of tularaemia are varied and depending upon the route of entry of the organism, six forms of the disease are seen in humans: ulceroglandular, glandular, oculoglandular, oropharyngeal, respiratory and typhoidal forms (WHO 2007, CFSPH 2009).

Ulceroglandular Form

Ulceroglandular tularaemia occurs when the infection is vector-borne or through skin or mucous membranes due to direct contact with the infected animals or the material contaminated with *F. tularensis*. This is the most common manifestation of the disease. The initial clinical signs are nonspecific and may include fever, chills, headache, body aches and malaise. This form of tularaemia can be recognized by the presence of a skin lesion and swollen glands. At the site of the entry of the bacteria, an inflamed papule usually develops which turns into a pustule that ulcerates (Figure 18.1). In some cases, the skin lesion may heal by the time the patient seeks medical care; in others, it persists. The regional lymph nodes become enlarged and painful, and may suppurate and drain profusely. Unusual cases with a vesicular skin rash

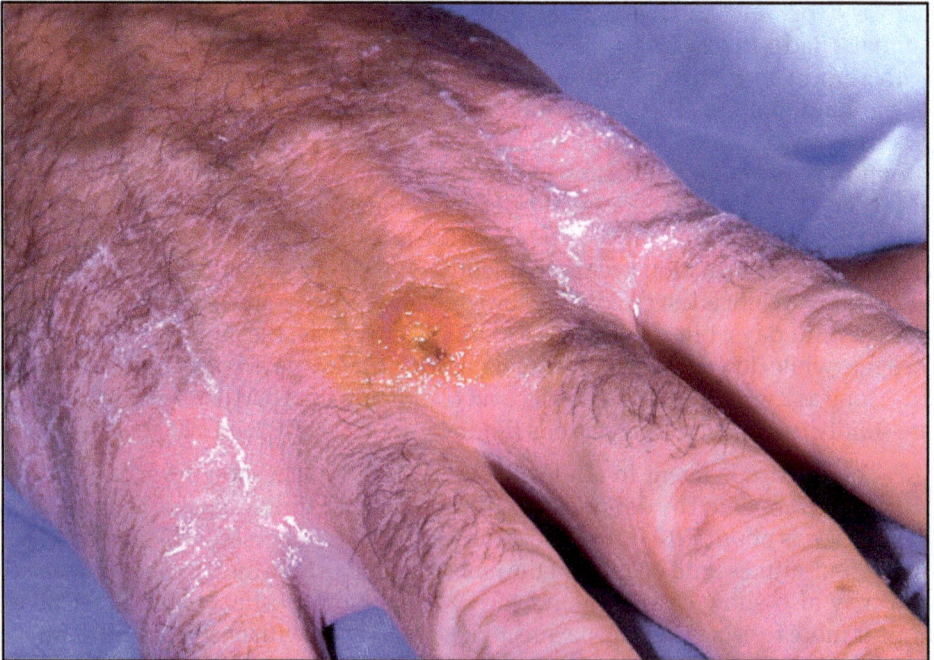

Figure 18.1: Tularaemia lesion
(Photo courtesy: CDC, Dr. Brachman)

have also been reported. The vesicles contain clear fluid that becomes turbid with time. Vesicles may also be found around an ulcerated eschar (WHO 2007, CFSPH 2009).

Glandular Form

The mode of occurrence and manifestation of glandular tularaemia is similar to the ulceroglandular form, but there is no lesion to indicate where the organism might have been inoculated.

Oculoglandular Form

It occurs through conjunctival infection occurring due to touching the eye with contaminated fingers or possibly from infective dust. The disease is characterized by unilateral, painful, purulent conjunctivitis with preauricular and/or cervical lymphadenopathy. In some cases, there may be chemosis, periorbital oedema and multiple small nodules or ulcerations on the conjunctiva. The possible complications include corneal perforation and iris prolapse (WHO 2007, CFSPH 2009).

Oropharyngeal Form

Ingesting contaminated food or water may lead to oropharyngeal tularaemia. In addition to the nonspecific symptoms such as fever and malaise, the organism may produce a throat infection, intestinal pain, diarrhoea and vomiting. Patients may develop exudative stomatitis and pharyngitis with pustules and ulcers. In some cases, the tonsils may also be inflamed. The lymph nodes in the neck are usually enlarged and tender, and abscesses may develop in these nodes (WHO 2007, CFSPH 2009).

Respiratory Form

Respiratory tularaemia occurs by inhaling contaminated dust or as a result of laboratory-acquired infection. When bacteria are inhaled, the infection may result in deep lymph-node enlargement. Inhalation of the organism may produce a fever alone or fever combined with a pneumonia-like illness. The severity of the disease varies with the causative organism. *F. tularensis* subsp. *tularensis* is more likely to causes serious disease. Severe forms of respiratory tularaemia may cause high fatality unless treated promptly. Inhalation of *F. tularensis* subsp. *holarctica* does not usually cause fulminating disease or pneumonia, and cases are not generally fatal. Pulmonary signs can also occur with any other form of tularaemia, when the lungs are affected via hematogenous spread (WHO 2007, CFSPH 2009).

Typhoidal Form

Typhoidal tularaemia is the term used for severe cases without an obvious route of exposure. Most cases are probably the result of inhalation, but this form can also develop after skin inoculation or ingestion. The symptoms may include nonspecific signs such as fever, prostration, headache, nausea and weight loss. Pneumonia is particularly common in the typhoidal form and can be severe (WHO 2007, CFSPH 2009).

Tularaemia may lead to complications such as meningitis, endocarditis, osteomyelitis, kidney failure, hepatitis, disseminated intravascular coagulation and acute respiratory distress. Person-to-person transmission has not been reported. However, infectious organisms can be found in the blood, lesions and tissues (CFSPH 2009).

Diagnostic Procedures

In Animals

Diagnosis of tularaemia infection in animals largely depends upon the detection of the organism or antigen in tissues and other biological materials. A number of conventional and modern techniques are available for this purpose. Serological tests are also available for diagnosis but they have limited application in animals. These tests have been described by the OIE (2012).

1. Identification of the Agent

F. tularensis can be demonstrated in smear preparations or in histological sections. It can also be identified by culture or animal inoculation. However, *F. tularensis* may be difficult to isolate from dead animals and carcasses due to overgrowth of other bacteria.

a) Smear Preparations

Impression smears of the liver, spleen, bone marrow, kidney, lungs or blood may be helpful for a presumptive diagnosis. Smear preparations are made on microscope slides. The bacteria are abundant in such smears, but can be missed because of their very small size (0.2-0.7 μm). The organism stains faintly with conventional stains such as Gram stain, and it can look like stain precipitates. Organisms are more likely to be found in impression smears with immunofluorescence. Demonstration of bacteria by direct or indirect fluorescent antibody staining is a safe, rapid and specific diagnostic method (CFSPH 2009, OIE 2012).

b) Histological Sections

Immunohistochemistry can detect antigens in tissue sections. Immunohistochemical methods, such as the fluorescent antibody test (FAT), may be performed on specimens from liver, spleen or bone marrow, fixed in neutral buffered formalin and paraffin embedded (OIE 2012).

c) Culture

F. tularensis can be isolated from enlarged lymph nodes, blood and tissues, including liver, spleen and bone marrow, but the overgrowth of other bacteria may prevent recovery from animals that are found dead. *F. tularensis* will not grow on ordinary media, although an occasional strain can, sometimes, on initial isolation, grow on blood agar. Incubation is at 37°C in ambient air or in 5 per cent CO_2. Heart blood, liver, spleen and bone marrow from moribund animals should be used for culture. It is necessary to use special culture media, such as Francis medium, McCoy and Chapin medium, modified Thayer-Martin agar, and glucose cysteine agar with thiamine. In addition to the non-selective media, the selective medium cystine heart agar with 5 per cent rabbit blood, and penicillin (100,000 units), polymyxin B sulphate (100,000 units), and cycloheximide (0.1 ml of a 1 per cent stock solution) per litre can be used (CFSPH 2009, OIE 2012).

F. tularensis colonies can be identified using slide agglutination, immunofluorescence or PCR. The organism is a Gram negative coccobacillus with bipolar staining in young cultures, but bacteria from older cultures may be pleomorphic (CFSPH 2009).

d) Capillary Tube Precipitation Test on Pathological Samples

The test can be performed with the antigen obtained from the tissues, such as spleen, liver or bone marrow. The antigen is allowed to react with tularaemia antiserum in capillary tubes at 37°C for 3 hours, and then kept at 4°C overnight. Formation of a ring of precipitate indicates a positive result (OIE 2012).

e) Animal Inoculation

Due to the hazards associated with the animal inoculation test, it is only recommended in cases when culture is negative and the agent identification is needed for epidemiological reasons (OIE 2012). It should only be undertaken where proper biosafety facilities and cages are available.

f) Molecular Techniques

PCR can also identify *F. tularensis* in clinical samples. Some PCR tests can distinguish the subtype. A real-time PCR has proven to be a highly sensitive and specific assay, and is a major diagnostic improvement (Tomaso *et al.* 2007). A DNA microarray has also been developed that is capable of distinguishing the highly virulent subspecies *F. tularensis* subsp. *tularensis* and the moderately virulent subspecies *F. tularensis* subsp. *holarctica* (Broekhuijsen *et al.* 2003). Several other molecular techniques can be used to identify strains for epidemiological studies (CFSPH 2009).

2. Serological Tests

Serological tests are useful diagnostic aids in human infection, but are of limited value in the more susceptible animal species that usually die before developing antibodies. Epidemiological surveys can be conducted in domestic animals in relatively resistant species and significant titers may be found in animals such as sheep, cattle, pigs and dogs. The most commonly used serological test is the tube agglutination test. Others tests include microagglutination and ELISA. Cross-reactions may occur with other bacteria including *Yersinia*, *Brucella* and *Legionella* (CFSPH 2009, OIE 2012).

In Humans

In humans, tularaemia is often diagnosed by serological tests, including tube agglutination, microagglutination and ELISA. Cross-reactions can occur with *Brucella* sp., *Legionella* sp., *Proteus* OX19, and *Yersinia* sp., usually at low titers. Tularaemia can also be diagnosed by isolating *F. tularensis* from blood or affected tissues/exudates such as sputum, pharyngeal or conjunctival exudates, ulcers, lymph nodes and gastric washings. PCR can also identify *F. tularensis* in clinical samples. Some PCR tests can distinguish the subtype. Identification of strains for epidemiological studies can also be done by other molecular techniques such as restriction fragment linked polymorphism (RFLP), Southern blot, pulsed-field gel electrophoresis (PFGE) and multi-locus variable number tandem repeat assays (MLVA). Immunohistochemical staining and histopathology are also useful (CFPSH 2009).

Treatment

F. tularensis is susceptible to a number of antibiotics such as tetracyclines and quinolones. Tularaemia can thus be treated with various antibiotics in humans and animals. Early treatment is most effective and avoids complications such as suppuration of the lymph nodes. Supportive therapy may also be required.

Prevention and Control Strategies

In animals, the risk of infection can be reduced by controlling ticks and other vectors. The disease in sheep can be rapidly controlled by spraying or dipping with an insecticide to kill the vector ticks. In areas where ticks are enzootic, sheep should be kept away from shrubby, infested pasture or sprayed regularly during the months when the tick population is greatest (Radostits *et al.* 2007). Cats and dogs should be controlled from hunting rodents and rabbits in areas where tularaemia is endemic. Early treatment of tularaemia with various antibiotics including tetracyclines and quinolones along with supportive therapy is effective. However, vaccines are not available for any species (CFSPH 2009).

In humans, safety against the bites from a variety of insects including ticks, tabanid flies and mosquitoes is very essential. Protective clothing and insect repellants may be helpful to ward off the vectors. Hunters and others who handle wildlife and their carcasses should use gloves and should properly cover the broken skin. Particular care should be taken to avoid bites and scratches. Washing of hands and cleaning of equipments is also important. Veterinarians and animal handlers should also take similar precautions. Game meat should be adequately cooked. In endemic areas, use of dust masks may be helpful in preventing inhalation of bacteria during dust-raising activities such as piling hay or mowing the lawn. Water should be filtered or treated before drinking. Mortality in rodents and lagomorphs should be reported and should raise suspicion of disease outbreak (CFSPH 2009). The disease should be controlled by expeditious treatment to avoid complications (CFSPH 2009).

Future Directions

Apart from being a causative agent of natural infections, *F. tularensis* is a potential biological weapon. Through inhalation, even a very low dose of the organism can result in high mortality which raises serious concerns about its misuse in terrorist activities. Vaccination is a proven strategy to mitigate a number of infectious diseases. An effective vaccine against tularaemia is required to reduce the number of naturally occurring tularaemia cases as well as to protect against a possible intentional release. In recent years, several vaccine candidates have been developed and tested in laboratory animals but many hurdles are yet to be resolved to come out with a safe and effective vaccine.

References

1.   Aiello SE. 2012. *The Merck Veterinary Manual*. Merck Sharp and Dohme Corp., Whitehouse Station, NJ, USA.

2.   Al Dahouk S, Nöckler K, Tomaso H, Splettstoesser WD, Jungersen G, Riber U, Petry T, Hoffmann D, Scholz HC, Hensel A and Neubauer H. 2005. Seroprevalence

of brucellosis, tularemia, and yersiniosis in wild boars (*Sus scrofa*) from north-eastern Germany. *J Vet Med B Infect Dis Vet Public Health* 52(10): 444-455.

3.   AWHN. 2012. Tularaemia and Australian wildlife Fact Sheet. Australian Wildlife Health Network. http://www.wildlifehealth.org.au/AWHN_Admin/ManageWebsite per cent 5CFactSheets per cent 5CUploadedFiles/121/Tularemia per cent 2014 per cent 20Jun per cent 202012 per cent 20(1.2).pdf. Accessed 13 November 2012.

4.   Broekhuijsen M, Larsson P, Johansson A, Byström M, Eriksson U, Larsson E, Prior RG, Sjöstedt A, Titball RW and Forsman M. 2003. Genome-wide DNA microarray analysis of *Francisella tularensis* strains demonstrates expensive genetic conservation within the species but identifies regions that are unique to the highly virulent *F. tularensis* subsp. *tularensis. J Clin Microbiol* 41: 2924-2931.

5.   CDC. 2012. Reported tularemia cases by state, United States, 2001-2010. Centres for Disease Control and Prevention. http://www.cdc.gov/tularemia/statistics/state.html. Accessed 13 November 2012.

6.   CFSPH. 2009. Tularemia. The Center for Food Security and Public Health, Iowa State University, Ames, IA 50011. http://www.cfsph.iastate.edu/Factsheets/pdfs/tularemia.pdf. Accessed 13 November 2012.

7.   Clark DV, Ismailov A, Seyidova E, Hajiyeva A, Bakhishova S, Hajiyev H, Nuriyev T, Piraliyev S, Bagirov S, Aslanova A, Debes AK, Qasimov M and Hepburn MJ. 2012. Seroprevalence of tularemia in rural Azerbaijan. *Vector Borne Zoonotic Dis* 12(7): 558-563. doi:10.1089/vbz.2010.0081.

8.   Deutz A, Guggenberger T, Gasteiner J, Steineck T, Bagó Z, Hofer E, Auer I and Böhm R. 2009. Investigation of the prevalence of tularaemia under the aspect of climate change. *Wiener Tierärztliche Monatsschrift* 96 (5/6): 107-113.

9.   ECDC. 2012. Epidemiological report on communicable diseases in Europe 2010. European Centre for Disease Prevention and Control, Stockholm.

10. European Commission. 2012. Tularaemia. http://ec.europa.eu/health/ph_information/dissemination/echi/docs/tularemia_en.pdf. Accessed 13 November 2012.

11. Francis E. 1921. Tularemia Francis. *Hygienic Laboratory Bulletin* No. 130, 1922.

12. Grunow R, Kalaveshi A, Kühn A, Mulliqi-Osmani G and Ramadani N. 2012. Surveillance of tularaemia in Kosovo, 2001 to 2010. *Euro Surveill* 17(28): pii=20217. http://www.eurosurveillance.org/ViewArticle.aspx?ArticleId= 20217. Accessed 13 November, 2012.

13. Gyuranecz M. 2011. Epizootic investigations of tularemia and the comparative characterization of *Francisella tularensis* strains. *Hungarian Veterinary Archive*, Faculty of Veterinary Science, Szent István University, PhD Dissertation. http://huveta.hu/handle/123456789/133.

14. Hotta A, Tanabayashi K, Yamamoto Y, Fujita O, Uda A, Mizoguchi T and Yamada A. 2012. Seroprevalence of tularemia in wild bears and hares in Japan. *Zoonoses Public Health* 59 (2): 89-95.

15. Kuehn A, Schulze C, Kutzer P, Probst C, Hlinak A, Ochs A and Grunow R. 2012. Tularaemia seroprevalence of captured and wild animals in Germany: The fox (*Vulpes vulpes*) as a biological indicator. *Epidemiol Infect* 17: 1-8.

16. Ladds P. 2009. Bacterial diseases in birds. In: *Pathology of Australian native wildlife*. CSIRO Publishing, Collingwood. p. 115-136.

17. McCoy GW and Chapin CW. 1912. Bacterium tularense, the cause of a plague-like disease of rodents. *Public Health Bull* 53: 17-23.

18. OIE. 2012. Chapter 2.1.18. Tularemia. *Manual of diagnostic tests and vaccines for terrestrial animals* 2012. World Organisation for Animal Health. http://www.oie.int/en/international-standard-setting/terrestrial-manual/access-online/. Accessed 14 November 2012.

19. PHAC. 2012. Tularemia – Frequently asked questions. Public Health Agency of Canada. http://www.phac-aspc.gc.ca/tularemia/tul-qa-eng.php. Accessed 13 November 2012.

20. Radostits OM, Gay CC, Blood DC and Hinchcliff KW. 2007. *Veterinary medicine: A textbook of the diseases of cattle, horses, sheep, pigs and goats*. 10th edn. WB Saunders, Philadelphia, USA.

21. Rydén P, Björk R, Schäfer ML, Lundström JO, Petersén B, Lindblom A, Forsman M, Sjöstedt A and Johansson A. 2012. Outbreaks of tularemia in a boreal forest region depends on mosquito prevalence. *J Infect Dis* 205(2): 297-304.

22. Tärnvik A, Priebe HS and Grunow R. 2004. Tularaemia in Europe: an epidemiological overview. *Scand J Infect Dis* 36: 350-355.

23. Tomaso H, Scholz HC, Neubauer H, Dahouk AL, Seibold E, Landt O, Forsman M and Splettstoesser WD. 2007. Real-time PCR using hybridization probes for the rapid specific identification of *Francisella tularensis* subspecies *tularensis*. *Mol Cell Probes* 21: 12-16.

24. Whipp MJ, Davis JM, Lum G, de Boer J, Zhou Y, Bearden SW, Petersen JM, Chu MC and Hogg G. 2003. Characterization of a novicida-like subspecies of *Francisella tularensis* isolated in Australia. *J Med Microbiol* 52: 839-842.

25. WHO. 2007. *WHO guidelines on tularaemia*. WHO/CDS/EPR/2007.7. World Health Organization, Geneva.

26. Yazgý H, Uyanýk MH, Ertek M, Kýlýç S, Kireçci E, Ozden K and Ayyilidz A, 2011. Tularemia seroprevalence in the risky population living in both rural and urban areas of Erzurum. *Mikrobiyol Bul* 45: 67-74.

2014, Zoonoses: Bacterial Diseases
Editor: Sudhi Ranjan Garg
Published by: DAYA PUBLISHING HOUSE, NEW DELHI

Pages 367–374

19

Streptococcal Zoonoses

S.R. Garg

Department of Veterinary Public Health and Epidemiology,
College of Veterinary Sciences, Lala Lajpat Rai University of
Veterinary and Animal Sciences, Hisar – 125 004

The genus *Streptococcus* includes a number of species of bacteria. These are widely distributed in nature. Some species occur as commensals and as normal microflora in a variety of animal species and human beings. Many species are pathogenic or may act as opportunistic pathogens while some have zoonotic implications as well. In animals, different species of streptococci cause a variety of diseases, for example, strangles in horses, pneumonia in pigs, mastitis in cattle, and septicemia in animals and poultry. Some species are pathogenic to fish also. In humans, *Streptococcus* species are known to cause a variety of diseases such as streptococcal pharyngitis (strep throat), scarlet fever, pneumonia, cellulites, endocarditis, polyarthritis septicemia, etc.

Geographical Distribution

The zoonotic streptococci can usually be found worldwide in their animal hosts but human infections have, in some cases, been reported only in limited geographic regions (CFSPH 2005). The prevalence of group C streptococci among equines has been demonstrated in organized equine farms as well as in unorganized sector in India (Malik and Kalra 2009). Environmental streptococci have been reported as a leading cause of both subclinical and clinical mastitis in animals throughout the world, including India (Sharma *et al.* 2006).

Etiological Agent

Streptococcus bacteria are Gram-positive cocci in the family Streptococcaceae. These grow in chains and are facultative anaerobes and catalase negative. Species of *Streptococcus* are generally classified into Lancefield groups. There are twenty Lancefield serogroups which include groups A through H and K through V. Some streptococcal species do not have Lancefield group antigens and some show newly identified antigens. Streptococci are also classified based on their ability to produce haemolysis or type of haemolysis on blood agar. While some species are non-haemolytic, others may either produce alpha haemolysis or beta haemolysis. Many streptococci are pathogenic for humans and animals. Streptococci are also subdivided into three large groups: pyogenic (where most pathogens reside), oral, and other. Some species are proven or suspected to be zoonotic (CFSPH 2005, Romich 2007).

S. equi subsp. *zooepidemicus* (beta-haemolytic; Lancefield group C) is an opportunistic pathogen that causes a variety of infections in many species and has been proven as zoonotic organism. *S. suis* (non-beta-haemolytic; Lancefield groups R, S and T), usually associated with pigs as commensal or pathogen, are also known to cause zoonotic infections in humans. *S. iniae* (beta-hemolytic; no Lancefield group antigens) associated with fish is also zoonotic in nature. *S. canis* (beta-haemolytic; Lancefield group G), an opportunistic pathogen found in dogs and other species, is also a proven zoonotic organism (CFSPH 2005).

In contrast to the above species, *S. pyogenes* (beta-haemolytic; Lancefield group A) is adapted to humans, but does not have natural animal reservoirs. It is a human pathogen that causes pharyngitis (strep throat), skin disease and many other infections in humans, but can cause reverse zoonoses in animals *i.e.*, transmission of infection may occur from infected persons to animals. However, subsequently, the colonized animals can re-transmit the infection to humans. *S. pyogenes* from humans can infect the bovine udder and may cause outbreaks of human illness through contaminated milk originating from such infected udders (CFSPH 2005).

S. agalactiae (beta-haemolytic; Lancefield group B) occurs both in humans and animals, but its zoonotic transmission is considered rare or insignificant. Some *Streptococcus* species are found in both animals and humans but their zoonotic significance is uncertain. These include the *S. bovis* group, *S. pneumoniae*, *S. dysgalactiae* subsp. *equisimilis*, *S porcinus* and the viridans group of streptococci (CFSPH 2005).

Host Range

Streptococcal species have been isolated from a variety of animals. *S. equi* subsp. *zooepidemicus* is a common commensal and opportunistic pathogen in horses. It has been isolated from a large spectrum of other animal hosts like cattle, sheep, goats, pigs, dogs, foxes, ferrets, guinea pigs, non-human primates and birds (CFSPH 2005). Similarly, *S. suis* is found mainly in pigs but has also been isolated from other species, including cattle, sheep, goats and bison. *S. canis* has been found in dogs and a variety of other animals, including cats, cattle, rats, mink, mice, rabbits and foxes. *S. iniae* has been found in freshwater dolphins and wild and farmed fish. *S. pyogenes* is a common human pathogen, but may be isolated from animals such as dogs, cattle, ducks and monkeys (CFSPH 2005, Romich 2007).

Transmission of Infection

Streptococci often constitute normal flora of animals and humans. The zoonotic streptococci can be directly transmitted to susceptible animals or humans in different ways. *S. equi* subspecies *zooepidemicus* can be transmitted through contact with purulent discharges, wound infections, fomites, aerosols and consumption of contaminated food products, particularly raw milk and dairy products. *S. suis* can be transmitted between pigs by direct contact, aerosols, fomites and possibly ingestion. To humans, it is transmitted by direct contact of broken skin or conjunctiva with infected materials from pigs or pork, or through fomites, aerosol, and ingestion of contaminated pork. Transmission of *S. canis* from contaminated skin and mucosa of dogs and other animals seems to be via close contact *i.e.,* open wound infection or through dog bites. The transmission of *S. iniae* also appears to be mainly through exposure of injured skin to infection while handling live or freshly killed fish or contaminated instruments. *S. pyogenes* is an airborne infection that spreads among humans by direct contact with infectious nasal and pharyngeal secretions, or with infected wounds. *S. pyogenes* infection from humans may be transmitted to cattle leading to mastitis in dairy cattle (reverse zoonosis). Consequently the microorganism can be found in dairy products which may become a source of infection for human beings (CFSPH 2005, Romich 2007).

Predisposing Factors

The factors such as overcrowding, inadequate ventilation, poor husbandry practices and others that cause stress predispose the animals to infections (CFSPH 2005).

Prevalence of Infection in Animals

Extensive information on the prevalence of streptococcal infection in animals is not available, however, streptococci are widely prevalent among the healthy as well as sick animals. The prevalence of group C streptococci among equines in India has been examined in various field samples from apparently healthy and clinically affected equines (Malik and Kalra 2009). The samples from apparently normal equines (n=185) yielded 16 isolates, while those from clinically affected equines (n=126) yielded 19 isolates. Over all from the total 311 samples, 35 streptococcal isolates were obtained, including *Streptococcus equi* (6), *S. zooepidemicus* (16) and *S. equisimilis* (13). Of 262 samples from organized farms, 25 yielded streptococci, while 10 isolates were recovered from 49 samples from the unorganized sector. All the streptococcal cultures, except 6 *S. equisimilis* isolates, were pathogenic to mice.

In Czech Republic, Lysková *et al*. (2007) isolated *S. canis* from specimens collected from healthy household pets and from pets with various opportunistic infections. Samples were taken from 124 dogs and 21 cats with proceeding infection. In addition, 199 dogs and 50 cats without clinical signs of infection were included in the study. Of the total 394 animals included in the study, 58 (18 per cent) dogs and 9 (12.7 per cent) cats were positive for *S. canis*. Of 145 samples taken from local infections, 31 (21.4 per cent) were positive for *S. canis* which was most frequently isolated from animals with otitis externa. On the other hand, the microorganism was not detected in the specimens taken from animals with rhinitis.

S. suis infection in pigs is reported worldwide, including North America (United States and Canada), South America (Brazil), Europe (United Kingdom, Netherlands, France, Denmark, Norway, Spain and Germany), Asia (China, Thailand, Vietnam and Japan), Australia, and New Zealand (Staats *et al.* 1997). *S. suis* can also be isolated from other animals, including ruminants, cats, dogs, deer, and horses, and is believed to be a commensal in the intestinal flora (Staats *et al.* 1997). In Vietnam, Hoa *et al.* (2011) determined the prevalence and diversity of *S. suis* carriage in the nasopharyngeal tonsils of healthy slaughterhouse pigs. They found that 41 per cent (222/542) of pigs carried *S. suis* of one or multiple serotypes. Eight per cent (45/542) animals carried *S. suis* serotype 2, which was the most common serotype detected. The studies demonstrated slaughterhouse pigs as a major reservoir of *S. suis* serotype 2 capable of causing human infection.

In a study of mastitis in India, out of 423 quarters of 109 apparently healthy lactating crossbred cows, 58 (13.71 per cent) showed subclinical mastitis and yielded 60 staphylococcal and streptococcal isolates. Out of these, 6 were *Streptococcus agalactiae*, 3 *Streptococcus dysgalactiae* and one *Streptococcus uberis* (Sharma *et al.* 2006).

Prevalence of Infections in Humans

S. pyogenes is a very common human pathogen, accounting for more than 10 million non-invasive infections, but the human infections with the other major zoonotic streptococci are generally rare. Most infections with *S. equi* subsp. *zooepidemicus* are sporadic but small to large outbreaks have also been reported (CFSPH 2005). *S. suis* also seems to be a rare zoonosis but over the past few years, the number of reported *S. suis* infections in humans has increased significantly, with most cases originating in Southeast Asia, where there is a high density of pigs (Wertheim *et al.* 2009). A recent study by Takeuchi *et al.* (2012) suggested a high incidence rate of *S. suis* infection in the general population in Thailand in 2010 and confirmed a cluster of three cases in 31 human cases.

Documented *S. canis* infections in humans are very rare, however, there is possibility of underestimating such infections because the bacterial species are not adequately identified or confirmed in many clinical infections (CFSPH 2005). *S. iniae* infections are also not very common. Most cases have been seen in very elderly people (CFSPH 2005).

Disease Manifestations in Animals

The streptococci constitute a vast group of organisms. Many species are commensals and may act as opportunistic pathogens in different species of animals. The incubation period varies with the form of the disease. In streptococcal toxic shock syndrome, healthy animals can become seriously ill rapidly and may die within few hours. Disease manifestations of the streptococcal species that are of major zoonotic significance are briefly described below.

S. equi subspecies zooepidemicus

This is an opportunistic pathogen found in a large number of animal species. The organism causes pneumonia, wound infections, endometritis, arthritis and mastitis in animals. Horses are the most commonly affected animals in which the organism is often isolated from secondary bacterial infections. It can invade the upper respiratory mucosa and lymph nodes in horses after a viral infection, sometimes mimicking strangles. The ascending infection from vagina can cause cervicitis, metritis, placentitis and abortions in susceptible mares. It causes mastitis in horses and has been associated with wound infections, septicemia in colts, and lower airway inflammation in young horses (CFSPH 2005, Romich 2007).

A serious epidemic due to the microorganism has also been recorded among pigs and monkeys in Bali, Indonesia. The syndromes reported in the affected animals included polyarthritis, bronchopneumonia, pleuritis, diarrhoea, epicarditis, endocarditis and meningitis. Most of the affected animals died within a few days. Sporadic cases of mastitis due to *S. equi* subsp. *zooepidemicus* occur in cattle and goats. Outbreak of mastitis in sheep has also been reported (CFSPH 2005). The microorganism has also been isolated from a variety of infections in other species, including septicemia, pneumonia and streptococcal toxic shock syndrome in dogs; abscesses, pneumonia, metritis and endocarditis in ferrets; and septicemia in poultry (CFSPH 2005).

S. suis

S. suis is a commensal microorganism in swine but more virulent strains can cause serious disease. It causes meningitis, acute meningitis with high mortality, septicemia and arthritis, especially in young, growing pigs. The disease in swine is manifested by meningitis, septicemia, paralysis, convulsion, cutaneous erythema and fever. *S. suis* sometimes causes pneumonia, endocarditis, myocarditis or abortions. The microorganism has also been isolated sporadically from other species (CFSPH 2005).

S. canis

S. canis is an opportunistic pathogen in dogs and cats. It causes otitis media, septicemia in neonates, lymphadenopathy in cervical lymph nodes, polyarthritis, and reproductive tract infections with abortion and mastitis in dogs. *S. canis* also causes various infections in cats. These include arthritis, wound infections, septicemia, pneumonia and streptococcal toxic shock syndrome. It also causes cervical lymphadenitis and sometimes neonatal septicemia in kittens (CFSPH 2005, Romich 2007).

S. iniae

These bacteria infect some fish asymptomatically and may cause abscesses in fish and dolphins. Sporadic disease is observed in some fish species and occasionally there may be serious outbreaks on fish farms. Some of the disease conditions include meningoencephalitis and panophthalmitis in trout and tilapia, and skin lesions and necrotizing myositis in red drum (CFSPH 2005, Romich 2007).

S. pyogenes
Cattle may acquire the infection from human and may suffer from mastitis.

Disease Manifestations in Humans

Different species of streptococci cause a variety of diseases in humans. The incubation period varies with the form of the disease (CFSPH 2005). Most human infections are associated with group A streptococci, which are usually *S. pyogenes*. A small proportion of infections is caused by other species.

S. pyogenes
S. pyogenes is a common human pathogen that accounts for more than 10 million infections, with approximately 9,000 cases occurring annually in the United States. Streptococcal pharyngitis (strep throat) is very common in human beings, which causes pain on swallowing, tonsillitis, high fever, headache, nausea, vomiting, malaise and rhinorrhea. The disease when accompanied by a rash is known as scarlet fever. *S. pyogenes* infections acquired from food can become symptomatic after 1 to 3 days. Acute bacterial meningitis usually appears within a few hours to a few days (CFSPH 2005, Romich 2007).

S. pyogenes also causes relatively mild skin infections such as pyoderma and impetigo, as well as otitis media, sinusitis, abscesses, cellulitis, osteomyelitis, arthritis, endocarditis and, rarely, serious infections such as pneumonia, meningitis, septicemia, necrotizing fasciitis (severe invasive disease) or streptococcal toxic shock syndrome (a severe and often fatal disease). The mortality rates are 10 per cent to 15 per cent for invasive disease, 20 per cent to 25 per cent for necrotizing fasciitis, and about 45 per cent for streptococcal toxic shock syndrome (CFSPH 2005, Romich 2007).

S. equi subsp. zooepidemicus
S. equi subsp. *zooepidemicus* has been isolated from humans with mild respiratory diseases, pneumonia, endocarditis, endophthalmitis, septic arthritis, meningitis, septicemia and streptococcal toxic shock syndrome. Mortality rates for group C bacteremia (to which this organism belongs) is about 20 per cent to 30 per cent and for group C meningitis is about 57 per cent (CFSPH 2005, Romich 2007).

S. suis
S. suis has mainly been associated with meningitis. The microorganism has also been isolated from cases of endocarditis, septicemia without meningitis, and septic shock. The mortality rate for *S. suis* meningitis is about 7 per cent with common sequelae of deafness and vertigo (CFSPH 2005, Romich 2007).

S. canis
S. canis has been associated with septicemia, meningitis and peritonitis.

S. iniae
S. iniae has usually been found in cases of cellulitis. Rare cases of osteomyelitis, septic arthritis, endocarditis, meningitis and discitis (infection of the vertebral discs) have also been reported. The incubation period for *S. iniae* infections ranges from less than 24 hours to about 2-3 days (CFSPH 2005).

Diagnostic Procedures

Microscopic examination of specimens for the presence of Gram positive cocci in pairs or chains can lead to presumptive diagnosis. Depending on the infection, bacteria may be found in pharyngeal secretions, wounds, blood, cerebrospinal fluid (CSF), placenta, tissues from aborted foetuses, or other sites in humans and animals. Streptococcal toxic shock syndrome is usually identified by the clinical signs. The definitive diagnosis of streptococcal infections depends on culturing and identifying the organism by examining their haemolysis patterns on blood agar, colony morphology, biochemical, and serological reactions. Lancefield grouping of the organism can be done by the capillary precipitation test or other serologic tests. In humans, immunofluorescence and enzyme-linked immunosorbent assays (ELISAs) can be used to identify *S. pyogenes* in throat swabs. Genetic tests such as DNA-DNA reassociation procedures, pulsedfield gel electrophoresis (PFGE) and 16S rRNA gene sequencing can also be used to identify *Streptococcus* species. Phage typing may also be used in epidemiologic studies (CFSPH 2005).

Treatment

Streptococcal infections can be treated with various antibiotics, including penicillin, amoxicillin, ampicillin, cephalosporins, vancomycin, clindamycin, etc. In streptococcal toxic shock syndrome and necrotic fasciitis, supportive treatment is important. Intravenous immunoglobulin therapy may also be used in humans. Surgical debridement of necrotic tissues is often necessary (CFSPH 2005).

Prevention and Control Strategies

Minimizing stress and other predisposing factors, adequate wound care and good hygiene can minimize the infections in animals. Good hygiene during milking and clean environmental conditions at the farms is essential to reduce the exposure to streptococci and to decrease the risk of mastitis in dairy animals. This is particularly important in case of *S. pyogenes*. It should also be ensured that infected farm workers and milker (suffering from strep throat or other streptococcal diseases) should not carry out milking operations and should not handle milk.

Transmission of streptococcal infections in humans through wounds and abrasions can be prevented by using protective clothing and gloves while handling pig, fish or animal products, etc. Foodborne infections can be prevented by avoiding raw, unpasteurized milk and dairy products. Good hygiene and other preventive measures should be observed when caring for horses with respiratory diseases and other *S. equi* subsp. *zooepidemicus* infections (CFSPH 2005).

Future Directions

Since streptococci constitute normal microflora of humans and animals, it is important to identify the mechanisms these bacteria use to cause infection and disease. The identification and classification of streptococci also require to be explored further. Research should be carried out to develop effective vaccines against streptococcal infections in humans and animals.

References

1. CFSPH. 2005. *Streptococcosis*. The Center for Food Security and Public Health, Iowa State University, Ames, IA 50011. http://www.cfsph.iastate.edu/Factsheets/pdfs/streptococcosis.pdf. Accessed 21 November 2012.

2. Hoa NT, Chieu TTB, Nga TTT, Dung NV, Campbell J, Anh PH, Tho HH, Chau NVV, Bryant JE, Hien TT, Farrar J and Schultsz C. 2011. Slaughterhouse pigs are a major reservoir of *Streptococcus suis* Serotype 2 capable of causing human infection in southern Vietnam. *PLoS ONE* 6(3): e17943. doi:10.1371/journal.pone.0017943.

3. Lysková P, Vydržalová M, Královcová D and Mazurová J. 2007. Prevalence and characteristics of *Streptococcus canis* strains isolated from dogs and cats. *Acta Vet Brno* 76: 619-625.

4. Malik P and Kalra SK. 2009. Prevalence of group C streptococci amongst equines in India. *Indian J Anim Sci* 79: 459-465.

5. Romich JA. 2007. *Understanding zoonotic diseases*. Chapter 3 Bacterial zoonoses. Part 5: Streptococcal infections. Delmar Cengage Learning.

6. Sharma A, Dhingra P, Pander BL and Kumar R. 2006. Bovine sub clinical mastitis: prevalence and treatment with homeopathic medicine. *International J Cow Sci* 2 (1): 40-44.

7. Staats JJ, Feder I, Okwumabua O and Chengappa MM. 1997. *Streptococcus suis*: past and present. *Vet Res Commun* 21: 381-407.

8. Takeuchi D, Kerdsin A, Pienpringam A, Loetthong P, Samerchea S *et al.* 2012. Population-based study of *Streptococcus suis* infection in humans in Phayao province in northern Thailand. *PLoS ONE* 7(2): e31265. doi:10.1371/journal.pone.0031265.

9. Wertheim HFL, Nghia HDT, Taylor W and Schultsz C. 2009. *Streptococcus suis*: An emerging human pathogen. *Clin Infect Dis* 48: 617-625.

2014, Zoonoses: Bacterial Diseases
Editor: Sudhi Ranjan Garg
Published by: DAYA PUBLISHING HOUSE, NEW DELHI

Pages **375–380**

20

Rat-bite Fever

S.R. Garg
Department of Veterinary Public Health and Epidemiology,
College of Veterinary Sciences, Lala Lajpat Rai University of
Veterinary and Animal Sciences, Hisar – 125 004

Rat-bite fever (RBF) includes two similar yet distinct disease syndromes caused by two bacterial species, *Streptobacillus moniliformis* and *Spirillum minus*. Both these organisms are found in the mouths of rodents. Humans may acquire infection from infected rodents. The RBF caused by *Streptobacillus moniliformis* is known as streptobacillary fever, streptobacilliosis, or epidemic arthritic erythema. The infection may also occur due to consumption of contaminated food or water; in such cases the disease is often called Haverhill fever. The disease caused by *Spirillum minus* infection is known as spirillary fever or sodoku which is more common in Asia (CFSPH 2006, CDC 2012).

Geographical Distribution

Streptobacillus moniliformis and *Spirillum minus* can be found worldwide. Reports of *S. moniliformis* have originated from the United States, Brazil, Canada, Mexico, Paraguay, United Kingdom, France, Norway, Finland, Germany, Spain, Italy, Greece, Poland, Denmark and Netherlands. Some cases have also been reported in Australia and Africa. *Spirillum minus* is common only in Asia and most reports from Asia document cases of sodoku, caused by *Spirillum minus*, however, human cases attributed to *Spirillum minus* have also been reported in Africa (CFSPH 2006, Elliott 2007).

Etiological Agent

Streptobacillus moniliformis is a Gram negative, non-motile, pleomorphic, rod-shaped organism. It often has spherical, oval, fusiform, or club-shaped swellings. In some cases, clumps of this bacterium may look like proteinaceous debris. It is an extremely fastidious organism that requires microaerophilic conditions to grow, making the laboratory culture difficult. Depending on the medium, the organism occurs singly or in chains (CFSPH 2006, Elliott 2007).

Spirillum minus, which was initially named *Spirocheta morsus muris* or *Sporozoa muris*, is a short, spiral-shaped, Gram negative rod (0.2-0.5 μm x 3-5 μm) with two to three coils and bipolar tufts of flagella. It has not been successfully grown on artificial media (CFSPH 2006, Elliott 2007).

Host Range

Rats are the reservoir hosts for *Streptobacillus moniliformis* and *Spirillum minus*, both of which also occur in mice. *Streptobacillus moniliformis* has also been found in hamsters, gerbils, guinea pigs, squirrels, non-human primates and birds. The organism also occurs in animals that eat rodents such as cats, dogs, ferrets and weasels. Disease has been reported in mice, birds, guinea pigs and non-human primates (CFSPH 2006).

Disease Transmission

Rats are considered the natural reservoir of RBF in which *Streptobacillus moniliformis* and *Spirillum minus* occur as the normal nasopharyngeal flora. Other animals that prey on rodents may acquire the infection from infected rodents. Mainly the causative organism is transmitted through bite wounds and scratches, but aerosol infection may also occur. Human infection may occur due to handling of infected rats or other animals or through animal bites or exposure to excreta. Food and water contaminated with excreta of rats may also transmit *Streptobacillus moniliformis* infection to humans resulting in Haverhill fever. Person-to-person transmission has not been reported (CFSPH 2006, CDC 2012).

Environmental and Other Predisposing Factors

Any person who is exposed to the causative organisms of RBF is at risk of infection. The persons living in rat-infested buildings or keeping rats as pets or those who work with rats in laboratories or pet stores are at greater risk of acquiring the disease (CDC 2012).

Prevalence of Disease in Animals

Streptobacillus moniliformis and *Spirillum minus* are most commonly found in rats. The former species is present in an estimated 50 per cent to 100 per cent of wild rats and the latter in up to 25 per cent of the wild rats in some countries. *Streptobacillus moniliformis* may also be found in laboratory rats. Sporadic infections with both types of the RBF agents have also been reported in other rodents, and epizootics of *Streptobacillus moniliformis* septicemia have been seen in wild and laboratory mice (CFSPH 2006).

Prevalence of Disease in Humans

A review of the cases of RBF reported in the English language literature by Elliott (2007) reveals 65 discrete case reports that provide full descriptions of the clinical presentation. Many additional cases are described within case series in which signs and symptoms specific to each case are not detailed. The 65 detailed cases were reported from 1938 to 2005 and were primarily from the United States, but the cases were also seen in the United Kingdom, Europe, Canada, Australia and Nigeria. Twenty-six (40 per cent) of the exposures occurred from a wild rat, eight (12 per cent) from a laboratory rat and three (5 per cent) from a pet rat shop. Twenty-two (34 per cent) of the patients described a non-bite or non-rat exposure. The remaining cases occurred in association with bites from a ferret (1), mouse (1), squirrel (2), gerbil (1) and dog (1). According to the CDC (2012), RBF is rare in the United States but accurate data about the incidence of the disease are unavailable because the disease may not be reportable to state health departments. However, recent case reports have highlighted the potential risk for RBF among persons having contact with rodents at home or in the workplace. According to CFSPH (2006), as of 2004, 200 cases of RBF had been reported. The incidence of RBF seems to be increasing with the increasing popularity of rats as pets.

In the United Kingdom also, RBF is rarely diagnosed in England and Wales where around 1-2 cases are reported per year (HPA 2012). Most cases are sporadic, but two large outbreaks of Haverhill fever have been reported, one in Haverhill, MA in 1926 and another in Essex, UK in 1983. Up to 10 per cent of rat bites may result in RBF (CFSPH 2006). In a retrospective study of 123 patients reporting with rat bites wounds during a 4 year period (2005-2009) in an Antirabies Center in Romania, 2 cases of RBF (sodoku) were diagnosed (Mirela 2009).

In India, Balakrishnan *et al.* (2006) reported a case of *Streptobacillus moniliformis* endocarditis in a patient with congenital heart disease. The patient had a history of living in a rat-infested area, and admitted having been bitten by a rat several months before the onset of symptoms. However, it was considered unlikely that disease contracted by a rat bite would take months to be manifested. Thus, it was considered more likely that the patient contracted the infection from food or water contaminated with rat excreta. De *et al.* (2010) reported the isolation of *Streptobacillus moniliformis* from the blood of a male child with acute lymphoblastic leukaemia. There was no history of rat bite but rats were present in the house. Food or water contaminated with rat excreta was considered the possible source of infection.

Disease Manifestations in Animals

Rats usually carry *Streptobacillus moniliformis* and *Spirillum minus* asymptomatically, but occasionally may demonstrate signs and symptoms of the disease. Occasionally, *Streptobacillus moniliformis* occurs as a secondary invader in subcutaneous abscesses. It has also been reported from the middle ear of rats with otitis media, as well as that of asymptomatic rats (CFSPH 2006). Symptomatic infections with *Streptobacillus moniliformis* have been reported in some other species too. Infected mice may develop polyarthritis, subcutaneous or hepatic abscesses, purulent lesions, and acute or subacute septicemia. The microorganism has also been associated with granulomatous pneumonia or cervical lymphangitis in guinea pigs. Nonhuman primates are reported to suffer from septic arthritis and endocarditis. Arthritis has been seen in naturally infected turkeys also (CFSPH 2006).

Disease Manifestations in Humans

The incubation period and clinical signs of streptobacillary RBF and spirillary RBF are different. The incubation period of streptobacillary RBF is 3-10 days but the appearance of symptom can be delayed as long as 3 weeks. The initial symptoms of streptobacillary RBF are non-specific and include fever, chills, myalgia, headache and vomiting. Patients may develop a maculopapular rash on the extremities 2-4 days after fever onset, followed by polyarthritis in approximately 50 per cent of patients (CDC 2012). In case of Haverhill, fever acquired through contaminated food or water, the symptoms differ slightly from those of RBF acquired through bites and/or scratches. Haverhill fever can be accompanied with more severe nausea/vomiting and pharyngitis (CDC 2012).

Symptoms of RBF due to *Spirillum minus* usually occur 7-21 days after exposure to an infected animal. Following partial healing of the rat bite, common symptoms and signs include fever, ulceration at the site, lymphangitis, lymphadenopathy, and a distinct rash of purple or red plaques.

The infection, if not appropriately treated, may result in endocarditis, myocarditis, meningitis, pneumonia or sepsis. The mortality rate for untreated RBF is 7-13 per cent (CDC 2012).

Diagnostic Procedures

In Animals

Streptobacillary RBF in animals can be diagnosed by isolation of the organism *Streptobacillus moniliformis*, serology, or molecular techniques. Because the organism does not grow well on conventional media, inoculation into rodents can also be used for diagnosis. Serologic tests, including enzyme-linked immunosorbent assay (ELISA,) indirect immunofluorescence and complement fixation, can be used for screening in rodents. Polymerase chain reaction (PCR) assays are also used to detect genetic material (CFSPH 2006).

Spirillary RBF can be diagnosed by demonstrating *Spirillum minus* organisms in darkfield or phase contrast preparations, or after Giemsa, Wright's, or silver staining. If microscopy is unsuccessful, animal inoculation into rodents may be used. Culturing the microorganisms in laboratory media is not successful (CFSPH 2006).

In Humans

The signs of rash, fever and arthritis, along with a known or suspected history of rodent exposure, should lead to suspicion of RBF in a patient. Streptobacillary RBF is usually diagnosed by isolating the organism from blood, joint fluid, or the wound but as it is difficult to grow *Streptobacillus moniliformis* in laboratory culture media, identification of pleomorphic Gram negative bacilli in appropriate specimens supports a preliminary diagnosis in the absence of a positive culture. Inoculation into rodents can also be used for diagnosis. Serology is not considered to be reliable in humans. PCR assays have occasionally been used in humans (CFSPH 2006, CDC 2012).

Spirillum minus does not grow in artificial media; so the diagnosis is made by identifying characteristic spirochetes in appropriate specimens using darkfield microscopy or differential stains. Specimens of blood, lymph node aspirates, the bite wound, or erythematous plaques can be used for examination. If microscopy is unsuccessful, laboratory animal inoculation can be useful. Blood or wound infiltrate can be inoculated into mice, guinea pigs or *Spirillum minus*-free rats for diagnosis (CFSPH 2006, CDC 2012).

Treatment

Streptobacillus moniliformis and *Spirillum minus* are susceptible to several antibiotics, including penicillin, erythromycin and tetracycline; hence RBF can be treated successfully with antibiotics in man and animals. Penicillin is most often prescribed, but erythromycin, tetracycline and other antibiotics are also used. Treatment results in a shorter clinical course and may prevent severe complications (CFSPH 2006). Cervical abscesses in guinea pigs, as well as other abscesses, may require surgical removal or incision and drainage (CFSPH 2006). The prognosis is excellent with early treatment, however, without treatment, RBF can be serious or potentially fatal. Severe illnesses can involve different organs and may lead to endocarditis, myocarditis, pericarditis, meningitis, pneumonia and abscesses in internal organs. While death from RBF is rare, it can occur if it goes untreated and the death rate can be as high as 25 per cent (CDC 2012, HPA 2012).

Prevention and Control Strategies

RBF in animal can be prevented by avoiding contact between domesticated animals and wild rodents, particularly rats. Laboratory rats and mice should be kept free from infection and should be monitored regularly. Laboratory rats, mice and guinea pigs should be kept in separate areas. Cats, dogs and ferrets should not be allowed to hunt wild rodents to prevent them from getting infected and becoming carriers of RBF organisms (CFSPH 2006).

In humans, the preventive measures include avoiding contact with rodents or rodent-infested places and avoiding consumption of contaminated milk, food or water. Safety precautions such as wearing protective gloves while handling rats or cleaning their cages, practising regular hand wash afterwards, and avoiding hand-to-mouth contact are also essential (CFSPH 2006, CDC 2012).

Future Directions

RBF is an under-reported disease. Further, the disease is given less attention despite the high mortality associated with it. Nonspecific nature of the clinical features and difficulties in the isolation and identification of the causative organisms make the task of disease diagnosis quite challenging. The difficulty in diagnosis usually leads to delay in the appropriate treatment and deterioration of the patient's condition though the disease can be treated with common antibiotics. The disease thus requires greater attention, particularly in those areas where exposure to rodents is a common occurrence.

References

1. Balakrishnan N, Menon T, Shanmugasundaram S and Alagesan R. 2006. *Streptobacillus moniliformis* endocarditis (Letter). *Emerg Infect Dis* 12: 1037-1038.

2. CDC. 2012. *Rat-bite fever.* Centers for Disease Control and Prevention, Atlanta, USA. http://www.cdc.gov/rat-bite-fever/. Accessed 23 November, 2012.

3. CFSPH. 2006. *Rat bite fever.* The Center for Food Security and Public Health, Iowa State University, Ames, IA 50011. http://www.cfsph.iastate.edu/Factsheets/pdfs/rat_bite_fever.pdf. Accessed 23 November, 2012.

4. De AS, Baveja SM, Salunke PM and Manglani MV. 2010. Isolation of *Streptobacillus moniliformis* from the blood of a child with acute lymphoblastic leukaemia. *Indian J Medical Microbiol* 28 (4): 387-389.

5. Elliott SP. 2007. Rat bite fever and *Streptobacillus moniliformis. Clin Microbiol Rev* 20: 13-22. doi: 10.1128/CMR.00016-06.

6. HPA. 2012. *Rat-bite fever.* Health Protection Agency, UK. http://www.hpa.org.uk/Topics/InfectiousDiseases/InfectionsAZ/RatbiteFever/. Accessed 23 November, 2012.

7. Mirela I. 2009. Rat bite fever: Sodoku disease and Haverhill fever. *Annals of University of Oradea.* Fascicle Environmental Protection 14: 835-839.

2014, Zoonoses: Bacterial Diseases
Editor: Sudhi Ranjan Garg
Published by: DAYA PUBLISHING HOUSE, NEW DELHI

Pages **381–389**

21

Cat Scratch Disease

V.M. Vaidya
Department of Veterinary Public Health and Epidemiology,
Bombay Veterinary College, Mumbai – 400 012

Cat scratch disease (CSD) is also known as bartonellosis, benign inoculation lymphoreticulosis, cat scratch fever, cat scratch adenitis, cat claw fever, cat scratch syndrome, teeny's disease, and parinaud's oculoglandular syndrome. It is usually a benign and self-limiting disease characterized by fever and regional lymphadenopathy. First discovered in 1889 by Henri Parinaud, the disease was first described by Robert Debre in 1950 (Chomel 2000). Caused by *Bartonella henselae*, CSD is a direct bacterial zoonosis (Pal 1997). Approximately 25,000 human cases diagnosed every year in the United States, CSD is the most common cat-associated zoonosis. The disease occurs when a person is bitten or scratched by an infected cat. People with CSD usually have swollen lymph nodes, especially around the head, neck and upper limbs. They may also experience fever, headache, sore muscles and joints, fatigue and poor appetite. The disease may progress to complications such as neuroretinitis, neurological signs, osteomyelitis and other conditions. The complications usually resolve without sequelae in healthy people, although they can be life-threatening in rare cases. Endocarditis is usually the most serious concern. Healthy adults generally recover with no lasting effects, but it may take several months for the disease to go away completely. The infection in immunocompromised individuals of all ages can be fatal without antibiotic treatment, although the majority (55-80 per cent) are under twenty years of age (Chomel 2000).

Geographical Distribution

CSD occurs worldwide and may be present wherever cats are found. Stray cats may be more likely than pets to carry Bartonella (CDC 2011). According to Heroman and McCurley (1982), more than 2,000 cases occur each year. Small epidemic outbreaks and familial clustering have been reported in several countries. Approximately 75 per cent of the cases may occur in children (Acha and Szyfres 2003). In humans and cats, two main *B. henselae* genotypic groups have been identified *i.e.* Houston-1 (type I) serotype and Marseille (type II) serotype. *B. henselae* type I is more common than type II in Japan and the Philippines. This genotype is also more prevalent in the eastern USA, where it represents approximately half of all isolates, than in the western states. *B. henselae* type II is more common than type I in the western USA, western Europe and Australia (CFSPH 2012).

Etiological Agent and Vectors

CSD is caused by *Bartonella henselae*. In 1983, researchers conducted histopathologic examination of lymph nodes of 39 patients and demonstrated in 34 of them the presence of small, Gram negative, pleomorphic bacilli located in capillary walls or near areas of follicular hyperplasia and inside microabscesses (Chomel 2000). The observed bacilli were intracellular in the affected areas (Acha and Szyfres 2003). The organisms are fastidious, aerobic and oxidase-negative, and belong to family Bartonellaceae (Chomel *et al.* 2004, CFSPH, 2012). The Bartonellaceae formerly belonged to the order Rickettsiales, but are now in the α-2 subgroup of the

Figure 21.1: Cat flea *Ctenocephalides felis*

Proteobacteria (CFSPH 2012). More than 20 species of *Bartonella* have been described in animals, and a number of these organisms are thought to be zoonotic or potentially zoonotic (CFSPH 2012). Among them, only *B. henselae* is well understood. Two species, *B. quintana* and *B. bacilliformis*, are maintained in human populations and cause illness in people. Other species have also been implicated rarely in clinical cases in people or animals. Cat flea *Ctenocephalides felis* appears to be the main vector for spread of *B. henselae* among cats (Figure 21.1).

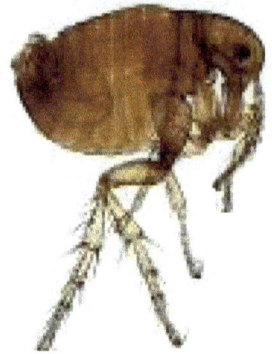

Host Range

The isolation of *B. henselae* from the blood of naturally infected cats and several months long bacteraemic phase in cats has suggested that cats are natural reservoir of *B. henselae* and can transmit the infection to humans by bite or scratch. Members of the genus *Bartonella* are maintained in many domestic and wild animal hosts. *B. henselae* infects house cats and other members of the Felidae. Additional species of *Bartonella* are found in cats, dogs, cattle, rodents, rabbits, bats, and other wild and domesticated animals.

Disease Transmission

Domestic cat is the main reservoir for *B. henselae*. Bacteraemia in cats was first reported in 1992 in the cat of a healthy owner (Regnery *et al.* 1992). Kittens are more likely to carry the bacteria in their blood, and may therefore be more likely to transmit the disease than adult cats. However, the results of experimental studies showed that fleas serve as a vector for transmission of *B. henselae* among cats (Chomel *et al.* 1996) and that viable *B. henselae* are excreted in the faeces of *Ctenocephalides felis*, the cat flea (Higgins *et al.* 1996). Another study showed that cats could be infected with *B. henselae* through intradermal inoculation using flea faeces containing *B. henselae* (Foil *et al.* 1998). As a consequence, it is believed that a likely means of transmission of *B. henselae* from cats to humans may be inoculation with flea faeces containing *B. henselae* through a contaminated cat scratch wound or across a mucosal surface. The organism is reported to survive for 3 days in flea faeces, which might result in contamination of the environment (CFSPH 2012). People mainly seem to acquire *B. henselae* in scratches and bites from cats. More than 90 per cent of clinical cases occur in people who have been in contact with cats, most often kittens, and the majority of these patients report having been scratched, bitten or licked (CFSPH 2012). A study reveals that bacteraemic cats are more likely to have *Bartonella* DNA in oral swabs, compared to nonbacteraemic cats. About 65 per cent of patients have the history of being scratched or bitten by cats and 90 per cent of the cases had some contact with cats (Acha and Szyfres 2003). The disease can be transmitted from cat to cat by the cat flea, but not by direct contact between animals. There is no human-to-human transmission.

Environmental and Other Predisposing Factors

The incidence of the disease varies by season, with most cases seen in autumn and winter. In the USA, CSD is most frequently seen between July and January, with the greatest number of hospitalizations between July and October. In France, one study found that the highest incidence was from September to April, with a peak in April. Seasonality in temperate regions is thought to be caused by the seasonal concentration of births, the acquisition of kittens as pets, and peaks in the flea population. Most cases have occurred in children in males than females, who have more contact with cats (Jackson *et al.* 1993, Acha and Szyfres 2003).

Prevalence of Disease in Animals

Asymptomatic infections with *B. henselae* are very common in cats. The prevalence of antibodies to *B. henselae* in cats varies widely in different geographic regions, with highest rates occurring in warm, humid areas where fleas are more common. In studies from the USA, Brazil, Ireland, UK, eastern Australia and other locations, the prevalence of bacteraemia in cats varied from 3 per cent to greater than 40 per cent, and was as high as 70-72 per cent in some populations by PCR (CFSPH 2012). In northern California, bacteraemia occurrences of 39.5 per cent (81/205) and 41 per cent (25/61) have been reported (Chomel *et al.* 1995). Several serosurveys have also been conducted in cat populations in the USA. In North Carolina, 21 per cent of 518 sick cats were seropositive for *B. henselae* (Breitschwerdt and Kordick 1995). Childs *et al.* (1995) reported a seropositivity in 28.2 per cent (370/1,314) cats from various parts of the USA.

In Chile, the general prevalence of IgG antibodies against *B. henselae* was 85.6 per cent. However, domestic cats had a lower prevalence when compared with stray cats *i.e.* 73 and 90 per cent, respectively. *B. henselae* was isolated in 41 per cent of blood cultures and all the isolated were confirmed as *B. henselae* by RFLP-PCR (Ferres *et al.* 2005). A report of a possible case of CSD caused by contact with a dog in Japan suggests that dogs could play a role in human *B. henselae* infection (Tsukahara *et al.* 1998).

Prevalence of Disease in Human Beings

A number of studies in several countries have shown CSD seroprevalence rates varying from less than 1 per cent to 25 per cent or more in the general population (CFSPH 2012). An estimated 22,000 to 24,000 human cases of CSD occurred in the USA in 1992, of which 2,000 required hospitalization (Jackson *et al.* 1993). In Connecticut, which is the only State in the USA where CSD is a reportable disease (since January 1992), 246 people met the case definition during the period 1992-1993. A prospective population-based surveillance system reported an average statewide annual incidence of 3.7 cases of CSD per 100,000 persons (Hamilton *et al.* 1995). In the Netherlands, the incidence of CSD was estimated to be 2,000 cases per year (Bergmans *et al.* 1997). In Chile, antibodies were detected in 13 per cent of children, and 10 per cent of technical and professional workers who cared for cats. Although antibodies to *B. henselae* are expected to be more common in people exposed to cats, the seroprevalence rates reported in high risk groups vary widely. In a study in Poland, antibodies to *B. henselae* were detected in 45 per cent of veterinarians, 53 per cent of cat owners and 48 per cent of homeless alcoholics (CFSPH 2012).

A study was performed over a 7-year period in which lymph node biopsy specimens or cytopunctures from 70 patients with lymph node enlargement were systematically tested by PCR for the presence of *B. henselae* DNA (*htrA* gene). The PCR assay for *B. henselae* was positive for 22 of the 29 definite CSD patients and 3 of the 15 possible CSD patients (Hansmann *et al.* 2005). Many isolated cases of CSD have been reported in different parts of the world (Arashima *et al.* 1993, Olejnik *et al.* 1996, Robson *et al.* 1999, Maruyama *et al.* 2000, Asensio-Sánchez *et al.* 2006).

In France, an immunocompetent 61-year old woman with a systemic CSD including a multifocal osteomyelitis was diagnosed and confirmed by PCR on the adenopathy. A literature review identified 51 other cases of osteomyelitis associated with CSD, 14 of those confirmed by PCR (Roubaud-Baudron *et al.* 2009). A survey of an occupational group potentially at risk for *Bartonella* infection found that the overall seroprevalence was 7.1 per cent (Noah *et al.* 1997). In Italy, an unusually high rate of reactors have been reported where antibodies to *B. henselae* were detected in approximately 62 per cent of children and adolescents who presented as outpatients to a clinic for health check-ups or minor illnesses, and who had no symptoms that might indicate bartonellosis. In this study, 8.5 per cent of the participants had high titers, suggesting recent or ongoing infections (CFSPH 2012).

Disease Manifestations in Animals

Cats that carry *B. henselae* do not show major clinical signs of CSD under natural conditions. Suspicions of lymphadenopathy caused by a CSD-like organism identified by silver-stained section have been reported (Chomel 2000). Some experimental studies and case reports have suggested some possibility of signs of disease. Development of inflammatory swellings or pustules at the inoculation site has been demonstrated in the experimentally infected cats. Other clinical signs reported were lymphadenopathy, myalgia and transient fever with lethargy and anorexia during febrile periods. Transient mild behavioural or neurological dysfunction, consisting of disorientation, nystagmus, hypersensitivity to stimuli, decreased responsiveness to environmental stimuli, or increased aggressiveness, as well as mild transient anemia, eosinophilia, and reproductive disorders were reported (CFSPH 2012).

Disease Manifestations in Humans

The disease characterized by a regional lymphadenopathy without lymphangitis is typically seen after 1-3 weeks. In 50 per cent of cases of CSD, a small skin lesion, often resembling an insect bite, appears at the inoculation site (usually the hand or forearm), and evolves from a papule to a vesicle and partially healed ulcers. These lesions heal within a few days to a few weeks. Lymphadenitis is generally unilateral and commonly appears in the epitrochlear, axillary or cervical lymph nodes (Carithers 1985). Manifestations of CSD range from the classic syndrome of regional lymphadenopathy to unusual manifestations such as severe, systemic or recurrent infection producing encephalitis, splenitis, mediastinal masses and pleurisy (Margileth *et al.* 1987). In a 1992-1993 Connecticut study, it was found that younger CSD patients (less than 15 years age) were more likely to have cervical adenopathy while older patients (15 years or more) were more likely to have inguinal and axillary adenopathy (Hamilton *et al.* 1995). Lymphadenopathy develops approximately three weeks after exposure (Chomel 1996). Swelling of the lymph node is usually painful and persists for several weeks to several months. In 25 per cent of the cases, suppuration occurs. The large majority of patients show signs of systemic infection: fever, chills, malaise, anorexia and/or headaches. In general, the disease is benign and heals spontaneously without sequelae.

Atypical symptoms of CSD occur in 5-10 per cent cases. The most common of these symptoms is Parinaud's oculoglandular syndrome which is a granulomatous conjunctivitis with concurrent swelling of the lymph node near the ear. Optic neuritis, involvement of retina, and neuropathy can also occur. Bacillary angiomatosis caused by *B. henselae* is primarily a vascular skin lesion that may extend to bone or be present in other areas of the body. Encephalopathy is one of the most serious complications of CSD, usually occurring two to six weeks after the onset of lymphadenopathy. However, patients usually make a complete recovery with a few or no sequelae. A cluster of five cases of children with acute encephalopathy associated with CSD was reported in south Florida (Noah *et al.* 1995). New clinical presentations associated with *B. henselae* infection were reported in immunocompetent people, such as neuroretinitis, and bacteraemia as a cause of chronic fatigue syndrome, as well as a case of aggressive *B. henselae* endocarditis in one cat owner (Chomel 1996). Bacillary peliosis is a condition that most often affects patients with HIV and other conditions causing severe immune compromise. The liver and spleen are primarily affected, with findings of blood filled cystic spaces on pathology (Perkocha *et al.* 1990). Rarely, CSD may present as mastitis or disguise itself as a solitary tumour of the breast (Gamblin *et al.* 2005)

Diagnostic Procedures

CSD is often diagnosed by the history and physical examination, with supporting evidence from laboratory tests. CSD can be clinically confused with other diseases that cause regional lymphadenopathies, such as tularemia, brucellosis, tuberculosis, pasteurellosis, infectious mononucleosis, Hodgkin's disease, venereal lymphogranuloma, lymphosarcoma, and lymphoma. All these diseases must be excluded before considering a diagnosis of CSD (Acha and Szyfres 2003).

1. Isolation

The isolation of the pathogen from blood is done by using a lysis-centrifugation technique. In cats, prolonged bacteraemia makes the isolation of the pathogen relatively easy. Isolation requires specialized media such as fresh chocolate agar or brain heart infusion agar enriched with blood. Visible colonies of *B. henselae* usually develop in 9 days to 6-8 weeks.

2. Microscopic Identification

Organisms may be detected in tissues with Warthin-Starry silver stain and Brown-Hopps Gram stains. Immunostaining has been used to identify the bacteria. Confirmatory diagnosis when atypical features are present includes demonstration of pleomorphic, Gram negative bacilli in the wall of capillaries and macrophages lining the sinuses near or in lymph node.

3. Serological Diagnosis

Serological tests for *B. henselae* include various indirect immunofluorescence assays and enzyme-linked immunosorbent assays. A four-fold rise in titer or the presence of IgM suggests a recent infection. IgM antibodies to *B. henselae* have been reported to persist for less than three months, while IgG may be detected for more than two years. Cross-reactions can occur between species of *Bartonella*. Cross-reactions have also been reported with other organisms such as *Chlamydia* spp. and *Coxiella burnetii*. Application of Hanger-Rose intradermal test is helpful. The Hanger-Rose antigen is prepared by suspending pus taken from an abscessed lymph node in a 1:5 saline solution and heating it for 10 hours at 60°C. The antigen is very crude and difficult to standardize. The test is carried out by intradermal inoculation with 0.1 ml of the antigen. The reaction may be read in 48 hours. Oedema measuring 0.5 cm and erythema of 1 cm are considered a positive reaction. The test is very useful, since 90 per cent of 485 clinically diagnosed cases gave positive results, while only 4.1 per cent out of 591 controls tested positive.

Polymerase Chain Reaction (PCR)

PCR is the best modern diagnostic method which can be performed in fresh sample or paraffin embedded formalin fixed tissue samples. A commercial PCR test for detecting *B. henselae* bacteraemia in cats directly from the blood is now available in the USA (Tobias *et al*. 1998).

Prevention and Control Strategies

General Hygiene and Sanitary Practices

Contact with cats should be avoided to prevent the disease. Hands should be washed thoroughly after playing with a cat. Bites and scratches from cats, particularly kittens, should be avoided and any bites or scratches should immediately be washed with soap and water. Cats should be discouraged from licking a person's skin, particularly the eyes, face and areas with abrasions or other open wounds. Cat's claws may be covered with soft nail cap to prevent injury. Suspected carrier cat should not come in contact with children or immunocompromised individuals. Flea control is helpful in reducing the risk of *B. henselae* transmission between cats.

Therapeutic Guidelines

Most cases of CSD in immunocompetent individuals are self-limiting. Treatment is usually supportive and symptomatic. Suppurating lymph nodes may be aspirated to remove pus and reduce the pain while the severely affected ones or persistent ocular granulomas are occasionally excised. Azithromycin, ciprofloxacin, doxycycline, and other multiple antibiotics have been used. Azithromycin is preferentially used in pregnancy to avoid the side effects of doxycycline. In immunocopromised individuals, intravenous administration of gentamicin, doxycycline and oral administration of erythromycin have been used successfully (Groves and Harrington 1994).

Future Directions

Development of a vaccine to prevent the spread of infection in cat population and to reduce the risk of infection of humans is although difficult due to the diversity of *Bartonella* species and types harboured by cats, and the lack of cross-protection between species and types, but there is a potential scope of research to develop a vaccine for CSD. Also, more study is required to know the exact route of cause of the disease, persistence of the etiological agent in the environment and for the development of rapid and reliable pen-side diagnostic test for effective control of the disease.

References

1. Acha PN and Szyfres B. 2003. Chlamydiosis, rickettsiosis and virosis. In: *Zoonoses and communicable diseases common to man*. 3rd edn. Pan American Health Organization, Washington, D.C.

2. Arashima Y, Kawano K, Baba S, Tezuka T, Munemura T, Asano R and Hokari S. 1993. A case of cat scratch disease: from the clinical pathological point of view. *Kansenshogaku Zasshi* 67: 81-84.

3. Asensio-Sanchez VM, Rodríguez-Delgado B, García-Herrero E, Cabo-Vaquera V and García-Loygorri C. 2006. Serous macular detachment as an atypical sign in cat scratch disease. *Arch Soc Esp Oftalmol* 81: 717-719.

4. Bergmans AMC, DeJong CMA, VanAmerongen G, Schot CS and Schouls LM. 1997. Prevalence of *Bartonella* species in domestic cats in the Netherlands. *J Clin Microbiol* 35: 2256-2261.

5. Breitschwerdt EB and Kordick DL. 1995. Bartonellosis. *J Am Vet Med Assoc* 206: 1928-1931.

6. Carithers HA.1985. Cat scratch disease. An overview based on a study of 1,200 patients. *Am J Dis Children* 139: 1124-1133.

7. CDC. 2011. *Bartonella-associated infections*. Centers for Disease Control and Prevention, Atlanta, GA 30333, USA.

8. CFSPH. 2012. *Cat scratch disease and other zoonotic Bartonella infections*. Centre for Food Security and Public Health, Ames, Iowa 50011. p. 20.

9. Childs JE, Olson JG, Wolf A, Cohen N, Fakile Y, Rooney JA, Bacellar F and Regnery RL. 1995. Prevalence of antibodies to *Rochaiimaea* species (cat scratch disease agent) in cats. *Vet Rec* 136: 519-520.

10. Chomel BB. 2000. Cat-scratch disease. *Rev Tech Off Int Epiz*. 19: 136-150.

11. Chomel BB, Abbott RC, Kasten RW, FloydHawkins KA, Kass PH, Glaser CA, Pedersen NC and Koehler JE. 1995. *Bartonella henselae* prevalence in domestic cats in California: risk factors and association between bactereamia and antibody titers. *J Clin Microbiol* 33: 2445-2450.

12. Chomel BB, Kasten RW, Floyd-Hawkins K. 1996. Experimental transmission of *Bartonella henselae* by the cat flea. *J Clin Microbiol* 34: 1952-1956.

13. Chomel BB, Boulouis HJ, Breitschwerdt EB. 2004. Cat scratch disease and other zoonotic *Bartonella* infections. *J Am Vet Med Assoc* 224: 1270-1279.

14. Ferres M, Abarca K, Godoy P, García P, Palavecino E, Mendez G, Valdés A, Ernst S, Thibaut J, Koberg J, Chanqueo L and Vial PA. 2005. Presence of *Bartonella henselae* in cats: natural reservoir quantification and human exposition risk of this and zoonoses in Chile. *Rev Med Chil* 133: 1465-1471.

15. Foil L, Andress E and Freeland RL. 1998. Experimental infection of domestic cats with *Bartonella henselae* by inoculation of *Ctenocephalides felis* (Siphonaptera: Pulicidae) feces. *J Med Entomol* 35: 625-628.

16. Gamblin TC, Nobles-James C, Bradley RA, Katner HP and Dale PS. 2005. Cat scratch disease presenting as breast mastitis. *Can J Surg* 48: 254-255.

17. Groves MG and Harrington KS. 1994. *Rochalimaea henselae* infections: newly recognized zoonoses transmitted by domestic cats. *J Am Vet Med Assoc* 204L: 267-271.

18. Hamilton DH, Zangwill KM, Hadler JL and Cartter ML. 1995. Cat-scratch disease - Connecticut, 1992-1993. *J Infect Dis* 172: 570-573.

19. Hansmann Y, DeMartino S, Piémont Y, Meyer N, Mariet P, Heller R, Christmann D and Jaulhac B. 2005. Diagnosis of cat scratch disease with detection of *Bartonella henselae* by PCR: a study of patients with lymph node enlargement. *J Clin Microbiol* 43: 3800-3806.

20. Heroman VM and McCurley WS. 1982. Cat-scratch disease. *Otolaryngol Clin North Am* 15: 649-658.

21. Higgins JA, Radulovic S, Jaworski DC and Azad AF. 1996. Acquisition of the cat scratch disease agent *Bartonella henselae* by cat fleas (Siphonaptera: Pulicidae). *J Med Entomol* 33: 490-495.

22. Jackson LA, Perkins BA and Wenger JD. 1993. Cat scratch disease in the United States: an analysis of three national data bases. *Am J Public Hlth* 83: 1707-1711.

23. Margileth AM, Wear DJ and English CK. 1987. Systemic cat scratch disease: report of 23 patients with prolonged or recurrent severe bacterial infection. *J Infect Dis* 155: 390-402.

24. Maruyama S, Kabeya H, Nogami S, Sakai H, Suzuki J, Suzuki H, Sugita H and Katsube Y. 2000. Three cases of cat scratch disease diagnosed by indirect immunofluorescence antibody assay and/or polymerase chain reaction of 16S rRNA gene of *Bartonella henselae*. *J Vet Med Sci* 62: 1321-1324.

25. Noah DL, Bresse JS, Gorensek MJ, Rooney JA, Cresanta JL, Regnery RL, Wong J, Del Toro J, Olson JG and Childs JE. 1995. Cluster of five children with acute encephalopathy associated with cat-scratch disease in South Florida. *Pediatr Infect Dis J* 14: 866-869.

26. Noah DL, Kramer CM, Verbsky MP, Rooney JA, Smith KA and Childs JE. 1997. Survey of veterinary professionals and other veterinary conference attendees for antibodies to *Bartonella henselae* and *B. quintana*. *J Am Vet Med Assoc* 210: 342-344.

27. Olejnik I, Wieczorek M, Ka³muk A and Gabriel A. 1996. Cat-scratch disease as a diagnostic problem of lymphadenopathy. *Pediatr Pol* 71: 363-366.

28. Pal M. 1997. Cat scratch disease. In: *Zoonoses*. R.M. Publishers and Distributors, Delhi, India.

29. Perkocha LA, Geaghan SM and Yen TS. 1990. Clinical and pathological features of bacillary peliosis hepatis in association with human immunodeficiency virus infection. *N Engl J Med* 323: 1581-1586.

30. Regnery RL, Martin M and Olson JG. 1992. Naturally occurring *Rochalimaea henselae* infection in domestic cat. *Lancet* 340: 557-558.

31. Robson JM, Harte GJ, Osborne DR and McCormack JG. 1999. Cat-scratch disease with paravertebral mass and osteomyelitis. *Clin Infect Dis* 28: 274-278.

32. Roubaud-Baudron C, Fortineau N, Goujard C, Le Bras P and Lambotte O. 2009. Cat scratch disease with bone involvement: A case report and literature review. *Rev Med Interne* 30: 602-608.

33. Tobias EJ, Noone KE and Garvey MS. 1998. Managing *Bartonella henselae* infection in cats. *Vet Med* 93:745-749.

34. Tsukahara M, Tsuneoka H, Lino H, Ohno K and Murano I. 1998. *Bartonella henselae* infection from a dog. *Lancet* 352: 1682.

2014, Zoonoses: Bacterial Diseases
Editor: Sudhi Ranjan Garg
Published by: DAYA PUBLISHING HOUSE, NEW DELHI

Pages 390–401

22

Rhodococcus equi Infection

S.K. Khurana
National Research Centre on Equines, Hisar – 125 001

Rhodococcus equi was first isolated in 1923 by Magnusson from the lung of a foal with pyogranulomatous pneumonia as *Corynebacterium equi* in Sweden. It was reclassified as *R. equi* in 1977 (Goodfellow and Alderson 1977). The organism primarily causes infections in grazing animals, mainly horses and foals (Giguere *et al.* 2011a, b). *R. equi* infection primarily causes chronic pyogranulomatous pneumonia in 1 to 4 month old foals. The disease does not occur in adult horses unless there is a severe immunodeficiency.

R. equi infection in a human patient was first reported in 1967 in a 29 year old man with plasma cell hepatitis receiving immunosuppressant medications. Since then, *R. equi* is emerging as an important opportunistic pathogen in immunocompromised patients, especially those with acquired immunodeficiency syndrome (AIDS). Though it rarely infects immunocompetent persons, *R. equi* infection is associated with significant mortality. As animals are the primary reservoirs of the infection, it is classified as a direct bacterial anthropozoonosis.

Geographical Distribution

The disease is present worldwide with many endemic areas and endemic equine farms as animal manure, especially horse manure, is suitable for growth of *R. equi* organisms in soil and environment. In temperate countries, the disease is common during summer season, whereas in tropical areas, it is more common in winter and spring season.

Etiological Agent and Vectors

R. equi is a Gram positive, facultative, intracellular, nonmotile, non-spore forming, pleomorphic, coccobacillus and a facultative intracellular pathogen of macrophages. It is nonfermenting (distinguishing it from pathogenic corynebacteria), gelatinase negative, catalase positive, usually urease positive, and oxidase negative. The organism is known as *Rhodococcus* because of its ability to form a salmon pink-coloured pigment. *R. equi* can be weakly acid fast. Previously called as *Corynebacterium equi*, the organism is currently grouped with the aerobic actinomycetes. Of the 40 genera in the actinomycetes group, genus *Rhodococcus* is placed among the nocardioform bacteria, along with *Mycobacterium, Nocardia, Gordonia, Tsukamurella* and *Corynebacterium*. *Rhodococcus* species include symbionts (*R. rhodnii*), animal pathogens animals (*e.g. R. equi*), plant pathogens (*R. fascians*), human pathogens (*e.g. R. equi, R. rhodochrous, R. erythropolis* and several unidentified *Rhodococcus* species) (Bell *et al.* 1998). *R. equi* is mainly a soil organism with simple nutritional and environmental requirements and grows well in animal manure.

Host Range

R. equi primarily affects equines, especially foals, aged between 1-4 months. Cattle, pigs and human beings are also affected. It has been isolated from a variety of land and water animals like cattle, crocodiles, wild birds and even blood sucking arthropods. The organism is present in soil all over the world except in Antarctica, and in fresh and sea water.

Disease Transmission

R. equi inhabits soil and gets entry to the respiratory tract of foals primarily by inhalation of airborne bacteria. Ingestion of the contaminated soil is another common mode of transmission where bacteria gains entry to the blood stream and seeds the lungs. The disease progresses slowly and large abscesses are formed throughout the lung. With the progression of the disease, the animal shows signs of pneumonia, fever, cough and depression. However, many of the symptoms are dormant with sudden fatal consequences. The organism replicates within the respiratory tract and then is coughed up and swallowed by the animal. Replication continues as the bacteria travel though the intestinal tract after which large numbers of the bacteria are defecated and subsequently regain entry into the soil. Although adults are only very rarely infected, it is known that mares also contaminate the soil by ingesting bacteria and passing these in large numbers in their faeces.

R. equi is found in the soil of most farms. Pneumonia caused by this organism can be endemic, sporadic or unrecognized in different farms. Several factors may influence the incidence of the disease, such as the degree of contamination on the farm, density of horses, climatic conditions and virulence of the isolate. *R. equi* frequently colonizes the gastrointestinal tract in grazing animals and thus becomes a source of infection for in-contact human beings.

In human beings, infection is acquired by inhalation from soil, inoculation into wound or mucous membranes or ingestion and sometimes through domesticated animals (Weinstock and Brown 2002). Other routes of R. equi infection including person-to-person transmission are poorly understood (Weinstock and Brown 2002).

Environmental and Other Predisposing Factors

R. equi is a robust soil organism and widespread in the environment. It grows well in the manure of animals, including horses, cattle, pigs and poultry. The organism potentially multiplies wherever there is horse manure. It may multiply 10,000 times in a few days in an area contaminated by horse manure. The growth of R. equi is optimum at 30°C. There seems to be a direct relationship between the number of R. equi in the environment of young foals and the number of cases of pneumonia. Because the organism reaches the lung by inhalation, manure-contaminated dusty environments are potentially lethal sources of infection.

R. equi can also multiply to dangerous levels in the gastrointestinal tract of a young foal. Foals establish a normal intestinal flora of bacteria from birth to 12 weeks. In this period, they seem to be susceptible to being infected by abnormally high levels of R. equi. Levels of up to 10,000 organisms per gram of manure are sometimes shed by the infected foals. After 12 weeks, the bacteria rapidly decline in number and no longer multiply in the intestine. The endemicity of infection on a farm is directly related to density of foals at a farm. The incidence of the disease peaks in foals at 6-12 weeks; this coincides with the period when maternal antibody, derived from colostral milk, has largely declined and before antibody produced by the foal has developed. It is thus important to ensure that foals get their full share of colostral antibody immediately after birth.

Little is known about the susceptibility of the organism to disinfectants. Because many horse breeding farms do not use concrete, and the disinfection procedures are often difficult to carry out in the stalls and impossible in the loafing paddocks, some farms may become heavily infected with the bacteria and develop problems each year with foal pneumonia.

Prevalence of Disease in Animals

The disease is prevalent in all countries that have significant horse breeding population. The disease is known to recur in many farms and thus becomes endemic there. It is most common in foals aged 1-4 months. The disease is possibly present in all states in India and there are many reports from Punjab, Haryana, Rajasthan and Maharashtra. Garg et al. (1985) found R. equi responsible for 84.6 per cent of foal deaths due to multiple pulmonary abscesses in parts of India. Khurana et al. (2009) have reported isolation of 14 isolates of R. equi from foals with respiratory problems from different parts of Haryana. The prevalence of R. equi has also been reported from several other countries, viz. Argentina, Australia, Canada, France, Hungary, Japan, Ireland and others (Alain et al. 2007).

Prevalence of Disease in Human Beings

The disease has been reported throughout the world, especially among the immunocompromised persons. Immunocompetent humans also suffer, but the prevalence is much less.

Disease Manifestations in Animals

The disease that usually occurs in foals aged 1-4 months is insidious in the beginning and clinical detection of the disease in its early stages requires considerable experience. It begins with increased diffuse bronchial sounds, often accompanied by a cough, which later develops into a wheeze or a rattling sound. Fever follows within 1-2 days, and is associated with an increased respiratory rate (over 40 per minute). Untreated foals develop progressive crackling sounds that can be heard all over the lung, and harsh inspiratory sounds audible even without a stethoscope.

Foals often become critically ill before anyone detects something seriously wrong. The disease is unique in that foals with fever appear quite active apparently. Sick foals remain bright and alert, sucking the mare well. The disease may also develop insidiously on a farm over many years, and before anyone is aware, the farm is heavily infected with the organism and thus becomes endemic. The clinical signs of the disease are not particularly distinctive from other types of pneumonia. However, experienced persons on the endemically affected farms may discern the signs of the disease. The disease should be suspected on farms where a number of foals, between 1-4 months, show mild fever and coughing. The more heavily a breeding farm is contaminated with the organism, the younger will be the age of the foals affected.

Mares with foals at foot from unaffected farms when sent for breeding to an infected farm bring back an infected foal on their return. Lack of experience with the disease on the unaffected farm and the insidious pneumonia in foals may mean that the disease will go undetected till it is very late to recover.

The disease is often confused with *Rotavirus,* a highly contagious virus that causes diarrhoea in foals. *R. equi* may occasionally cause disease in other body systems. Inflammation of the eye (uveitis), bone infection, inflammation of the joints, diarrhoea and abdominal abscess formation are noted infections outside of the lungs.

Disease Manifestations in Humans

Necrotizing pneumonia is the most common manifestation of *R. equi* infection in humans. Extrapulmonary *R. equi* infections include wound infection, subcutaneous abscess, brain abscess, thyroid abscess, retroperitoneal abscess, peritonitis, meningitis, pericarditis, osteomyelitis, endophthalmitis, lymphadenitis, lymphangitis, septic arthritis, osteitis, bloody diarrhoea, and fever of unknown origin among others. Bacteraemia and dissemination of infection follow from the primary infection site, usually lungs. Physical findings depend on the site of infection and include fever, tachypnea, crackles and other common physical findings of pneumonia, lymphadenopathy, septic arthritis, hyperemia, decreased visual acuity, evidence of anterior chamber involvement, meningitis, soft tissue masses, induration, fluctuance in localized infections consistent with abscesses.

In patients infected with human immunodeficiency virus (HIV), *R. equi* has been isolated from the stool, with or without evidence of pneumonia, suggesting that the infection may also occur from the alimentary tract. Most of the information about the pathogenesis of *R. equi* infections in humans is derived from animal isolates. However, the infection in humans seems to differ from that in foals. The differences between human and animal *R. equi* infections may suggest that the conclusion drawn from animal models may not be entirely applicable to the pathogenesis of *R. equi* infections in humans (Makrai *et al.* 2002, Takai *et al.* 2003).

Morbidity and Mortality in Humans

Morbidity is related to complications and chronicity of the infection. *R. equi* pneumonia may be complicated by many conditions including abscesses, empyema, pleural effusion, hemoptysis, direct chest wall involvement and pneumothorax. Pericardial tamponade may result from purulent pericarditis. Bacteremia leading to overwhelming sepsis has been reported, more often in immunocompromised patients. About 47 per cent of patients infected with HIV and 17 per cent of patients with non-HIV-associated immunocompromised conditions have been reported to have chronic *R. equi* infection (Verville *et al.* 1994). Relapses are also common after discontinuation of antibiotics. An important site of extrapulmonary relapse is the central nervous system.

R. equi infections carry an overall mortality rate of about 25 per cent. In two different reports, the mortality rate was 50-55 per cent in patients infected with HIV and 20-25 per cent in patients with non-HIV-associated immunocompromised conditions (Harvey and Sunstrum 1991, Cornish *et al.* 1999). In contrast, the mortality rate was only about 11 per cent in immunocompetent patients; though lower but it is still significant. Lower mortality rates in this subgroup of patients may be due to the fact that localized infections represent about 50 per cent of the cases reported. However, these numbers may not hold true in the era of highly active antiretroviral therapy. Long-term remissions have been reported in patients infected with HIV (Vladusic *et al.* 2006). High mortality rates in *R. equi* infections may be mainly due to following factors.

1. *R. equi* may be misidentified among both immunocompetent and immunocompromised populations.

2. Patients with *R. equi* infection may receive inappropriate initial antibiotic therapy because of misdiagnosis. *R. equi* pneumonia does not respond to standard empirical treatment with beta-lactams (other than imipenem and meropenem) and tetracyclines.

3. Simultaneous opportunistic infections are common, especially in patients infected with HIV. In this subgroup of patients, the mortality rate directly attributed to *R. equi* infection alone may be less. Capdevila *et al.* (1997) reported patients infected with HIV who had *R. equi* pneumonia; in this group of patients, the mortality rate directly attributed to *R. equi* infection was only 15.4 per cent.

R. equi infections have no reported racial predilection. In all *R. equi* infections, the male-to-female ratio is about 3:1. The reason for this is not clear; however, among immunocompromised patients, the predilection in males may be explained by the higher prevalence of HIV infection among males. *R. equi* infections have been described in all age groups, from infants to elderly persons. Studies have revealed a mean age of infection of 34-38 years.

Infection in Children

R. equi infections in children differ from those in adults. Immunocompromising conditions, including hematopoietic malignancies, immunosuppression associated with chemotherapy, and HIV infection account for only about one-third of *R. equi* infections in children. Pediatric *R. equi* infections account for approximately one third of all cases among immunocompetent individuals, perhaps because of the increased prevalence of trauma among children, predisposing them to localized *R. equi* wound infections. *R. equi* infections in immunocompetent children have been shown to have a favourable prognosis.

Clinical Symptoms in Adults

The onset of *R. equi* infections is generally insidious, and presenting symptoms vary according to the infection site. Symptoms in immunocompetent patients do not differ from those in immunocompromised patients. In *R. equi* infections secondary to trauma, such as endophthalmitis, septic arthritis and traumatic meningitis, symptoms may present within 24 hours of the trauma.

Pulmonary *R. equi* infections include fever and cough, malaise, chest pain, dyspnoea, hemoptysis, loss of weight, etc. Some other manifestations of *R. equi* infection include lymphadenopathy, eye drainage and pain, joint pain, altered level of consciousness and bloody diarrhoea. Anaemia caused by colonic polyps infected with *R. equi* has also been reported. *R. equi* infections can also be acquired nosocomially.

Exposure to soil contaminated with manure is the most likely route of infection. About 80-90 per cent of patients with *R. equi* infection are immunocompromised. About 50-60 per cent of the patients have HIV infection, 15-20 per cent patients have hematopoietic and other malignancies, and 10 per cent are transplant recipients.

R. equi infections can also occur in immunocompetent persons. Infections in these patients include pneumonia, endophthalmitis, septic arthritis, traumatic meningitis, brain abscess, fever of unknown origin, lymphangitis, and lymphadenitis. A history of trauma should be obtained because about 50 per cent of *R. equi* infections described in immunocompetent patients are due to trauma.

Diagnostic Procedures

Diagnosis in Foals

Clinical Signs

Clinical signs can be characteristic of the disease and include fever, cough, depression, laboured breathing and tracheal rattle. Most of the foals do not show symptoms of the infection until the disease becomes very severe and fatal. The disease may remain in mild form or unnoticed in initial phases, and the condition of the foal may deteriorate abruptly often resulting in death. Confirmation of the disease is dependent on history of occurrence of disease on a farm or endemicity at the farm. The specimens of nasal swab, bronchial aspirate and faecal matter may be collected for laboratory examination and confirmation. These should be transported in ice and submitted as soon as possible to the laboratory.

Gross Post-mortem Lesions and Histologic Findings

Pyogranulomatous pneumonia and lymphadenitis of bronchial lymph nodes are main lesions. The ability of *R. equi* to persist in and destroy macrophages is the basis of its pathogenesis. The typical pattern is a necrotizing granulomatous reaction dominated by macrophages filled with granular cytoplasm that shows positive results on periodic acid-Schiff stain and contains large numbers of coccobacilli.

Differential Diagnosis

The pneumonic form should be differentiated from viral respiratory infections caused by rhinovirus, herpesvirus and influenza virus, *Streptococcus zooepidemicus*, parasitic pneumonia by migrating stages of *Parascaris equorum* and *Pneumocystis carinii* pneumonia. Diarrhoea should be differentiated from infection due to *Salmonella*, parasitism due to cythostomes and antibiotic induced diarrhoea. Joint infection should be differentiated from septic arthritis due to *Streptococcus zooepidemicus*, *Salmonella* and some other bacteria.

Cultural Examination

R. equi infections are conventionally diagnosed by culturing, biochemical tests and morphological characteristics of the causative organism. The organism is cultured easily in ordinary nonselective media. Large, smooth, irregular, mucoid colonies appear within 48 hours. Salmon-coloured pigment appears later. However, the colony characteristics, cellular morphology, and reaction to acid fast staining differ between *R. equi* isolates. Identification of *R. equi* by classical bacteriological techniques is difficult and misclassification of an isolate may occur. Further, the conventional cultural methods take a long time for diagnosis.

PCR Assays

A specific PCR assay for rapid and reliable identification of *R. equi* is based on the amplification of a fragment of the *ChoE* gene encoding cholesterol oxidase, an enzyme believed to be a major virulence factor of *R. equi*. Mutational analysis indicated that *ChoE* is the membrane-damaging factor responsible for the typically shovel-shaped synergistic hemolysis (CAMP-like) reaction elicited by *R. equi* in the presence of sphingomyelinase C-producing bacteria, such as *Listeria ivanovii*, *Bacillus cereus*, and *Staphylococcus aureus*. This CAMP-like reaction can be used as a phenotypic marker for the rapid presumptive identification of *R. equi*.

Indirect Fluorescent Antibody Technique (IFA)

Indirect fluorescent antibody technique (IFA) using a monoclonal antibody against the 15- to 17-kDa virulence-associated antigens (VapA) of *R. equi* and PCR targeting the structural gene of VapA detected bacteria in tracheal aspirates less sensitively than the isolation technique although they were more rapid. Therefore a combination of tracheal aspiration and bacterial isolation may be the most valuable method for routine diagnosis of *R. equi* pneumonia in foals.

Diagnosis in Human Beings

Clinical Examination

Clinical examination with detailed medical background may help in diagnosis. *R. equi* infections in human beings should be differentiated from lung abscesses due to other etiologies, metastatic cancer, mycobacterium avium-intracellulare, nocardiosis, *Pneumocystis carinii* pneumonia, pneumonia of fungal/viral origin, tuberculosis, etc.

Laboratory Tests

The following laboratory tests are used for diagnosis of *R. equi* infections:

☆ **CBC count:** This is important for evaluation of leukocytosis, anaemia, and neutropenia.

☆ **HIV screening tests:** All patients with *R. equi* infection should undergo screening for HIV because more than half of reported cases involve patients infected with HIV.

☆ **Blood and stool culture:** Cultural examination is done to isolate *R. equi* from the clinical specimens from infected sources, such as abscess, eye drainage, and cerebrospinal fluid.

Imaging Techniques

Chest radiography is done even in patients with extrapulmonary *R. equi* infections. Multiple nodular infiltrates are the usual findings of *R. equi* infection. In patients infected with HIV, *R. equi* infection has a preference for the upper lobes, but in immunocompetent patients, it has no definite predilection for any particular lobe. If untreated, nodular infiltrates are followed by cavitation. More than half of all patients have cavitation. Cavitation is more common in patients infected with HIV. Other findings of *R. equi* infection on chest radiography include interstitial pneumonia, abscesses and pleural effusion. Cavities observed with *R. equi* infection are thick-walled and may demonstrate air-fluid levels, indicating progression to abscess formation. CT scanning of the thorax is more sensitive and may show more nodules and cavitation than are observed on a plain radiograph.

Cultural Examination

Cultural examination of clinical samples may be done as described in case of animal infections.

Histologic Findings

The ability of *R. equi* to persist in and destroy macrophages is the basis of its pathogenesis. The typical pattern is a necrotizing granulomatous reaction dominated by macrophages filled with granular cytoplasm that shows positive results on periodic acid-Schiff stain and contains large numbers of coccobacilli. Pulmonary *R. equi* infections in immunocompromised hosts may have typical histopathological finding of malakoplakia, which is an unusual inflammatory disorder with accumulation of characteristic histiocytes with calcified lamellar cytoplasmic bodies (Michaelis-Gutman bodies). It gives strong indication of *R. equi* infection. Although malakoplakia in immunocompromised patients is mostly found in pulmonary infections, it also has been demonstrated in subcutaneous infections and abscesses.

Prevention and Control Strategies

Prevention in Equines

R. equi has proven itself quite resistant to various approaches of preventing the disease. Preventing disease by proper management and sanitation at farms is very important. In the areas where foals are kept, regular disposal of manure and dust control have proven effective in reducing the levels of bacteria. It is also important to ensure that loafing paddocks are well grassed, and not totally grazed, reducing them to dusty sandpits.

No suitable vaccination is available due to many complicated immunological reasons. The most likely time of infection is during the first 2 weeks of life. If foals are administered antibiotics during the first 2 weeks of life, infection may be reduced in them. Another approach is the administration of hyperimmune plasma to foals during the first few days of life and then again at 3 weeks of age. This is reported to reduce the incidence of the disease by up to one-third and it may also lower the severity of pneumonia if the foals are infected. Early detection of the disease also helps in preventing its spread.

Treatment in Foals/Equines

R. equi pneumonia is reported to have a poor to grave prognosis due to lack of a good treatment protocol. Erythromycin, clarithromycin and azithromycin are individually paired with rifampin to treat foals. While potent and effective, these drug combinations do carry the risk of serious side effects to the mare and foal. These drugs have the tendency to cause diarrhoea in older foals in hot climates or seasons; hence only confirmed *R. equi* cases should be given such medication. Treatment durations vary depending on the stage of the pneumonia, but generally run 2 to 8 weeks.

Prevention and Treatment in Human Beings

Special care and hygiene for immunocompromised humans is very essential in prevention of *R. equi* infection. Increased awareness about the infection may help in its early diagnosis and timely treatment. Treatment may require prolonged combination antibiotic therapy, sometimes in combination with surgical therapy. A study based on animal experiments demonstrated that the effective single agents against *R. equi* include vancomycin, imipenem and rifampin. In the same study, combinations of antibiotics were not more effective than vancomycin alone. However, combinations of antibiotics may limit the emergence of *in vivo* antibiotic-resistant mutants. Munoz *et al.* (2008) reported successful linezolid monotherapy for pulmonary *R. equi* infection in a heart transplant recipient. However, the initial therapy in this patient consisted of a combination of parenteral antibiotics.

Use of combination antibiotics is recommended. Some studies have reported using at least one antibiotic with intracellular penetration (*e.g.* erythromycin, rifampin). *In vitro* synergy studies demonstrated 4 combinations of antibiotics to be effective against *R. equi* infection (*i.e.* rifampin-erythromycin, rifampin-minocycline, erythromycin-minocycline, imipenem-amikacin) (Nordmann *et al.* 1992). In the same study, the combination of macrolides and aminoglycosides was found to be antagonistic. However, it has been used with success clinically (Nordmann *et al.* 1992).

In a case report by Scotton *et al.* (2000), a patient with nosocomial meningitis was initially treated with vancomycin and rifampin. However, because of continued fever, monotherapy with levofloxacin was instituted with success. Clinical success with vancomycin monotherapy followed by oral sulfamethoxazole-trimethoprim has been documented in an immunocompetent patient with brain abscess, in addition to neurosurgical treatment. However, monotherapy should generally not be used to treat systemic *R. equi* infections. In general, pulmonary infections should be treated for a minimum of 2 months. Treatment should initially consist of parenteral antibiotics, followed by oral combination therapy.

In immunocompetent patients, a shorter duration of therapy is recommended. Selective local *R. equi* infections with no systemic involvement can be treated with shorter courses. Topical antibiotics have also been used in *R. equi* endophthalmitis, in combination with systemic antibiotics and surgical therapy.

The prognosis of *R. equi* infection depends on the underlying immunosuppressive conditions and other concurrent infections. Early diagnosis and treatment may prevent chronicity and relapses. Prognosis is favourable in most local *R. equi* infections and among immunocompetent children.

Future Directions

R. equi is very versatile organism and goes across species. Since it resides intracellularly, treatment with conventional antibiotics is not possible. There seems to be a need for early and accurate diagnostic tests so that both foals and human patients may be saved. Though molecular diagnostic tools are available, these are required at grassroot level for human patients and in field for foals and other animals.

Development of suitable vaccines, especially for foals, is also very essential, but the age at which the disease occurs coupled with its complex immunological nature makes the proposition very challenging. In the first 4 months of life, foals must be protected from potentially serious damage to the lung caused by inhalation of the *R. equi* bacteria. Why young foals are so prone to this organism, compared to other offspring (such as piglets or calves) is not known, but they grow out of this susceptibility by 5-6 months of age. Probably, most foals develop antibodies to the organism, without the disease being apparent. The development of natural immunity appears to be slow and is inadequate in the young foals in the face of heavy challenge by inhalation. The scientists have a challenge to find ways to promote antibody production by mares or foals to protect them during what is apparently a time of critical susceptibility to the disease.

References

1. Alain A, Ocampo-Sosa, Lewis DA, Navas J, Quigly F, Collejo R, Scortti M, Leadon DP, Forgarty U, Jose A and Vazquez-Boland. 2007. Molecular epidemiology of *Rhodococcus equi* based on traA, vapA and vapB virulence plasmid markers. *J Infect Dis* 196: 763-769.

2. Bell KS, Philp JC, Aw DW and Christofi N. 1998. The genus *Rhodococcus*. *J Appl Microbiol* 85: 195-210.

3. Capdevila JA, Bujan S, Gavalda J, Ferrer A and Pahissa A. 1997. *Rhodococcus equi* pneumonia in patients infected with the human immunodeficiency virus. Report of 2 cases and review of the literature. *Scand J Infect Dis* 29: 535-541.

4. Cornish N and Washington JA. 1999. *Rhodococcus equi* infections: Clinical features and laboratory diagnosis. *Curr Clin Top Infect Dis* 19: 198-215.

5. Garg DN, Manchanda VP and Chandiramani NK. 1985. Etiology of post natal foal mortality. *Indian J Comp Micribiol Immunol Infect Dis* 6: 29-35.

6. Giguère S, Cohen ND, Keith Chaffin M, Hines SA, Hondalus MK, Prescott JF and Slovis NM. 2011a. *Rhodococcus equi*: Clinical manifestations, virulence and immunity. *J Vet Intern Med* 25: 1221-1230.

7. Giguère S, Cohen ND, Keith Chaffin M, Slovis NM, Hondalus MK, Hines SA and Prescott JF. 2011b. Diagnosis, treatment, control and prevention of infections caused by *Rhodococcus equi* in foals. *J Vet Intern Med* 25: 1209-1220.

8. Goodfellow M and Alderson G. 1977. The actinomycete - Genus *Rhodococcus*: A home for "rhodochrous complex". *J Gen Microbiol* 100: 99-122.

9. Harvey RL and Sunstrum JC. 1991. *Rhodococcus equi* infection in patients with and without human immunodeficiency virus infection. *Rev Infect Dis* 13: 139-145.

10. Khurana SK, Malik P, Virmani N and Singh BR. 2009. Prevalence of *Rhodococcus equi* infection in foals. *Indian J Vet Res* 18: 20-22.

11. Makrai L, Takai S, Tamura M, Tsukamoto A, Sekimoto R, Sasaki Y, Kakuda T, Tsubaki S, Varga J, Fodor L, Solymosi N and Major A. 2002. Characterization of virulence plasmid types in *Rhodococcus equi* isolates from foals, pigs, humans and soil in Hungary. *Vet Microbiol* 88: 377-384.

12. Munoz P, Palomo J, Guinea J, Yanez J, Gianella M and Bouza E. 2008. Relapsing *Rhodococcus equi* infection in a heart transplant recipient successfully treated with long-term linezolid. *Diagn Microbiol Infect Dis* 60: 197-199.

13. Nordmann P, Kerestedjian J J, Ranco E. 1992. Therapy of *Rhodococcus equi* disseminated infections in nude mice. *Antimicrob Agents Chemother* 36: 1244-1248.

14. Scotton PG, Tonon E, Giobbia M, Gallucci M, Rigoli R and Vaglia A. 2000. *Rhodococcus equi* nosocomial meningitis cured by levofloxacin and shunt removal. *Clin Infect Dis* 30: 223-224.

15. Takai S, Tharavichitkul P, Takarn P, Khantawa B, Tamura M, Tsukamoto A, Takayama S, Yamatoda N, Kimura A, Sasaki Y, Kakuda T, Tsukaki S, Maneekarn N, Sirisanthana T and Kirikae T. 2003. Molecular epidemiology of *Rhodococcus equi* of intermediate virulence isolated from patients with and without acquired immune deficiency syndrome in Chiang Mai, Thailand. *J Infect Dis* 88: 1717-1723.

16. Verville TD, Huycke MM, Greenfield RA, Fine DP, Kuhls Tl and Slater LN. 1994. *Rhodococcus equi* infections of humans. 12 cases and a review of the literature. *Medicine (Baltimore)* 73(3): 119-132.

17. Vladusic I, Krajinovic V and Begovac J. 2006. Long term survival after *Rhodococcus equi* pneumonia in a patient with human immunodeficiency virus infection in the era of highly active antiretroviral therapy: case report and review. *Acta Med Croatica* 60: 259-263.

18. Weinstock DM and Brown AE. 2002. *Rhodococcus equi*: An emerging pathogen. *Clin Infect Dis* 34: 1379-1385.

2014, Zoonoses: Bacterial Diseases
Pages 402–416
Editor: Sudhi Ranjan Garg
Published by: DAYA PUBLISHING HOUSE, NEW DELHI

23

Zoonotic Mycoplasmoses

D.N. Garg
Department of Veterinary Public Health and Epidemiology,
Lala Lajpat Rai University of Veterinary and Animal Sciences,
Hisar– 125 004

Mycoplasmas have been accorded an elite status as the 'next generation' prokaryotic pathogens. Being the smallest bacteria, devoid of cell-wall peptidoglycans, these are unique amongst the microbes known to mankind. Sporadic infections of immunocompromised persons with mycoplasmas that originated from domestic animals have been reported and the susceptibility of the human population has increased. The reports of the instances of human infections with animal mycoplasmas and *vice versa* are leading to the recognition of zoonotic potential of mycoplasmas. However, slow reporting of zoonotic mycoplasmoses may be attributed mainly to the difficulty in measuring zoonotic potential of mycoplasma organisms, particularly in some complex diseases such as multiple sclerosis, rheumatoid arthritis, leukemia, atherosclerosis, chronic fatigue syndrome, CJD and Cohn's colitis etc.

Zoonotic mycoplasmoses may be best defined as sharing of mycoplasmal infections, which are naturally transmitted between vertebrate animals and humans. This kind of sharing occurs under natural conditions permitting bidirectional movement of mycoplasmas and is not based on the experimental evidence of occurrence. Animals acting as reservoir for human mycoplasmosis lead to anthropozoonotic mycoplasmosis. Similarly, human reservoirs may transmit infection to animals, leading to zooanthroponotic mycoplasmosis. Amphizoonotic mycoplasmosis can be transmitted in either direction between man and animals.

Zoonotic Mycoplasmosis

The concept of zoonotic mycoplasmosis is new and only some mycoplasmas and acholeplasmas belonging to *Mycoplasmatales* and *Achopleplasmatales* show the existence of zoonotic mycoplasmoses (Garg 2009). So far, eleven species of *Mycoplasma* under family *Mycoplasmataceae* and two species of *Acholeplasma* under family *Acholeplasmataceae* can be claimed as zoonotic. No other mollicute including ureaplasma, aneroplasma and asteroplasma has yet been recorded as zoonotic. Thus far recorded zoonotic mycoplasmas/acholeplasmas classified on the basis of reservoir hosts and the direction of zoonotic transmission are listed below.

☆ **Zooanthropozoonotic organisms:** *Mycoplasma pneumoniae*, *M. salivarium*

☆ **Anthropozoonotic organisms:** *M. arthritidis*, *M. arginini*, *M. bovis*, *M. canis*, *M. edwardii*, *M. felis*, *M. lipofaciens*, *M. phocacerebrale*, *Mycoplasma* sp. (Strain M 7806)

☆ **Amphizoonotic organisms:** *Acholeplasma laidlawii*, *A. oculi*

Zooanthropozoonotic Mycoplasmoses

In humans, mycoplasmas, acholeplasmas and ureaplasmas are mucosally associated, residing predominantly in the respiratory and urogenital tract. Till now, 14 species in the genus *Mycoplasma*, two in the genus *Acholeplasma* and one in the genus *Ureaplasma* have been isolated from humans. Out of these, only two *Mycoplasma* species, *viz.* M. *pneumoniae* and M *salivarium*, have been recorded as zooanthropozoonotic.

Mycoplasma pneumoniae

M. *pneumoniae* was first identified and described in the early 1960s. The primary site of colonization of the organism is oropharynx and it is pathogenic to humans. The pneumonia is designated 'primary atypical' to distinguish it from 'typical' pneumonia due to *Streptococcus pneumoniae*, the pneumococcus. A detailed definition of epidemiology of M. *pneumoniae* disease has emerged (Lind and Bentzon 1988). Attack rate is low and disease spreads slowly in communities with close contact among their members. Transmission by droplets requires a rather high inoculum unless the host is immunocompromised. M. *pneumoniae* organisms are usually confined to humans, which is the only natural host of these microorganisms.

A zoonotic outbreak of M. *pneumoniae* infection in a class of a Budapest secondary school has been reported (Mikola *et al.* 1997). Thirty students became feverish with upper respiratory tract catarrh. In two patients hospitalized with pneumonia, a mycoplasma infection was confirmed by cold agglutination test. Their classroom contained a terrarium with hamsters, from which M. *pneumoniae* was isolated. However, the zoonotic transmission was not conclusively proven in this report.

Mycoplasma salivarium

M. *salivarium* exists as commensal in the oropharynx of human and non-human primates. Pathogenicity of M. *salivarium* is not known. However, isolation of the organism with a significantly higher incidence from the gingival sulci of individuals with periodontal disease (87 per cent) than in persons with healthy periodentium

(32 per cent) has stimulated interest in its possible role in periodental pathology (Engel and Kenny 1970, Forest 1979). Also, it is the predominant mycoplasmal species in dental plaques. Likewise, this organism has been identified as a causative agent of septic arthritis in hypogammaglobulinaemia (So *et al.* 1983).

M. salivarium, usually confined to humans or other primates, was repeatedly isolated from nasal and pharyngeal secretions of 14 of 284 swine (Erickson *et al.* 1988) suggesting that the organism can establish persistently in swine without evidence of overt disease. It has also been isolated from tonsils of three horses (Poland and Lemcke 1978). *M. salivarium* is neither a true equine mycoplasma nor a porcine mycoplasma; thus it might have been acquired by them from their environmental companion. No further report came on this account. These reports were also preliminary and further investigations are required to draw a conclusion on the zooanthroponotic nature of the microorganisms.

Anthropozoonotic Mycoplasmoses

Numerous mycoplasmas are known to exist as animal pathogen (Ross 1993), opportunist and as non-pathogen. However, only a few, as described briefly below, have taken a zoonotic route.

Mycoplasma arginini

M. arginini is a mammalian parasitic microorganism with an apparently wide host range. It is found with a high frequency in the respiratory tract and less frequently in the conjunctivae and genital tract of sheep, goats and cattle. Occasionally, isolations have also been reported from pigs, horses, dogs, domestic and wild cats. It is also frequently isolated from bovine serum used in tissue culture.

A septicaemic infection caused by *M. arginini* in a patient with advanced Hodgkin's disease and marked immunodeficiency emphasizes the potentially heightened risk to immunocompromised hosts with mollicutes of both human and non-human organisms (Yechouron *et al.* 1992). This organism was repeatedly isolated from the blood and bronchial washings of a patient during the course of the disease. *M. arginini*, commonly found in a variety of species of large animals, was thought to have been acquired by the patient during his work as a slaughterhouse worker. Prayson *et al.* (2008) have reported isolation of *M. arginini* from an unusual case of deep infection of open femur fracture in a 56-year hunter. The hunter was attacked by an African lion causing crushing bite leading to open femur fracture and after some time bone culture was found positive for *M. arginini*. The researchers put the theory that the lion had fed on a goat prior to the attack on hunter resulting in deep tissue inoculation of *M. arginini*. Quirk *et al.* (2001) detected mycoplasma DNA in 6 (13.0 per cent) of the 46 tumor-DNA samples. Nucleotide sequence similarity searches of nested PCR products revealed that 1 *M. salivarium* and 5 *M. arginini* DNA sequences were amplified from the ovarian tissues.

Evidence that *M. arginini* is pathogenic for animals and humans is yet to be presented. However, the circumstantial evidence in the cases reported so far suggests the zoonotic transfer and pathogenic role of the microorganism (Yechouron *et al.* 1992, Quirk *et al.* 2001, Prayson *et al.* 2008).

Mycoplasma arthritidis

M. arthritidis is a common pathogen of laboratory and wild rats causing purulent polyarthritis; sometimes occurring as localized outbreaks in stocks of laboratory rats and also isolated repeatedly from apparently healthy rats. Between 1948 and 1965, several mycoplasma strains were isolated from human urogenital tract and synovial tissues in laboratories at widely diverse geographical locations. These organisms did not relate to other known human mycoplasmas but were designated as *M. hominis* type 2. Later these strains were reclassified as *M. arthritidis* (Lemcke 1965). The justification of this reclassification was subsequently confirmed by nucleic acid homology, by enzyme analysis and by examination of electrophoretic patterns of the cell proteins of these strains. No subsequent isolates of the organism from human tissues have been reported and no adequate role in human disease has been documented.

M. arthritidis is the causative agent of chronic proliferative arthritis of rodents and possess superantigen called MAM (*M. arthritidis* mitogen). It is the only mycoplasma species which has been demonstrated to induce differentiation of mouse lymphocytes into cytotoxic/suppressor effector cells. Arthritogenic potential of some mycoplasmas led to speculations concerning their role in human rheumatoid arthritis. Besides, the earlier isolations of *M. arthrtidis* from synovial tissues, numerous other investigations over the past 4 decades have either failed to isolate mycoplasmas or have resulted in inconsistent findings.

Mycoplasma bovis

M. bovis is a potential bovine pathogen causing mastitis, pneumonia, arthritis and genital disorders (Pfutzner and Sachse 1996). Once established in a herd, the organism persists and may become endemic producing mastitis in dairy herd.

Madoff *et al.* (1979) have reported the isolation of *M. bovis* from sputum of a woman with lobar pneumonia, psychosis, and probable myocarditis, nephritis, and hemolytic anaemia. Except for an exposure to cowdung 3 weeks before her illness, there was no history suggesting that the patient had any contact with cattle. Thus, the exact source from which the patient acquired this organism will remain an enigma. This has been the only report of isolation of *M. bovis* from human so far and no further isolation from human has been reported.

The significance of frequent isolations of *M. bovis* from respiratory tract of cattle with respiratory disease is yet to be resolved though the incidence of pneumonia due to *M. bovis* can be as high as 100 per cent (Pfutzner 1990). *M. bovis* infection may spread by aerosol, which may be a logic way of its transmission to human in the case reported by Madoff *et al.* (1979). The possibility of transfer of *M. bovis* via consumption of infected bovine milk is another possibility as this organism appears more invasive and may cause septicaemia.

Mycoplasma canis

M. canis is found as a common inhabitant of the mucous membrane of upper respiratory tract, conjunctivae and genitals of dogs. It has also been isolated occasionally from non-human primates. Several strains of *M. canis* were isolated from the oropharynx of members of a family at the time when their dog had an acute respiratory disease infection (Armstrong *et al.* 1971). Although one member of the family was taking immunosuppressant, the pathogenic role of *M. canis* in human disease was neither established nor the dynamics of zoonotic transfer of *M. canis* studied. Pathogenicity of *M. canis* is not known under natural and experimental conditions. Experimentally, a strain isolated from the pericardium of a dog did not produce any lesion in dogs and small rodents. Thus, the zoonotic transfer of *M. canis* may be regarded as yet of no public health significance.

Mycoplasma edwardii

M. edwardii is frequently found in the upper respiratory and urogenital tracts of male and female dogs. It has also been isolated from shrews. *M. edwardii* has recently been reported from septicaemic infection in a patient with advanced acquired immunodeficiency syndrome (AIDS). This has been reported by Baseman and Tully (1997) as a personal communication to them by M. K. York. The pathogenicity of the microorganism is not known although occasionally it has been isolated from pneumonic lungs of dogs, and recently has been reported from an advanced case of AIDS.

Mycoplasma felis

M. felis is a common inhabitant of the upper respiratory and lower genital tract of asymptomatic and diseased cats. Strains related to *M. felis* have also been isolated frequently from tonsils and other regions of respiratory tract of healthy and diseased horses. *M. felis* is associated with conjunctivitis in cats, being isolated significantly more frequently from the conjunctivae of diseased cats than in the convalescent or healthy cats.

Bonilla *et al.* (1997) described first documented case of *M. felis* infection in a woman who had common variable immune deficiency and presented with septic arthritis of the left hip and right knee. *M. felis* was isolated from both the joints. She had been exposed to cat before the diagnosis of *M. felis* septic arthritis was made. Evidences so far that *M. felis* is pathogenic are inconclusive though conjunctivitis can be experimentally induced in cats. Prevalence of *M. felis* in oligoathritis patients exposed to cats, the possible role of *M. felis* as arthritogenic agent and the routes as well as risk of transmission are under investigation.

Mycoplasma lipofaciens

M. lipofaciens was first isolated in 1983 from the infraorbital sinus of an adult chicken. Its pathogenicity is not known as yet but experimental infection with the type strain has caused some chicken and turkey embryo mortality and can be transmitted by air. A human infection with an avian *M. lipofaciens* has been reported (Lierz *et al.* 2008). A clinical trial to investigate the capability of the organism to spread horizontally between infected and non-infected turkey poults in an incubator demonstrated airborne transmission of the pathogen within 24 hours. During this trial, the veterinarian conducting the study, a 36 years old man, was monitored for infection. Each day, 2 swabs were taken from both nostrils, starting from the day before the infected poults hatched (day 0) through day 7 after the poult hatching date.

When handling eggs and poults, the veterinarian wore gloves but not a protective mask. Two days after the poults hatched, the veterinarian reported throat pain and a slight rhinitis indicating a respiratory disease. The next day only the rhinitis with minor nasal pain was present. The veterinarian handling the infected poults was free of nasal *Mycoplasma* organisms a day before contact. His infection occurred concurrent with demonstration of airborne transmission among poults. It suggested that *M. lipofaciens* can be transmitted successfully to humans and may cause clinical symptoms. The study documents non-artificial human infection with an avian *M. lipofaciens*.

Mycoplasma phocacerebrale

M. phocacerebrale was first reported in 1991 from seals died because of an acute disease characterized by pneumonia, skin lesions, diarrhoea, polyarthritis, nervous signs, and abortions in pregnant females during 1988 mass mortality (more than 18,000) that spread among the harbour seals (*Phoca vitulina*) of the Baltic Sea and the North Sea. The fact that they were isolated in large numbers from several pathologically altered internal organs (lung, heart, brain) of the diseased seals suggests that even if they are not the primary cause, they might have been involved in the production of these pathological changes and in the general disease, leading to the death of the seals. This probability is enhanced by other observations which showed that the seal mycoplasmas displayed strong cytotoxic capacities, as observed in investigations with tracheal organ cultures in which the organisms caused inhibition of the ciliary activity and extensive damage to the multilayered epithelium.

Seal finger monoarticular infection of the fingers of seal hunters and those who handle pinnipeds, resulting from bites or direct contact with seal bones or untreated seal pelts has been described in coastal Scandinavia and Canada. Its aetiology is yet unknown, but most probably it might be due to *M. phocacerebrale*, which was isolated from an epidemic of seal finger in the Baltics. The affected digits are usually amputated once they become unusable; seal finger is also treatable with tetracycline.

Two strains of *M. phocacerebrale* have been isolated from finger of an aquarium female trainer who was bitten by a seal and from the front teeth and pharynx of the biting seal (Baker *et al*. 1998). The trainer sustained a painful seal bite on her right forefinger; six days following the bite, the patient's finger was swollen, painful, erythematous, and there was a serous discharge. This was the first reported case in which mycoplasma has been associated with seal finger. The fact that the *Mycoplasma* strains isolated from the patient and the seal belong to the same species suggests a possible role of mycoplasma in seal finger.

Mycoplasma sp. (srain M 7806)

McCabe *et al*. (1987) described a mycoplasmal infection in the hands of a veterinarian acquired through a cat bite. The infection resulted in severe soft tissue cellulitis with tissue destruction sufficient to require a tendon graft. The organism (strain M 7806) was identified as an unclassified, glucose fermenting *Mycoplasma* species and resolved after vibramycin was administered. Although other serologically related strains were later isolated from the oropharynx of feline, the organisms remained unclassified. This unclassified strain of *Mycoplasma* has never been isolated subsequently either from human or animals. Therefore, no comment can be offered about its pathogenic, epidemiologic and zoonotic potential.

Amphizoonotic Acholeplasmoses

Currently, 13 species of *Acholeplasma* exist in nature, which are apparent parasites of vertebrate hosts. Only two species, *A. laidlawii* and *A. oculi*, are commonly isolated from the diseased sites of various animals and humans.

Acholeplasma laidlawii

A. laidlawii was originally isolated from sewage, manure, humus and soil, but its designation as true saprophyte was challenged later by its occurrence in mammals and birds. It is frequently recovered from oral cavity, respiratory and genital tract secretions, eye, lymphnodes, semen and serum. Its pathogenicity has not been well established, although some strains are lethal for chicken embryos. A limited number of *A. laidlawii* strains have been isolated from oral cavity (Razin *et al.* 1964) and from flora of human burns (Markham and Markham 1969).

As *A. laidlawii* is abundantly isolated from a wide variety of syndrome in humans and animals, it looks like a perfect example of amphizoonosis; being freely transmitted between humans and animals. A fresh look to *A. laidlawii* is required for its possible potential role when it is found associated with diseased hosts.

Acholeplasma oculi

A. oculi has been isolated from conjunctivae of goats with keratoconjuntivitis, nasal secretions of porcines, nasopharynx, lungs, spinal fluid, joints, and semen of equines and external genitals of guinea pigs. Experimentally, it has been shown to produce conjunctivitis, pneumonia and death in goats. A single isolate of *A. oculi* was cultured from amniotic fluid at 19 weeks of gestation (Waites *et al.* 1987). Its identification by direct immunofluorescence appeared to preclude contamination. The remainder of pregnancy in question was unremarkable and a full term infant was delivered without any complication. Pathogenicity of *A. oculi* cannot be ignored. However, in absence of any record of proven virulence, it is not wise to draw any conclusion as to its zoonotic potential.

Zoonotic Hemoplasmoses

Uncultured wall-less pathogenic prokaryotes infecting erythrocytes are now classified in Mollicutes and are called hemoplasmas. Hemoplasmas are small epicellular parasites that adhere to the host's erythrocytes. Initially, hemoplasmas were classified in the order *Rickettsiales* based on morphology and response to antibiotic therapy but based on the 16S rRNA, the genera *Haemobartonella* and *Eperythrozoon* are found closely related to pneumonic group of mycoplasmas. Haemoplasmas are now classified within the genus *Mycoplasma*, based on 16S rRNA gene and RNase P RNA gene phylogeny. They are divided, based on phylogeny (rather than pathogenicity or host specificity), into two groups: a haemominutum group and a haemofelis group. Haemoplasmas have been identified in many domestic and wild animal species and can cause haemolytic anaemia. Infected humans may be asymptomatic or have various clinical signs, including acute fever, anaemia and severe haemolytic jaundice, especially in infected neonates. Pregnant women and newborns are more vulnerable to the disease and show more severe clinical signs after infection.

Hemotrophic mycoplasma infection is a neglected zoonotic disease, though it poses a threat to public health and the animal industry, especially in China (Hu *et al.* 2009). So far, six hemotrophic *Mycoplasma* species have been identified in rodents and mammals (Yang *et al.* 2000). Hemoplasmas might also act as a cofactor in HIV infection, contributing to acceleration of the course of the disease. A case of a *Mycoplasma haemofelis*-like infection in an HIV-positive patient co-infected with *Bartonella henselae* in Brazil (Santos *et al.* 2008) has been reported.

Hemoplasma DNA is present in saliva and faeces of cats and *M. haemofelis*-DNA has also been detected in cat fleas (*Ctenocephalides felis*), which suggests that aggressive interactions among cats involving biting may lead to transmission of the organism. Thus, cat bites, cat scratches and flea-infested cats are considered to be involved in transmission of *M. haemofelis* and/or *B. henselae* infection to AIDS infected humans. Further studies are needed to establish the role and prevalence of hemoplasma infection in AIDS patients, as well as the zoonotic potential of *M. haemofelis*.

A novel hemotropic hemoplasma species in a human patient with chronic moderate neutropenia, acute hemolysis, and pyrexia has also been detected most recently (Steer *et al.* 2011), which represents the first report of hemolysis in association with confirmed hemoplasma infection in a human. The origin of the hemoplasma infection in this patient is not known, but it quite possibly represents a zoonotic infection, probably acquired from blood-sucking arthropods such as ticks, fleas, and lice or by direct transmission from another mammalian host through parenteral or oral inoculation of blood. However, no such transmission events were recalled by the patient before the onset of disease. It is also possible that this novel hemoplasma represents a species of which the primary host is humans and the name *Candidatus Mycoplasma haemohominis* is proposed for it.

Current Status of Zoonotic Mycoplasmoses in India

There is no recorded evidence of any episode of zoonotic mycoplasmoses, whether zooanthrropozoonotic or anthropozoonotic or amphizoonotic, in India till date. Likewise, no confirmed case of zoonotic hemoplasmosis has been reported from India.

Detection of Zoonotic Mycoplasmas

Details of the culture media, cultural conditions and other requirements for isolation and identification of mollicutes are available in the literature (Razin and Tully 1995, Tully and Razin 1995). The current biotechnological approach to laboratory diagnosis of mollicutes infection has also been amply described (Garg 1993).

Differential characteristics of zoonotic species of genus *Mycoplasma* and *Acholeplasma* so far recorded are given the Table 23.1. Final identification, however,

Table 23.1: Differential characteristics of zoonotic species of *Mycoplasma* and *Acholeplasma*

Organism	Gl	Ma	Arg	Phos	Film, Spot	Tet A/An	Gel	SD	CD	H	G+C Mol per cent
Mycoplasma											
M. arginini	−	−	+	−	−	−/+		−	−	−	27.6-28.6
M. arthritidis	−	−	+	−	−	−/−	+	−	−	−	30.0-32.6
M. bovis	−	−	−	+	d	+/+		−	−	X	27.8-32.9
M. canis	+	−	−	−	−	−/+	d	−	−	+	28.4-29.1
M. edwardii	+	−	−	−	+	−/+	−	−	−	−	29.2
M. felis	+	−	−	+	+	−/+	−	−	−	−	25.2
M. lipofaciens	+	−	+	−	+	−/+		−	−	−	24.5
M. phocacerebrale	−	−	+	+	+	−/−		−	−	−	25.9
M. pneumoniae	+	+	−	−	−	+/+	−	−	−	+	38.6-40.8
M. salivarium	−	−	+	−	+	−/w	−	−	−	−	27.3-31.4
Mycoplasma sp. (Strain M 7806)	+										
Acholeplasma											
A. laidlawii	+	−	−	−	−	w/+	d	−	−	−	31-36
A. oculi	+	−	−	−	−	+/+					26.0

+: 90 per cent or more strains are positive; −: 90 per cent or more strains are negative; d: 11 to 89 per cent strains are positive; x: Not definitely setteled; Gl: Glucose catabolization; Ma: Mannose; Arg: Arginine deimination; Phos: Phosphatase; Tet: Tetrazolium reduction; A/An: Aerobic/anerobic; Gel: Gelatine hydrolysis; SD: Coagulated serum digestion; CD: Casein digestion; H: Haemadsorption, the test performed with RBC from species from which mycoplasma originated plus guinea pig (sometimes with a variety of RBC).

depends on determination of serological relatedness, for which methods such as growth inhibition, direct or indirect immunofluroscence and metabolic inhibition tests are available. Rapid detection of the organisms and quick diagnosis is essential to prevent severe consequences in affected humans and animals. Laboratory culture is well adopted to most, though not all, species of zoonotic mycoplasmas. Many techniques for direct (DNA probes, rDNA) and amplified detection of mycoplasmas (PCR) are currently in use (Razin *et al.* 1987, Johansson 1993, Razin 1994).

Rapid detection of *M. pneumoniae* antigen (total antigen, P1 adhesin or glycolipids) includes direct immunofluorescence, counter-immunoelectrophoresis, immunoblotting and antigen capture enzyme immunoassay (EIA). Detection of *M. pneumoniae* is also possible by probing or specific nucleotide sequence by primers of P1 or cytoadhesin gene with product identified by dot-blot-hybridization (DBH) following PCR amplification (P1-PCR-DBH-Assay).

For other zoonotic mycoplasmas, rapid detection systems include: a PCR diagnosis for *M. arthritidis* (Van Kupperveld *et al.* 1992); an antigen capture ELISA using monoclonal antibodies (Heller *et al.* 1993), a DNA probe (Hotzel *et al.* 1993) and PCR system based on amplification of 16S rRNA gene for *M. bovis* (Chavez-Gonalez *et al.* 1995); an immunoenzyme assay using MOAb and streptavidin-biotin detection for *M. arginini*, *M. salivarium* and *A. laidlawii* (BRL, Maryland, USA); and DNA probes to detect either mycoplasmal DNA or mycoplasmal rRNA and 16S rRNA based PCR for *M. arginini*, *M. salivarium* and *A.laidlawii* (Gobel and Stanbridge 1984, Razin *et al.* 1984, Mattson and Johansson 1993 Teysson *et al.* 1993). An oligonucleotide array has also been recently developed to detect and genotype mollicutes based on internal transcribed spacer (ITS) (Jang *et al.* 2009). This method proposed oligonucleotide array effective for 23 mycoplasmas including many zoonotic mycoplasmas in a single hybridization. While 16S rDNA identity of mycoplasma ranges from 62.7 to 98.8 per cent, the ITS sequence identity of mycoplasma ranges from 32 to 99 per cent. Thus, it is more rapid and accurate method for species specific identification using sequence-based molecular technique. However, such fast detection methods are neither available for *M. canis* nor for *A. oculi* providing scope for research on this account.

Detection of the Hemoplasma infection has historically relied on the evaluation of Romanowsky stained blood smears. Hemoplasmas are mostly found attached to the erythrocytes; some organisms may also be found in the plasma because of detachment from the cells. Due to cyclic parasitemia and chronic infection, this technique is neither sensitive nor specific for infection diagnosis. Consequently, molecular diagnostic techniques, such as standard conventional and real-time PCR, have been extensively and successfully used to identify acute and chronically infected humans and animals (Messick 2004, Sykes *et al.* 2007, Yu *et al.* 2007, Hoelzle 2008, Kamrani *et al.* 2008, Santos *et al.* 2008, Steer *et al.* 2011, Boes *et al.* 2012). The quantitative real-time (q)PCR assays have also been developed to detect all known haemoplasma species, and a human housekeeping gene in order to demonstrate both successful DNA extraction from clinical samples and to test for sample inhibition, and to apply these qPCRs to human blood samples and blood smears (Tasker *et al.* 2010). The high sensitivity and specificity of generic haemoplasma qPCR assays make them ideal for screening large numbers of human samples and extremely useful for detection of previously recognized haemoplasma species in samples as well as for identification of novel haemoplasma species.

Future Directions

Emerging zoonotic mycoplasmosis is a direct challenge to veterinary and medical profession. So far, there has been no episode of zoonotic mycoplasmosis in the form of an outbreak; only a few sporadic incidences have been reported, all being occupational zoonosis. The involvement of a veterinarian in the life cycle of mycoplasma is incidental or accidental as dead end host. Though the available reports on zoonotic mycoplasmoses are limited but it appears that it is just waiting to happen at the verge of emergence. The recognition of zoonotic mycoplasmoses faces several hurdles. These include: (i) absence of specific or distinguishable clinical sign in the affected individual(s) or difficulty in diagnosing healthy or convalescent carriers capable of transmitting zoonotic mycoplasmas, (ii) lack of training to recognize the pathogenic mycoplasmas, and (iii) recognition of uncultivable hemoplasmas. Our human or veterinary clinical or diagnostic laboratories are not yet ready to diagnose mycoplasmal infections, either in humans or in animals. Laboratory training of medical and veterinary diagnosticians in mycoplasma work may flip the reporting of zoonotic occurrence of mycoplasmas.

Fastidious nature of mollicutes results in their slow cultivation, identification and diagnosis. Recently developed molecular methods certainly obviate the need to cultivate and identify these organisms, but such sophisticated methods are generally restricted to the research laboratories only. Unless these methods are routinely used by diagnosticians, slow reporting of zoonotic mycoplasmoses may continue. Development of diagnostic kits and their commercial availability is essential to assess the real extent of occurrence of zoonotic mycoplasmas and also for better understanding of their epidemiological features such as transmission path, geographic locations, etc.

Garg and Singh (2009) in a recent paper have dwelled upon the questions as to whether zoonotic mycoplasmoses are emerging, and whether microbiological examination of foods of animal origin *i.e.* meat, milk and their products, should include their screening for mycoplasmas. It is assumed that mycoplasma, a Gram-negative organism, occurring in large numbers in animal products, could lead to some sporadic cases or outbreaks of foodborne gastroenteritis, which remain undiagnosed etiologically. Mycoplasmas do not grow on ordinary plating media; rather special culture techniques are required. Further, limulus amoebocyte lysate (LAL) is not used routinely for screening for endotoxin. It will, therefore, be a wise step to introduce such tests and procedures in food analysis and public health laboratories to define the possible association of mycoplasma with contaminated foods.

Till now, a few reports on zoonotic mycoplasmas provide only incomplete and inconclusive information. Generation of competent human resource capable to work with mycoplasma-biotechnology is a prerequisite to increase the capability of diagnostic laboratories for diagnosis of mycoplasmoses. The diagnostic and research work should always be directed to gather information on transmission vehicles and routes, ecological parameters as well as on spatial and temporal distribution of mycoplamsmas and mycoplasmoses to explore their hidden aspects.

References

1. Armstrong D, Yu BH, Yagoda A and Kagnoff MF. 1971. Colonization of humans by *Mycoplasma canis*. *J Infect Dis* 124: 607-609.

2. Baker AS, Ruoff KL and Madoff S. 1998. Isolation of *Mycoplasma* species from a patient with seal finger. *Clin Infect Dis* 27: 1168-1170.

3. Baseman JB and Tully JG. 1997. Mycoplasmas: Sophisticated, reemerging and burdened by their notoriety. *Emerg Infect Dis* 3: 21-32.

4. Boes KM, Goncarovs KO, Thompson CA, Halik LA, Santos AP, Guimaraes AMS, Feutz MM, Holman PJ, Vemulapalli R and Messick JB. 2012. Identification of a *Mycoplasma ovis*-like organism in a herd of farmed white-tailed deer (*Odocoileus virginianus*) in rural Indiana. *Vet Clin Path* 41: 77-83.

5. Bonilla HF, Chenoweth CE, Tully JG, Blythe LK, Robertson JA and Kauffman CA. 1997. *Mycoplasma felis* septic arthritis in a patient with hypogamma-globulinemia. *Clin Infect Dis* 24: 222-225.

6. Chavez-Gonzalez YR, Bascunana CR, Bolske G, Mattsson JG, Molina CF and Johansson K-E. 1995. *In vitro* amplification of the 16S rRNA genes from *Mycoplasma bovis* and *Mycoplasma agalactiae* by PCR. *Vet Microbiol* 47: 183-190.

7. Engel LD and Kenny GE. 1970. *Mycoplasma salivarium* in human gingival sulci. *J Periodontal Res* 5: 1-9.

8. Erickson BZ, Ross RF and Bove JM. 1988. Isolation of *Mycoplasma salivarium* from swine. *Vet Micrbiol* 16: 385-390.

9. Forest N. 1979. Characterisation de *Mycoplasma salivarium* dans les parodontopathies. *J Biol Buccale* 7: 321-330.

10. Garg DN. 1993. Current biotechnological approaches to laboratory diagnosis of Mollicutes infection. In: Srivastava RN, Prasad G, Kalra SK and Gupta Y (eds.) *Animal health biotechnology*. Arvin Printing Press, New Delhi. p.164-174.

11. Garg DN. 2009. Mycoplasmas of zoonotic significance. Indian veterinary community. http//www.indianveterinarycommunity.com/images/IVC_Garg_mycoplasmosis.pdf. Accessed 15 March 2012.

12. Garg DN and Singh Y. 2009. Zoonotic mycoplasmoses: Waiting to happen. In: Singh SP, Funk J, Tripathi SC and Joshi N (eds.) *Food safety quality assurance and global trade: concerns and strategies*. International Book Distributing Co., Lucknow, India. p. 167-172.

13. Gobel UB and Stanbridge EJ. 1984. Cloned mycoplasma ribosomal RNA genes for detection of mycoplasma contamination in tissue cultures. *Science* 226: 1211-1213.

14. Heller M, Berthold E, Pfutzner H, Leirer R and Sachse K. 1993. Antigen capture ELISA using a monoclonal antibody for detection of *Mycoplasma bovis* in milk. *Vet Microbiol* 37: 127-133.

15. Hoelzle LE. 2008. Haemotrophic mycoplasmas: recent advances in *Mycoplasma suis*. *Vet Microbiol* 130: 215-226.

16. Hotzel H, Dermuth B, Sachse K, Pflitsch A and Pfutzner H. 1993. Detection of *Mycoplasma bovis* using *in-vitro* deoxyribonucleic acid amplification. *Rev Sci Tech OIE* 12: 581- 591.

17. Hu Z, Yin J, Shen K, Kang W and Chen Q. 2009. Outbreaks of hemotrophic *Mycoplasma* infections in China. *Emerg Infect Dis* 15: 1139-1140.

18. Jang H, Kim H, Kang B, Kim C and Park H. 2009. Oligonucleotide array-based detection and genotyping of mollicutes (*Acholeplasma*, *Mycoplasma* and *Ureaplasma*). *J Micrbiol Biotechnol* 19: 265-270.

19. Johansson K-E. 1993. Detection and identification of mycoplasmas with diagnostic DNA probes complimentary to ribosomal RNA. In: Khane I and Adoni A (eds.) *Rapid diagnosis of mycoplasmas*. Plenum, New York. p. 139-154.

20. Kamrani A, Parreira VR, Greenwood J and Prescott JF. 2008. The prevalence of *Bartonella*, hemoplasma and *Rickettsia felis* infections in domestic cats and in cat fleas in Ontario. *Can J Vet Res* 72: 411-419.

21. Lemcke RM. 1965. A serological comparison of various species of *Mycoplasma* by an agar gel double diffusion technique. *J Gen Micrbiol* 38: 91-100.

22. Lierz M, Jansen A and Hafez HM. 2008. Avian *Mycoplasma lipofaciens* transmission to veterinarian. *Emerg Infect Dis* 14: 1161-1163.

23. Lind K and Bentzon MW. 1988. Changes in the epidemiological pattern of *Mycoplasma pneumoniae* infections in Denmark: a 30 years survey. *Epidemiol Infect* 101: 377-386.

24. Madoff S, Pixley BQ, Del Guidice RA and Modellering Jr. 1979. Isolation of *Mycoplasma bovis* from a patient with systemic illness. *J Clin Micrbiol* 9: 709-711.

25. Markham JG and Markham NP. 1969. *Mycoplasma laidlawii* in human burns. *J Bacteriol* 98: 827-828.

26. Mattson JG and Johnasson K-E. 1993. Oligonucleotide probes complimentary to 16S rRNA for rapid detection of mycoplasma contamination in cell cultures. *FEMS Microbiol Lett* 107: 139-144.

27. McCabe SJ, Murray JF, Ruhnke HL and Rachlis A. 1987. *Mycoplasma* infection of the hand acquired from a cat. *J Hand Surg* 12A: 1085-1088.

28. Messick JB. 2004. Hemotrophic mycoplasmas (hemoplasmas): a review and new insights into pathogenic potential. *Vet Clin Path* 33: 2-13.

29. Mikola I, Balogh G, Nagg A, Matyas M, Rady M, Glavits R and Stipkovits L. 1997. Zoonotic outbreak of *Mycoplasma pneumoniae* infection. *Magyar allatorvosok Lapja* 119: 403-405.

30. Pfutzner H. 1990. Epizootiology of *Mycoplasma bovis* infection in cattle. *Zbl Bakt suppl* 20: 394-399.

31. Pfutzner H and Sachse K. 1996. *Mycoplasma bovis* as an agent of mastitis, pneumonia, arthritis and genital disorders in cattle. *Rev Sci Tech OIE* 15: 1477-1494.

32. Poland J and Lemcke R. 1978. Mycoplasmas of the respiratory tract of horses and their significance in upper respiratory tract disease. Proc. 4th Int. cong on equine infect dis Lyon. *J Equine Med Surg suppl* 1: 437-446.

33. Prayson MJ, Venkatarryappa I, Srivastava M, Northern I and Burdette SD. 2008. Deep infection with *Mycoplasma arginini* in an open femur fracture secondary to an African lion bite: A case report. *Injury Extra* 39: 243-246.

34. Quirk JT, Kupinski JM and DiCioccio RA. 2001. Detection of *Mycoplasma* ribosomal DNA sequences in ovarian tumors by nested PCR. *Gynecol Oncol* 83: 560-562.

35. Razin S. 1994. DNA probes and PCR in diagnosis of *mycoplasma* infections. *Mol Cell Probe* 8: 497-511.

36. Razin S and Tully JG. 1995. *Molecular and diagnostic procedures in mycoplasmology. Vol. I. Molecular characterization*. Acdemic Press, New York.

37. Razin S, Michmann J and Shimshoni Z. 1964. The occurrence of *mycoplasma* pleuropneumonia like organisms (PPLO) in the oral cavity of dentulous and edentulous subjects. *J Dent Res* 43: 402-405.

38. Razin S, Gross M, Wormser M, Pollack Y and Glaser G. 1984. Detection of mycoplasmas infecting cell cultures by DNA hybridization. *In Vitro* 20: 404-408.

39. Razin S, Hyman HC, Nur I and Yogev D. 1987. DNA probes for detection and identification of *mycoplasmas*. *Israel J Med Sci* 23: 735-741.

40. Ross RF. 1993. Mycoplasmas: Animal pathogens. In: Khane I and Adoni A (eds.) *Rapid diagnosis of mycoplasmas*. Plenum, New York. p. 69-109.

41. Santos AP, Santos RP, Biondo AW, Dora JM, Goldani LZ, de Oliveira ST, de Sa Guimaraes AM, Timenetsky J, de Morais HA, Gonzalez FHD and Messick JB. 2008. *Hemoplasma* infection in HIV-positive patient, Brazil. *Emerg Infect Dis* 14: 1922-1924.

42. So AKL, Furr PM, Taylor-Robinson D and Webster ADB. 1983. Arthritis caused by *Mycoplasma salivarium* in hypogammaglobulinaemia. *Brit Med J* 286: 762-763.

43. Steer JA, Tasker S, Barker EN, Jensen J, Mitchell J, Stocki T, Chalker VJ and Hamon M. 2011. A novel hemotropic mycoplasma (Hemoplasma) in a patient with hemolytic anemia and pyrexia. *Clin Infect Dis* 53(11): e147-e151. doi:10.1093/cid/cir666. Accessed 16 March 2012.

44. Sykes JE, Drazenovich NL, Ball LM and Leutenegger CM. 2007. Use of conventional and real-time polymerase chain reaction to determine the epidemiology of hemoplasma infections in anemic and nonanemic cats. *J Vet Intern Med* 21: 685-693.

45. Tasker S, Peters IR, Mumford AD, Day MJ, Gruffydd-Jones TJ, Day S, Pretorius A-M, Birtles RJ, Helps CR and Neimark H. 2010. Investigation of human haemotropic *Mycoplasma* infections using a novel generic haemoplasma qPCR assay on blood samples and blood smears. *J Med Microbiol* 59: 1285-1292.

46. Teysson R, Poutiers F, Saillard C, Grau O, Laigret F, Bove JM and Bebear C. 1993. Detection of mollicute contamination in cell cultures by 16S rRNA amplification. *Mol Cell Probe* 7: 209-216.

47. Tully JG and Razin S. 1995. *Molecular and diagnostic procedures in mycoplasmology. Vol. II. Diagnostic procedures*. Academic Press, New York.

48. Van Kupperveld FJM, van der Logt JTM, Angulo AF, van Zoest MJ, Quint WGV, Neisters HGM, Galama JMD and Melchers WJG. 1992. Genus and species specific identification of *mycoplasmas* by 16S rRNA amplification. *Appl Environ Micrbiol* 58: 2006-2615.

49. Waites KB, Tully JG, Rose DL, Marriot PA, Davis RO and Cassell GH. 1987. Isolation of *Acholeplasma oculi* from human amniotic fluid in early pregnancy. *Curr Microbiol* 15: 325-327.

50. Yang D, Tai X, Qui Y and Yun S. 2000. Prevalence of *Eperythrozoon* spp. infection and congenital eperthrozoonosis in humans in Inner Mangolia. *Epidemiol Infect* 125: 421-426.

51. Yechouron A, Lefebvre J, Robson HG, Rose DL and Tully JG. 1992. Fatal septicemia with *Mycoplasma arginini*: a new human zoonosis. *Clin Infect Dis* 15: 434-438.

52. Yu D-H, Kim H-W, Desai AR, Han I-A, Liy H, Lee M-J, Kim In-S, Chae J-S and Park J. 2007. Molecular detection of feline hemoplasmas in feral cats in Korea. *J Vet Med Sci* 69: 1299-1301.

2014, Zoonoses: Bacterial Diseases
Editor: Sudhi Ranjan Garg
Published by: DAYA PUBLISHING HOUSE, NEW DELHI

Pages **417–435**

24

Bacterial Zoonoses Associated with Poultry

Kuldeep Dhama[1], Ruchi Tiwari[2], Mohd. Yaqoob Wani[3] and Senthilkumar Natesan[4]

[1]*Division of Pathology,* [3]*Division of Veterinary Biotechnology, Indian Veterinary Research Institute, Izatnagar – 243 122*
[2]*College of Veterinary Sciences, Pandit Deen Dayal Upadhyaya Veterinary University, Mathura – 281 001*
[4]*Institute of Human Virology and Department of Medicine, University of Maryland School of Medicine, 725 West Lombard Street, Baltimore, MD 21201, USA*

Poultry industry in India has emerged as a dynamic venture due to commercialization of intensive poultry farming in the past three decades. Infections and disease outbreaks not only cost poultry producers and related industries millions of rupees a year as lost revenue but also pose public health risk due to transmission of zoonotic avian diseases by direct and indirect contact, vectors, contaminated inanimate objects, oral ingestion or inhalation of aerosolized materials. Many people work or live with birds daily, facing the risk of exposing themselves to numerous pathogens. Pet bird keepers and consumers of poultry products also face similar risk.

It is important to know about the zoonotic potential of avian diseases to plan remedial measures. Veterinarians, public health physicians and researchers make the first line of defense against these zoonotic diseases by creating public awareness and devising effective prevention and control strategies. Some important bacterial zoonotic diseases of poultry, their public health significance and salient measures to reduce public health hazards are discussed below.

Salmonellosis

Salmonellosis is caused by *Salmonella*, Gram-negative bacteria, which have more than 2400 serologically distinct variants. The organism has been responsible for serious economic losses to the poultry producers worldwide. Infection of poultry with salmonellae causes localized or systemic infections, in addition to a chronic asymptomatic carrier state. Infections can progress to dehydration, weakness, and sometimes, especially in the very young or very old, high mortality may occur. Infected birds become lethargic, anaemic, have watery droppings, show poor appetite, polydypsia and may develop arthritis. Focal infections may occur in any organ, including heart, kidney, joints, meninges and the periosteum. In young chicks unabsorbed yolk is a common finding.

Poultry is considered a major reservoir of *salmonella*, and poultry products have been identified as important source of the organism that causes human illness. Salmonellosis is among the leading foodborne infection throughout the world (Singh 2005, Kabir 2010). *Salmonella* Enteritidis and *S*. Typhimurium are the predominant serotypes having more public health concern. However, poultry born zoonoses are mainly due to *S*. Enteritidis (El-Tras *et al.* 2010). *S*. Gallinarum (fowl typhoid) is of little public health significance and rare cases of *S*. Pullorum (pullorum disease) infection in humans have been reported following the ingestion of contaminated poultry products (Shivaprasad 2000, Kabir 2010). Only massive exposure following the ingestion of contaminated foods causes development of a rapid onset of acute but transient enteritis, followed by prompt recovery without treatment. *S*. Enteritidis, excreted in avian faeces, is capable of penetrating eggshells and can be transmitted through food chain to humans due to consumption of uncooked eggs, causing severe gastrointestinal disease, particularly in small children, elderly and immunocompromised individuals. It produces endotoxins which cause foodborne illness. During the past two decades, *S*. Enteritidis has emerged as a leading cause of human infections in many countries, with hen eggs being a principal source of the pathogen.

Faecal oral route and direct transmission from infected bird (occupational exposure) are responsible for human disease incidences. Avian strains of *Salmonella* are not a risk for healthy adults but can infect small children, older and immunosuppressed individuals. The incubation period in humans is 6-72 hours. In humans, the foodborne salmonellosis mainly causes enterocolitis characterized by nausea, vomiting, bloody diarrhoea, dehydration and fever. In severe cases, there can be a high fever, bacteremia/septicemia, sepsis, headache and abdominal cramps. Recovery may occur in 2-4 days. Organisms may settle in organs such as liver, kidney and heart or in the joints and may cause inflammation (osteomyelitis). Although mortality is very low (probably less than 1 per cent), death rate is higher in infants and old people usually on account of septicemia. Salmonellae can be transmitted from person to person and humans carrying *salmonella* can also infect the birds. Widespread distribution of contaminated foods, from *Salmonella* outbreaks in poultry, can sometimes involve huge numbers of consumers. *Salmonella* infections are becoming major economic and public health risks due to the emergence of multi-drug resistant (MDR) *Salmonella* species.

The effective control of avian *salmonellosis* is possible by strict biosecurity measures. In birds, vertical transmission and egg contamination play dominant role in the transmission of this infection. Therefore, preventive measures should ensure that only eggs from flocks known to be free from salmonellae be introduced into hatcheries. The comprehensive control measures, including production of salmonella-free feed, rigorous biosecurity practices for eliminating biological vectors (pests), effective cleaning, sanitation and disinfection of poultry houses, and prophylactic treatment of poultry, can reduce the spread and incidences of this disease of public health concern to a large extent (Linam and Gerber 2007).

Preventing free flying birds, carriers, rodents, etc. from entering the poultry rearing area need to be practised strictly in order to minimize the exposure to salmonellae. Pellet feeding can also minimize the introduction of this disease in poultry flocks. The importance of hygienic conditions in the poultry farm and also at personal level, proper handling and packaging of food materials should always be kept in mind. Rapid investigation of the disease situation at the farm level should be taken care of. Frequent inspection of abattoirs, food processing plants, feed mills, egg grading stations and meat vendors should be done by the competent authority for assessing the sanitary and hygienic practices.

Campylobacteriosis

Campylobacteriosis is the most reported foodborne gastroenteritic disease and poses a serious health burden in industrialized countries. The infection in poultry is principally caused by *Campylobacter jejuni*, a Gram-negative bacterium. *Campylobacters* can affect humans, animals and a variety of birds (chickens, turkeys, parrots, canaries, pigeons etc.) and are considered to be the major foodborne pathogens in the poultry industry, especially in the developed countries (Chattopadhyay *et al.* 1991). *C. jejuni* lives in the small intestine and colon and may be isolated from clinically ill as well as normal birds. It is generally non-pathogenic in mature poultry. However, some strains of this species are invasive and/or toxic, which can cause enteritis and death in newly hatched chicks and poults (below 2 weeks). Majority of birds shed large numbers of organisms ($>10^6$ colony forming units/g faeces). The diseased birds develop distension of the intestine, hepatitis, lethargy, loss of appetite, weight loss, yellowish diarrhoea or watery droppings. Egg production is markedly lowered. Highly pathogenic isolates derived from people with enterocolitis may induce up to 32 per cent mortality in chicks.

Epidemiological studies have revealed a major association between *Campylobacter* infection in humans and handling and consumption of raw or undercooked poultry meat (Saleha *et al.* 1998, Hariharan 2002, Lindmark *et al.* 2009). Recent genetic typing studies revealed that chicken isolates can frequently be linked to human clinical cases of campylobacter enteritis (Hermans *et al.* 2011). Chickens may account for 50-70 per cent of infections. Colonized broiler chicks are the primary vector for transmitting this pathogen to humans. The disease can spread to humans by direct infection of the poultry farm workers or during processing plant operations through handling infected poultry, or through contact with their droppings, or by eating *C. jejuni* contaminated raw/undercooked poultry products, causing enterocolitis. Almost all people are susceptible to the infection but usually young children and young adults, pregnant women, debilitated and immunocompromised individuals are affected more.

After an incubation period of 2-5 days, clinical signs in humans include watery to bloody diarrhoea, fever, abdominal cramps, malaise, headache, myalgia, nausea and vomiting, which may last for more than 24 hours. This is a self-limiting disease, which generally lasts 3-5 days or up to 10 days at the most, but convalescent shedding of the organisms occurs in excreta. Pregnant women, debilitated and the immunocompromised individuals such as AIDS patients are at a greater risk with more severe and prolonged disease. *Campylobacter* enteritis has been associated with development of complications like septicaemia, Guillain-Barre syndrome (GBS), reactive arthritis, meningitis, thrombotic thrombocytopenic purpura, etc. High incidence of clinical disease associated with this organism, its low infective dose in humans, and its potentially serious sequelae, confirm its importance as a significant public health hazard (Hariharan 2002, Dhama *et al.* 2011c).

World Health Organization (WHO) report on diarrhoeal zoonosis has indicated that direct or indirect contact of human beings with poultry and other domesticated animals gives rise to higher risk of campylobacteriosis among rural communities. Surveys of farm animals in a number of countries have revealed an incidence of up to 70 per cent in sheep, cattle and pigs, and up to 100 per cent in chickens. Worldwide, much of the poultry meat presented at retail shops/slaughterhouses is contaminated with *C. jejuni* and/or *C. coli*, largely a result of the preferential colonisation of the avian digestive tract by these organisms, and many unhygienic operations there present opportunities for cross-contamination, thus creating possible public health hazards. The isolation of antibiotic resistant strains of *Campylobacter* species from poultry represents a major public health problem. Therefore, proper antibiotic sensitivity test should be conducted before starting a treatment programme. The significance of *Campylobacter* as diarrhoegenic agent in human beings is now considered with more attention in developing countries like India (Dhama *et al.* 2011b).

Measures for reducing avian campylobacteriosis and its public health hazards should be strictly implemented (Lin 2009, Dhama *et al.* 2011b). Strict biosecurity measures and good management practices, including cleanliness, hygiene, sanitation and disinfection, 'all in all out' stocking policy are important. Providing clean water and/or effective water sanitation treatment such as chlorination and effective vermin (mice and rats), insect and housefly control programme are also necessary. Dead birds and body discharges/wastes should be disposed of properly and hygienically. The entry of visitors on farm should be limited and their direct contact with the flock should be restricted. Other domestic animals should not be kept in the same premises. All houses should be bird-proofed and free flying birds should be kept away from the poultry farm. Insects and houseflies should also be controlled. Purchasing birds from disease free sources, strict observance of personal hygiene and appropriate safety measures, washing or sanitizing hands and feet frequently with antibacterial soap or suitable disinfectants, and changing outer protective clothing, especially after handling any bird or litter are some other preventive and control measures. Disinfectant boot dips should be placed to reduce the probability of introducing and spreading the infection. *Campylobacter* contamination and cross-contaminations should be avoided during poultry processing at slaughterhouses and processing plants, bird markets, retail shops and kitchens. Good kitchen hygiene and proper cooking are important at consumers' household level. Poultry products should be stored and transported at a temperature of 4°C or lower to prevent the proliferation of *Campylobacters*. Refrigeration of raw chicken promptly, never leaving it outside at room temperatures or at hot and humid climate of tropical countries should be observed. Decontamination of utensils and work surfaces should be taken care of. In case of infection, it is important to seek timely and appropriate medical treatment.

Hazard analysis critical control point (HACCP) approach covering the full chain of poultry production and processing is needed to combat the *Campylobacter* problem. Corrective measures, including end-product decontamination when necessary, must be introduced at defined critical points confronted with the hazard of campylobacter introduction. Methods reducing *Campylobacter* contamination in poultry processing need proper attention; improved washing of carcasses, the use of counter flow scalding, water immersion chilling, etc. can reduce the levels of *C. jejuni* contamination in poultry meat and its products. Irradiation (Gamma irradiation, cold pasteurization) is an effective method to eliminate *C. jejuni* from poultry meat and products. Irradiation using Cobalt-60 and electron beam generation are cost-effective procedures. Control measures at farm level and meat processing plants should be combined into an integrated approach for pathogen reduction. Strict biosecurity measures should be adopted. Additionally, it is important to educate people about the risks associated with handling raw poultry meat, and consuming undercooked or contaminated products which will help in preventing spread of this zoonotic disease (Dhama *et al.* 2011b).

Colibacillosis

Colibacillosis is caused by *Escherichia coli*, a Gram negative bacterium that normally inhabits the intestinal tract of all animals. There are a number of different strains and many are species specific. Multiple serogroups are associated with disease, especially O1, O2 and O78 among many others. *E. coli* affects birds of all ages. Faeco-oral route is the main route of infection. Contaminated food and water consumption and faecal contamination of the egg surface cause infection in poultry (Barnes *et al.* 2003). The incubation period in most of the species affected is 12-72 hours. In birds, pathogenic *E. coli* infections may cause septicemia, chronic respiratory disease, air sacculitis, omphalitis, synovitis, pericarditis, necrotic enteritis and salpingitis, resulting into significant economic losses in poultry industry worldwide (Barnes *et al.* 2003). Transmission in humans is via the faecal-oral route. Colibacillosis is often considered as foodborne or waterborne zoonotic disease. Earlier, the avian strains of *E. coli* were considered not to cause any important disease in humans and animals, hence attracted not much zoonotic significance. However, the growing evidence suggests a possible role of avian pathogenic *E. coli* (APEC) in human disease, and it has been suggested recently that among the poultry diseases transmissible to human, avian colibacillosis and avian salmonellosis are the leading ones (Manges *et al.* 2007, Kabir 2010). The presence of pathogenic microorganisms in poultry meat and its products remains a significant concern for public health worldwide. Colibacillosis in humans usually manifests diarrhoea, and other complications, including fever, dysentery, shock, and purpura.

Management practices designed to minimize the exposure level of these types of organisms in the bird's environment are necessary in any preventive programme. Birds also need to be protected against other pathogens that promote infections with APEC. Unravelling the molecular basis of virulence of APEC in their natural hosts would provide the basis for the development or improvement of strategies to control such infections in the food producing avian species (Dziva and Stevens 2008). Good management practices, avoiding overcrowding, proper ventilation, good sanitation and hygiene, along with effective treatment with suitable antibiotics are highly effective in the prevention and control of colibacillosis in birds and thereby reducing its zoonotic incidences. Prevention of egg contamination by fumigating them within two hours after lay, and by removing cracked eggs or the eggs soiled with faecal material is an important step. Vaccination is also effective in avian colibacillosis.

Chlamydophilosis (Chlamydiosis)

Avian chlamydophilosis is a widespread zoonotic disease caused by a Gram negative obligate intracellular bacterium *Chlamydophila psittaci*, previously known as *Chlamydia psittaci* (Order: Chlamydiales, Family: Chlamydiaceae). It is also known as psittacosis/parrot fever (in psittacines and humans) or ornithosis (in all other birds). All avian species are potential hosts for *Chlamydophila* but the disease more commonly affects parakeets, cockatoos, lovebirds, pigeons and canaries. As far as the poultry industry is concerned, turkeys are extremely susceptible, while chickens are more resistant (Andersen and Vanrompay 2000, 2003, Dhama *et al.* 2008). The infectious agent is the elementary body. Human infections usually occur by inhalation of contaminated aerosol, as the infected birds shed elementary bodies in their faeces, urine, saliva, ocular secretions, nasal exudates, and feather dust. Poultry farmers, bird keepers, workers and veterinarians involved in breeding and selling birds, and commercial poultry processors are at greater risk of exposure. Most human cases are contracted from psittacines, pigeons, and turkeys. Psittacosis, due to contact with poultry, probably occurs more often than is thought and the infection can be asymptomatic or symptomatic. Zoonotic transmission from processed poultry meat, especially turkeys, is important.

Psittacosis is a systemic zoonotic infection. The incubation period is 5 to 14 days and the mean duration of illness is 9-10 days. Infection in humans varies from a mild flu-like infection to a serious atypical pneumonia with neurological signs and possibly death, if not treated. Patients typically present with 1 week of fever, headache, myalgias, and a non-productive cough. Pneumonia is the most common manifestation, though all organ/systems can be involved. Due to a low awareness of the disease and a variable clinical presentation, psittacosis is often not recognised as such by general practitioners (Beeckman and Vanrompay 2009).

Since it is a contagious disease, implementation of strict biosecurity measures is essential. These involve isolation of infected, unhealthy and dead birds and keeping them in quarantine for observation and treatment; movement control of visitors, vehicles, contaminated farm materials, feed, water, vectors (wild or feral birds), reservoirs, etc.; and sanitation of farm premises, feeders, waterers with appropriate procedures. The elementary bodies can survive in the environment for several days but may remain viable in dried debris for months. They are rapidly inactivated in faeces and are susceptible to sunlight and heat (above 20°C). Chlamydophila are susceptible to most disinfectants including quaternary ammonium compounds, alcohol, benzalkonium chloride, bleach and hydrogen peroxide (Johnston *et al.* 2000). Dry and dusty areas in and around poultry or pet bird premises can be disinfected with household bleach or a commercial disinfectant. Carcasses must be wetted with detergent and water or disinfectant during necropsy. While examining the carcasses or any infected material, it should be handled in a biohazard cabinet. Strict trade regulations are very essential to check the disease.

The knowledge about the distribution and infectious cycle of chlamydiae is still insufficient, and laboratory diagnosis of chlamydial zoonoses remains unsatisfactory in both human and veterinary medicine. Acute chlamydial infections are usually treated with macrolides, tetracyclines, or quinolones (Rhode *et al.* 2010). Doxycycline is the treatment of choice. Persistent varieties are not treated by standard therapy. The possible emergence of *C. psittaci* antibiotic-resistant strains has also been suggested (Beeckman and Vanrompay 2009). Diagnostic monitoring and reporting of *C. psittaci* infections in poultry workers, along with development of an efficient veterinary vaccine and information campaigns on zoonotic risk and preventive measures against its transmission, would be beneficial to public health (Dickx and Vanrompay 2011). There is a considerable need for research, especially with regard to the diagnosis and treatment of persistent varieties.

Listeriosis

Listeriosis, caused by *Listeria monocytogenes,* a Gram positive bacterium, is quite rare in birds and usually occurs as a septicemia or sometimes localized encephalitis. Signs of infection, if seen, are suggestive of septicemia and may include depression and listlessness, emaciation, diarrhoea and sudden death at times. Depression, incoordination, ataxia, torticollis, opisthotonous and other nervous signs are seen in the encephalitic form. The organism is shed in the nasal secretions and faeces of the bird. It can infect almost all domestic animals besides many species of rodents, wild animals and poultry (Wesley 2007). Chickens are thought to be the carriers and also the prime reservoirs for the disease (Njagi *et al.* 2004). Transmission and subsequent infections occur by ingestion of contaminated feed, water, litter and soil. Cold, wet conditions causing excessively moist litters are associated with outbreak of listeriosis (Barness 2003, Kahn 2005). Human beings predominantly get infection from raw broiler meat due to *Listeria* contamination and unhygienic conditions of the processing area, rather than acquiring direct infection from birds (Kosek-Paszkowska *et al.* 2005).

Human listeriosis is prevalent all over the world, causing several foodborne outbreaks, particularly in the developed countries (Mahmood *et al.* 2003). Consumption of raw meats, poorly cooked and ready-to-eat poultry products is responsible for the disease. It should be kept in mind that this organism will continue to grow in some prepared foods kept at low temperatures. Mortality among neonates, very young and immunocompromised individuals is very high (up to 30-40 per cent) making it a serious public health hazard. Person-to-person spread, though recognized, is uncommon. Direct contact with birds is of little importance in transmission except in case of highly susceptible persons. Incubation period varies from 1 day to 3 weeks. The disease may present itself as meningitis or, more rarely, encephalitis. Meningitis is characterized by high temperature, stiffness of neck, often ataxia, tremors, seizures, and fluctuating consciousness. The onset is sudden and death may follow within 24-48 hours. The infection can also cause septicaemia, abortion in pregnant woman, stillbirth or infection of the newborn. The symptoms such as headache, vomiting, fever, malaise, pneumonia and conjunctivitis have also been observed. In workers at avian processing plants, conjunctivitis due to *L. monocytogenes* has been linked to handling of apparently normal but infected chickens.

Prevention of listeriosis depends on identifying and eliminating sources of infection. Control of disease in poultry is largely dependent on avoiding potential sources of infection and practising a high standard of management. Rigid sanitation and disinfection procedures with culling and isolation of affected birds are helpful. Widespread use of antibiotics in feed may have had prophylactic value in listeriosis prevention in poultry. The disease can be prevented in humans by avoiding consumption of contaminated foodstuffs and avoiding cross-infections. Food should be properly stored and cooked, and good kitchen hygiene should be practised to avoid cross-contamination of other foods (Rebagliati *et al.* 2009). Thorough and proper cooking of raw meat is necessary. Packed and frozen meat products require heating in accordance with manufacturer's instructions. Hands, knives and cutting boards should be washed and disinfected after handling uncooked meat. Disinfectants like 1 per cent sodium hypochlorite, 70 per cent ethanol or glutaraldehyde should be used in processing units to inactivate the organism. Moist heat (121°C for a minimum of 15 min) or dry heat (160-170°C for 1 h) can kill the organism present in the utensils. Disease can be satisfactorily prevented by wearing protective clothing while handling infected birds or their tissues.

So far no vaccination is available. The organism is often resistant to most of the commonly used antibiotics; high levels of tetracyclines are recommended for treatment, which are efficacious in both acute and subacute forms of the disease. Treatment of the chronic form is usually unsuccessful. With the generalized use of antibiotics in poultry feed, for growth promotion, the cases of listeriosis in poultry have decreased to few. Post-packaging treatments are to be implemented to destroy *L. monocytogenes* in food products (USFDA 2011). Short-term refrigerated storage of cooked perishable foods is important since this organism will continue to grow in some prepared foods kept at low temperatures (Gillespie *et al.* 2006). At the level of consumers, identification of high risk foods and education of high risk individuals should be given due consideration (Rebagliati *et al.* 2009). Limiting listeriosis requires implementation of effective food safety control measures and ensuring that these control strategies are consistently met.

Tuberculosis

Avian tuberculosis (ATB), caused by acid fast bacteria *Mycobacterium avium* and *M. genavense*, is a contagious, debilitating disease affecting a wide range of birds (domestic poultry, pet, captive or zoo birds). ATB affects chickens, pheasants, quail, guinea fowl, turkeys, parrots, budgerigars, ducks, goose, doves, patridges, pigeons and other captive and wild game birds and has also been reported in ostriches, emus and rheas in many zoological parks. With a long incubation period and a protracted slow course, the symptoms can prolong for weeks or months. Being established through a longer exposure, ATB affects adult birds, and is less prevalent in young fowls and lesions are less severe in them. The disease has a worldwide distribution. *M. avium* subsp. *avium* (MAA) is considered as the most important pathogen causing tuberculosis in domestic birds. *Mycobacterium avium* complex (MAC) has 28 serotypes; 1, 2, and 3 are considered virulent for chickens while serovar 3 is recovered sporadically from wild birds.

The clinical manifestations in birds include emaciation, depression and diarrhoea along with marked atrophy of breast muscle, all these contributing to decreased production and mortality (Fulton and Thoen 2003, Dhama *et al.* 2011c). Unlike tuberculosis in animals and humans, lesions in lungs are rare and if present, are not detrimental. Tubercular nodules can be seen in liver, spleen, intestine and bone marrow. Granulomatous lesion with central necrosis and multinucleate giant cells without calcification is a prominent feature of avian tuberculosis. The disease is a rarity in organized poultry sector due to improved farm practices but birds kept in zoo aviaries are still susceptible. It can readily infect swine and also cause sensitivity in cattle to the tuberculin test. The disease is not common in wild birds, but may develop upon contact with infected chickens. *M. tuberculosis* less commonly causes infection in birds often acquired from pet bird owners. Infected birds with advanced lesions excrete the organism in their faeces. *M. avium* is highly resistant to environmental challenges and can survive in soil for up to 4 years. It is spread by ingestion of food or water contaminated with faeces from birds, which shed the organism.

M. avium, M. intracellulare and *M. genavense* are of public health concern, particularly in those with immunocompromised diseases such as HIV/AIDS or patients under transplant therapy. Humans are considered relatively resistant to *M. avium*, but infection usually occurs in immunocompromised individuals with multiple or larger exposure to the bacteria causing a progressive disease that is refractory to antibiotic treatment. Transmission in humans via aerosol usually results in pulmonary infections which can be difficult to treat. *M. avium* can also cause local wound infections with swelling of regional lymph nodes. Localized primary lymphadenitis and a disseminated form of infection occur. In adults, it frequently affects the lungs, producing respiratory signs. In children, the cervical lymph nodes are often involved, while the immunosuppressed individuals have the disseminated form of the infection (Une and Mori 2007, Kriz *et al.* 2010, OIE 2010). Gastrointestinal tract is the main route for *M. avium* infection in AIDS patients.

Eradication of avian tuberculosis is difficult due to the chronic carrier state and intermittent shedding of a large number of organisms by the affected birds. Effective measures to eliminate the disease and establishing and maintaining tuberculosis-free flocks need to be followed. If a positive bird is identified, it should be separated from the flock. All contact birds should be quarantined and tested at 6-12 week intervals. In commercial poultry flocks, relatively rapid turnover, coupled with improved sanitation, has largely eliminated this infection. In humans, all research and diagnostic operations involving the handling of open live cultures of *M. avium*, or material from infected birds, must be carried out with adequate biohazard containment. Since both tuberculin and agglutination tests are not confirmatory, there is possibility of dissemination of disease, as long as one infected bird remains in a flock. Entire infected flocks should be slaughtered and their carcasses burnt. All the equipments and housing materials should be destroyed by burning. Contaminated soil is a continual source of infection and should be removed along with litter adopting appropriate hygiene procedures and followed by disinfection. Frequent removal of faecal material is the single most important factor in preventing transmission. The older flocks should be removed gradually and repopulated with day-old chicks from reliable sources. Strict sanitation of contaminated premises, buildings and equipments are to be followed. Disinfection of the poultry houses should be done frequently. The practice of managing poultry in free range system and keeping the breeders for several years is conducive for the spread of tuberculosis among them; therefore should not be encouraged. Exotic birds should be quarantined for 60 days and tuberculin testing should be done (Fulton and Thoen 2003, OIE 2010).

Clostridial Infections

Rare but deadly disease botulism occurs when the bacterium *Clostridium botulinum* grows and produces a powerful paralytic toxin in foods. Another species, *C. perfringens*, is also an emerging threat for public health that can cause two different foodborne diseases in humans; diarrhoea and necrotic enteritis, caused respectively by enterotoxin positive type A strains and type C strains of the organism (Filip *et al.* 2004). Food that is heated, cooled down too slowly when freezing, and reheated insufficiently is of great risk as it facilitates bacterial spores to form vegetative colonies rapidly. Symptoms of the self limiting type A food poisoning start about 6 to 24 h after consumption of contaminated food and last about 24 hours, characterized by abdominal pain, nausea and diarrhoea. Symptoms of type A usually are relatively mild and limited to the elderly and very young children but occasionally death may occur due to dehydration. Type C food poisoning in humans has only been reported on very rare occasions (Filip *et al.* 2004). *C. difficile* is new (re)emerging zoonotic pathogen (Indra *et al.* 2009, Gould and Limbago 2010). It has been found relatively commonly in retail chicken meat, albeit at low levels (Weese *et al.* 2010). Contamination of meat with *C. difficile* strains implicated in human infections raises concerns about food as a source of its infection

Erysipelas

Infection with the zoonotic bacterium *Erysipelothrix rhusiopathiae*, a facultative, non spore forming, non acid fast, small, Gram positive bacillus, causes severe disease outbreaks (erysipelas) in poultry flocks (Eriksson *et al.* 2010). It is a pathogen or a commensal and may cause erysipelas in farmed turkeys, chickens, ducks and emus. The disease can be transmitted to humans through direct contact with birds, tissues and droppings (Wang *et al.* 2010). The bacterium has been isolated from the poultry red mite (*Dermanyssus gallinae*), hence, the mite has been suggested as a possible means of transmission of *E. rhusiopathiae* on and between poultry farms.

The risk of infection increases in a person having unprotected cuts or abrasions on hands. Disease in humans may present as cellulitis, bacteremia, endocarditis, encephalitis and arthritis. Veterinarians and farmers, in particular, should be aware of the symptoms. Symptoms in humans include itching that usually begins several days after the infection. The itching becomes painful and the skin becomes red and swollen. When not treated, the infection can spread through the lymph nodes and in some rare cases can cause a heart valve infection (Eriksson *et al.* 2010). Infection by this organism is possibly under-diagnosed due to the resemblance it bears to other infections, and also due to the problems encountered in isolation and identification. Control by sound poultry husbandry, flock management, good sanitation and immunization procedure is recommended.

Arizonosis

This is an egg-transmitted disease, primarily of turkey poults, which is of public health concern mainly in western countries. The disease is caused by any of the 300 or more serotypes of *Salmonella arizonae*. Transmission of arizonosis occurs by the faecal-oral and also by vertical routes. Affected poults show unthriftness and many will develop opacity of eye and blindness, the involvement of eyes being unique. Incoordination and paralytic signs may also develop. Some birds develop diarrhoea, dehydration, pasting around vent, peritonitis, salpingitis, or local ovarian infections, but infections of the intestinal tract are more common. In humans, diarrhoea is the most common finding and most of the infections are subclinical and septicemia may occur only in immunocompromised individuals. As it is a vertically transmitted disease, the primary stock kept for breeding should be free of *S. arizonae*. It is somewhat less hardy than most salmonellae, but can survive for months in soil, feed and water. Fumigation of hatching eggs, treatment with antibiotics, and rigorous hatchery sanitation also aid in reducing the transmission.

Other Diseases

Apart from the above pathogens, wild and migratory birds are also known to spread *Borrelia burgdorferi* (Lyme disease) which could affect animals as well as human beings (Rizzoli *et al.* 2007). Recently, methicillin-resistant *Staphylococcus aureus* (MRSA) has been reported to occur in domestic animals, chickens and humans, and showed transmission in both directions *i.e.*, from animals and birds to man, and vice versa (Saleha and Zunita 2010). *Enterococcus faecalis* has been linked to severe extra-intestinal infections in poultry and is a major cause of nosocomial infections in humans. Recently, *E. faecalis* of human and poultry origin have been shown to share virulence genes which suggests and supports the possibility of zoonotic risk associated with this organism (Olsen 2011).

Disease Monitoring and Surveillance

The changed scenario of the disease incidences in poultry is posing new challenges for poultry management and healthcare professionals and disease specialists. As the situation unfolds, it is highly imperative that disease epidemiology, serosurveillance, monitoring and network approaches are intensified (Kataria *et al.* 2005, Dahlhausen 2010). Majority of the conventional techniques in poultry disease diagnosis have been perceived to be time consuming, laborious and even requiring *in vivo* systems. With the advent of biotechnological tools, detection of poultry pathogens has been strengthened in terms of reliability and rapidity, and also for characterizing and monitoring various pathogens. Molecular techniques have been used very commonly for the detection, differentiation and characterization of various avian pathogens in the recent past and have tremendous application potentials for current and future diagnostics. Progress is also underway with respect to making these approaches more suitable to the diagnostic laboratories all over the country and worldwide. Such tools and techniques include PCR and allied techniques *viz.* PCR-RE/RFLP, RT-PCR, PCR-ELISA, Q-PCR, RRT-PCR, LUX-RRT-PCR, m-PCR, real-time PCR/HRM curve analysis, RFLP, RAPD-PCR, PFGE, and various others like hybridization, LAMP, NASBA, nucleotide sequencing, phylogenetic analysis, and recent tools of biochips, microarrays and nanotechnology (Munster *et al.* 2008, Dahlhausen 2010). These emerging molecular tools have comfortably overtaken the conventional diagnostic methods, but their wholesome practical implementation is still awaited.

The newer concepts of poultry disease management cover advanced monitoring, surveillance strategies and biosecurity measures. For serosurveillance, the poultry enterprises use automated technology for rapid processing of large number of serum samples for antibody detection, which facilitates flock profiling and provides useful information regarding the health status of birds. The surveillance has been escalated in view of several global threats and international organizations have together established Global Livestock Early Warning and Response System (GLEWS). Global Avian Influenza Network for Surveillance (GAINS) regularly monitors and keeps vigil on various migratory birds (Brown and Stallknecht 2008). FAO, through Emergency Prevention System for Transboundary Animal Diseases, EMPRES collects, records, and analyses the data on major diseases in both wild birds and domestic poultry. Nowadays, satellite-based tracking systems are being exploited for improving the surveillance of wild birds. Global positioning system (GPS) transmitters can be implanted on wild birds to track their migratory routes or breeding grounds.

Biosecurity

Basically it refers to the methods adopted to secure a disease free environment by preventing the exposure of the birds/flock to disease causing organisms derived from multiple sources by reducing the introduction and spread of pathogens into and between farms. It involves an understanding of the principles of epidemiology and economics and requires teamwork to maximize benefits. It should be the cheapest and at the same time most effective means of disease control *i.e.* "one size fit all" approach holds good. It is a first line of defense. The main objective of biosecurity is to prevent diseases, for which there are three key principles of biosecurity - isolation-confinement of birds within a controlled environment; traffic control- controlling traffic onto and within farms; and sanitation and hygiene- disinfecting materials, equipments or poultry houses (Mulder 1997, Anandh *et al.* 2003, Dhama *et al.* 2003, Nathaniel and Nelson 2004, Grunkemever 2011). The animal-human interface and infectious disease in industrial food animal production requires rethinking about biosecurity and biocontainment (Graham *et al.* 2008).

General Biosecurity Measures

Biosecurity is the most important tool in preventing clinical and/or sub-clinical diseases in poultry. It works on the basis that an "ounce of prevention is worth a pound of cure." It is an integral part of total poultry management, provides safety of living things and thus has to be a way of life and means for optimal poultry production. It always keeps an eye on the critical control points (CCPs) like people, poultry, vehicles, equipments, feed, litter, water, vermin, insects and wild birds in the farm environment. Implementation of biosecurity measures involves the use of common sense, economics, and the relative risk, and it should be adopted differently according to the environment with the main objective of preventing disease transmission. Biosecurity measures are required in poultry farms, hatcheries, egg processing plants, live bird markets, and poultry meat processing plants to prevent the disease transmission among the birds and the transmission of zoonotic diseases from birds to human beings.

Biosecurity takes up proper control of most likely ways of disease entrance on the farm. Some general principles of ascertaining biosecurity at the farm are listed below. Biosecurity programme is like a chain and any weak link or even a single failure or breakdown in the programme can lead to disaster. Hence, the rules should be followed all the 365 days of the year.

☆ Follow business with suppliers who practise good biosecurity standards.

☆ Practise 'all in all out' (one age group per farm) breeding practices.

☆ Implement cleaning and disinfection programme regularly.

☆ Promptly remove and properly dispose of dead birds.

☆ Keep free flying birds and wild birds away from the flock.

☆ Limit visitors on the farm.

☆ Store feed in a building, which is bird proof, insect proof, and rodent proof.

☆ Have a check on water quality, as it is a potential source of contamination.

General Measures for Disease Prevention and Control

☆ Follow good husbandry practices, sanitation, hygiene and disinfection measures along with suitable vaccination programme at poultry farm and hatchery.

☆ Place disinfectant boot dips to reduce the probability of introducing and spreading infections.

☆ Judiciously use suitable antibiotics and anti-fungal agents for treating diseases.

☆ Follow vector control strategy and avoid carriers of disease causing agents. Keep free flying birds and wild animals away from the farm premises.

☆ Follow adequate quarantine measures, culling and disposal of dead or affected birds, contaminated material, excreta, discharges and wastes.

☆ Learn safe and humane bird handling techniques and use proper safety measures to avoid accidental exposures to infectious agents.

☆ Follow proper personal hygiene, sanitation and biosafety measures while handling poultry or its products. Wash hands frequently with a good soap, detergent and antideptics and wear protective clothing (gloves, face mask, glasses/goggles, apron, shoe cover, coveralls) when handling birds, eggs or tissues to prevent exposure from sick birds or contaminated environment. Do not eat or drink while handling poultry or in the poultry housing areas.

☆ Use potable, clean and sanitized (chlorinated) water.

☆ Follow good laboratory and biohazard containment practices while handling poultry pathogens.

☆ Take suitable precautions for foodborne pathogens and avoid cross-contaminations during poultry processing at slaughterhouses, processing plants, bird markets, retail shops and food preparations of poultry meat.

☆ Follow appropriate cleanliness, hygiene and disinfection procedures at slaughterhouses and food processing plants.

☆ The consumers of poultry or its products should practise good kitchen hygiene and cooking practices to avoid infection through foods. Avoid cross-contamination of foods such as salads etc. that are eaten raw. Use separate utensils and carefully clean all utensils with soap and hot water after use.

☆ Good toilet practices should be followed by persons, especially the children, with an acute diarrhoeal illness or food poisoning to reduce the risk of spreading the infection.

☆ Treat human excreta appropriately before its disposal.

☆ Seek timely and appropriate medical treatment in case of any infection. Tell your physician that you work closely with birds, and visit them regularly, and discuss appropriate precautions.

☆ Young, elderly, pregnant or immunocompromised individuals need to take extra precautions.

☆ Stay current on appropriate vaccinations and follow prophylactic immunization and specific therapeutic measures.

☆ Veterinary clinicians and health experts should follow timely and appropriate disease prevention and control measures.

☆ Backyard poultry farming practices need special attention towards biosecurity measures.

☆ Notifiable diseases should be immediately reported to the regulatory authorities and officials.

☆ Follow appropriate sampling and dispatch procedures for disease diagnosis. Confine live birds being submitted to the laboratory in safe boxes and make sure they will not return to farm.

☆ Plan precautionary measures in advance for timely and appropriate restriction on poultry trade in the event of an outbreak or zoonotic threat.

☆ Be aware of the avian diseases transmissible to humans. Public awareness on zoonotic diseases should be created using mass media.

Conclusion and Future Directions

The presence of pathogenic microorganisms in poultry birds, poultry meat and other products and by-products remains a significant concern for public health worldwide. The situation warrants strategies for curbing the diseases in birds and guarantee of supply of safe products. Necessary measures are required to be taken by farmers and industry to prevent infections and contaminations at all stages of production, processing and marketing. Science-based education and risk communication strategies aimed at susceptible populations and focused on high risk foods should be delivered through healthcare providers or other credible sources of information. High risk individuals should be provided with guidance on safe and healthy eating practices. It also includes educating the workers about the risk of occupational exposures and measures for protecting themselves, their families and the community. Public health implications of zoonotic infections should be elaborated to such workers. Any zoonotic disease outbreak, small or large, needs to be reported transparently for maintaining the safety of people, birds, animals and animal products. Proper methods of disease prevention and containment should be followed according to the regulations and guidelines in the event of an outbreak to protect the public health. In large scale outbreaks, international coordination and cooperation may be required. All the international trade of poultry and poultry products should be monitored carefully.

Research and development activities should be promoted for evolving effective and newer antibiotics, along with alternative approaches to counter the emerging multiple drug resistant strains of pathogens. In this context, new modalities of treatment like bacteriophages, avian egg antibodies, probiotics, cytokines and others need to be explored to their full practical potentials. Efficient drugs and vaccines need to be developed to reduce the disease incidences in poultry. Multi-disciplinary approach utilizing the competence of veterinarians, wildlife specialists, ornithologists, avian disease experts and life science researchers along with collaborations of department of animal husbandry and zoological, forestry, wildlife and environmental authorities etc. is of utmost importance in the changing global scenario.

References

1. Anandh MA, Lakshmanan V and Muthukumar SP. 2003. Control of zoonosis by biosecurity. *Livestock International* 4: 19-21.

2. Andersen AA and Vanrompay D. 2000. Avian chlamydiosis. *Rev Sci Tech Off Int Epiz* 19: 396-404.

3. Andersen AA and Vanrompay D. 2003. Chlamydiosis (psittacosis, ornithosis). In: Saif YM, Barnes HJ, Fadly AM, Glisson JR, McDougald LR and Swayne DE (Eds.) *Diseases of poultry*, 11th edn, Iowa State University Press, Ames, IA.

4. Barnes HJ. 2003. Miscellaneous and sporadic bacterial infections. In: Saif YM, Barnes HJ, Fadly AM, Glisson JR, McDougald LR and Swayne DE (Eds.) *Diseases of poultry*, 11th edn, Iowa State University Press, Ames, IA.

5. Barnes HJ, Vaillancourt J and Gross WB. 2003. Colibacillosis. In: Saif YM, Barnes HJ, Fadly AM, Glisson JR, McDougald LR and Swayne DE (Eds.) *Diseases of poultry*, 11th edn, Iowa State University Press, Ames, IA.

6. Beeckman DS and Vanrompay DC. 2009. Zoonotic Chlamydophila psittaci infections from a clinical perspective. *Clin Microbiol Infect* 15: 11-17.

7. Chattopadhyay UK, Rathore RS, Das MS, Pal D and Dey NK. 1991. Poultry as a reservoir and source of human campylobacteriosis. *Indian Vet J* 68: 911-914.

8. Dahlhausen B. 2010. Future veterinary diagnostics. *J Exotic Pet Medicine* 19: 117-132.

9. Dhama K, Senthil Kumar N, Kataria JM and Dash BB. 2003. Biosecurity- an effective tool to prevent many threats to poultry health: Part I and II. *Indian Poultry Review* 35 (3): 35-41, 35 (4): 39-44.

10. Dhama K, Mahendran M and Tomar S. 2008. Avian chlamydiosis (psittacosis/ornithosis) and its public health significance. *Poultry Fortune* 9: 26-30.

11. Dhama K, Tiwari R and Basaraddi MS. 2011a. Avian diseases transmissible to humans. *Poultry Technology* 6: 28-32.

12. Dhama K, Hansa A and Anjaneya. 2011b. Campylobacter infection in poultry and its zoonotic significance. *Poultry Punch* 27: 50-60.

13. Dhama K, Mahendran M, Tiwari R, Singh SD, Kumar D, Singh SV and Sawant PM. 2011c. Tuberculosis in birds: Insights into the *Mycobacterium avium* infections. *Vet Med Int* doi:10.4061/2011/712369.

14. Dickx V and Vanrompay D. 2011. Zoonotic transmission of *Chlamydia psittaci* in a chicken and turkey hatchery. *J Med Microbiol* 60: 775-779.

15. Dziva F and Stevens MP. 2008. Colibacillosis in poultry: unravelling the molecular basis of virulence of avian pathogenic *Escherichia coli* in their natural hosts. *Avian Pathol* 37: 355-366.

16. El-Tras WF, Tayel AA and Samir A. 2010. Potential zoonotic pathways of *Salmonella enteritidis* in laying farms. *Vector Borne Zoonotic Dis* 10: 739-742.

17. Eriksson H, Brannstrom S, Skarin H and Chirico J. 2010. Characterization of *Erysipelothrix rhusiopathiae* isolates from laying hens and poultry red mites (Dermanyssusgallinae) from an outbreak of erysipelas. *Avian Pathol* 39: 505-509.

18. Filip VI, Jeroen DB, Frank P, Gerard H, Freddy H and Richard D. 2004. *Clostridium perfringens* in poultry: an emerging threat for animal and public health. *Avian Pathol* 33: 537-554.

19. Fulton RM and Thoen CO. 2003. Tuberculosis. In: Saif YM, Barnes HJ, Fadly AM, Glisson JR, McDougald LR and Swayne DE (Eds.) *Diseases of poultry*, 11th edn, Iowa State University Press, Ames, IA.

20. Gillespie IA, McLauchlin J, Grant KA, Little CL, Mithani V, Penman C, Lane C and Regan M. 2006. Changing pattern of human listeriosis, England and Wales, 2001-2004. *Emerg Infect Dis* 12: 1361-1366.

21. Gould LH and Limbago B. 2010. *Clostridium difficile* in food and domestic animals: a new foodborne pathogen? *Clin Infect Dis* 51: 577-582.

22. Graham JP, Leibler JH, Price LB, Otte JM, Pfeiffer DU, Tiensin T and Silbergeld EK. 2008. The animal-human interface and infectious disease in industrial food animal production: rethinking biosecurity and biocontainment. *Public Health Rep* 123: 282-299.

23. Grunkemeyer VL. 2011. Zoonoses, public health, and the backyard poultry flock. *Vet Clin North Am Exot Anim Pract* 14: 477-490.

24. Hariharan. 2002. *Campylobacter jejuni*: Public health hazards and potential control methods in poultry: a review. *Vet Med Czech* 49: 441-446.

25. Harkinezhad T, Geens T and Vanrompay D. 2009. *Chlamydophila psittaci* infections in birds: a review with emphasis on zoonotic consequences. *Vet Microbiol* 135: 68-77.

26. Indra A, Lassnig H, Baliko N, Much P, Fiedler A, Huhulescu S and Allerberger F. 2009. *Clostridium difficile*: a new zoonotic agent? *Wien Klin Wochenschr* 121: 91-95.

27. Johnston WB, Eidson M, Smith KA, Stobierski MG, Besser RE, Conti LA, Flammer K, Reilly KF, Ritchie B and Tully TN. 2000. Compendium of measures to control *Chlamydia psittaci* infection among humans (psittacosis) and pet birds (avian chlamydiosis). *Morbidity and Mortality Weekly Report* 14: 1-17.

28. Kabir SML. 2010. Avian colibacillosis and salmonellosis: a closer look at epidemiology, pathogenesis, diagnosis, control and public health concerns. *Int J Environ Res Public Health* 7: 89-114.

29. Kahn CM. 2005. Listeriosis. *Merck Veterinary Manual*. 9th ed. Merck and Co. NJ.

30. Kataria JM, Dhama K, Dey S and Tomar S. 2005. Emerging poultry diseases: occupational risk to poultry farm workers. *IV Conference and National Symposium of Indian Association of Veterinary Public Health Specialists on Newer Strategies for the Diagnosis and Control of Zoonosis*, November 11-12, 2005, ICAR Research Complex for NEH Region, Meghalaya.

31. Kosek-Paszkowska K, jacekbania, Bania J, Bystroñ J, Molenda J and Czerw M. 2005. Occurrence of *Listeria* sp. in raw poultry meat and poultry meat products. *Bull Vet Inst Pulawy* 49: 219-222.

32. Kríz P, Slaný M, Shitaye JEJE and Pavlík I. 2010. Avian mycobacteriosis in humans remains a threat in the Czech Republic. *Klin Mikrobiol Infekc Lek* 16: 10-17.

33. Lin J. 2009. Novel approaches for *Campylobacter* control in poultry. *Foodborne Pathog Dis* 6: 755-765.

34. Linam WM and Gerber MA. 2007. Changing epidemiology and prevention of *Salmonella* infections. *Pediatr Infect Dis J* 26: 747-748.

35. Lindmark H, Boqvist S, Ljungstrom M, Agren P, Bjorkholm B and Engstrand L. 2009. Risk factors for campylobacteriosis: an epidemiological surveillance study of patients and retail poultry. *J Clinc Microbiol* 47: 2616-2619.

36. Mahmood MS, Ahmed AN and Hussain I. 2003. Prevalence of *Listeria monocytogenes* in poultry meat, poultry meat products and other related inanimates at Faisalabad. *Pakistan J Nutr* 2: 346-349.

37. Manges AR, Smith SP, Lau BJ, Nuval CJ, Eisenberg JN, Dietrich PS and Riley LW. 2007. Retail meat consumption and the acquisition of antimicrobial resistant *Escherichia coli* causing urinary tract infections: A case-control study. *Foodborne Pathog Dis* 4: 419-431.

38. Mulder RWAW. 1997. Transmission of avian pathogens. Biosecurity in poultry production in the next century. *Acta Vet Hungarica* 45: 307-315.

39. Nathaniel T and Milt Nelson T. 2004. *Biosecurity for poultry*. University of Maryland.

40. Njagi LW, Mbuthia PG, Bebora LC, Nyaga PN, Minga Uand Olsen JE. 2004. Carrier status for *Listeria monocytogenes* and other *Listeria* species in free range farm and market healthy indigenous chicks and ducks. *East African Med J* 8: 529-534.

41. OIE. 2010. Avian tuberculosis. Manual of diagnostic tests and vaccines for terrestrial animals. *OIE*.

42. Olsen RH, Schønheyder HC, Christensen H and Bisgaard M. 2011. *Enterococcus faecalis* of human and poultry origin share virulence genes supporting the zoonotic potential of *E. faecalis*. *Zoonoses Public Health* doi: 10.1111/j.1863-2378.2011.01442.x.

43. Rebagliati V, Philippi R, Rossi M and Troncoso A. 2009. Prevention of foodborne listeriosis. *Indian J Pathol Microbiol* 52: 145-149.

44. Rizzoli A, Rosà R, Rosso F, Buckley A and Gould E. 2007. West Nile virus circulation detected in northern Italy in sentinel chickens. *Vector Borne Zoonotic Dis* 7: 411-417.

45. Rohde G, Straube E, Essig A, Reinhold P and Sachse K. 2010. Chlamydial zoonoses. *Dtsch Arztebl Int* 107: 174-80.

46. Saleha A and Zunita Z. 2010. Methicillin resistant *Staphylococcus aureus* (MRSA): An emerging veterinary and zoonotic pathogen of public health concern and some studies in Malaysia. *J Anim Vet Adv* 9: 1094-1098.

47. Saleha AA, Mead GC and Ibrahim AL. 1998. *Campylobacter jejuni* in poultry production and processing in relation to public health. *World's Poultry Sci J* 54: 49-58.

48. Shivaprasad HL. 2000. Fowl typhoid and pullorum disease. *Rev Sci Tech Off Int Epiz* 19: 405-424.

49. Singh BR. 2005. *Prevalence of Salmonella serovars in animals in India.* (www.aclisassari.com)

50. Une Y and Mori T. 2007. Tuberculosis as a zoonosis from a veterinary perspective. *Comp Immunol Microbiol Infect Dis* 30: 415-425.

51. USFDA. 2011. Guidance for industry: Control of *Listeria monocytogenes* in refrigerated or frozen ready-to-eat foods; Draft Guidance. USFDA.

52. Wang Q, Chang BJ and Riley TV. 2010. *Erysipelothrix rhusiopathiae*. *Vet Microbiol* 140: 405-417.

53. Weese JS, Reid-Smith RJ, Avery BP and Rousseau J. 2010. Detection and characterization of *Clostridium difficile* in retail chicken. *Lett Appl Microbiol* 50: 362-365.

54. Wesley IV. 2007. Listeriosis in animals. In: Ryser ET and Marth E H (Eds.) *Listeria, listeriosis and food safety*, 3rd edn. CRC Press.

2014, Zoonoses: Bacterial Diseases
Editor: **Sudhi Ranjan Garg**
Published by: **DAYA PUBLISHING HOUSE, NEW DELHI**

Pages **436–441**

25

Bacterial Zoonoses in Pakistan

Muhammad Athar Khan and Abdul Rehman
Department of Epidemiology and Public Health,
Faculty of Veterinary Science, University of Veterinary and
Animal Sciences, Lahore, Pakistan

The emergence of zoonotic diseases has dramatically increased in Pakistan during the last decade probably because of the lack of awareness, financial and resource constraints, poorly developed surveillance and monitoring systems, the absence of food safety regulations, etc. As the country largely depends on agriculture and livestock husbandry, such diseases not only cause morbidity and mortality but also affect the economy of the country by creating barriers to international trade. The occurrence of some newly emerging diseases has also been recorded. Major bacterial zoonoses encountered in Pakistan are briefly summarized below.

Anthrax

Anthrax, also known as wool sorter's disease, is caused by *Bacillus anthracis*. As it is a soil-borne disease, the grazing animals, particularly sheep and goats, are at a higher risk of contracting it than other animals. Sporadic cases in human beings and outbreaks of anthrax in animals do occur in various parts of Pakistan every year (Ahmad 2004). In humans, the disease is communicable through contaminated meat, skin abrasions, or wool processing. During wool handling, the spores are inhaled and deposited in the lungs causing a fatal outcome. Butchers are also at a greater risk of contracting the diseases during slaughtering procedures and processing of meat, primarily via skin abrasions. The disease has three forms: pulmonary, gastrointestinal and cutaneous forms, depending on the entry of infectious agent. Of these, cutaneous form is the most common (up to 95 per cent), whereas pulmonary form is considered to have the highest case fatality rate, which is more than 90 per cent. Humans are considered as the dead-end host. Anthrax is a notifiable disease in Pakistan and treatment of the diseased animal, either suspected or confirmed, is not allowed to be practised due to the zoonotic potential of the pathogen. The burial of a suspected or confirmed case is mandatory. Although no comprehensive data regarding the prevalence and incidence of anthrax are available in Pakistan, but still sheep and goat population is vaccinated at least once a year. About 60,000 doses of anthrax spore vaccine is being produced annually in the Veterinary Research Institute, Lahore, Pakistan. The individuals engaged in high risk occupations (veterinarians, butchers, individuals working in fiber and leather industry and livestock farmers) should be focused for awareness campaigns about the mode of transmission of the disease to reduce the incidence.

Brucellosis

Brucellosis is one of the important chronic diseases of animals and humans (undulant fever) caused by *Brucella*. In Pakistan, the disease in ruminants has become an endemic disease. The reported incidence of brucellosis in cattle and buffaloes in different areas of the country is 3.25 to 4.4 per cent (Naeem *et al.* 1990). If untreated, the disease may last for years but prolonged administration of tetracycline and streptomycin is effective (Gul and Khan 2007).

Seroprevalence of brucellosis in humans (abattoir workers) was estimated as 21.7 per cent by ELISA in Lahore, 5.5 per cent in Swat by ELISA and Rose Bengal plate test (RBPT) (Fatima and Farkhanda 2008). The estimated prevalence was 14 and 18 per cent in cattle and 15.38 and 35 per cent in buffalos at different livestock farms in Punjab province (Nasir *et al.* 2004) and 4.75 per cent in Swat (Hamidullah *et al.* 2009). In case of horses, the estimate was 20.7 per cent by RBPT and 17.7 per cent by standard agglutination test (SAT) in Faisalabad (Abubakar *et al.* 2010).

Glanders

Glanders, a bacterial disease of equines caused by *Burkholderia mallei*, is one of the oldest known diseases. It is characterized by ulcerating nodules in the upper respiratory tract, lungs and skin and may occur in three forms *i.e.* nasal, pulmonary and the cutaneous. Farcy is synonym for cutaneous form. The most common source of infection of glanders is contaminated food and water. In Pakistan, glanders is diagnosed by indirect method in ISO certified labs by complement fixation test (CFT) and mallein test. A trace of cases has been reported in equines since 1999 to 2007 from Punjab province but no human cases have been reported (Niaz 2010). There is no reliable treatment and no vaccine available for glanders. Glanders can be controlled and eradicated by strict testing, surveillance system, quarantine, culling of suspected cases, awareness of people, improving sanitary conditions and adopting hygienic measures (CDC 2012).

Leptospirosis

Serological evidence or isolation of leptospires has been reported in animals and humans in Pakistan and world over. It is an important disease of cattle, buffalo, sheep, goat, horse, other animals (dog, wild and zoo animals) and humans (Chaudhry *et al.* 1996). It is a cosmopolitan, zoonotic disease and has now been grouped under the diseases causing abortion syndrome (up to 30 per cent) in cattle. In livestock, the disease is characterized by reproductive disorders including abortion. The main symptoms as observed by many workers are hematuria, hemoglobinurea, jaundice, decreased milk production, septicemia and abortion in females. Although the mortality rate is quite low (5 per cent) (Guidugli *et al.* 2001), the morbidity rate is high. Its incidence in the tested herds varies from 3-6 per cent in sheep and goats and up to 30-48 per cent in bovines depending on herd immunity. The incidence varies with the intensity of the spread of the disease as detected from apparent symptoms and confirmation by serology and isolation of the organism. Generally, buffaloes have higher rate of infection than cattle, probably due to their wallowing habit (Jamil 1998).

Frequently high serum antibody titers have been observed against *L. canicola*, *L. pomona* and *L. hardjo*, which indicates that these three serovars infect cattle and buffaloes and remain viable in their urogenital tract. The other serovars, although more commonly found, may not leave as high titers. These do not have cattle as their primary host and although they infect cattle and buffaloes, but probably do not remain viable in these beyond a limited time (Khan and Khan 1988). The occurrence of leptospirosis in the farm animals is of significant public health importance as the organisms are shed via urine.

Salmonellosis

Salmonellosis is an important disease of digestive tract of mammals and fowls caused by *Salmonella*. *Salmonella enteritidis*, *S. typhimurium*, *S. dublin* and *S. newport* are common among the salmonellae. There is ample evidence for the transmission of *Salmonella* to humans via food, poultry meat, eggs and contaminated water. Recently increase has been observed in the number of food poisoning outbreaks due to *S. enteritidis* in poultry. Gross pathological observations after natural and experimental infections with *Salmonella enteritidis* revealed that the organism may cause infection in chicks and laying hens and cause prolonged faecal shedding. Isolation of the organism from ovaries, oviducts and egg contents indicates the possibility of transovarian transmission of the pathogen (Younus *et al.* 2006a). Variation in mortality rates, clinical signs, faecal shedding and frequency of production of contaminated eggs has been observed in chicks and hens experimentally infected with *S. enteritidis*. The choice of bacterial strain, phage type, age of bird and inoculum size might also affect the outcome of an infection (White *et al.* 1997). The prevalence of salmonellosis in Pakistan is low as most of the people follow the primordial level of prevention to stop the occurrence of foodborne diseases through maintaining hygienic conditions, as their religious practices guide them to do so. Secondly, most of the people do not use salad which is a high source of salmonella infection. Thirdly, the food, especially eggs, is adequately cooked traditionally, which results in the destruction of the pathogen (Younus *et al.* 2006b).

Tuberculosis

The World Health Organization (WHO) ranks Pakistan at 6th place in the global burden of the disease (Pakistan Today 2012). Tuberculosis (TB) is a chronic disease of humans and animals, caused majorly by *Mycobacterium tuberculosis* and rarely by *Mycobacterium bovis. Mycobacterium bovis* acts as the principal zoonotic agent. Cattle are the most important source of infection that can be transmitted to humans through infected or contaminated milk and meat, aerosol, and contact with infected animals in slaughterhouse leading to infection via direct injury to the skin and mucous membrane. Animal can also be reinfected by infected humans. In humans, TB can occur in different forms like extra pulmonary TB, miliary TB, bone and joint TB, and cutaneous TB, while lungs, intestine and regional lymph nodes are affected in case of animals (Ashraf 2008). Tuberculin allergic test is used for diagnosis of TB in livestock population. In a study, 10 out of 334 animals were found positive for TB in Punjab province of Pakistan in 2007-2008 (Usman 2003). According to WHO Global Surveillance and Monitoring Project, TB incidence is estimated at 181 per 1,00,000 people in Pakistan and an estimated 40 per cent of the prisoners are infected with TB. In 2006-2007, bovine tuberculosis in livestock, especially in large ruminants, was reduced through tuberculin testing and culling of suspected animals in Pakistan. About 2,50,000 new cases occur every year in Pakistan of which 75 per cent patients are in the 15-59 years age group. Punjab contributes 57 per cent, Sindh 23 per cent, Khyber Pakhtunkhwa 13 per cent and Balochistan 5 per cent cases (Metzger *et al.* 2010).

Future Perspectives

A wide range of zoonotic diseases exist in Pakistan. The lack of efficient monitoring and surveillance system, unavailability of staff trained in zoonotic diseases control both in medical and veterinary graduates, lack of coordination between animal health and human health sectors, communication gap between different sectors that matter in the control and eradication process of the diseases, low literacy rate, lack of funding and other resources, poorly developed food safety regulations and absence of political interest and commitment to control the zoonotic diseases are the major challenges in Pakistan.

Efficient monitoring system and control of zoonotic diseases need to include animal and human health issues into the new public health strategy plans. The government may take the following actions to improve the situation:

1. Calculate the disease load at national level, and find out which disease should be given preference according to infection rate and national GDP.
2. Monitoring and control systems for animal borne diseases should be improved and integrated with other infectious disease monitoring and control systems. Data related to animal and human health should be recorded on routine basis.

3. Multi-sectorial committees should be formed, which should be efficient enough to ensure cooperation for animal-borne diseases, control at national level and for that purpose enough resources should be provided to them. These committees must include members from all the concerned animal and public health departments.

4. High level of communication must be ensured between animal and human health departments for the timely exchange of disease information and reports on the disease control.

5. Joint training of the staff of the animal and public health departments should be conducted to facilitate the planning, implementation and evaluation of preventive and control actions.

6. Active participation of general public should be encouraged through public health education.

7. The syllabus of education for animal and human health professionals should be updated according to the recent knowledge and practical needs for the control of animal-borne diseases.

8. Political cooperation and coordination with all the concerned stakeholders must be ensured.

9. Decision-makers and policymakers should be well aware of the particular disease load in the community.

10. Networking among the concerned reference centres at both regional and national level is very essential for effective control of animal-borne diseases.

References

1. Abubakar M, Arshed MJ, Hussain M, Ehtisham-ul-Haq and Ali Q. 2010. Serological evidence of *Brucella abortus* prevalence in Punjab province, Pakistan - A cross-sectional study. *Transboundary Emerg Dis* 57(6): 443-447.

2. Ahmad K, Dil AS, Kazi BM, Saba NU, Ansari J and Nomani K. 2004. Pakistan's experience of a bioterrorism-related anthrax scare. *East Mediterr Health J* 10(1-2): 19-26

3. Ashraf W. 2008. Studies on bovine tuberclosis, paratuberclosis and brucellosis in cattle and buffalo at a local abattoir in Faisalabad. *Thesis*, University of Agriculture, Faisalabad.

4. CDC. 2012. Glanders (*Burkholderia mallei*) general information. Centers for Disease Control and Prevention. http://www.cdc.gov. Accessed 26 January 2012.

5. Chaudhry JI, Khan MA, Akhtar T, Khan AG and Chaudhry MS. 1996. Seroprevalence of leptospirosis in buffaloes. *Buffalo J* 1: 65-71.

6. Fatima Mukhtar and Farkhanda Kokab. 2008. Brucella serology in abattoir workers. *J Ayub Med Coll Abbottabad* 20(3): 57-61.

7. Guidugli F, Castro AA and Atallah NA. 2001. Antibiotics for preventing leptospirosis (Cochrane Review). In: *The Cochrane Library*. Issue 3. Oxford: Update Software. 2001.

8. Gul ST and Khan A. 2007. Epidemiology and epizootology of brucellosis A review. *Pakistan Vet J* 27(3): 145-151.

9. Hamidullah M, Khan R and Khan I. 2009. Seroprevalence of brucellosis in animals in district Kohat NWFP and comparison of two serological tests. *Pakistan J Sci* 61(4): 242-243.

10. Jamil M. 1998. Epidemiological survey and serogrouping of type strains of leptospirosis in vertebrate animals in Pakistan. Technical research report of Pakistan Science Foundation, University of Agriculture, CVS, Lahore.

11. Khan MA and Khan MS. 1988. Serological response against *Leptospira enterrogans* in the aborting animals. *Proc. World Buffalo Congress*, December 12-16, 1988, New Delhi, India. p. 352.

12. Metzger P, Baloch NA, Kazi GN and Bile KM. 2010. Tuberculosis control in Pakistan: Reviewing a decade of success and challenges. *East Mediterr Health J* 16 Suppl: S47-53.

13. Naeem K, Akhtar S and Ullah N. 1990. The serological survey of bovine brucellosis in Rawalpindi and Islamabad. *Pakistan Vet J* 10(4): 154-156

14. Nasir AA, Parveen Z, Shah MA and Rashid M. 2004. Seroprevalence of brucellosis in animals at government and private livestock farms in Punjab. *Pakistan Vet J* 24(3): 144-146.

15. Niaz N. 2010. A study of incidence of glanders in Lahore. *M.Phil. Thesis in Microbiology*, University of Veterinary and Animal Sciences, Lahore.

16. Pakistan Today. 2012. http://www.pakistantoday.com.pk/2012/04/21/city/islamabad/pakistan-ranks-6th-among-high-tb-risk-countries/. Accessed 2 March 2012.

17. Usman M. 2003. Prevalence and pathological studies on tuberculosis in buffaloes at various livestock farms. *M.Phil. Thesis in Pathology*, University of Veterinary and Animal Sciences, Lahore.

18. White PL, Baker AR and James WO. 1997. Strategies to control *Salmonella* and *Campylobacter* in raw poultry products. *Rv sci tech Off int Epiz* 16(2): 525-541.

19. Younus M, Chaudhary ZA, Khan MA, Munir MA, Shakori AR, Mehmood N and Tipu MY. 2006a. Prevalence of *Salmonella* species in poultry eggs in Lahore through polymerase chain reaction. *Sci int (Lahore)* 18(3): 229-233.

20. Younus M, Chaudhary ZI, Khan MA, Munir MA, Shakori AR, Mehmood N and Tipu MY. 2006b. Prevalence of *Salmonella* species in poultry meat in Lahore through polymerase chain reaction. *Sci int (Lahore)* 18(4): 309-312.

2014, Zoonoses: Bacterial Diseases   *Pages* **442–449**
Editor: **Sudhi Ranjan Garg**
Published by: **DAYA PUBLISHING HOUSE, NEW DELHI**

26

Bacterial Zoonoses in Ethiopia

Asefa Deressa, Fasil Mengistu and Abraham Ali
Ethiopian Health and Nutrition Research Institute,
Zoonoses Research Team, Addis Ababa, Ethiopia

The burden of zoonotic pathogens in Ethiopia is not yet fully quantified and documented but a number of zoonoses are commonly encountered, mainly in rural settings where health systems are not well established and where people are not sedentary and move along river basins in search of green fodder for the livestock. Agriculture is the means of livelihood for 90 per cent of the people in the country, largely by mixed farming; the highlanders with 20 per cent of livestock resource and the lowlanders with the remaining 80 per cent. Some communities, be it in pastoral or in mixed farming systems, due to their socio-cultural traditions are closely associated with food animals and pets, sharing the same housing that favours the spill over of zoonotic infections from animals to humans and vice versa. Fragmented research work has been carried out and some data have been generated for endemic and prioritized diseases in different parts of the country. The situation of some major bacterial diseases in the country is briefly discussed below.

Brucelloses

Seroprevalence study of bovine brucellosis in 2005 to 2007 in Ada'a Liben Dairy Cooperatives revealed an individual prevalence of 2.3 per cent and herd prevalence of 1.8 per cent to 3.3 per cent (Abraham *et al.* 2008). Another study in cattle (n=1782) estimated 5.8 per cent seroprevalence in Sululta, 4.55 per cent in Muke turi, 2.33 per cent in Debre Zeit and 3.38 per cent in Wondo Genet areas (Mulugeta *et al.* 2008). In yet another study during 2008-2009, the examination of 435 blood samples from cattle of 6 or more months of age in four districts of Jijiga zone in eastern Ethiopia demonstrated 1.38 per cent overall prevalence of bovine brucellosis (Hailu *et al.* 2011). A study in 500 small ruminants during 2008-2009 in and around Bahir Dar in northwest Ethiopia revealed 0.4 per cent seroprevalence of brucellosis (Yeshwas *et al.* 2011).

Bovine Tuberculosis

The reported prevalence rate of bovine tuberculosis (BTB) ranges from 3.4 per cent in small holder production system) to 50 per cent in intensive dairy productions. In the slaughterhouses, it ranged from 3.5 per cent to 5.2 per cent at various places in the country (Shitaye *et al.* 2007). *Mycobacterium bovis* strains isolated from tuberculosis (TB) lesions from 1138 cattle slaughtered at Kombolcha abattoir in northeastern Ethiopia were characterized. At least one TB lesion was observed in 57 (5 per cent) cattle, of which 27(47 per cent) yielded mycobacterium isolates. Of these, 25 isolates were identified as *M. bovis* and two as *M. tuberculosis* (Gobena *et al.* 2010). Comparative intradermal tuberculin test in cattle in northwest Ethiopia revealed an overall 15.7 per cent disease prevalence (192/1220) considering doubtful reactors as negative (Fetene *et al.* 2011). A similar study in 2,216 cattle in four regions of Ethiopia revealed 3 per cent overall prevalence of cattle bovine tuberculosis, the highest (7.9 per cent) being in Meskan Mareko in central Ethiopia and the lowest (1.2 per cent) in Woldia in the northeast edge of the Rift Valley (Tschopp *et al.* 2009).

The status of BTB in wildlife populations that often share habitat with livestock is unknown. Blood and tissue samples from 133 mammals of 28 species collected from 2006 to 2008 revealed that serum samples from 20 of 87 (23 per cent) animals were positive for BTB by the rapid test while acid fast bacilli were isolated from 29 of 89 (32.5 per cent) animals (Tschopp *et al.* 2010).

Salmonollosis

On examination of raw and medium cooked 'kitfo' samples from ten food establishments in 1996-1997 in Addis Ababa, *Salmonella* were isolated from 21 of the 50 raw kitfo samples (Mezgbu and Mogessie 1998). A study over the period 1997 to 2002 in apparently healthy slaughtered cattle and camel retail meat products revealed *salmonella* in 4.2 per cent cattle, 16.2 per cent camel, 12.1 per cent minced beef and 23.6 per cent chicken meat and giblets (Bayleyegne *et al.* 2003).

In a study undertaken to determine the prevalence and distribution of *Salmonella* in apparently healthy slaughtered camels (n=119) in eastern Ethiopia, samples of faeces, mesenteric, lymph nodes, spleen, liver, abdominal and diaphragmatic muscles were analyzed. Salmonellae were detected in total 16.2 per cent (n=714) samples including 15.1 per cent samples of faeces, 15.9 per cent of mesenteric lymph nodes, 11.8 per cent of livers, 14.3 per cent of spleen, and 20.1 per cent of the abdominal and diaphragmatic muscle specimens (Molla *et al.* 2004).

Campylobacteriosis

Sheep carcasses (n=218) and goat carcasses (n=180) at a private export abattoir in Debre-Zeit in 2007- 2008 were examined for *Campylobacter* spp. Campylobacters were isolated from 10.1 per cent carcasses. Of the 40 thermophilic campylobacter isolates, *C. jejuni and C. coli* accounted for 29 (72.5 per cent) and 11 (27.5 per cent), respectively (Tefera *et al.* 2009). In another study, campylobacters were isolated from 192 (39.6 per cent) faecal specimens collected from 485 various food animals (cattle 205, poultry 191, pigs 18, sheep 71) in urban and rural farm animal settings in Jimma in southwest Ethiopia during January-April 2004 (Tesfaye *et al.* 2005). The highest isolation rate was recorded in chickens (68.1 per cent), followed by pigs (50 per cent), sheep (38 per cent) and cattle (12.7 per cent). Out of 192 isolates, 135 (70.3 per cent) were *C. jejuni*, 51 (26.6 per cent) *C. coli* and 6 (3.1 per cent) *C. lari*. *C. jejuni* was the most prevalent species in chickens (80.8 per cent), followed by sheep (59.3 per cent) and cattle (53.8 per cent). All isolates from pigs were *C. coli* (100 per cent).

Listeriosis

Examination of 316 samples of retail meat and milk products in Addis Ababa during 2003-2004 revealed 103 (32.6 per cent) samples positive for *Listeria* (Molla *et al.* 2004). Listeria were isolated from 69.8 per cent samples of pork, 47.5 per cent of minced beef, 43.5 per cent of ice cream, 18.6 per cent of fish, 15.4 per cent of chicken and 1.6 per cent of cottage cheese. *L. monocytogenes* was detected in 5.1 per cent samples. The isolates belonged to serotype 1/2b, 4b and 4e. Other Listeria included *L. innocua* (65 per cent), *L. seeligeri* (8.7 per cent), *L. welshimeri* (6.8 per cent), *L. murrayi, L. ivanovii* and *L. grayi* (each 0.9 per cent). In another study in Addis Ababa in 2008-2009, 102 (26.1 per cent) out of 391 food samples of animal origin were positive for *Listeria* (Simon *et al.* 2010).

Leptospirosis

In a study, serum specimens collected from cattle, goats and sheep in three areas revealed leptospira titers of 1:16 or greater in 26.3 per cent cattle and 16.7 per cent goats but none in sheep. The most frequently indicated serotype was *autumnalis*. Fifteen of 36 (41.7 per cent) positive cattle had antibodies to more than 1 serotype. Wild rodents were trapped in five provinces of southwest Ethiopia. Serologic testing for leptospirosis, using the macroscopic slide agglutination test, failed to reveal any serum positive to the 18 antigens represented in the test. Negative kidney cultures reinforced the negative serologic findings (Obeck and Birhanu 1976).

Anthrax

An investigation of sudden death in a goat in Wabessa village in the Dessie Zuria district of Ethiopia in 2002 revealed outbreak mortality rates of 7.7 per cent in cattle, 32.7 per cent in goats and 47.1 per cent in donkeys. It is observed that the disease occurs annually in this area in May and June (Shiferaw 2004).

Zoonotic Disease Surveillance System

Surveillance is considered the best strategy for strengthening the prevention and control of zoonotic diseases. It provides evidence and basis to arrive at decisions on public and veterinary public health interventions for better preparedness and response to epidemics of zoonotic diseases. The collection, analysis and dissemination of information in the surveillance system help in proper planning, implementation and evaluation of intervention strategies.

The major problems and challenges related to prevention and control of many diseases observed by the WHO-Africa in the region are listed below:

☆ Breakdown or weakness of the health infrastructure,

☆ Weak disease surveillance systems,

☆ Absence of integration of animal and public health systems,

☆ Insufficient information on the burden of zoonotic diseases,

☆ Insufficient human and financial resources, and

☆ Weakness or absence of collaboration and coordination between public health, veterinary, agricultural and wildlife sectors.

Strengthening of Disease Prevention and Control Strategies

The existing challenges need to be overcome by strengthening and establishing linkages between animal health and public health sectors and by the following measures:

☆ Conducting studies to provide evidence for priority setting and guiding action,

☆ Building up existing data,

☆ Strengthening the capacity of diagnostic and research laboratories,

☆ Sharing information, and

☆ Setting priorities based on both health impact and the agricultural market.

During the occurrence of a zoonotic disease in the human population in areas where that disease is endemic in the animal population or when an epidemic situation occurs in the animal or human population, the priority would be to address the needs of the patients and the community in general. It is also important to address issues concerning animal health due to the threat it is posing to the human population as well as due to economic reasons. Moreover, the threat of transmission of infections from humans to animals should also be kept in mind. It is often necessary to understand the natural ecology of a zoonotic pathogen and its host in order to control disease transmission and prevent future outbreaks. Some general guiding principles for studying zoonotic diseases are listed below:

☆ Identify the source of infection to determine whether it is from wildlife, domestic or peri-domestic animals, or from multiple sources.

☆ Establish the mode of transmission to determine whether it is by direct contact, vector-borne, environmental contamination, or a combination of different modes of transmission.

☆ Identify potential host species and the natural reservoirs of the zoonotic pathogen. Current molecular and epidemiological knowledge can be used to identify target species for surveys.

☆ Conduct preliminary surveys of target species and follow up, when indicated, with long-term ecological and epidemiological studies of identified reservoir species in the wild and/or in an experimental setting where appropriate.

In the event of an epidemic, generally there is no time for training the public health and veterinary professionals or even to assemble all necessary supplies. Therefore, preparedness is the key to successful intervention for mitigation and control of an epidemic. There should be some guidelines intended to provide a detailed description of the interventions of the various players at various levels. There must also be standard guidelines for all levels of the public health and animal health systems involved in disease surveillance and reporting.

Future Perspectives

Zoonoses control programme is a complex process that has to consider several issues in Ethiopia. Various factors like diverse agro-ecology, biological, socio-cultural, technical, economic considerations, as well as financial and operational aspects are the key issues. A national zoonoses control programme in Ethiopia should broadly focus on improving the general state of health in rural and urban settings, strengthening infrastructure and development of services and overall well being of the public at large. In order to have a sound and effective control programme, meaningful statistical and epidemiological data is a prerequisite.

Ethiopian Health and Research Nutrition Institute (EHNRI) serves as the National Public Health Laboratory in the country. A strong collaboration between EHNRI and the Ministry of Agriculture and Rural Development (MoARD) and the Ministry of Health (MoH) is required so that coordinated interventions can take place. At the regional level, laboratories involved in public health and animal health must also share information regarding confirmation of any zoonotic disease. These laboratories must be able to conduct diagnostic tests for confirmation of zoonotic diseases and also be able to help in monitoring the drug susceptibility patterns. Strong collaboration between public health and animal health laboratories must be ensured at all times and information shared as soon as there is suspicion of zoonotic disease. It is important that standard techniques are used in laboratories. The clinical and public health laboratory referral linkages need to be established and strengthened across the country.

It seems worthwhile to launch zoonoses surveillance with specific targets of pathogen infections across the country. A fully operational and effective zoonoses surveillance system coordinated activities at the various levels of the public health and animal health systems as well as coordination and information sharing between Ministry of Health and Ministry of Agriculture are necessary. However, activities at community level are fundamental for successful implementation of activities (Luis 2006). In line with this, the role of inter-sectoral institutions among MoARD, MoH and EHNRI should be clearly defined. Accordingly, the MoH and MoARD should provide integrated zoonoses diseases surveillance support for the system, including:

☆ Development of comprehensive technical guidelines for each level;

☆ Adopting guidelines at each level within the country;

☆ Advocacy for resource mobilisation;

☆ Monitoring and detection of zoonotic diseases across regions; and

☆ Interacting with the MoARD/EHNRI in the planning and interventions deemed necessary to control/eradicate a specific disease.

EHNRI is the reference centre for zoonotic diseases in the country. Surveillance activities and trainings are carried out and coordinated by this institution. It also provides the linkage between the MoARD and MoH. The institute also works as a reference centre for diagnosis of many of the zoonotic diseases in the country. It is through its close collaboration with the MoARD and the MoH that the new system for surveillance, control and eradication of zoonotic diseases may achieve the expected results. Inclusion of other partners (national and international) in the system is also necessary for a strong programme for zoonoses prevention and control.

References

1. Abraham Abebe, Tesfu Kassa, Tamirat Degefa and Edris Hansar. 2008. Investigation on seroprevalence of bovine brucellosis and related major reproductive health disorders in Ada'a Liben dairy co-operatives. *Eth Vet J* 12(2): 19-35.

2. Abraham Ali, Fasil Mengistu, Kedir Hussen, Garoma Getahun, Asefa Deressa, Eshetu Yimer and Kassahun Tafese. 2010. Overview of rabies in and around Addis Ababa in animals examined in EHNRI Zoonoses Laboratory between 2003 and 2009. *Eth Vet J* 14: 91-101.

3. Abunna Fufa, Getachew Tilahun G, Megersa B, Regassa A and Kumsa B. 2008. Bovine cysticercosis in cattle slaughtered at Awassa Municipal Abattoir, Ethiopia: Prevalence, cyst viability, distribution and its public health implication. *Zoonoses Public Health* 55: 82-88.

4. Abyot Desta, Solomon Shiferaw, Andargachew Kassa, Techalew Shimelis and Simachew Dires. 2005. Leishmaniasis, Diploma Program, Debub University for the Ethiopian Health Center Team in collaboration with the Ethiopia Public Health Training Initiative, the Carter Center, the Ethiopia Ministry of Health, and the Ethiopia Ministry of Education, 2005.

5. Ameni G, Desta F and Firdessa R. 2010. Molecular typing of *Mycobacterium bovis* isolated from tuberculosis lesions of cattle in north eastern Ethiopia. *Vet Record* 167: 138-141.

6. Asefa Deressa, Workenesh Ayele, Amenu Gisso, Temesgen Lemi, Million Wendabeku and Rosemary Sang. 2007. Assessing the threat of Rift Valley fever (RVF). Ethiopian Medical Association XLIII, Annual Medical Conference and Medical Exhibition, United Nation Conference Center (UNCC), May 30-June1, 2007.

7. Bayleyegne Molla, Daniel Alemayehu and Woubit Salah. 2003. Sources and distribution of Salmonella serotypes isolated from food animals, slaughterhouse personnel and retail meat products in Ethiopia: 1997-2002. *Ethip J Health Dev* 17(1): 63-70.

8. Bayleyegn Molla, Roman Yilma and Daniel Alemayehu. 2004. *Listeria monocytogenes* and other *Listeria* species in retail meat and milk products in Addis Ababa, Ethiopia. *Ethiop J Health Dev* 18(3): 208-212.

9. de Castro L. 1908. Medicina vecchiae medicinia nuova in Abissinia. Soc Geo *Italiana Boll Ser* 49: 1087.

10. Dinka Ayana, Getachew Tilahun and Abebe Wossene. 2009. Study on *Eimeria* and *Cryptosporidium* in sheep and goats at ELFORA export abattoir, Debre- zeit, Ethiopia. *Turk J Vet Anim Sci* 33 (5): 367-371.

11. Fetene T, Kebede N and Alem G. 2011. Tuberculosis infection in animal and human populations in three districts of Western Gojam, Ethiopia. *Zoonoses Public Health* 58: 47-53.

12. Fiseha G. 1995. Zoonotic diseases in Ethiopia. *ESAP*, 1.

13. Hailu Degefu, Mohamed Mohamud, Mussie Hailemelekot and Moti Yohannes. 2011. Seroprevalence of bovine brucellosis in agropastoral areas of Jijiga zone of Somali National regional state, Eastern Ethiopia. *Ethiop Vet J* 15: 37-47.

14. Luis Fillipe Lde JLoureiro. 2005. *Establishment of a surveillance system for zoonotic diseases in Ethiopia*. Addis Ababa, Ethiopia.

15. Megersa B, Tesfaye E, Regassa A, Abebe R and Abunna F. 2010. Bovine cysticercosis in cattle slaughtered at Jimma Municipal Abattoir, South western Ethiopia. Prevalence, cyst viability and its socio-economic importance. *Vet World* 3 (6): 257-262.

16. Mezgbu Tegegne and Mogessie Ashenafi. 1998. Microbial load and incidence of *Salmonella* spp. in 'Kitfo', a traditional Ethiopian spiced, minced meat dish. *Ethip J Health Dev* 12(2): 135-140.

17. Molla B, Mohamed A and Salah W. 2004. *Salmonella* prevalence and distribution of serotypes in apparently healthy slaughtered camels (*Camelus dromedaries*) in Eastern Ethiopia. *Trop Anim Health Prod* 36 (5): 451-458.

18. Mulugeta Tefera, Tesfu Kassa, Kelay Belehu, Abraham Abebe, Melaku Negash and Ahmed Ali. 2008. Studies on the occurrence of brucellosis in cattle in selected sites of Ethiopia and indications on its zoonotic significance. *Ethiop Vet J* 12 (1): 161-181.

19. Nigatu Kebede. 2008. Cysticercosis of slaughtered cattle in northwestern Ethiopia. *Res Vet Sci* 85: 522-526.

20. Obeck DK and Birhanu GM. 1976. *Leptospirosis survey of rodents and domestic animals in Ethiopia*. Report No. TR 14. Naval Medical Research Unit No. 5, New York.

21. Rea Tschopp, Esther Schelling, Jan Hattendrof, Abraham Aseffa and Jakob Zinsstag. 2009. Risk factors of bovine tuberculosis in cattle in rural livestock production systems of Ethiopia. *Preventive Vet Med* 89: 205-211.

22. Shitaye JE, Tsegaye W and Pavlik I. 2007. Bovine tuberculosis infection in animal and human populations in Ethiopia: a review. *Veterinarni Medicina* 52 (8): 317-332.

23. Tefera Woldemariam, Daniel Asrat and Girma Zewde. 2009. Prevalence of thermophilic *Campylobacter* species in carcasses from sheep and goats in an abattoir in Debre Zeit area. Ethiopia. *Ethiop J Health Dev* 23(3): 229-233.

24. Tesfaye Kassa, Solomon Gebre-selassie and Daniel Asrat. 2005. The prevalence of thermo tolerant *Campylobacter* species in food animals in Jimma Zone, southwest Ethiopia. *Ethiop J Health Dev* 19(3): 225-229.

25. WHO/FAO/OIE, 2004. Report of the WHO/FAO/OIE joint consultation on emerging zoonotic diseases in collaboration with the Health Council of the Netherlands, 3–5 May 2004 – Geneva, Switzerland.

26. Yeshwas Ferede, Desalegne Mengesha, Gebreyesus Mekonen and Mussie H Melekot. 2011. Study on the seroprevalence of small ruminant brucellosis in and around Bahir Dar, North West Ethiopia. *Ethiop Vet J* 15 (2): 35-44.

2014, Zoonoses: Bacterial Diseases
Editor: **Sudhi Ranjan Garg**
Published by: **DAYA PUBLISHING HOUSE, NEW DELHI**

Pages **450–458**

27

Community-based Approach for Zoonoses Control

K. Vrinda Menon and B. Sunil
Department of Veterinary Public Health,
College of Veterinary and Animal Sciences, Mannuthy, Thrissur – 680 651

Zoonoses act as a double-edged weapon causing fatal disease in humans and undermining animal health and productivity. Zoonoses are fundamental determinants of community health (Stephen *et al.* 2004). One of the principal causes of the emergence and re-emergence of such diseases is the increased risk of exposure to certain pathogens. This is because of several factors such as animal and human diet changes, increased densities of production animals or wildlife population, human and animal population displacement, increased contacts with wildlife reservoirs due to various outdoor leisure activities, accelerated degradation of natural environment and global warming (Wilcox and Gubler 2005). Moreover, the host defence is also showing increasing breakdown due to immunosuppression and emergence of drug-resistant bacterial strains. Evidently, apart from the prevention and control of the pathogens in animals and humans by movement restrictions, vaccination, improved diagnosis and treatment, raising awareness in the community about the disease and its control measures is very important in the present-day scenario. Community participation is, therefore, widely promoted as an important feature in different projects concerning disease control. However, the definition, applications and expectations of community participation vary considerably among the professionals including veterinarians.

The grassroots involvement of local communities in a zoonotic disease prevention and control programme depends on ensuring that these communities understand the disease in its varied dimensions. The commitment of the community towards the control of disease is possible only through efficient awareness campaign which will allow better disease reporting and vaccination coverage. Community involvement in health generates in individuals a sense of responsibility for their own health and welfare. Higher the level of self reliance and social awareness, more the individuals will accept responsibility for protecting their animals and themselves from disease hazards transmitted directly or through foods of animal origin or vectors. Community-based zoonotic disease surveillance system may be helpful in making better assessment of the impact of disease in the population as the reliance on official reports alone may lead to gross underestimation of the problem (Mariner 2002). Hence, epidemiological surveys should include the resources such as livestock owners, community animal health workers, para-veterinarians, abattoir workers, etc.

Community Participation

Community is defined as a social group of people sharing an environment, interest, belief, resources, preferences, needs, risks, common leadership and a number of other conditions and degree of cohesiveness (McMillan and Chavis 1986). The community involvement in a programme is necessary for the sustainability of a community initiation. It also makes it easy to access the local constraints and opportunities. The extension programmes initiated by the government also require involvement of the community for sustainability. Hence, empowering the community to participate in the health issues is essential for the success of such projects.

Community participation in health development has been identified and adopted as one of the fundamental strategies for accomplishing the priority objectives of the primary health care. According to the declaration of the International Conference in Primary Health Care (1978, Alma-Ata), "Community participation is the process by which individuals and families assume responsibility for their own health and welfare and for those of the community, and develop the capacity to contribute to their and the community development." Community participation results in involving people to solve their own problems. Moreover, such participation is a basic human right and fundamental principle to democracy.

The participation process in a community includes initiation, preparation, participation and continuation. Community assessment is the most important activity in caring for people in the community (Wilcox 1994). Knowing the health situation of the people living in the community can lead and help the public health personnel in implementing appropriate and effective interventions to improve the health status of people.

Tools for Community Assessment

There are seven tools that are effectively used for community assessment. These are convenient to use and can help public health personnel understand the community (Chuengsatiansup 2002).

1. **Community mapping:** Mapping is the only method which helps the public health personnel to rapidly see the whole community. While mapping, they can observe the community surroundings, environment, people's way of life and living situation. It helps them understand the physical and social aspects of the community. Community mapping helps the health personnel not only in getting the picture of the community area but also to learn about the community.

2. **Kinship mapping:** It is the method to describe the genetic relationship of people in the community and how they care for their relatives. It helps the public health personnel develop a better understanding about their way of living.

3. **Community organisation structure:** A community structure means the relationship of people in a community. Community organisations should have memberships, same goals of life and management. Knowing the community organisations helps the public health personnel know where to start working with the community.

4. **Community health system:** Usually public health personnel use their thinking to dominate people in a community without understanding about their way of life and beliefs. The community health system assists in changing the endless beliefs and faith of the community and also helps in understanding the potential in them to tackle various health issues.

5. **Community calendar:** Understanding everyday life of people in the community is a tool to know about the diseases related to people's behaviour and nature of work performed during the period.

6. **Cultural activity:** It shows the practical relationship and lifestyle of people in the community. If the public health personnel understand it, they will understand the role of the community.

7. **Life history:** The assessment on the occurrence of a disease in a community can be made by knowing the life history of people. For example, rearing of cattle by feeding the animals on the wet floor under their house, especially in the north-east regions of India, makes the community more prone to TB.

Community Animal Health Workers

The main reason for failure in disease control programme is the inability to instill confidence in the beneficiaries about the value of the campaign. This can be generally overcome by community dialogue. Once a common understanding is reached, an agreement that is termed as 'community contract' (FAO 2000) can be made where each side clearly states what can be done. The community remembers the broken promises, intended or not, usually long; so the veterinary team may not get a second chance.

As each community is characteristic by unique traditional institutions, customs and experiences, there is no one ideal approach in practice that can be recommended. One general model that has worked well in animal health care in broad spectrum of culture is the community animal health workers. Animal health projects have been relatively successful in many developing countries, which involved community participation as the guiding principle for project design and implementation. These projects worked with local people to describe and analyse the animal health concerns and to identify the solutions. The communities selected people who would work as community-based animal health workers (CAHW). The concept of CAHW probably arose from experiences in the human health sector. CAHW systems have a key role in strengthening the capabilities (epidemiological surveillance, disease control, animal disease reporting systems) of veterinary services in remote areas (Catley 2002). Community-based animal health delivery systems can also assist in animal identification systems, traceability systems, and animal movement control systems. In remote areas of developing countries where infrastructure and enforcement of regulations is weak, CAHWs have an important, but as yet untapped, role to play in raising awareness on the need for these capabilities. They have already proven to be excellent entry-points for human health, relief and conflict management issues in many areas of Africa. They should be personally accountable during animal health programme and campaign.

Kudumbashree (meaning prosperity of the family) is a women-oriented community based mission launched by the Government of Kerala in 1998 for wiping out absolute poverty from the state through concerted community action under the leadership of Local Self Governments and is today one of the largest women empowering projects in India (Government of Kerala 2007). Women self help group movement has gathered momentum as a powerful instrument for socio-economic transformation of poor people in India. The attitude of these self help groups towards livestock raising was high (Anand 2002). The Kudumbashree workers have been successfully engaged in collection of waste from households on a nominal charge and the improvement of the communities in the disposal of this waste has been successful in certain areas of Ernakulam and Thiruvananthapuram corporation areas in Kerala. Hence, these workers can be mobilised for animal health care activities by training them and making them accountable for the work done by them through regular monitoring by the veterinarians of the area.

Role of Community

Health programmes are unlikely to succeed if community involvement is not an integral part of the structure and execution at local level. Laws, regulations and veterinary policy measures alone will not bring the desired results. Moreover, the individuals and the community must be willing to acquire new knowledge, and to translate it into wholesome habits and constructive behaviour patterns. Human, animal health and veterinary public health (VPH) systems are responsible for providing clear information and explaining the favourable and adverse consequences of various intervention measures being proposed, as well as their relative costs. The core functions of VPH involve diagnosis, surveillance, epidemiology, control, prevention and elimination of zoonoses. VPH activities can improve human health by reducing exposure to hazards arising from interaction with animals and animal products. Hence, there is an urgent need to expand the links between human and animal medicine.

Community participation plays an integral role in the implementation of VPH programmes. It can be achieved by encouraging the participation of all stakeholders, including women and children in decision-making at the local level so as to increase the ownership accountability and sustainability. The strategies to get community participation in VPH programmes include the use of trained auxiliaries to deliver VPH services locally, involvement of communities in the development and management of VPH programmes, using participatory field research to identify community priorities, evaluating the impact of VPH programmes and making appropriate adjustments, involving NGOs already working in the area in both human and animal health, increasing the outreach to women in rural areas and coordinating with human health services in the region (WHO 2002).

Evolving Community Participation

Community members should be completely involved as participants in the health programmes in their communities. They have the important advantages of speaking the local dialect, knowing how to reach people, enjoying social acceptance and they also know the local situations or local needs. Both in rural and urban areas, community groups are all important in the planning and implementation of programmes. They provide the resources needed for adapting plans to local conditions, carrying out tasks at little or no cost, and overcoming constraints. They must be informed about their approach and their role in achieving the aims of the programme. By the process of education and by acquiring experience and knowledge, individuals and communities learn to understand their own situation and be motivated to solve their problems. Community participation enables the individuals and communities to become agents of their own development, rather than passive receivers of information and assistance. Community involvement in health generates in individuals a sense of responsibility for their own health and welfare. To be successful, they have to acquire the capacity to evaluate the situations, choose options and determine their contributions. In other words, individuals and families, and the community as a whole, are not obliged to accept the otherwise conventional solutions that may be imposed, but are not suitable.

In the early phases of a control programme, the general public, especially of communities in endemic areas, have to be made aware of the danger to health as well as the economic importance of zoonoses and foodborne diseases. One of the most effective methods has been found to be the discussions in small groups. In such discussions, the health worker (educator) suggests some kind of concrete action, for example, formation of working committees, which may be constituted soon after the discussions. Such committees have proved to be extremely useful in the early phases of several control programmes. The most common teaching aids and media are posters, documents, pictures, slides, films, radio and television programmes. Communicating the health message is very important, and different methods and techniques need to be combined to accomplish the educational purpose. However, the information must be correct, complete and acceptable to the people. The language of the messages must be understandable.

Types of Community Participation

Participation of the community can be sought in different ways that influence its effectiveness (Table 27.1). The organizational principles of national zoonoses control programmes should depend on the epidemiological pattern of the diseases and on the availability and structure of health care services. They are interrelated with farming practices, habits and levels of urbanization, as well as trade in animals and animal products. It is important that health education and community participation should be included in a zoonoses control project or food hygiene programme from the start and should be closely linked to and coordinated with all changes to it.

Table 27.1: Seven types of community participation

Type of Participation	Description
Manipulative participation (Co-option)	Community participation is simply a pretence, with people's representatives on official boards who are unelected and have no power.
Passive participation (Compliance)	Communities participate by being told what has been decided or already happened. It involves unilateral announcements by an administration or project management without listening to people's responses. The information belongs only to external professionals.
Participation by consultation	Communities participate by being consulted or by answering questions. External agents define problems and information gathering processes, and so control analysis. Such a consultative process does not concede any share in decision-making, and professionals are under no obligation to take on board people's views.
Participation for material incentives	Communities participate by contributing resources, such as labour, in return for material incentives (*e.g.* food, cash). This type of participation is quite common. However, people have no stake in prolonging these practices when the incentives end.
Functional participation (Cooperation)	Community participation is seen by external agencies as a means to achieve project goals. People participate by forming groups to meet predetermined project objectives; they may be involved in decision-making, but only after major decisions have already been made by external agents.
Interactive participation (Co-learning)	People participate in joint analysis, development of action plans and formation or strengthening of local institutions. Participation is seen as a right, not just the means to achieve the project goals. The process involves inter-disciplinary methodologies that seek multiple perspectives and make use of systematic and structured learning processes. As groups take control over local decisions and determine how available resources are used, they have a stake in maintaining structures or practices.
Self mobilization (Collective action)	People participate by taking initiatives independently of external institutions to change systems. They develop contacts with external institutions for resources and technical advice they need, but retain control over how resources are used. Self-mobilisation can spread if governments and NGOs provide an enabling framework of support. Such self-initiated mobilisation may or may not challenge existing distributions of wealth and power.

Source: Adapted from Pretty (1994) and Cornwall (1996).

Communicable Disease Control

Many studies document the benefits of using a community participatory approach to relief in emergency settings and to development in post-emergency phase for controlling communicable diseases. Community participatory relief programme can deliver aid in a timely manner, ensure that resources reach the most vulnerable and poorest individuals, enhance rather than weaken the existing health structures and empower communities to take more control of their lives (Table 27.2).

Table 27.2: Steps to promote community participation in prevention and control of communicable diseases

Community Level	Level of Community Participation
Consultation and sharing of information	Community volunteers and community health workers to be trained to recognise disease, report cases and work on prevention of disease, surveys, data reviews and group interviews.
Decision-making	Inviting community representatives and hold relief coordination meetings in the community.
Initiating programmes	Creating awareness in the community about the resources available for potential community driven activity.

Zoonoses Control Programmes

Planning at local level regarding the control of zoonotic diseases depending on the need of the community and organisation of resources is required for the successful implementation of project. The implementation requires:

1. Selection of the community;
2. Mapping of the risk groups in the community;
3. Identification of risk hotspots;
4. Participatory community risk assessment; and
5. Participatory community risk assessment planning.

The community public health education requirements include:

1. Sensitization of trainers of trainees (public) on all relevant public health matters;
2. Public awareness in schools, religious and political fora;
3. Creating awareness among the decision makers especially village leaders, stakeholders in local government/councils;
4. Retraining of meat inspectors and other service providing cadres involved in meat inspection; and
5. Sensitizing consumers or general public using television and radio programmes.

For implementation of zoonoses control programmes, advocacy is needed to influence the people, policies and systems to bring about widespread changes in the community. Zoonoses control programmes should be included sectorally as well as within the institution in order to sustain the efforts for a longer period of time. This can be achieved by involving various stakeholders and by organising workshops at various levels. For example, community participation has been recognised as a key factor in the effectiveness of rabies prevention and control. Stray dogs in developing countries pose a major threat of rabies, so this segment of the dog population needs to be particularly targeted for rabies control and prevention campaigns. However, such dogs are frequently considered as community or neighbourhood dogs in many developing countries including India. As a result, any drastic dog population control measures often generate resentment among the community. It is, therefore, crucial to build trust between the community and the personnel engaged in the dog population management work. Success can be achieved by educating the schoolchildren about rabies, who in turn help in creating further awareness about the disease in their family. They can also assist by reporting rabies suspected dog cases in their community.

Conclusions

The success in the prevention and control of major zoonoses depends on the capability to mobilize the community participation and on coordination and intersectoral approaches, especially between veterinary and public health services. For example, the avian influenza outbreaks in many countries especially Indonesia and Thailand, could be controlled due to active participation of the communities and coordinated efforts of the health and the veterinary sectors and the government authorities. Similarly, rabies control in Kisumu district of Kenya was initiated in 2010 by the combined effort of veterinary and medical personnel due to alarmingly high cases of dog bite. The programme was started by creating awareness among the village leaders about the need of vaccination against rabies in dogs who in turn motivated the other members of the community. The community members were mobilised to identify households raising dogs and doing door-to-door vaccination. People who refused to vaccinate the dogs were penalised by taking the expense of post-bite vaccination (Omemo 2010).

These and many other examples have shown that community participation is vitally important in order to achieve good results (Deborn *et al.* 2001). We need to understand the community potential, community perspective on zoonoses and zoonoses mainstreaming. There should be no passing of responsibilities but sharing of responsibilities for control of diseases. The desired level of improvement may not take place unless people want change and intend to make it happen.

References

1. Anand JS. 2002. *Self-help groups in empowering women. Case study of selected SHGs and NHGs.* Discussion Paper 38. Kerala Research programme on local level development. Center for Development Studies, Thiruvananthapuram, Kerala, India.

2. Catley A. 2002. Monitoring and assessment of community based animal health projects. In: Catley AC, Blakeway S and Leyland T (eds) *Community based animal*

health care: A practical guide to improve veterinary services. ITDG Publishing, London, UK.

3. Chuengsatiansup K. 2002. *Community life approach: Learning manual that make community work easy, effective and fun*. Society and Health Institute, Nonthaburi, Thailand.

4. Cornwall A. 1996. Towards participatory practice: participatory rural appraisal (PRA) and the participatory process. In: de Koning K and Martin M (Eds) *Participatory research and health: Issues and experiences*. Zed Books Ltd., London. p. 94-107.

5. Daborn C, Njau P, Wood S and Martin M. 2001. *Delivery of animal health services in eastern Africa*. Proceedings of the workshop held at the MS Danish Center, Usa river, Arusha, 14-18 December, 1998. Available at www.vetaid.org/publications/arusha98/index.htm.

6. FAO. 2000. FAO Animal Health Manual 10. *Manual of participatory epidemiology*, Food and Agriculture Organisation, Rome. Available at http://www.fao.org.

7. Government of Kerala. 2007. Kudumbashree mission, state poverty eradication mission of Kerala. *Handbook for resource persons (Malayalam)*, Department of local self government Thiruvanathapuram.

8. Mariner JC. 2002 Community based animal health workers and disease surveillance. In: Catley AC, Blakeway S and Leyland T (eds) *Community based animal health care*. ITDG Publishing, London, UK.

9. McMillan DW and Chavis DM. 1986. Sense of community: A definition and theory. *J Commun Psychol* 14(1): 6-23.

10. Omemo P. 2010. Community based rabies control in Kenya. *Global alliance for rabies control*. www.rabiescontrol.net.

11. Pretty JN. 1994. Alternative systems of inquiry for a sustainable agriculture. *IDS Bulletin* 25(2): 37-47.

12. Rifkin SB. 1996. Paradigms lost: Toward a new understanding of community participation in health programmes. *Acta Tropica* 61: 79-92.

13. Stephen C, Artsob H, Bowie WR, Debot M, Fraser E, Leighten T, Morshed M, Ong C and Patrick D. 2004. Perspectives on emerging zoonotic disease research capacity building in Canada. *Can J Infect Dis Med Microbiol* 15(6): 331-334.

14. WHO. 2002. *Future trends in veterinary public health*. WHO Technical Report Series 907. World Health Organization, Geneva. 96 pp.

15. Wilcox D. 1994. Community participation and empowerment: putting theory into practice. *RRA Notes*, Issue 21, IIED, London. p. 78-82.

16. Wilcox BA and Gubler DG. 2005. Disease ecology and the global emergence of zoonotic pathogens. *Environ Health Prev Med* 10: 263-272.

Index